# Mechanobiology in Health and Disease

# Mechanobiology in Health and Disease

Edited by

**Stefaan W. Verbruggen**
*Department of Biomedical Engineering, Columbia University, New York, NY, United States*
*Institute of Bioengineering, School of Engineering and Materials Science, Queen Mary University of London, London, United Kingdom*

Academic Press is an imprint of Elsevier
125 London Wall, London EC2Y 5AS, United Kingdom
525 B Street, Suite 1650, San Diego, CA 92101, United States
50 Hampshire Street, 5th Floor, Cambridge, MA 02139, United States
The Boulevard, Langford Lane, Kidlington, Oxford OX5 1GB, United Kingdom

© 2018 Elsevier Ltd. All rights reserved.

No part of this publication may be reproduced or transmitted in any form or by any means, electronic or mechanical, including photocopying, recording, or any information storage and retrieval system, without permission in writing from the publisher. Details on how to seek permission, further information about the Publisher's permissions policies and our arrangements with organizations such as the Copyright Clearance Center and the Copyright Licensing Agency, can be found at our website: www.elsevier.com/permissions.

This book and the individual contributions contained in it are protected under copyright by the Publisher (other than as may be noted herein).

**Notices**

Knowledge and best practice in this field are constantly changing. As new research and experience broaden our understanding, changes in research methods, professional practices, or medical treatment may become necessary.

Practitioners and researchers must always rely on their own experience and knowledge in evaluating and using any information, methods, compounds, or experiments described herein. In using such information or methods they should be mindful of their own safety and the safety of others, including parties for whom they have a professional responsibility.

To the fullest extent of the law, neither the Publisher nor the authors, contributors, or editors, assume any liability for any injury and/or damage to persons or property as a matter of products liability, negligence or otherwise, or from any use or operation of any methods, products, instructions, or ideas contained in the material herein.

**Library of Congress Cataloging-in-Publication Data**
A catalog record for this book is available from the Library of Congress

**British Library Cataloguing-in-Publication Data**
A catalogue record for this book is available from the British Library

ISBN 978-0-12-812952-4

For information on all Academic Press publications
visit our website at https://www.elsevier.com/books-and-journals

*Publisher:* Mara Conner
*Acquisition Editor:* Fiona Geraghty
*Editorial Project:* John Leonard
*Production Project Manager:* R. Vijay Bharath
*Cover Designer:* Christian Bilbow

Typeset by SPi Global, India

# Contents

Contributors .................................................................................................. xiii
Foreword to Mechanobiology in Health and Disease ........................................ xvii
Preface ........................................................................................................... xix
Acknowledgments .......................................................................................... xxi

**CHAPTER 1 Techniques for Studying Mechanobiology ................... 1**
Eimear B. Dolan, Stefaan W. Verbruggen and
Rebecca A. Rolfe
1 Introduction to Mechanobiology ...................................................... 2
2 Animal Models and Tissue Engineering to Study
  Mechanobiology ............................................................................... 3
  2.1 Analysis of a Single Cell ............................................................ 3
  2.2 Cellular Interactions With Their Local Environment ............. 6
  2.3 Bioreactors to Mimic the in vivo Environment ...................... 8
  2.4 Animal Loading Models ........................................................... 11
  2.5 Fluorescent Proteins (FPs) and Imaging Techniques ........... 14
3 Molecular and Genetic Techniques to Study Mechanobiology .... 17
  3.1 Analysis of mRNA Expression ............................................... 17
  3.2 Analysis at the Protein Level ................................................. 23
  3.3 Techniques for Editing Gene Function and Altering the
      Mechanical Environment ........................................................ 25
4 Computational Techniques in Mechanobiology ........................... 26
  4.1 Computational Modeling ........................................................ 27
  4.2 Image Analysis ........................................................................ 31
5 Future Perspectives ......................................................................... 33
  References ........................................................................................ 34

**CHAPTER 2 Cell Geometric Control of Nuclear Dynamics and Its
Implications ................................................................. 55**
Abhishek Kumar, Ekta Makhija, A.V. Radhakrishnan,
Doorgesh Sharma Jokhun and G.V. Shivashankar
1 Introduction ..................................................................................... 55
  1.1 Physical Link Between Nucleus and Cytoskeleton ............... 56
  1.2 Microrheology of the Nucleus ................................................ 57
  1.3 Boundary Conditions .............................................................. 57
2 Nuclear Translational Motion ......................................................... 58
  2.1 Image Processing and Computational Methods ................... 58

|     | 2.2 Geometric Control of Nuclear Translation ........................... 60 |
|     | 2.3 Role of Cytoskeleton in Nuclear Translation ...................... 60 |
| 3   | Nuclear Rotational Motion ................................................................. 62 |
|     | 3.1 Image Processing and Computational Methods.................... 62 |
|     | 3.2 Cell Geometric Regulation of Nuclear Rotation ................. 63 |
|     | 3.3 Role of Cytoskeleton in Nuclear Rotation............................ 63 |
|     | 3.4 Implications of Nuclear Rotation in Cellular Functions ...... 65 |
| 4   | Nuclear Envelope Fluctuations ........................................................... 66 |
|     | 4.1 Image Processing and Computational Methods.................... 66 |
|     | 4.2 Cell Geometric Regulation of Nuclear Fluctuations ............ 66 |
|     | 4.3 Role of Cytoskeleton in Nuclear Fluctuations...................... 66 |
| 5   | Discussion............................................................................................... 68 |
|     | 5.1 Summary of Nuclear Translation, Rotation, and Envelope Fluctuations ....................................................... 68 |
|     | 5.2 Implications in Gene Expression ........................................... 69 |
|     | 5.3 Implications in Physiology .................................................... 70 |
|     | 5.4 Conclusions............................................................................. 71 |
|     | References............................................................................................... 71 |

## CHAPTER 3 Mechanobiology Throughout Development ................ 77
Jason P. Gleghorn and Megan L. Killian

| 1 | Introduction........................................................................................... 77 |
| 2 | Mechanobiology in Early Development ............................................. 78 |
|   | 2.1 Neural Tube Morphogenesis ................................................. 80 |
|   | 2.2 Lung Morphogenesis .............................................................. 81 |
|   | 2.3 Appendicular Limb Development ......................................... 82 |
| 3 | Elucidating the Mechanics of Developing Tissues ............................ 85 |
|   | 3.1 Tracking Cell Fate .................................................................. 85 |
|   | 3.2 Mechanical Behavior of Developing Tissues ....................... 86 |
| 4 | Conclusion ............................................................................................. 88 |
|   | Acknowledgments ................................................................................. 88 |
|   | References............................................................................................... 89 |

## CHAPTER 4 Cartilage Mechanobiology: How Chondrocytes Respond to Mechanical Load ..................................... 99
Sophie J. Gilbert and Emma J. Blain

| 1 | Introduction........................................................................................... 100 |
| 2 | Structural Overview.............................................................................. 100 |
|   | 2.1 The ECM ................................................................................. 100 |
|   | 2.2 Aging of Articular Cartilage .................................................. 105 |
|   | 2.3 The Response of Articular Cartilage to Physiological Mechanical Loading ............................................................... 105 |

| | 3 | Cartilage Loading at Cellular Level .................................................. 106 |
|---|---|---|
| | | 3.1 Importance of Micromechanical Environment ................... 106 |
| | | 3.2 Chondrocyte Mechanoreceptors ............................................ 108 |
| | | 3.3 Chondrocyte Mechanotransduction ....................................... 111 |
| | 4 | Concluding Summary ................................................................... 115 |
| | | Acknowledgment ........................................................................... 116 |
| | | References ...................................................................................... 116 |

## CHAPTER 5 Advances in Tendon Mechanobiology ..................... 127
James H.-C. Wang and Bhavani P. Thampatty

| | 1 | Introduction .................................................................................. 128 |
|---|---|---|
| | 2 | Discovery of TSCs ........................................................................ 129 |
| | 3 | The Properties of TSCs vs. Tenocytes ......................................... 130 |
| | 4 | The Factors That Influence the Fate of TSCs .............................. 132 |
| | 5 | The Study of TSC Mechanobiology ............................................ 133 |
| | | 5.1 Mechanobiological Responses of TSCs In Vitro ................ 133 |
| | | 5.2 Mechanobiological Responses of TSCs In Vivo ................. 137 |
| | 6 | The Role of TSCs in the Development of Loading-Induced Tendinopathy ................................................................................. 140 |
| | 7 | The Mechanobiological Studies in Aging Tendons .................... 141 |
| | | 7.1 Features of Aging Tendons ................................................... 141 |
| | | 7.2 The Role of TSCs in the Development of Aging-Induced Tendinopathy ........................................................................... 144 |
| | 8 | Mechanosignaling in TSC Differentiation .................................. 145 |
| | 9 | Concluding Remarks ................................................................... 147 |
| | | Acknowledgments ........................................................................ 148 |
| | | References ...................................................................................... 148 |

## CHAPTER 6 Bone Mechanobiology in Health and Disease ......... 157
Stefaan W. Verbruggen and Laoise M. McNamara

| | 1 | Bone .............................................................................................. 158 |
|---|---|---|
| | | 1.1 Bone Biomechanics ............................................................... 158 |
| | | 1.2 Hierarchical Structure of Bone ............................................. 159 |
| | | 1.3 Bone Porosity and Fluid Flow .............................................. 159 |
| | 2 | Bone Cells ..................................................................................... 161 |
| | | 2.1 MSCs and Osteoprogenitors ................................................. 161 |
| | | 2.2 Osteoblasts ............................................................................. 161 |
| | | 2.3 Osteocytes .............................................................................. 163 |
| | | 2.4 Osteoclasts ............................................................................. 165 |
| | 3 | Bone Growth and Adaptation ...................................................... 165 |
| | | 3.1 Bone Growth .......................................................................... 165 |
| | | 3.2 Bone Modeling and Remodeling ......................................... 166 |

| | | |
|---|---|---|
| **4** | Bone Mechanobiology | 167 |
| | 4.1 Mechanosensation and Mechanosensors | 169 |
| | 4.2 Mechanotransduction and Signaling Pathways | 173 |
| | 4.3 Mechanical Environment of Bone Cells | 176 |
| **5** | Bone Mechanobiology and Disease | 178 |
| | 5.1 Bone Mechanobiology and Osteoporosis | 178 |
| | 5.2 Bone Mechanobiology During Osteoarthritis | 185 |
| | 5.3 Mechanobiology, Osteosarcoma, and Metastatic Bone Disease | 185 |
| | 5.4 Mechanobiological Therapy for Bone Diseases | 187 |
| | 5.5 Mechanobiology for Bone Regeneration | 187 |
| **6** | Conclusion | 188 |
| | Acknowledgments | 189 |
| | References | 189 |
| | Further Reading | 214 |

**CHAPTER 7 Vascular Mechanobiology, Immunobiology, and Arterial Growth and Remodeling ............ 215**

Alexander W. Caulk, George Tellides and Jay D. Humphrey

| | | |
|---|---|---|
| **1** | Introduction | 216 |
| **2** | Vascular Structure and Function | 216 |
| | 2.1 Intima | 217 |
| | 2.2 Media | 218 |
| | 2.3 Adventitia | 219 |
| **3** | Quantification of Arterial Mechanics | 220 |
| | 3.1 In Vivo Methodologies | 221 |
| | 3.2 Ex Vivo Methodologies | 222 |
| | 3.3 Application | 226 |
| **4** | Cellular Sensing of Its Mechanical Environment | 227 |
| | 4.1 Mechanosensing in Endothelial Cells | 227 |
| | 4.2 Mechanosensing in Vascular Smooth Muscle Cells | 228 |
| **5** | Regulation of the Inflammatory Response—Cellular Function and Phenotype | 230 |
| | 5.1 Monocytes/Macrophages | 231 |
| | 5.2 T Cells | 234 |
| | 5.3 Dendritic Cells | 235 |
| **6** | ECM and Inflammation | 236 |
| | 6.1 Matrix Metalloproteinases | 236 |
| | 6.2 Other Proteases | 238 |
| | 6.3 Tissue Inhibitors of Metalloproteinases | 238 |

| | 7 | Matrix G&R in Cardiovascular Disease | 238 |
|---|---|---|---|
| | | 7.1 Mechanics and Inflammation in Hypertension | 240 |
| | | 7.2 Aging, Inflammation, and Vascular Mechanics | 242 |
| | 8 | Conclusions and Future Directions | 243 |
| | | Acknowledgment | 243 |
| | | References | 244 |

## CHAPTER 8 Mechanobiology of the Heart Valve Interstitial Cell: Simulation, Experiment, and Discovery ........... 249

Alex Khang, Rachel M. Buchanan, Salma Ayoub, Bruno V. Rego, Chung-Hao Lee, Giovanni Ferrari, Kristi S. Anseth and Michael S. Sacks

| 1 | Introduction | 250 |
|---|---|---|
| 2 | Major Questions and Challenges | 251 |
| 3 | Advances in Investigating VIC Mechanobiology | 252 |
| | 3.1 Isolated Cell Studies | 252 |
| | 3.2 In Situ Tissue Level Evaluation of VIC Contraction Behaviors | 260 |
| | 3.3 Uniaxial Planar Stretch Bioreactors | 269 |
| | 3.4 Model-Driven Experimental Design | 273 |
| 4 | Future Directions | 277 |
| | Acknowledgments | 280 |
| | References | 280 |

## CHAPTER 9 Platelet Receptor-Mediated Mechanosensing and Thrombosis ....................................................... 285

Lining A. Ju, Yunfeng Chen, Zhenhai Li and Cheng Zhu

| 1 | Introduction | 285 |
|---|---|---|
| 2 | Ultrasensitive Force Techniques | 287 |
| 3 | Force-Induced VWF Activation | 289 |
| 4 | VWF-GPIbα Binding Kinetics | 290 |
| 5 | Force-Induced VWF Cleavage by ADAMTS13 | 291 |
| 6 | Force-Induced GPIbα Domain Unfolding | 293 |
| 7 | Coupling Between Unbinding Kinetics and the Unfolding Mechanics | 293 |
| 8 | MSD Unfolding and GPIb Mechanotransduction | 295 |
| 9 | Four-Step Model for Receptor-Mediated Mechanosensing | 296 |
| 10 | Conclusions | 298 |
| | Notes and Acknowledgments | 299 |
| | References | 299 |

## CHAPTER 10 Mechanobiology of Primary Cilia in the Vascular and Renal Systems ......305
Surya M. Nauli, Ashraf M. Mohieldin, Madhawi Alanazi and Andromeda M. Nauli

1 Introduction ......305
2 Primary Cilia as Mechanical Sensors ......306
3 Ciliopathies ......308
4 Mechanobiology ......308
5 Cilia Length Regulation ......309
   5.1 Ciliotherapy in Vascular Hypertension ......310
   5.2 Mechanosensory Cilia in Vascular System ......311
   5.3 Mechanosensory Cilia in Renal System ......314
6 Conclusion ......315
   Acknowledgements ......316
   References ......316

## CHAPTER 11 Neuromechanobiology ......327
William J. Tyler

1 Introduction ......327
2 Viscoelastic Plasma Membranes Govern Neuronal Function ......328
3 Electrodeformation and Flexoelectricity: A Mechanism of Neuronal Signaling? ......330
4 The Role of Mechanosensitive Ion Channels in Neuronal Function ......333
5 Cytoskeletal Elements Mediate Force Sensing and Transduction in Neurons: Mechanisms for Plasticity? ......335
6 Methods and Tools for Mechanical Interfacing with the Nervous System ......339
7 Future Directions and Outlook for Neuromechanobiology ......342
   References ......343

## CHAPTER 12 Mechanobiology of the Eye ......349
Ashutosh Richhariya, Nikhil S. Choudhari, Ashik Mohamed, Derek Nankivil, Akshay Badakere, Vivek P. Dave, Sunil Punjabi and Virender S. Sangwan

1 Introduction ......349
2 The Anterior Segment ......350
   2.1 The Cornea and the Tear Film ......350
   2.2 Optics of the Cornea and the Role of Mechanobiology ......353
   2.3 Mechanobiology of the Lens, Capsule, and Zonules ......355
   2.4 The Forces of Accommodation ......357

      2.5  Senescence and Pathophysiology of the Accommodative Apparatus .................................................................358
      2.6  The Aqueous Humor and Glaucoma ..................................359
  3  The Posterior Segment ...........................................................362
  4  The Extra Ocular System .......................................................364
      4.1  Muscle Functions...........................................................364
      4.2  Related Disease Conditions ...........................................366
  5  Conclusion .............................................................................367
      References................................................................................367
      Further Reading .......................................................................375

## CHAPTER 13 Gastrointestinal Mechanosensory Function in Health and Disease .............................................. 377
Amanda J. Page and Hui Li

  1  Introduction .............................................................................378
  2  Mechanosensitive Components of the Gastrointestinal Tract ...378
      2.1  Gastrointestinal Mechanosensation .................................378
      2.2  Extrinsic Innervation of the GI Tract .............................380
      2.3  Intrinsic Innervation of the GI Tract ..............................392
      2.4  Interstitial Cells of Cajal ................................................394
      2.5  Smooth Muscle Cells .....................................................397
      2.6  Endocrine Cells..............................................................398
  3  Conclusion .............................................................................399
      References................................................................................400

## CHAPTER 14 Mechanobiology of Skin Diseases and Wound Healing .................................................................... 415
Sun Hyung Kwon, Jagannath Padmanabhan and Geoffrey C. Gurtner

  1  Introduction .............................................................................415
  2  Mechanobiology of the Skin ..................................................417
      2.1  Mechanobiology in Skin Homeostasis ...........................417
      2.2  Mechanoresponsive Skin Cells .......................................418
      2.3  Molecular Pathways Mediating Skin Mechanobiology ......419
      2.4  Skin Mechanobiology and the ECM ..............................421
      2.5  Altered Mechanobiology in Skin Diseases and Wound Healing ...........................................................................422
  3  Mechanotransduction in Skin Fibrosis and Fibrotic Wound Healing ....................................................................................422
      3.1  Hypertrophic Scars .........................................................424
      3.2  Keloids ............................................................................425
      3.3  Scleroderma (Systemic Sclerosis) ..................................426

|   |   | 3.4 Dupuytren's Contracture ....................................................... 427 |
|---|---|---|
|   |   | 3.5 Other Fibrotic Skin Disorders ............................................. 428 |
|   | 4 | Mechanobiology of Degenerative Cutaneous Wound Healing ................................................................................ 428 |
|   |   | 4.1 Diabetic Skin Ulcers ............................................................ 429 |
|   |   | 4.2 Wound Healing in Aging .................................................... 433 |
|   | 5 | Therapeutic Strategies for Modulation of Dysfunctional Mechanobiology .............................................................................. 434 |
|   |   | 5.1 Preoperative Surgical Techniques ...................................... 435 |
|   |   | 5.2 Mechanomodulatory Devices ............................................. 435 |
|   |   | 5.3 Biochemical Inhibitors ........................................................ 436 |
|   | 6 | Future Research and Directions .................................................... 438 |
|   | 7 | Conclusion ...................................................................................... 438 |
|   |   | References ....................................................................................... 438 |

### CHAPTER 15 Mechanobiology of Metastatic Cancer .................... 449

Martha B. Alvarez-Elizondo, Rakefet Rozen and Daphne Weihs

|   | 1 | Introduction .................................................................................... 449 |
|---|---|---|
|   | 2 | Cell Mechanics, Structure, and Interactions ................................. 452 |
|   |   | 2.1 Mechanical Structure of the Cell—The Cytoskeleton ........ 452 |
|   |   | 2.2 The Cytoskeleton in Motility and Force Application ......... 453 |
|   |   | 2.3 Role of the Tumor Microenvironment ................................ 456 |
|   |   | 2.4 Mechanotransduction in Cancer Cells ................................ 457 |
|   | 3 | Cell Motility and Migration .......................................................... 458 |
|   |   | 3.1 Motility Modes .................................................................... 458 |
|   |   | 3.2 Stimuli for Directional Motility or Migration of Cells ...... 460 |
|   | 4 | Steps of Metastasis ........................................................................ 461 |
|   |   | 4.1 Local Invasion Through Tissue ........................................... 461 |
|   |   | 4.2 Cell Mechanics in the Intravasation Phase ........................ 463 |
|   |   | 4.3 Metastatic Cells in the Blood Stream ................................. 464 |
|   |   | 4.4 Extravasation ....................................................................... 464 |
|   | 5 | Methods for Cytomechanobiology in Cancer Cells ..................... 465 |
|   |   | 5.1 Cell Mechanics Assays ....................................................... 466 |
|   |   | 5.2 Cell Migration Assays ........................................................ 469 |
|   |   | 5.3 Cell Strength Evaluation .................................................... 472 |
|   | 6 | Evaluating Metastatic Potential through Mechanical Interactions ..................................................................................... 476 |
|   |   | 6.1 Adhesion and Adhesive Forces .......................................... 477 |
|   |   | 6.2 Mechanical Invasiveness .................................................... 479 |
|   | 7 | Summary ........................................................................................ 481 |
|   |   | References ....................................................................................... 482 |

Index ................................................................................................................. 495

# Contributors

**Madhawi Alanazi**
Department of Biomedical & Pharmaceutical Sciences, Chapman University, Irvine, CA, United States

**Martha B. Alvarez-Elizondo**
Faculty of Biomedical Engineering, Technion—Israel Institute of Technology, Haifa, Israel

**Kristi S. Anseth**
Department of Chemical and Biological Engineering, BioFrontiers Institute; Howard Hughes Medical Institute, University of Colorado, Boulder, CO, United States

**Salma Ayoub**
Willerson Center for Cardiovascular Modeling and Simulation, Institute for Computational Engineering and Sciences, Department of Biomedical Engineering, The University of Texas at Austin, Austin, TX, United States

**Akshay Badakere**
L V Prasad Eye Institute, Kallam Anji Reddy Campus, Hyderabad, India

**Emma J. Blain**
Arthritis Research UK Biomechanics and Bioengineering Centre, Biomedicine Division, School of Biosciences, Cardiff University, Cardiff, United Kingdom

**Rachel M. Buchanan**
Willerson Center for Cardiovascular Modeling and Simulation, Institute for Computational Engineering and Sciences, Department of Biomedical Engineering, The University of Texas at Austin, Austin, TX, United States

**Alexander W. Caulk**
Department of Biomedical Engineering, Yale University, New Haven, CT, United States

**Yunfeng Chen**
Department of Molecular Medicine, MERU-Roon Research Center on Vascular Biology, The Scripps Research Institute, La Jolla, CA, United States

**Nikhil S. Choudhari**
L V Prasad Eye Institute, Kallam Anji Reddy Campus, Hyderabad, India

**Vivek P. Dave**
L V Prasad Eye Institute, Kallam Anji Reddy Campus, Hyderabad, India

**Eimear B. Dolan**
School of Pharmacy, Royal College of Surgeons in Ireland, Dublin; Discipline of Anatomy, School of Medicine, College of Medicine Nursing and Health Sciences, National University of Ireland, Galway, Ireland; Institute for Medical Engineering Science, Massachusetts Institute of Technology, Cambridge, MA, United States

**Giovanni Ferrari**
Department of Surgery, Columbia University, New York, NY, United States

**Sophie J. Gilbert**
Arthritis Research UK Biomechanics and Bioengineering Centre, Biomedicine Division, School of Biosciences, Cardiff University, Cardiff, United Kingdom

**Jason P. Gleghorn**
Department of Biomedical Engineering, University of Delaware, Newark, DE, United States

**Geoffrey C. Gurtner**
Hagey Laboratory, Division of Plastic Surgery, Department of Surgery, Stanford University School of Medicine, Stanford, CA, United States

**Jay D. Humphrey**
Department of Biomedical Engineering; Vascular Biology and Therapeutics Program, Yale University, New Haven, CT, United States

**Doorgesh Sharma Jokhun**
Mechanobiology Institute (MBI), National University of Singapore, Singapore

**Lining A. Ju**
Heart Research Institute; Charles Perkins Centre, The University of Sydney, Camperdown, NSW, Australia

**Alex Khang**
Willerson Center for Cardiovascular Modeling and Simulation, Institute for Computational Engineering and Sciences, Department of Biomedical Engineering, The University of Texas at Austin, Austin, TX, United States

**Megan L. Killian**
Department of Biomedical Engineering, University of Delaware, Newark, DE, United States

**Abhishek Kumar**
Yale Cardiovascular Research Center (YCVRC), Department of Internal Medicine, Yale School of Medicine, Yale University, New Haven, CT, United States

**Sun Hyung Kwon**
Hagey Laboratory, Division of Plastic Surgery, Department of Surgery, Stanford University School of Medicine, Stanford, CA, United States

**Chung-Hao Lee**
School of Aerospace and Mechanical Engineering, The University of Oklahoma, Norman, OK, United States

**Hui Li**
Adelaide Medical School, University of Adelaide, Adelaide, SA, Australia

**Zhenhai Li**
Molecular Modeling and Simulation Group, National Institutes for Quantum and Radiological Science and Technology, Kyoto, Japan

**Ekta Makhija**
Mechanobiology Institute (MBI), National University of Singapore; BioSystems and Micromechanics Group, Singapore-MIT Alliance for Research & Technology, CREATE, Singapore

**Laoise M. McNamara**
Biomedical Engineering, College of Engineering and Informatics, National University of Ireland Galway, Galway, Ireland

**Ashik Mohamed**
L V Prasad Eye Institute, Kallam Anji Reddy Campus, Hyderabad, India

**Ashraf M. Mohieldin**
Department of Biomedical & Pharmaceutical Sciences, Chapman University, Irvine, CA, United States

**Derek Nankivil**
Johnson & Johnson Vision Care Inc., Jacksonville, FL, United States

**Andromeda M. Nauli**
Department of Pharmaceutical Sciences, Marshall B. Ketchum University, Fullerton, CA, United States

**Surya M. Nauli**
Department of Biomedical & Pharmaceutical Sciences, Chapman University; Department of Medicine, University of California Irvine, Irvine, CA, United States

**Jagannath Padmanabhan**
Hagey Laboratory, Division of Plastic Surgery, Department of Surgery, Stanford University School of Medicine, Stanford, CA, United States

**Amanda J. Page**
Adelaide Medical School, University of Adelaide; South Australian Health and Medical Research Institute (SAHMRI), Adelaide, SA, Australia

**Sunil Punjabi**
Department of Mechanical Engineering, Ujjain Engineering College, Ujjain, India

**A.V. Radhakrishnan**
Mechanobiology Institute (MBI), National University of Singapore, Singapore; Raman Research Institute, Bangalore, India; Interdisciplinary Institute for Neuroscience, UMR 5297, CNRS, Bordeaux, France

**Bruno V. Rego**
Willerson Center for Cardiovascular Modeling and Simulation, Institute for Computational Engineering and Sciences, Department of Biomedical Engineering, The University of Texas at Austin, Austin, TX, United States

**Ashutosh Richhariya**
L V Prasad Eye Institute, Kallam Anji Reddy Campus, Hyderabad, India

**Rebecca A. Rolfe**
Department of Zoology, School of Natural Sciences, Trinity College Dublin, Dublin, Ireland

**Rakefet Rozen**
Faculty of Biomedical Engineering, Technion—Israel Institute of Technology, Haifa, Israel

**Michael S. Sacks**
Willerson Center for Cardiovascular Modeling and Simulation, Institute for Computational Engineering and Sciences, Department of Biomedical Engineering, The University of Texas at Austin, Austin, TX, United States

**Virender S. Sangwan**
L V Prasad Eye Institute, Kallam Anji Reddy Campus, Hyderabad, India

**G.V. Shivashankar**
Mechanobiology Institute (MBI), National University of Singapore, Singapore; Institute of Molecular Oncology (IFOM), Italian Foundation for Cancer Research, Milan, Italy

**George Tellides**
Department of Surgery; Vascular Biology and Therapeutics Program, Yale University, New Haven, CT, United States

**Bhavani P. Thampatty**
MechanoBiology Laboratory, Department of Orthopaedic Surgery, University of Pittsburgh School of Medicine, Pittsburgh, PA, United States

**William J. Tyler**
School of Biological and Health Systems Engineering, Arizona State University, Tempe, AZ, United States

**Stefaan W. Verbruggen**
Department of Biomedical Engineering, Columbia University, New York, NY, United States; Institute of Bioengineering, School of Engineering and Materials Science, Queen Mary University of London, London, United Kingdom

**James H.-C. Wang**
MechanoBiology Laboratory, Department of Orthopaedic Surgery, University of Pittsburgh School of Medicine, Pittsburgh, PA, United States

**Daphne Weihs**
Faculty of Biomedical Engineering, Technion—Israel Institute of Technology, Haifa, Israel

**Cheng Zhu**
Coulter Department of Biomedical Engineering, Georgia Institute of Technology, Atlanta, GA, United States

# Foreword to mechanobiology in health and disease

Although mechanobiology is a term that has been in the literature for more than 50 years, two periods are particularly noteworthy. Darcy Thompson, although he didn't use the term, certainly defined the fundamental issues in his seminal book "On Growth and Form" first published in 1917, in which he argued how developmental processes are influenced by physical force. These concepts, while widely appreciated in the intervening years, also laid the groundwork for the rapid expansion of interest years later in how forces shape cell behavior. Interest gained momentum in the 1960s and 1970s with a progressive appreciation of the roles that shear stresses in arteries play in the pathogenesis of atherosclerosis, as well as a variety of other disease processes ranging from arthritis to connective tissue damage and repair.

Just after 2000, the term "mechanotransduction" came into wide usage, as investigations began to delve into the fundamental mechanisms by which cells sense mechanical force and then convert ("transduce") it to a biochemical signal. This work included studies on the role of tissue level stress and strain as viewed at the nanoscale in terms of alterations in protein conformation leading to changes in molecular binding properties and chemical signaling, as well as the activation of mechanosensitive ion channels. Subsequent studies identified many of the mechanisms of force sensation, yet the field continues to grow as does our appreciation of its biological consequences. It is now fair to say that there is a wide recognition among the bioengineering/biophysics/biology communities that force transmission and transduction into a biochemical signal constitutes one of the major determinants of biological function with important consequences in development, physiology, and disease.

The current text comes at a time, therefore, at which we understand many of the underlying physical factors governing mechanobioology, and we also appreciate its extensive role in health and disease. Our field, however, continues to grow, and has experienced a "second wave" of interest just in the past 4 years, as evidenced by nearly a doubling in the number of papers listed in Pubmed using the term "mechanobiology" between 2013 and 2017. Thus, now is an excellent time to capture and compile some of these recent advances, many of which delve deeper into the disease processes and, importantly, how physical forces can influence, and even used in some instances to control disease progression, maintain normal physiological function, or regulate natural developmental processes.

This current text effectively reflects the growing collection of physiological systems in which mechanobiology plays a central role. Notably, chapters on "Neuromechanobiology" (Chapter 11) and "Cell Nuclear Mechanobiology" (Chapter 2) represent entirely new fields. "Cancer Mechanobiology" (Chapter 15) is somewhat more established, but has also experienced considerable recent growth as we seek to understand how mechanics influences not only proliferation of the primary tumor, but especially the various processes the occur during the metastatic cascade. Some others, such as "Vascular Mechanobiology" (Chapter 7) and "Cartilage Mechanobiology" (Chapter 4), constitute more mature fields, for which we now have a sufficient understanding of the underlying principles that one can begin to couple experimental studies with computational models, either deterministic or data-driven, as a means of predicting outcomes of various interventions or changes in the mechanical environment of a tissue.

While the chapters are written by leading experts in their respective fields and dig deeply into each topic, the text remains accessible to those from life sciences or physical sciences. It serves equally well either as a current update for researchers in mechanobiology or as an introduction to those new to the field. It marks an important milestone, and is highly recommended to all.

**Roger D. Kamm**
*Departments of Biological Engineering and Mechanical Engineering,*
*Massachusetts Institute of Technology*

# Preface

Mechanobiology is a nascent interdisciplinary area of research that has recently emerged from the closely related field of classical biomechanics. While biomechanics is largely concerned with the physical interactions between the body and its surrounding environment, mechanobiology explores the biological responses by tissues and cells when exposed to physical stimuli. Biomechanics has been studied, in one guise or another, since the time of Aristotle. However, the first glimpse of mechanobiology was seen in Darwin's theories of the adaptation of biological structures to the surrounding physical environment. Following Darwin's evolutionary theory, Wilhelm Roux determined that skeletal structures adapted not only to the environment but also to physical interactions with it. Similar theories were proposed for other tissues, culminating in the idea, put forth by D'Arcy Thompson in his seminal text *On Growth and Form*, that "cell and tissue, shell and bone, leaf and flower" are molded "in obedience to the laws of physics." However, while the ability of biology to adapt to its physical surroundings has been apparent for centuries, it is only relatively recently that scientists have dedicated themselves entirely to studying precisely how and why these changes occur.

Thus, the field of mechanobiology in general is beginning to take shape and is growing rapidly as research in other areas leads to scientific or technological advances that impact the field, drawing new scientists into researching mechanobiology. Indeed, the concept for this book arose from conversations at a meeting of the European Congress of Biomechanics, during which delegates commented on how mechanobiology had, over the years, gone from being discussed tangentially as tissue remodeling or mechanotransduction in a handful of talks to having its own dedicated parallel sessions held throughout the conference program. It was clearly apparent that mechanobiology was developing from traditional biomechanics to a fully fledged field of study in its own right.

As an inherently interdisciplinary field of study, mechanobiology is investigated by multiple disparate groups of biologists, engineers, clinicians, and scientists, encompassing a broad span of interrelated multidisciplinary fields. This presents unique challenges and opportunities. Research findings in the subject are frequently published on niche topics in specialist journals in numerous distinct areas, often using terminology and technologies specific to their respective fields. Identifying research techniques (both established and developing), information sources, and, indeed, research groups that may overlap or influence an area of mechanobiological research is particularly challenging for established investigators. This can stunt the development of collaborations, as scientists in similar areas may not attend the same conferences or publish in similar journals. Similarly, for a student or early career researcher entering the field, there is little material available to give an overview of mechanobiology before focusing on a specific area. This collection aims to provide a broad, yet thorough, introduction for these researchers and a detailing of the state of the art in various exciting areas of study.

My hope is that readers find this book, which takes a system-by-system approach to the human body, both illuminating and educational, and that it encourages researchers into this interesting new field of research.

# Acknowledgments

I would like to extend my thanks to each of the authors for their excellent contributions to this book. These are a group of early pioneers, current trailblazers, and future leaders in the field of mechanobiology, and I am extremely grateful for their time and effort. I would also like to thank Prof. Roger Kamm for his excellent foreword, which I consider a testament to the high standard of writing in these pages. I would like to gratefully acknowledge Fiona Geraghty, who first suggested to me to compile this collection, as well as Edward Payne, Carla Lima, and John Leonard for their assistance in managing the editorial process. Lastly, I thank Jack Stenson for encouraging me to begin and supporting me to finish this work.

**Stefaan W. Verbruggen**

# CHAPTER 1

# Techniques for studying mechanobiology

Eimear B. Dolan[*,†,‡], Stefaan W. Verbruggen[§,¶], Rebecca A. Rolfe[∥]

School of Pharmacy, Royal College of Surgeons in Ireland, Dublin, Ireland[*] Discipline of Anatomy, School of Medicine, College of Medicine Nursing and Health Sciences, National University of Ireland, Galway, Ireland[†] Institute for Medical Engineering Science, Massachusetts Institute of Technology, Cambridge, MA, United States[‡] Department of Biomedical Engineering, Columbia University, New York, NY, United States[§] Institute of Bioengineering, School of Engineering and Materials Science, Queen Mary University of London, London, United Kingdom[¶] Department of Zoology, School of Natural Sciences, Trinity College Dublin, Dublin, Ireland[∥]

## ABBREVIATIONS

| | |
|---|---|
| 2D | two-dimensional |
| 3D | three-dimensional |
| AFM | atomic force microscopy |
| $Ca^{2+}$ | calcium ion |
| cDNA | complementary DNA |
| CFD | computational fluid dynamics |
| CRISPR | clustered regularly interspaced short palindromic repeats |
| DIC | digital image correlation |
| ELISA | enzyme-linked immunosorbent assay |
| ECM | extracellular matrix |
| FE | finite element |
| FP | fluorescent protein |
| FRAP | fluorescent recovery after photobleaching |
| FRET | fluorescent resonance energy transfer |
| FSI | fluid-solid interaction |
| GPCR | G-protein-coupled receptor |
| IHC | immunohistochemistry |
| mRNA | messenger RNA |
| PIV | particle image velocimetry |
| qRT-PCR | quantitative real-time polymerase chain reaction |
| TFM | traction force microscopy |

Mechanobiology in Health and Disease. https://doi.org/10.1016/B978-0-12-812952-4.00001-5
© 2018 Elsevier Ltd. All rights reserved.

# 1 INTRODUCTION TO MECHANOBIOLOGY

Mechanobiology is a field at the forefront of biomedical investigation, situated at the interface between the fields of engineering and biology. While new examples of the human body adapting or responding to mechanical loading are regularly being discovered, this phenomenon has long been observed in multiple tissue types and across numerous anatomical locations. Examples of tissue adaptation in response to changes in loading include bone, cartilage, tendon, vessels, heart, lung, and skin [1–7]. Each of these cases involves cell-driven responses by tissues and organs to loading, requiring translation of loading that occurs at the whole-organ scale down to mechanical stimulation of individual cells. The resulting changes in cell activity are then manifested back up through the scales, causing adaption at the tissue or organ level [8].

While intricately related to what could be termed "classical" biomechanics, mechanobiology can be thought of as its mirror opposite. Biomechanics largely concerns the study of the physical effects and interactions induced by biological activity (e.g., the forces imparted onto the ground during running), whereas mechanobiology describes the biological response to an applied mechanical stimulus (e.g., the loss of the bone in low-gravity environments). Therefore, while mechanobiological effects can be observed at the scale of an organ or organism, they are fundamentally the result of changes wrought by cells in response to mechanical stimuli [9]. In fact, it has been shown that most eukaryotic cells themselves exert force on their surrounding tissues, even in the absence of any external mechanical stimulus [10,11]. Furthermore, it has been proposed that all cells are mechanosensitive [12], as forces are essential for basic cellular functions like mitosis and migration [13,14]. Thus, mechanobiology is fundamentally a multiscale phenomenon, spanning the length scales from the very smallest molecules to whole organs and presenting unique challenges to researchers attempting to further our understanding. This complex relationship across multiple scales is illustrated in Fig. 1.

The objective of this chapter is to introduce researchers from various backgrounds to some of the wide range of experimental and computational techniques being applied to advance the study of mechanobiology. The first section examines investigative methods at the organ and tissue level, including animal models and tissue-engineering techniques. The second section moves toward the cell and molecular levels, introducing imaging methods, biochemical assays, and molecular analysis techniques to determine the biological responses to mechanical stimuli. The final section describes computational methods, which have been applied at multiple scales to analyze imaging data, quantify loading experienced by biological tissues, and predict structural responses to mechanical stimuli.

Mechanical stimulation is transferred down from organ to molecular scales, with various animal models (e.g., the rat ulnar loading model [15]), tissue-engineering bioreactors (e.g., spinner-flask bioreactor [16]), cell culture techniques (e.g., stretching individual cells [17]), and cytoskeletal disruption [18] used to

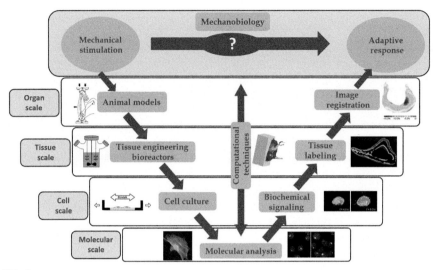

**FIG. 1**

Mechanobiology describes the adaptation of the body to mechanical stimulation and occurs across multiple scales, with researchers using a range of techniques to apply loading and measure the response at different scales.

replicate this experimentally. These stimuli are then transduced into biochemical and structural responses, with a range of techniques such as RNA assays (e.g., in situ hybridization on RNA analysis of gene expression [19]), biochemical assays (e.g., calcium signaling [20]), tissue labeling (e.g., tetracycline-alizarin staining for bone tissue growth [21]), and imaging techniques (e.g., image registration of knee menisci [22]) applied to measure these. Computational techniques (such as FSI modeling [23]) can operate across multiple levels, acting as a bridge across the length scales to model in vivo mechanobiology.

## 2 ANIMAL MODELS AND TISSUE ENGINEERING TO STUDY MECHANOBIOLOGY

### 2.1 ANALYSIS OF A SINGLE CELL

Single-cell investigations are advantageous for understanding cell behavior in response to specific stimuli (mechanotransduction). The results of single-cell investigations can be used to guide the development of mechanical environments that elicit favorable cell responses and inform tissue-engineering approaches [24,25]. Single-cell investigations are also used to investigate cell material properties, vital information that is required for computational investigations [26]. Force-application techniques are used to investigate single-cell mechanics, whereby the cell is deformed in some way by a known force or stress and its mechanical and/or biochemical response is measured. Typically, the surface of the cell is indented or extended [26,27]. There are a number of force-application techniques available, as discussed in detail by Rodriguez et al. [26] and summarised in Fig. 2. Optical tweezers, atomic force microscopy (AFM), and micropipette aspiration are commonly used tools for single-cell investigation, shown in Fig. 2.

### 2.1.1 Force application techniques to analyze a single cell
#### 2.1.1.1 Optical tweezers
Optical tweezers (often referred to as optical trap) are one method often used to apply a known force to a cell. This technique was developed by Arthur Ashkin in 1970 [28] and was originally used to trap individual atoms, viruses, and bacteria [29]. In this method, nano- to micron-sized beads are attached to the cell membrane. Displacement of the cell membrane is controlled by directing infrared lasers at the transparent beads. When photons pass through the beads, there is a change in their direction. The change in direction causes a change in momentum, resulting in a force on the bead. This change is dependent on the refractive index of the beads. Optical tweezers can exert forces in excess of 100 pN on particles ranging in size from nanometers to microns while simultaneously measuring the 3D displacement of the trapped particle with subnanometer accuracy and submillisecond time resolution [30].

#### 2.1.1.2 Atomic force microscopy
AFM was first developed to probe nanoscale features of solid materials using its high sensitivity to intermolecular forces ($\sim$pN) and spatial resolution ($\sim$nm). More recently, AFM has been used throughout the literature to measure the apparent elasticity of living cells. An AFM system generally consists of a probing tip attached to a flexible cantilever that is lowered onto the cell, and the deflection of the cantilever is monitored. The local Young's modulus ($E$) of a living cell can be measured by recording the force acting on the AFM tip while it is indented into a cell, which results in a force-displacement curve. This force-displacement curve can be used to calculate the force-indentation curve by fitting it with the Hertz model (contact mechanics) allowing the estimation of the local $E$; a detailed description is provided in [31,32]. The following two conditions must be met for an accurate measurement: (a) The indentation depth is not more than $\sim$10% of the sample thickness [33,34], and (b) the indentation depth is $>$200 nm [35]. Additionally, the variable shape of a typical AFM probe will determine the nature of the force-deformation curve [27]. AFM indentation is typically performed on highly localized regions of the cell, probing individual structures and determining the heterogeneity of cell.

#### 2.1.1.3 Micropipette aspiration
A micropipette is a small glass capillary with an internal diameter smaller than that of a cell. In this technique, the micropipette is extended to the surface of a cell, and a small negative pressure is applied to create a tight seal between the cell and the tip

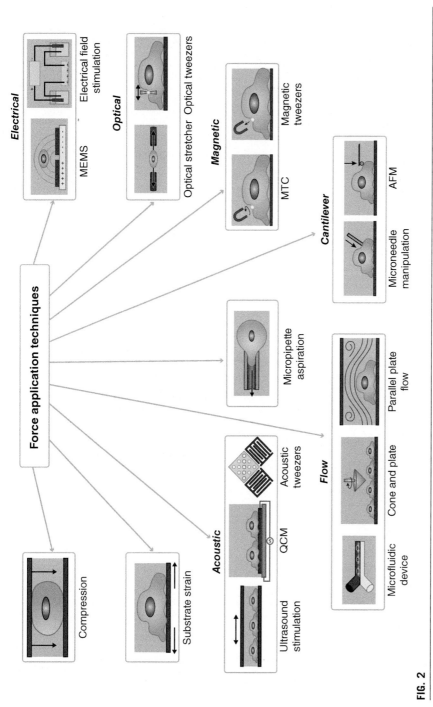

FIG. 2

Force-application techniques for single cells. *QCM*, Quartz Crystal Microbalance; *AFM*, Atomic Force Microscopy; *MTC*, Magnetic Twisting Cytometry; *MEMS*, Microelectromechanical Systems.

of the micropipette. Once this seal is formed, a known negative pressure is applied inducing cell deformation or "aspiration." Micropipette aspiration is used to study whole-cell mechanics by investigating how much cellular material is pulled into a glass pipette in response to the known negative pressure applied. Video microscopy is used to monitor the volume of cell material outside the pipette by tracking the change in radius of material and the height of cellular material inside the micropipette [25,26]. The Young's modulus of the cell can be calculated from the applied vacuum pressure, the length of the cell inside the pipette, and the inner radius of the pipette, if the cell is assumed to be a solid homogenous material [27]. If the cell is assumed to behave as a viscous solid, the cell viscosity can be calculated from these aforementioned values, the radius of the spherical portion of the cell outside the pipette, and the lengthening rate of the cellular material within the pipette [26,27]. The device can measure piconewton-level forces [27]. This technique has been used to determine the elastic modulus and viscoelastic properties of various cell types throughout the years [26,36,37]. It has been extensively used to measure cells in suspension [38–43] but more recently has been used to measure cells adhered to a substrate [44,45]. This technique has also been used to investigate the stiffness of nuclear mechanics by gently isolating the nucleus from the cell cytoplasm [41,46].

## 2.2 CELLULAR INTERACTIONS WITH THEIR LOCAL ENVIRONMENT

In addition to the investigation of cells themselves, forces generated by the cell in their local environment are key in the study of mechanobiology and tissue engineering. Forces are produced by cells during development, contraction, migration, and other common cell processes [26]. Contractile cellular forces (cellular tractions) are transmitted to other cells via cell-cell interactions and to the local extracellular matrix (ECM) through cell-matrix interactions. These forces generated by cells drive the bending, stretching, alignment, and repositioning required for tissue development and homeostasis, and they also regulate cell functions ranging from receptor signaling and transcription to differentiation and proliferation. Cell tractions are in the range of pico- to nanonewtons and occur across small-length scales (nano- to micrometers), making direct measurement a particularly challenging task. Exciting research in the nascent fields of microfluidics and organ-on-a-chip technologies provide the promise of studying cell mechanobiology in a tailored 3D microenvironment, more closely replicating the physiological and mechanical environment in vivo. These interesting new techniques combine much of the methods described in this section, and detailed reviews can be found elsewhere [46a,46b,46c,46d]. Methods for measuring cellular forces include collagen contraction, tissue pillars, two-dimensional (2D) and three-dimensional (3D) traction force microscopy (TFM), and micropillar arrays. For a review of these methods, refer to Polacheck and Chen [47]; TFM and micropillar arrays are discussed briefly below.

### 2.2.1 *Techniques to analyze cellular tractions*
#### 2.2.1.1 Traction force microscopy

Cellular TFM developed by Dembo and Wang [48] remains the most widely used method to measure cell forces. Traction forces generated by cells can be decomposed into a component acting parallel to the substrate surface and a normal component, which acts perpendicular to the substrate surface. The forces that act parallel to the substrate surface generate deformations in the optical viewing plane and can be visualized using wide-field microscopy. TFM involves tracking synthetic elastic polymer substrates as they move in response to cellular forces [47]. Briefly, standard 2D TFM involves mixing small fluorescent beads (<1 µm) into a substrate and seeding cells on/in the substrate. The substrate used for this application must be a flat, deformable material that has well characterized mechanical properties. The material must behave as an isotropic linear elastic material under deformations that are likely to occur. In addition to this, the substrate must be resistant to degradation in order to decouple force measurements from changes in mechanical properties of the substrate. The fluorescent beads are optically imaged in a stressed state, and then, the cell traction forces are released by cell lysis, detachment, or myosin inhibition, and the beads are tracked in space and time to determine their position in an unstressed state. Computational algorithms are then used to determine the displacement of the beads from the images and the force required to cause such displacements. This technique allows cellular forces to be mapped at a subcellular resolution as the size of the beads is much smaller than the size of the cells. However, complicated computation calculations are required to determine bead displacements and forces [47]. Various computational techniques are discussed in detail in the following publications [49–51].

Tracking substrate deformation in a 2D plane (as described above) is not representative of a 3D environment as contractile forces generated by cells are distributed throughout the 3D space. For this reason, TFM techniques have been modified to track bead displacement in 3D with confocal microscopy. However, computing traction forces in 3D requires considerable computational resources. Measuring tractions of cells in 3D is difficult for two main reasons: (a) the experimental/computational complexities and (b) the mechanical properties of biologically relevant 3D culture substrates are much more complicated than those of well-characterized nondegradable synthetic materials used for 2D TFM [47].

TFM and related techniques have enabled characterization of the force dynamics involved in a variety of cell biological processes such as adhesion maturation [52,53], migration [48,54–56], differentiation [57], and malignant transformation [58]. Although great progress has been made in this field over the past number of years, it still remains unclear how forces measured in vitro on mechanically simplified materials relate to forces in living tissues. Current methods measure the forces between a cell and a single material, but in vivo, cells are connected to a host of materials and other cells, all of which contribute to the generation and propagation of cellular forces [47]. However, the ever-growing community of engineers, mathematicians, and scientists are working on the continual development of solutions to overcome the shortcomings of current methods.

### 2.2.1.2 Micropillar arrays
Micropillar arrays are another method to measure cellular traction forces. In this technique, single cells are seeded onto an array of micron-sized evenly distributed pillars/cantilevers. The tops of the cantilevers serve as the cell substrate, which results in a high density of force sensors beneath a single cell. Cellular- or subcellular-scale pillars are typically 0.5–10 µm. The displacements of each cantilever in an array can be tracked, and the observed displacements can be used to calculate the tissue traction forces using beam theory [25,47]. Furthermore, these posts can be fabricated in a cost-effective manner, as described by Rodriguez et al. [26]. Micropillars are also known as micropost arrays or microfabricated postarray detectors (mPADs) [26,47]. Micropillar arrays have been used to investigate cell spreading [59,60], migration [61–63], contractility [60,64,65], focal adhesion strength [66], and cadherin junction tractions [67,68].

## 2.3 BIOREACTORS TO MIMIC THE IN VIVO ENVIRONMENT
In the body, the forces experienced by tissues and cells vary in both type and magnitude depending on the physiological location. As a result, each type of tissue construct (skin, bone, cartilage, tendon, blood vessel, etc.) has different requirements, making bioreactor design a complex task. For this reason, tissue-specific bioreactors have been developed based on a thorough understanding of biological and engineering aspects, to generate loading conditions in vitro similar to those experienced by cells in their native niche [69]. A tissue-engineering bioreactor can be defined as a device that uses mechanical means to influence biological processes [70]. Bioreactors are generally designed to perform at least one of the following functions: (a) provide a spatially uniform cell distribution, (b) maintain the desired concentration of gases and nutrients in culture medium, (c) facilitate mass transport to the tissue, (d) expose the construct to physical stimuli, and/or (e) provide information about the formation of 3D tissue [71–73]. Numerous studies have demonstrated that the application of mechanical cues assists in the differentiation and growth of stem cells and the production of functional ECM, such as aligned tendon [74–77], cartilage [78–81], and mineralized bone [52,82–84]. Bioreactor studies are often combined with computational/mathematical modeling to advance the understanding of the dynamic environment.

### 2.3.1 Types of bioreactors
Bioreactors range from advanced commercial systems to custom-built systems developed and built by researchers. Bioreactors have been specifically developed to apply mechanical stimulation via compressive loading, tensile strain, hydrostatic pressure, shearing fluid flow, or indeed a combination of these elements. These types of bioreactors are shown in Fig. 3. For a more thorough review, refer to Pörtner et al. [69].

Flow perfusion bioreactors are most commonly used as they replicate a dynamic environment by allowing 3D structures to obtain nutrients and eliminate waste. Flow

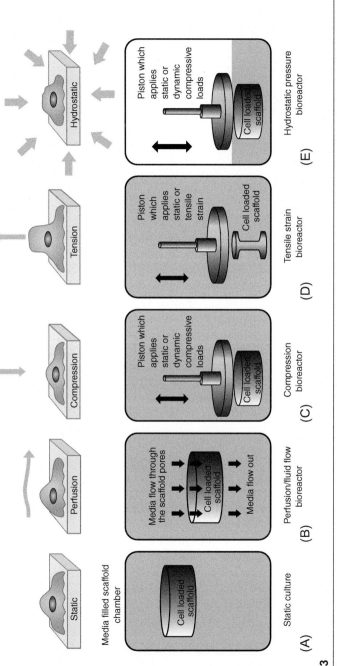

**FIG. 3**

Static culture (A) and perfusion (B), compression (C), tensile (D), and hydrostatic (E) forces applied to cell loaded scaffolds using bioreactors.

perfusion bioreactors generally consist of a pump that forces the media through a scaffold (located in a scaffold chamber) at a quantifiable flow rate. Media can be easily changed in this experimental setup; however, results may largely depend on the flow rate. Microfluidics systems typically consist of cells grown on a porous scaffold or flat surface, and fluid is pumped across the cell layer, whereas in microcarrier systems, cells are seeded on a scaffold that is placed in an agitated solution [69,71].

Compressive forces in vivo generate shear stress and strain as fluid is forced from the compressed area to the interstitial spaces. Both static and dynamic forces occur in vivo. Compression bioreactors are particularly important in the musculoskeletal system; specifically, osteocytes and chondrocytes are particularly sensitive to compressive forces. Generally, compression bioreactors consist of a motor that provides a linear motion and a controlling mechanism providing displacement regimes. The compressive force is transferred to the construct by flat platens that distribute the load evenly. Mass transfer is usually improved in dynamic compression bioreactors as compression causes fluid flow through the scaffold [71,85].

Tensile forces are commonly experienced in tendons, ligaments, and muscles. In order to grow these tissues in vitro it is necessary to align the cell growth along the appropriate axis. Once the cells are aligned, the intracellular cytoskeleton and ECM deposition will also be aligned parallel to the strain axis. Many tensile strain bioreactors have very similar design to compression bioreactors, differing only in the direction in which the load is applied. In this case, the scaffold is clamped in position using nonslip grips, and tensile strain is applied [69,71].

Hydrostatic pressure bioreactors can be used to apply mechanical stimulus to cell-loaded constructs and are commonly used in cartilage tissue engineering. Hydrostatic pressure bioreactors generally consist of a scaffold chamber that can withstand the pressure applied and a means to apply the pressure, such as an actuator-controlled piston. In this case, the piston must apply the pressure via an impermeable membrane so as not to sacrifice sterility of the experimental setup [71].

The four basic steps of bioreactor design are (a) identifying the needs and technical requirements, (b) defining and evaluating the related concepts, (c) designing and drawing the device, and (d) building and validating the device. Furthermore, the design has to be adapted to the specific purpose of the research and how the tissues will be used [86]. A description of bioreactor design requirements is provided by Partap et al. [71].

### 2.3.2 Future of bioreactors

Static culture conditions do not accurately represent the dynamic in vivo environment and are being gradually replaced by bioreactor culture systems. A better understanding of the mechanobiological environment of cells in 3D is required for the successful fabrication of functional engineered tissue. Bioreactors are a vital cog in the transition to the next generation of cell research, whereby readily available, easy-to-use systems will allow researchers to apply appropriate mechanical loading to their experiments and hence mimic the native cell environment [87]. However, currently, most bioreactors are specialized devices with a low-volume output. Many exhibit operator-dependent variability, and their assembly is time-consuming and labor-intensive [71]. First, bioreactors are required to enable us to study this complex 3D environment, and following on from this, scaled-up automated bioreactors are required to produce this engineered tissue.

## 2.4 ANIMAL LOADING MODELS

Animal loading models are often required to elucidate the mechanobiology of a living tissue under normal and altered mechanical conditions. These models are commonly used to study bone mechanobiology, as loading is particularly important for bone development, remodeling, and regeneration. The bone is constantly remodeled by the coordinated action of bone-resorbing osteoclasts and bone-forming osteoblasts. During physical activity, mechanical forces are exerted on bones through ground reaction forces and by the contractile activity of muscles [88,89]. These physical forces result in a maintenance or gain of bone mass and adaptive bone remodeling. The lack of physical activity/mechanical loading results in resorption of the bone [90]. Numerous animal loading models have been developed throughout the years to test specific hypothesis about bone modeling and remodeling. Animal loading models are used to apply forces at the organ scale in order to generate responses at the cellular level in an effort to determine what mechanical signals elicit specific cellular responses.

In a controlled experimental environment, the force required to generate these mechanical signals can come from intrinsic sources, such as voluntary muscle contraction during a vigorous exercise session (noninvasive), or from normal activity following the surgical removal of a nearby bone that formerly shared the load (invasive). Alternatively, the load can originate from extrinsic sources, such as pressure applied to the skin adjacent to the bone (noninvasive) or loads applied to surgically implanted pins (invasive) [91]. For a review of some of the most widely used animal loading models for bone, refer to Robling et al. [91]. While we focus here on skeletal tissues, animal models have been applied in mechanobiological studies of a number of other organs, for example the tendons in mouse treadmill running (described in Chapter 5), the vasculature of hypertensive mice (described in Chapter 7) and scar mechanotransduction in pig skin tissue (described in Chapter 14).

### 2.4.1 Noninvasive extrinsic skeletal loading models

Early models enabling extrinsic control of load levels provided a significant insight into bone remodeling; however, they typically employed invasive surgical procedures, which can present complications (e.g., infection and inflammation) in experiments and interpretation of results. This led to the development of noninvasive animal loading models that are capable of applying a well-defined mechanical signal to the bone without the potential complications of surgery. Noninvasive models are technically simpler, less expensive and do not rely on healing processes, as compared with the surgical models [91]. The two most commonly used noninvasive animal

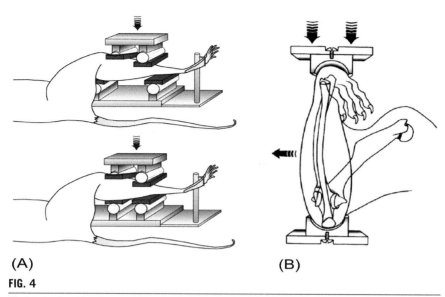

**FIG. 4**

Skeletal animal loading models. (A) Tibial four-point bend model [94]; (B) ulnar compression model [15].

loading models are the tibial four-point bending approach developed by Turner and coworkers [92] and the ulnar compression model of Lanyon and coworkers [93].

### 2.4.1.1 Tibial four-point bend model

The tibial four-point bend model was first described by Turner [92], whereby the rat tibia is subjected to four-point bending in the mediolateral direction. The right hind limb of an anesthetized animal is placed between pairs of upper and lower padded load points. A downward force is applied to the upper points, and the load is transmitted to the tibia through the skin, fascia, muscle, and periosteum, resulting in the production of a bending moment in the region between the two upper points. The bending moment imposes a compressive strain on the lateral tibial surface and tensile strain on the medial surface, as shown in Fig. 4A (top). On the contralateral leg of the animal, a sham configuration is implemented, whereby the upper and lower points directly oppose each other, as shown in Fig. 4A (bottom). In this configuration, the sham leg is squeezed, but the bone does not deform [95]. This model has since been scaled down for a mouse model [96].

### 2.4.1.2 Ulnar compression model

In the ulnar compression model [93], the forearm of an anesthetized animal is secured between two small metal cups that are mounted on the platens of a materials testing machine or other actuator. The elbow is secured with one cup, and the dorsal surface of the volar-flexed wrist is secured in the second cup. Compressive forces applied to the platens are transmitted to the ulnar diaphysis through the skin, fascia, articular cartilage (at the distal end), and ulnar metaphyseal bone, as shown in Fig. 4B.

The natural curvature of the ulnar diaphysis translates ~90% of the axial compression into a mediolateral bending moment [91]. The ulnar loading model was also initially developed for the rat and has been modified for the mouse [97] and rabbit [98].

In both the tibial four-point bend model and the ulnar compression model dynamic loads are applied, and the load magnitude, rate, number of cycles, and duration are well controlled. These noninvasive models, combined with increased computing power, higher resolution imaging, and new molecular techniques, will enable systematic evaluation of loading parameters to understand the nature of the osteogenic stimuli and pathways [99]. Additionally, the explosive growth of transgenic animal technology will undoubtedly lead to a more comprehensive understanding of the process of mechanically induced bone formation.

### 2.4.2 Embryonic animal models with an altered mechanical environment

Evidence from animal models has been key to help our understanding of the importance of movement as a regulatory tool in sculpting skeletal development. In animal models, the mechanical environment can be altered in a number of ways including the in vivo immobilization of the musculature (in ovo immobilization) or the use of mutant mouse embryos in which the skeletal rudiments develop with reduced, absent, or noncontractile muscle (reviewed in Ref. [100]). Both the chick and mouse are valuable vertebrate models used to investigate the effect of mechanical stimulation on embryonic skeletal development, due to their similarities with human musculoskeletal development.

#### 2.4.2.1 In ovo immobilization

An advantage of the chick embryonic model is that it can be physically manipulated in ways that are impossible in the mammalian embryo. The chick shares many features of embryonic development with mammals and has a huge advantage of development external to the mother, in ovo (in the egg), which allows procedures and alterations to the embryos and resulting effects to be examined (reviewed in Ref. [101]). During chick development, innervations of chick myotomes occur at approximately embryonic day 3 (E3) [102], and it has been reported that spontaneous limb movements occur from E3.5 to hatching [103]. Immobilization can be achieved in the developing chick embryo either surgically, by extirpation of the spinal nerves, or by application of pharmaceutical agents that block neuromuscular signals (e.g., Refs. [104–106]). Immobilization studies on the chick have shown that biophysical stimuli are required for correct initiation of ossification [104], several aspects of joint morphogenesis [105,107–110], and correct spine development [106].

#### 2.4.2.2 Mammalian models

Essential information about regulatory genes and the role of environmental stimuli for skeletal development has emerged using the developing chick model; however, for appropriate comparison with human development, the mammalian murine model has been utilized. Another benefit of the murine model is the substantial knowledge of the genome and the similarity in gene regulation mechanisms with the human. Muscle contractions begin relatively early in development, at approximately the same time as the cartilage template is taking shape, after approximately E12.5 in the mouse [111]. Genetic manipulation has produced mouse models that can be used to study the effect of mechanical stimulation from movement on skeletal development; they include mice with reduced ($Myf5^{nlacZ/+}$:$MyoD^{-/-}$ [112]), immobile ($Mdg^{-/-}$ [113]), or absent (Splotch [114,115], Splotch delayed [116,117], $Myf5^{nlacZ/nlacZ}$:$MyoD^{-/-}$ [118,119], $Six1^{-/-}$:$Six4^{-/-}$ [120]) skeletal muscle. These mouse models that lack normal muscle contraction show similar skeletal phenotypes to those observed in the chick immobilization studies including joint fusions and alterations in the ossification pattern [121].

### 2.4.2.3 Zebrafish models

The recent emergence of the zebra fish as a model for mechanoregulation of the skeletal system builds on the work of the previously described chick and mouse. The zebra fish is a system, however, in which many transgenic lines are available, specifically those that mark the various cell types of the musculoskeletal system [122]. This system has aided the understanding of cellular behavior following the manipulation of the mechanical environment [123]. Paralysis of the zebra fish exhibits a reduction in the size of all pharyngeal cartilage, establishing muscle loading in this model as a regulator of chondrocyte intercalation [124]. Similarly, zebra-fish mutants that lack neuromuscular nicotinic receptors (nic b107) and are therefore immobile display jaw morphology abnormalities, such as smaller and wider elements [124]. Both flaccid and rigid paralysis of the zebra fish have been shown to show similar changes to the morphology and function of the jaw joint [125]. It has recently been demonstrated using live zebra-fish joint imaging that cell behavior such as proliferation, migration, intercalation, and cell morphology changes required to shape the jaw joint are altered under reduced biomechanical conditions [126]. The malleable nature of this model could potentially hold promise for joint malformation recovery studies following periods of immobilization, as may occur in utero.

## 2.5 FLUORESCENT PROTEINS (FPs) AND IMAGING TECHNIQUES

The discovery of green fluorescent protein (GFP) in 1962 [127] has led to the development of a number of FPs with various hues. FPs are members of structurally similar class of proteins that share the unique property of emitting fluorescence at a specific wavelength when excited by a specific wavelength. FPs can be fused to virtually any protein of interest and genetically encoded into cells to analyze protein geography, movement, and chemistry in living cells [128]. FPs have been widely used for live-cell imaging over the past 20 years and have advanced our understanding of many important molecular and cellular functions in live cells. For a thorough review on the various FPs, refer to Wang et al. [129]. As a result of the innovation in FPs, new imaging technologies that utilize FPs have also been developed. Techniques utilizing novel FPs and imaging technology have been making a substantial impact on mechanobiology research over the past number of years.

### 2.5.1 FPs as markers in mechanobiology

Mechanical forces can activate a number of signaling molecules located in the cell membrane and other subcellular compartments. As FPs are genetically encoded, they are well suited for the imaging of the spatiotemporal localization and activation of signaling molecules and structures in live cells in response to mechanical stimuli. A large number of signaling molecules have been labeled with FPs, and as such, the position and movement of these molecules can be visualized with high spatiotemporal resolution techniques [129,130]. FPs and live-cell imaging can be used to visualize organelles, cytoskeleton, signaling molecules, and gene expression in mechanobiology, as discussed in detail by Wang et al. [129].

Briefly, at the organelle level, FPs can be fused to signaling molecules that localize to subcellular organelles to monitor where the organelle resides. FPs can highlight organelles to serve as reference points for the determination of the global mechanical properties [131–133]. FPs can also be fused to cytoskeleton molecules such as actin, microtubules, and intermediate filaments making the cytoskeleton fluorescent whereby morphology and deformations of the cytoskeleton can be monitored in a dynamic fashion [134,135]. Labeling and monitoring the dynamics and intercompartmental traffic of signaling molecules have been the most successful use of FPs in mechanobiology to date. FPs have been used to observe the molecular dynamics in terms of intracellular mechanical tension/stress [136–138], extracellular mechanical environment [139,140], external mechanical loading [141–144], and the mechanical impact the cells exert on the extracellular environment [52,53]. FPs have also been used to investigate the translocation of specific target molecules among different subcellular organelles [145–148]. In gene expression, FPs are fused to the promotor region of the gene of interest; when cells are exposed to various types of mechanical stimulation, the up-/downregulation of the gene can be monitored by the levels of expressed FPs [129].

### 2.5.2 Imaging technologies using FPs
#### 2.5.2.1 Live cell imaging

Time-lapse imaging is used to observe and capture cellular dynamics by imaging live cells at regular time intervals using fluorescent or indeed light microscopy. In this technique, a camera captures sequences of images that are later viewed at faster speed to track cellular responses over time. The two main experimental challenges in collecting robust live-cell imaging data are to minimize photodamage while retaining a useful signal-to-noise ratio (specifically for fluorescent imaging techniques) and to provide a suitable environment for cells or tissues to replicate physiological cell dynamics. Living cells will only behave normally in a physiological environment, and control of factors (temperature and cell culture medium) using an environmental chamber is therefore critically important. The single most important factor to successful live-cell imaging and meaningful data is to limit excitation light as photobleaching is inevitable with this technique, as discussed by Ettinger et al. [149].

### 2.5.2.2 Fluorescent resonance energy transfer (FRET)

Fluorescent resonance energy transfer (FRET) is a phenomenon of quantum mechanics that involves the nonradiative transfer of energy from a donor to an acceptor fluorophore (molecule that fluoresces) [129]. A fluorophore can serve as a FRET donor if its emission spectrum overlaps the excitation spectrum of another fluorophore, the acceptor fluorophore. When the donor and the acceptor are in close proximity to one another (<10nm) at the correct orientation, the excitation of the donor can elicit an energy transfer, inducing emission of the acceptor. FRET efficiency is defined as the proportion of the donor molecules that have transferred excitation state energy to the acceptor molecules and is dependent on the distance and orientations between the fluorophores. FRET is a reversible reaction and occurs instantaneously [129]. Genetically encoded FRET biosensors can be easily introduced into cells making this technique well suited to molecular live-cell imaging to monitor mechanotransduction with high spatiotemporal resolutions. FRET-based techniques have been employed to visualize signal transduction in response to mechanical stimulation, as discussed by Wang et al. [129]. Briefly, Chachisvilis et al. [150] fused ECFP and EYFP (fluorescent proteins) to human $B_2$ bradykinin receptor, a G-protein-coupled receptor (GPCR), to detect the activation of GPCR. Using FRET, they showed that shear stress activated $B_2$ bradykinin in bovine aorta endothelial cells, and this effect can be inhibited by $B_2$-selective antagonist. These results suggest that the membrane $B_2$ bradykinin GPCRs are involved in mediating primary mechanochemical signal transduction in endothelial cells [150]. More recently, FRET has been reported in 3D where Zhao et al. demonstrate that $Ca^{2+}$ and cAMP levels of live embryos expressing dual FRET sensors can be monitored simultaneously at microscopic resolution [151].

### 2.5.2.3 Fluorescent recovery after photobleaching (FRAP)

Fluorescent recovery after photobleaching (FRAP) is a technique where fluorescent signals are selectively photobleached within a subcellular region, and the recovery of the fluorescence is monitored in that region over time. The fluorescent intensity of the bleached area will recover at different rates, depending on the levels of diffusion and active transportation of fluorescent molecules [129]. FRAP has been widely used to investigate molecular dynamics in mechanobiology. Using FRAP, Vereecke et al. [152,153] assessed the speed of intracellular $Ca^{2+}$ wave propagation during mechanical stimulation in rat retinal pigment epithelial cells. FRAP has also been used to show that mechanical stress controls the focal adhesion assembly by modulating the kinetics of zyxin in bovine adrenal capillary endothelial cells [136].

### 2.5.2.4 Confocal and two-photon microscopy

Cells in their native 3D environment behave very differently to in vitro 2D cultures [154–156]. Investigation into cellular responses requires 3D images at the cellular, subcellular, and ultrastructural levels. Imaging structures in 3D is an inherently difficult task as the contribution of a signal from above and below the focal plane produces background fluorescence, affecting the quality of the image. Depths and scattering effects [156] require new imaging techniques to achieve high-resolution images of cells and indeed FPs in 3D environments.

In confocal laser scanning microscopy, a point-source laser light excites a fluorophore in the sample that either is autofluorescent or has been stained with specific fluorescent dyes. The sample is imaged at sequential focal planes, and a pinhole detector excludes out-of-focus background fluorescence from detection. A stack of 2D optical sections is acquired, which enables production of 3D representations of internal structures [157]. However, in this technique, as the excitation light generates fluorescence, it also produces photobleaching and phototoxicity throughout the specimen (even though the signal is only collected from the plane of focus). The penetration depth is also limited by absorption of excitation energy throughout the beam path and by specimen scattering of the photons [157–159].

Two-photon excitation microscopy has been developed as an alternative to conventional single-photon confocal microscopy. In two-photon excitation microscopy, a fluorophore is excited by the simultaneous absorption of two long wavelength (low energy) photons. In this case, their combined energy induces excitation of a fluorophore, which normally requires the absorption of a high energy to become excited. This can only occur at a very focused area with limited volume (femtoliter scale) [129], and as such, noise originating from the areas outside the focal region is eliminated. As a result of the enhanced signal-to-noise ratio, the penetration depth of imaging is improved (several hundred micrometers) without significant photobleaching [129,160,161]. As this technique enables increased depth penetration and can be less phototoxic to live specimens, it has been widely used at the molecular, cellular, tissue, and animal levels [162–165,165a,165b].

# 3 MOLECULAR AND GENETIC TECHNIQUES TO STUDY MECHANOBIOLOGY

## 3.1 ANALYSIS OF mRNA EXPRESSION

To understand how a biological system works, researchers seek to comprehend the functioning of the systems' component parts. As all cells in a given organism possess an identical genetic makeup, it is the unique phenotypes or observable characteristics directed by differential gene expression that guide the system's complexity. The first step to understanding this complexity in assorted cell types is to discover which genes are expressed by the cells of interest, thus guiding cellular differentiation into specific tissue types and then into functioning systems. Great progress has been made over the recent years toward understanding the role that mechanical stimulation has on the development and maintenance of tissues and the impact it has in guiding cell differentiation [100,166,167]. Much of this understanding of the integration of mechanical forces and cellular responses has been possible through the analysis of messenger RNA (mRNA) profiles and changes in gene expression following alterations in the mechanical environment. The central dogma of molecular genetics is that DNA codes for protein not directly but indirectly through processes called transcription and translation. This indirect route of information transfer involves an intermediate ribonucleic acid (RNA) molecule that relays the message. This so-called mRNA carries the genetic information transcribed from DNA and is used to translate a template for protein synthesis. Through the analysis of mRNA by different means, as discussed below, it is possible to understand the types of proteins that are being guided to be produced by this molecular information transfer.

The objective of this section is to provide researchers from traditional engineering backgrounds with the theoretical principles and practical techniques of experimental molecular biology, to utilize in the field of mechanobiology.

The first step in selecting a method of mRNA expression analysis is to assess whether a hypothesis can be tested using known specific genes that may be responsive to experimental mechanical manipulation. In these cases, methods of detection of specific individual/single genes would be appropriate. Techniques for individual-/single-gene expression include in situ hybridization (spatial expression) and quantitative real-time polymerase chain reaction (qRT-PCR). In cases when specific gene changes would be unknown, a more high-throughput screening approach using techniques such as microarray or RNA sequencing that focus on genome-wide patterns of gene expression would be more appropriate. Utilizing high-throughput methods of mRNA detection offers the benefit of simultaneously capturing changes in interacting groups of genes, with the potential of illustrating novel mechanisms of mechanotransduction. This section will present the basic principles underlying the molecular and genetic techniques of high-throughput and individual-level detection of mRNA expression that have been utilized for mechanobiology studies.

### 3.1.1 Microarray analysis

DNA microarray or chip-based detection was the first of its type to take a large-scale screening approach to collect large data sets to allow data mining and reveal intricate functions. The methodological approach was originally used for sequence analysis but then became widely adopted to quantitatively measure changes in gene expression (reviewed in Ref. [168]). Microarray technology works on the principle of nucleic acid site-specific sequence binding or hybridization onto synthetic sequences present on a chip. A microarray (or chip) is a flat surface in which 10,000–100,000 distinct oligonucleotide (short number of nucleotides) probes are present. These probes represent unique sequences for individual genes that will allow for complementary binding of mRNA from cell/tissue samples. For both in vitro and in vivo experiments involving alterations of the mechanical environment, separate control and experimental groups are formed. The method includes total RNA extraction, reverse transcription of the RNA using oligo-dT primers, and inclusion of a promoter sequence. In vitro transcription is then performed to form complementary DNA (cDNA) incorporating a fluorescent label. The fluorescently labeled cDNA is hybridized with the microarray (chip), to allow the complementary sequence-specific binding of the sample cDNA with the oligonucleotide probe sets on the chip. Following rinsing and digital scanning of the chip, the abundance of RNA (bound labeled

# 3 Molecular and genetic techniques to study mechanobiology

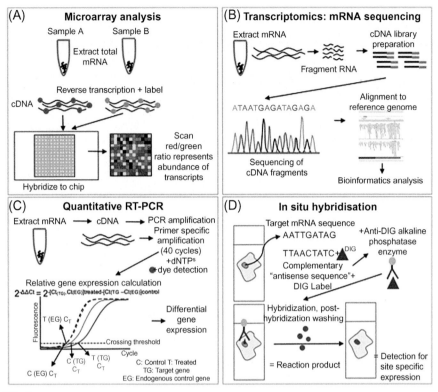

**FIG. 5**

Molecular techniques using mRNA expression to analyze mechanobiology. Basic flow through of each technique as described in the text for both high-throughput identification of changes in gene expression using microarray analysis (A) and transcriptomic mRNA sequencing (B) and more individual- or single-based gene expression changes using quantitative RT-PCR (C) and in situ hybridization (D).

cDNA) is determined by measuring fluorescent density (Fig. 5A). Data are then normalized among replicates for control and experimental groups and the statistical analysis performed. Differential gene expression is generally indicated with a fold change of $\geq 2.0$ or $\leq 2.0$.

The molecular response following alterations in the mechanical environment has been reported in various in vitro and in vivo studies in an attempt to understand in more detail the means by which mechanical stimuli modulate the cellular response during cellular differentiation. Profiling of genome-wide changes under altered mechanical environments has been carried out using in vitro culture systems in conjunction with microarray technology, including osteoblast cell lines subjected to weightlessness or microgravity conditions [169] and chondrocytes subjected to anabolic loading [170], dynamic compression [171], or hydrostatic pressure [172].

Analysis of genetic responses to altered mechanical environment during in vivo conditions has also been performed, including expression changes to an absence of movement during embryonic limb development [173].

A limiting factor of the hybridization methodology is its high background, because it is unable to distinguish RNA molecules sharing high sequence similarity. Microarrays also rely on hybridization with a labeled probe in which sequence is known, while RNA-sequencing technology doesn't depend on the genomic sequences being known that allows the potential to identify novel gene sequences.

### *3.1.2 Transcriptomics: Total RNA and mRNA sequencing*

A transcriptome is the whole set of RNAs transcribed by the genome from a specific tissue or cell type at a particular developmental stage and/or under a certain physiological condition. Following the sequencing of the genome, transcriptome analysis allowed researchers to understand further information on gene structure and regulation of gene expression. This technique has been utilized in multiple aspects of biology to reveal the regulation network of biological processes and guidance on aspects of diseases and drug discovery [174–176]. Transcriptome sequencing is a major advance in the study of gene expression because it allows a snapshot of the whole transcriptome rather than a predetermined subset of genes. Direct comparisons between RNA-sequencing-based approaches and microarray technologies to reveal alterations in gene expression between tissues report that RNA-seq identifies a great number of differentially expressed genes [173,177–179] and is more sensitive in reproducibly detecting alterations in gene expression at lower quantitative levels [173].

The steps for RNA sequencing begin in the same way as for a microarray, and total RNA is extracted. This RNA is then converted into a library of cDNA fragments. Sequence adaptors are added to each cDNA fragment, and a short sequence is obtained from each fragment from one end (single-end sequencing) or both ends (paired) using high-throughput sequencing technology (reviewed in Ref. [180]). The resulting sequence reads are aligned with the reference genome or transcriptome, or in the case where there is a limited reference genome, they can be assembled to produce a genome-scale transcription map that consisted of level of expression for particular genes (Fig. 5B).

This technique has advanced greatly over the last 10 years and is overcoming challenges with respect to cDNA library construction, bioinformatics, and sequence coverage versus cost (reviewed in Ref. [180]). A factor to consider when utilizing this technique for expression analysis is the quantity of high-quality RNA available for analysis. Recent work has optimized a protocol to extract high-quality RNA from human articular cartilage and performed RNA-seq; this advancement could be valuable to understand more about expression changes in osteoarthritic patients [181]. More recent advancements in the next-generation sequencing field have seen the emergence of single-cell RNA-sequencing technology (scRNA-seq) that is designed to overcome population-averaged RNA-seq that may mask rare subpopulations of cells (such as stem cells). Single-cell RNA-seq attempts to investigate expression profiles at the cell level, and comparisons between tube- and microfluidic-based extraction methods are being explored [182,183]. The advancement of single-cell genomics had the advantage of exploring cellular process with a more accurate resolution and thus may be of benefit to understanding mechanotransduction events in multiple contexts.

### 3.1.3 Quantitative real time PCR

qRT-PCR is a technique that is comprehensively used to analyze the expression levels of individual gene transcripts in a particular tissue or cell population following environmental manipulation. PCR was first devised in 1985, and it has had a major impact on biological research and genetic engineering. Through which, it is now possible to analyze 40,000-year-old DNA, DNA from fingerprints, blood, and tissue found at crime scenes, and analyze single embryonic cells for prenatal diagnosis of genetic disorders and virally infected cells. It is no doubt that this technique has been invaluable to our understanding of mechanobiology and the analysis of cellular changes following a change in the mechanical environment. qRT-PCR begins by converting sample mRNA into cDNA with corresponding sequences (using reverse transcriptase and DNA polymerase). PCR amplification is then performed encompassing a denaturation step (to separate DNA strands), an annealing step (to allow known sequence-specific primers for a particular gene to bind to the ends of the target sequence), and an extension phase (when DNA polymerase adds free nucleotides to the end of each primer). These steps are then repeated up to 40 cycles, which results in an exponential growing population of identical DNA molecules. Inclusion of a fluorescently bound dye during the annealing phase that fluoresces only when bound to a double-stranded PCR product is read computationally, and the levels of expression or $C_t$ (cycle threshold) value for a particular gene can be quantified. The cycle threshold method [184] is an example of the relative quantification approach that compares a gene of interest between experimental and control samples, following normalization to an endogenous control gene (Fig. 5C).

Changes in or identification of mechanosensitive genes following the application or removal of mechanical stimulation commonly utilizes qRT-PCR to confirm high-throughput data output [169,171,173]. The use of qRT-PCR is also valuable in assessing changes in the cellular phenotype following the manipulation of the mechanical environment. Work on mechanisms of chondrocyte differentiation shows that the application of hydrostatic pressure on embryonic cells and adult-derived progenitor cells results in a "stable" cartilage phenotype [185–187]. Work on wound healing identified that mechanical strain results in the upregulation of matrix remodeling genes and the production of more matrix [188]. This technique has also been valuable in revealing changes in gene expression due to changes in the mechanical microenvironment in glaucomatous cells [189] and changes in the thermal environment of bone cells during surgical cutting [190,191]. The value of this approach supports the quantification of the molecular changes in a tissue or cell population; however, it does not show the exact location in which these changes are taking place; this can be addressed with the technique called in situ hybridization.

### 3.1.4 In situ hybridization

In situ hybridization is a powerful tool for detecting DNA or RNA sequences in intact cells, tissues of whole organisms. Mary Lou Pardue and Joseph Gal pioneered the technique of in situ hybridization by using a radioactive test DNA to label stationary DNA of a cytological preparation [192] (Fig. 5D). This approach allowed for the first time the spatial localization of genetic information. This technique has continued to advance, and the method of complementary binding of a nucleotide probe to a specific target sequence is still applicable, with probes being labeled radioactively, colorimetrically, or fluorescently [193]. In situ hybridization is extensively used in research and clinical applications, especially for diagnostic purposes. Use of this technique has aided interpretation of phenotypic changes following the manipulation of the mechanical environment during skeletal development (Fig. 6A) [104,110,173,194,196] to elucidate the expression profile of genes in mechanosensitive regions [197].

Gene expression profiling at an individual-/single-gene level that has altered the mechanical environment in vivo has investigated candidate genes for altered expression, in order to assess the molecular response to alterations in biophysical stimuli. Both qualitative and quantitative approaches, as described above, to investigate changes in these genes have been performed [104,173,194,196,198,199].

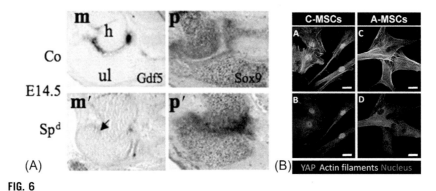

**FIG. 6**

(A) In situ hybridization shown by Kahn et al. [194] in the developing mouse humeroradial joint at 14.5 days of embryonic development. Sections of control (m, p) and Spd muscleless mutant (m', p') show no Gdf5 expression in Spd (*arrow* in (m')) in contrast to the control joint region (m). *Arrows* indicate joint loss in the mutant as visualized by Sox9 (p') gene expression (*ul*, ulna; *h*, humerus). (B) Immunofluorescent detection of nuclear or cytoplasmic YAP, shown by [195]. This study investigates the effect of age and substrate stiffness on MSCs on nuclear-to-cytoplasmic location of YAP as a measure of osteogenic mechanotransductive signaling. (A and B) YAP is located in the nucleus of children-derived MSCs (C-MSCs), while (C and D) it is located in the cytoplasm of adult-derived MSCs (A-MSCs) when cultured in control conditions.

Application of various forms of mechanical stimulation during in vitro culture regimes is a major goal of bioengineering techniques in order to create tissue suitable for regeneration applications. Through the methods described, it has been possible to reveal molecular responses to mechanical stimulation and identify tissue compositions due to the known molecular identity profiles.

## 3.2 ANALYSIS AT THE PROTEIN LEVEL

How cells perceive and relay dynamic mechanical signals to illicit an intracellular response and an alteration at the mRNA transcript level still remains unclear. Other avenues to aid understanding of these changes are at the protein level. It is credible that biomechanics impacts on proteins that guide cell matrix, cell-cell adhesion, and cytoskeletal and ultimately nuclear interactions [200]. This perspective in analyzing mechanobiology has been driven by the recent developments in functional proteomics and the ongoing advances in mass spectrometry quantitation (reviewed in Ref. [200]). More traditional approaches to understanding or observing changes at the protein level include the use of antibodies through spatial localization (immunohisto-/cytochemistry) or enzyme-linked immunosorbent assay (ELISA) or the identification of proteins based on molecular weight (western blotting [201]). The basis of these three techniques relies on a particular antigen-antibody complex binding.

### 3.2.1 Immunohistochemistry

To identify specific proteins in a tissue or cell type of interest, antibody molecules for specific target molecules are exposed to the sample. The binding of these molecules is detected by incubating the sample with a secondary antibody specific for immunoglobulin molecules and conjugated to a fluorophore (for fluorescent detection). This provides both a visible signal and amplification of the signal that can be visualized using a fluorescent microscope. Immunohistochemistry (IHC) provides information about the spatial localization of protein expression and qualitative evaluation of expression levels. The general steps for the procedure involve fixation, embedding, and sectioning (for tissue samples); detergent permeabilization of cell membranes; antigen retrieval (commonly used for paraffin-embedding sections to increase specificity of binding); and blocking and incubation with appropriate primary and secondary antibodies. Double or triple labeling of antigens can be performed in a single sample, as long as each primary antibody is either a different isotype or raised in a different species so that each can be recognized by distinct secondary antibodies with different labels. Appropriate negative controls are required during the procedure to confirm specificity of staining. A common approach is to use "no primary antibody controls," in which the primary antibody is omitted but the secondary is placed on the sample; this will give insight into nonspecific binding.

The use of IHC in mechanobiology research has facilitated investigation into phenotypic changes in cell populations following the application of mechanical stimulation in vitro [187,202,203]. This technique also provides value for the investigation of mechanisms of mechanotransduction, for example, through the evaluation of primary cilia following the application of mechanical stimulation [204–206]. The nuclear localization of specific signaling pathway components similarly utilizes this technique, in an attempt to understand how cells sense and adapt to external forces and physical constraints. An example of this is the analysis of the role of the YAP/TAZ (Hippo pathway components) as nuclear relays of mechanical signals exerted by ECM rigidity and cell shape (Fig. 6B) [195,207].

### 3.2.2 Western blotting
Western blotting (or sodium dodecyl sulfate-polyacrylamide gel electrophoresis (SDS-PAGE)) is a technique that identifies specific proteins based on separation by molecular weight through gel electrophoresis. The theory of the procedure is as follows: sample preparation (most commonly cell lysates), gel electrophoresis (two types of agarose gels: stacking and separating), blotting (electric transfer from gel to a membrane), washing, blocking, and antibody incubation [201]. The signal is then detected from the bound antibody, usually with an enzyme; this then corresponds to the target protein. Normalization among samples and experimental groups is based on the loading of each sample lane with an equal amount of total protein. For further validation, it is common practice to reprobe the membrane for a putatively constitutively expressed protein, such as beta-actin.

### 3.2.3 ELISA
The most sensitive and quantitative technique for protein analysis, ELISA, allows high specificity, even in complex solutions such as blood. The technique uses a biochemical assay to detect the presence of an antigen in a liquid sample. Since its first description by Engvall et al. in the 1970s [208], ELISA has experienced rapid adoption as a diagnostic tool in medicine, a valuable investigative method in scientific research and a quality control check in various biotech industries. While several different variations of the technique have been developed, the "sandwich" ELISA is the most pertinent and useful for analyzing soluble proteins in scientific research and will therefore be discussed in detail here. A sandwich ELISA operates by using two separate antibodies that recognize different epitopes, which can be either two different monoclonal antibodies or a polyclonal antibody solution. This method allows the measurement of growth factors and/or cytokine levels in cultures or biological liquids, providing convenient assessment of the biological responses to stimuli such as mechanical loading. ELISA kits are commercially available (typically as a 96-well plate), though they can be expensive. It is recommended that for high-throughput experiments for repeated analysis of particular antigens, a custom kit be developed in-house. However, custom sandwich ELISAs used for research purposes should be validated due to the risk of false-positive results [209].

The initial step in a sandwich ELISA involves coating the surface of the wells such that a known quantity of the capture antibody adsorbs onto its surface. Following blocking of any nonspecific binding sites, the wells should be incubated with serially diluted standards of known concentration and experimental samples.

A "blank" group, one without samples or standards, should be included to allow measurement of the background signal in the assay. Next, an enzyme-conjugated detection antibody should be added, followed by a substrate that forms a soluble colorimetric, fluorescent, electrochemical, or chemiluminescent product when cleaved. Between each step, extensive washing should be performed, adding a "stop" buffer at the end of the assay to terminate the enzyme reaction. A plate reader is then employed to collect the raw data, and a standard curve is generated upon removal of the background signal. Given appropriate conditions, ELISA can accurately measure sample concentrations in the low ($<10$) pictogram/millimeter range, and ELISA data are usually normalized to total protein or DNA (e.g., pictogram antigen/nanogram DNA) in order to account for potential variability in cell number among samples and experimental groups [210]. ELISA kits have been applied to study mechanobiology in varied tissues and organs, for example, exploring the effect of loading on intervertebral disk degeneration [211], examining correlation between pressure and cell stiffness in heart valve cells [212], and incorporating into microfluidic devices to characterize mechanotransduction in vitro [213].

## 3.3 TECHNIQUES FOR EDITING GENE FUNCTION AND ALTERING THE MECHANICAL ENVIRONMENT

### 3.3.1 In vitro mutagenesis—Mice

A technique called in vitro *mutagenesis* encompasses specific mutations being introduced into a cloned gene, and the mutated gene is returned to a cell in such a way that it disables or "knocks out" the normal cellular copies of the same gene. If the introduced mutations alter or destroy the function of the gene product, the phenotype of the mutant cell may help reveal the function of the missing normal protein. Using molecular and genetic techniques worked out in the 1980s, researchers can generate mice with any given gene disabled in order to study the role of that gene in development and in the adult. Multiple mouse models have been utilized to examine and investigate the role of the mechanical environment on cellular function, some of which are detailed below.

### 3.3.2 CRISPR

The novel molecular technique that has become increasingly popular over the past 5 years based on the identification of the functions of clustered regularly interspaced short palindromic repeats (CRISPR) and CRISPR-associated (Cas) genes and the manipulation of these for genome editing. The functions of CRISPR and Cas genes are essential for adaptive immunity, enabling organisms to eliminate invading genetic material [214,215]. The relative simplicity of the CRISPR nuclease system makes it amenable to adaptations for genome editing, which was realized in 2012 [216]. Manipulation of this system as a tool in molecular biology allows for either gene silencing or activation. To date, this tool has been used in multiple systems, including human, bacteria, zebra fish, and mice (reviewed in Ref. [214]). This tool has recently been utilized to eliminate gene function in a model of tendon biology in rats, to understand the cellular responses to mechanical stress [217]. CRISPR has the potential to expand our knowledge of the molecular mechanisms that are involved in mechanobiology, and the potential of its use will unquestionably be demonstrated over the next 5–10 years.

### 3.3.3 In ovo/ex ovo manipulation—Chick

A powerful tool for unraveling the molecular mechanisms involved in developmental processes is to ectopically express a gene or signaling pathway of interest and examine the effect. The avian embryo offers many advantages for developmental studies over mammalian embryos, due to the ease of access for in ovo (in the egg) manipulations. Different types of manipulations can be performed in ovo including surgical and chemically induced immobilization (discussed below). Retroviral transmission has been used very successfully to deliver genes into tissue locations in chick embryos (reviewed in Ref. [218]). Retroviral independent gene transfer can be achieved in chick embryos using in ovo electroporation, a more successful technique for targeting specific embryonic tissues/cells compared with microparticle bombardment and lipofection, offering a positive alternative to broad retroviral infection [219,220]. The basis of the electroporation technique relies on the transient generation of pores in the plasma membrane, to allow macromolecules to penetrate the cytoplasm and DNA to enter due to its negative charge [221]. Multiple electroporation systems have been described with respect to targeting different tissues in the developing chick, for example, the neural tube [222], the somites [223], and the eye [224]. In general, in ovo electroporation has been most commonly applied to chick embryos at early stages of development, younger than Hamburger and Hamilton stage 20 (HH20/~E3.5). Therefore, an alternative to carrying out the DNA transfer in ovo for older embryos is to use shell-less culture techniques [225–229]. Such ex ovo methods have described using petri dishes [226], plastic cups [229], and drinking glasses [228]. These ex ovo methods provide additional accessibility that may be required in order to target a specific tissue at older stages of development. The use of this targeted technique may prove advantageous for investigating the molecular mechanisms involved in mechanoregulation, as it is possible to combine this technique with that of an altered mechanical environment.

## 4 COMPUTATIONAL TECHNIQUES IN MECHANOBIOLOGY

Computational techniques for probing research questions in the field of mechanobiology have developed alongside experimental investigations, as these techniques can both inform experimental design and glean new information from experimental observations. As computational power increases exponentially, these methods have become ever more sophisticated and enlightening. Indeed, many advances in mechanobiology have been spurred by computational investigation, shedding new light on problems ranging from the mechanical response, to loading of individual cells, to predicting tissue differentiation in response to loading.

While many experimental techniques exist to study mechanobiology, as outlined in the previous sections, almost all involve some sort of destructive interference with the tissue or cellular mechanical environment. Therefore, most mechanobiological problems represent excellent examples of Heisenberg's uncertainty principle, wherein it is effectively impossible to observe an intact in vivo mechanobiological environment without interfering with its native behavior. Examples of this include osteocytes in the bone [23], skeletal development in utero [230,231,231a,231b], cell migration in the intestinal epithelium [232], and growth of aortic aneurysms [233].

This section will outline a number of computational methods that have proved invaluable to the study of mechanobiology and will give a perspective on its future development as a field.

## 4.1 COMPUTATIONAL MODELING

Computational, or in silico, modeling comprises interdisciplinary methods that apply mathematics, physics, and computer science to replicate and analyze the behavior of complex systems through the use of computer simulation. By characterizing a system using numerous variables, the simulation can adjust these variables and predict the resulting effects on the system. In silico modeling of physical behaviors has developed from theoretical origins in the early 20th century into a powerful engineering tool to assess the mechanical behavior of physical structures; mechanical systems; and, more recently, biological processes. Rapid advances in computational power over the past two decades have brought computational modeling to the fore as a key tool to test prevailing theories or develop entirely new ones. The primary methods by which this is achieved are finite element (FE) method and finite volume method, whereby the system is broken down into a mesh of smaller, simpler regions, allowing modeling of solid or fluid behaviors, respectively. While FE modeling involves treating these elements like simple structures obeying physical laws, finite volume modeling calculates the change in flow of a fluid through the simple volume and into the next discrete volume. The standard physical equations solved in the elements or volumes are then assembled into a larger system of equations, allowing modeling and analysis of the entire problem [8]. The use of these techniques both complements and enhances the development of new and existing theoretical models in the field of mechanobiology. These techniques have been applied to a range of different tissues and diseases, a selection of which will be described in the following sections.

### 4.1.1 Computational fluid dynamics

Computational fluid dynamics (CFD) as a technique is readily applicable to the cardiovascular system, given its key role as a fluid transport system for the body. It has been applied for some time to investigate a wide range of vascular diseases, in disparate locations in the body. Since the early application of CFD methods to aneurysms in 1992 [234], they have developed rapidly to gain the confidence of clinicians as a strong diagnostic tool for predicting risk of rupture [235,236].

Similarly, the first application of CFD to coronary artery disease was published in 2000 [237] and has since been combined with significant advancements in medical imaging to develop realistic patient-specific models in 3D [238,239]. A recent and intriguing development is the study of mechanobiology of blood cells themselves using CFD [240], with multiple research groups simulating the interactions of crowded blood cell clusters in 3D [241–243]. Similar use of CFD in mechanobiology allows modeling the vitreous humor of the eye, predicting concentrations of shear stress on the chamber wall [244].

The other major mechanobiological application of CFD has been to predict flow of marrow or interstitial fluid within the bone. Early computational models were developed to characterize loading-induced fluid flow across whole bones [245]. Similar techniques were used to analyze an idealized lacunar-canalicular system, predicting abrupt changes in the drag forces within the canaliculi arising from changes in geometry or proximity to bone microporosity and haversian canals [246]. CFD techniques facilitated analysis of models of bone cells, in particular osteocytes, with idealized models predicting high shear stresses within the canaliculi [247]. More recently, CFD studies have demonstrated the importance of local geometry on fluid flow in the pericellular space, with geometries obtained from transmission electron microscopy (TEM) and ultrahigh-voltage electron microscopy (UHVEM) images suggested [248,249]. Additionally, numerical models have explored the effect of the pericellular matrix on flow through the canaliculus, investigating the permeability [250–252], fluid movement [253,254], and electrochemomechanical effects [254,255]. On a larger scale, shear stress within bone marrow under macroscopic loading has been characterized using CFD [256,257], predicting important mechanical stimuli for tissue engineering of the bone [258].

### *4.1.2 FE analysis*

Given that the bone is a stiff, mechanically active, adaptive tissue, FE models have been employed for decades to investigate the biomechanics of the bone. Application of FE to orthopedic tissues began in 1972 [259] and initially was largely focused on either the design of prostheses or fundamental research into structural biomechanics [260]. More recently, FE has been used to determine adaptation of tissue structure in response to loading and mechanical stimuli at the tissue and cell levels under macroscale mechanical loading. Adaption is largely modeled through either tissue growth or tissue differentiation algorithms (reviewed in Ref. [8]), with these approaches being used to successfully model fracture healing [261], skeletal morphogenesis [230,262], and regeneration [263]. A range of mechanical stimuli at the tissue or cell level can be computed from models, predicting stress and strain [264,265], marrow shear stress [266–268], and even thermal stimuli [269] at the tissue and cellular scales. At the cellular level, first complete 3D idealized FE model of the bone cell environment predicted that strains in the lacunar walls are amplified by the local matrix geometry [270]. These findings were corroborated by recent FE studies applying accurate 3D geometries of osteocytes using scans from confocal laser scanning microscopy and X-ray nanotomography, predicting that geometry alone can amplify strain transfer to the osteocyte in vivo [271,272]. FE models have also been applied to investigate mechanosensation of bone cells in vitro, allowing exploration of the stimulatory effects of cell morphology, focal adhesion density [273], and substrate material properties [274], as well as the translation of mechanical stimulation to the nucleus via the cytoskeleton [275].

The complex process of modeling of heart valve mechanics at the organ scale began as structural models, applying blood pressure as static loads [276–280]. Dynamic loading developed later and incorporated realistic geometries [281–283] and anisotropic [284] and nonlinear [285,286] material properties. At the tissue scale, research has concentrated on developing constitutive models to capture the mechanical behavior of heart valve tissue, with FE modeling recruited to implement these models (reviewed in detail elsewhere [287]). Modeling at the cell scale has developed recently and advanced rapidly, applying FE methods to either model the cell itself as a continuum [288,289] or characterize the structural behavior of the cytoskeleton [288,289].

Similar to the bone, cartilage is a mechanically responsive tissue for which FE analysis has provided many insights, including articular cartilage thickness distributions, skeletal morphology, and endochondral ossification patterns (reviewed in Ref. [290]). FE models have also been applied to investigate the expansion and growth of skin tissue, allowing the development of algorithms to predict mechanically controlled skin growth in health and disease [291–294]. Promising research into mechanics and mechanobiology in the vocal folds [295] and the vocal ligament [296] is also being carried out using FE analysis, demonstrating the wide breadth of research topics that benefit from these methods.

### 4.1.3 Multiscale and multiphysics modeling

While the various models outlined above focus on research questions confined to individual loading cases of a specific tissue structure or cell type, mechanobiology in vivo occurs across multiple scales, with translation of loading to cell and molecular levels followed by transduction into responses expressed at tissue and organ scales. Therefore, researchers across varying fields of study have recently applied multiscale modeling techniques to investigate this phenomenon.

In bone tissue, multiscale modeling techniques have been applied alongside periodic boundary conditions to determine that the strain experienced by osteocytes under the same macroscopic loading varies significantly and strongly depends on their location relative to microstructural porosities [297]. Furthermore, it was found that orientation of tissue structures such as lamellae can have a significant effect on strain experienced at the level of individual bone cells [297]). A similar multiscale FE approach has been applied to cells suspended in bone marrow, demonstrating the importance of cell-cell attachments for mechanosensation within the bone marrow under macroscopic bone loading [298].

Multiscale modeling has also been applied in more disparate cases, such as modeling fluid flow and matrix deformation in the liver, allowing optimization of

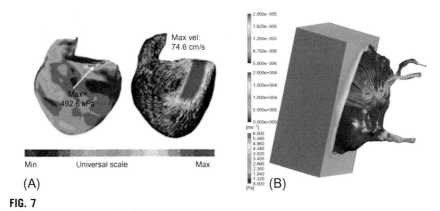

**FIG. 7**

Examples of fluid-structure interaction (FSI) models: (A) the contracting heart [301], showing stress distribution (left) and velocity profile (right); (B) an individual bone cell environment, with extracellular matrix strain, interstitial fluid velocity, and shear stress on the osteocyte surface shown [23].

perfusion conditions for tissue engineering [299]. In the study of morphogenesis, multiscale models are used both to investigate tissue-level effects in response to cellular- and molecular-scale events and to study cell arrangement in developing tissue [300].

As has been discussed, cells are exposed to various types of interrelated physical stimuli and therefore reside in a multiphysics environment. Multiphysics modeling represents a novel and developing array of methods that couple the effects of several physical phenomena in a single simulation or system of coupled simulations (see examples in Fig. 7). The type of multiphysics modeling most applicable to the study of mechanobiology is fluid-structure interaction (FSI) techniques, which couple classic CFD and FE modeling by relaying results between solvers in an iterative manner until a solution to both is converged upon. These new methods have been applied to models of in vitro systems, allowing determination of the mechanical stimulation applied to cells by experimental settings [302] and the stimulation experienced by individual cells at different locations in a tissue-engineering scaffold [303,304].

In the bone, FSI models have elucidated the function of the primary cilium as a mechanosensor on bone cells, determining the importance of cilia length [305]. Furthermore, FSI has been applied to the complex multiphysics environments within the bone, recently predicting that stimulatory magnitudes of shear stress result from macroscopic loading-induced fluid flow in accurate 3D models of osteocytes [23,306]. In an attempt to definitively compare these various mechanosensors, a comprehensive study of bone cell mechanosensation both in vitro and in vivo used FSI to predict that both integrin attachments and primary cilia are highly stimulated in vitro but that the primary cilia is less stimulated in vivo unless embedded in the surrounding matrix [307].

Multiscale techniques have been deployed to investigate in-stent restenosis alongside agent-based and cellular automata-based FE models of cell behavior [308–313], multiphysics modeling incorporating blood flow shear stress stimuli [314], and most recently mechanical/damage stimuli to individual cells [311], significantly advancing our understanding of this complex problem. Early advances were made in tying together mechanobiological stimuli across multiple scales and capturing multiphysics behavior in the aortic heart valve, with different forms of FSI simulations in this area developing over a decade of research (reviewed in detail elsewhere [315]). The use of these methods has shed new light on a range of different cardiovascular conditions, allowing analysis of transient, three-dimensional behavior over a range of length scales.

These multiscale and multiphysics models demonstrate the value of computational mechanobiology models for providing information on biophysical parameters that cannot be measured experimentally and the localized effects of multiple types of mechanosensors and complex patterns of physiological loading.

## 4.2 IMAGE ANALYSIS

One of the key problems in the study of mechanobiology is quantifying physical effects caused or experienced by cells, which is particularly challenging without directly interfering with them. While computational modeling attempts to recreate these effects, image analysis allows researchers to calculate the mechanics of mechanobiological behavior from experimental observations. These techniques can be developed in different manners and for various cells or tissues, with a selection of methods particularly useful in the field of mechanobiology discussed here.

### 4.2.1 Digital image correlation

Digital image correlation (DIC) is an optical technique that combines image registration and tracking methods for accurate 2D measurements of changes in images. Correlation theories for the measurement of alterations in data were first applied to digital images in 1975 [316]. These theories have been optimized in the recent years to apply to numerous applications [317], including confocal microscopy [318]. DIC is based upon the calculation of a correlation coefficient that is determined from pixel intensity array subsets on multiple corresponding images and extracting the deformation mapping function that relates the images. In this manner, the displacements of individual regions in an image are tracked over a series of images, with the resulting strain calculated.

DIC can thus be applied in mechanobiology to characterize strain at the tissue level or within individual cells under loading. The various tissues for which the technique has provided detailed strain maps and material properties include the skin [319,320]; the gallbladder [321]; the vasculature, such as the aorta [322–324]; the tympanic membrane of the ear [325]; and individual trabecular struts within bone [326]. DIC techniques have been recruited to diagnose cancer, as demonstrated by the detection of a basal cell carcinoma via strain mapping [327]. DIC can also

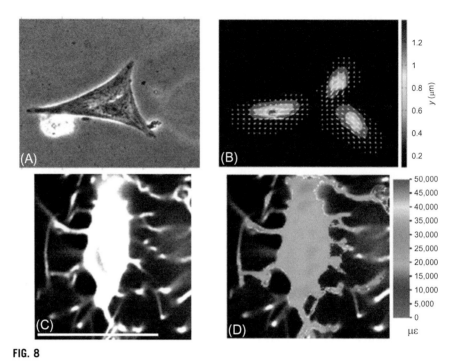

**FIG. 8**

Contour plots of cell mechanobiology developed using DIC: imaging of cell (A) contractility by (B) mapping substrate deformation [328] and imaging of (C) ex vivo osteocytes to generate contour plots of (D) cellular strains [329].

be applied at the scale of cells and can quantify the displacement field in a substrate under cell contraction [328] (Fig. 8A) or the velocity of cell migration [330]. A recent study applied this technique to osteocytes and osteoblasts in vivo, allowing cellular strains to be observed for the first time in bone tissue and providing direct evidence that loading of whole bones is amplified at the cell level [329] (Fig. 8B). DIC was also recently applied to analyze the beating of individual human cardiomyocytes, measuring both beating time and phases [331].

### 4.2.2 Particle image velocimetry

Developed over the past three decades, particle image velocimetry (PIV) has become a standard tool in experimental fluid mechanics. Given its ability to measure the instantaneous velocity field simultaneously at many points, it is possible to compute fluid vorticity and strain in rapidly evolving flows [332]. The technique has evolved from theoretical origins [333], with significant increases in computing power facilitating the development of digital PIV [334], while the advent and proliferation of standard digital cameras provided inputs perfectly suited to PIV [332].

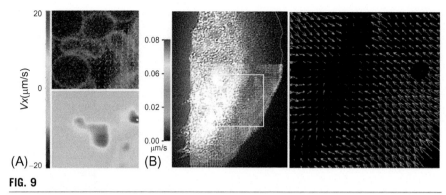

**FIG. 9**

The use of PIV allows imaging of (A) movement of individual red blood cells under flow [344] and (B) movement of the cytoskeleton within a cell [346].

PIV has been used extensively to investigate vascular biomechanics and mechanobiology, shedding light on the complex flow around heart valves and heart valve replacement devices [335–338]. Further development has facilitated the use of PIV to investigate flow inside bioreactors and scaffolds used for tissue engineering [339–341], including for the study of heart valve tissue mechanobiology [342]. Higher resolution capabilities have facilitated the investigation of cell level flow, including the dynamics of individual red blood cells [343]. Indeed, micro-PIV techniques are capable of capturing the flow around individual red blood cells [240,344]. In a fascinating application of the technology, it has recently been used at the cellular scale to calculate the shear stress affecting cell cytoskeletons [345] and to detect deformation of the cytoskeleton itself [346], as shown in Fig. 9. Finally, PIV has also been applied to capture the guidance of collective cell migration by substrate geometry [347] and the mechanical waves generated during the expansion of tissue, as occurs in both development [348] and cancer [349].

## 5 FUTURE PERSPECTIVES

Mechanobiology is a nascent field that, as is evident from the methods discussed, has benefited from rapid advancements in molecular analysis, imaging, and computational techniques. Our understanding of mechanisms of mechanosensing and mechanotransduction is deepening, alongside growing recognition in many overlapping fields of the importance of mechanical effects in cell behavior, tissue development, and various diseases. In particular, the potential of mechanobiological tools to augment tissue engineering by replicating in vivo mechanical environments provides an important avenue of study. The precision through which experimental manipulation of mechanical stimuli has advanced, in addition to improvements in measurements of cell mechanics, provides opportunities to investigate mechanotransduction in ever greater detail. Furthermore, the power and sophistication of computational tools have improved significantly and will likely spur further discoveries in the future. As engineering techniques become more intertwined with cellular and molecular analysis techniques, novel insights into the fundamental mechanisms by which cells appraise their mechanical environment will be gleaned. This will likely shed new light on the pathways by which cells transduce these stimuli into mechanical signals, presenting new therapeutic targets.

## REFERENCES

[1] Arnoczky SP, Lavagnino M, Egerbacher M. The mechanobiological aetiopathogenesis of tendinopathy: is it the over-stimulation or the under-stimulation of tendon cells? Int J Exp Pathol 2007;88(4):217–26.

[2] Beaupre GS, Stevens SS, Carter DR. Mechanobiology in the development, maintenance, and degeneration of articular cartilage. J Rehabil Res Dev 2000;37(2):145–51.

[3] Chicurel ME, Chen CS, Ingber DE. Cellular control lies in the balance of forces. Curr Opin Cell Biol 1998;10(2):232–9.

[4] Eckes B, Krieg T. Regulation of connective tissue homeostasis in the skin by mechanical forces. Clin Exp Rheumatol 2004;22(3 Suppl. 33):S73–6.

[5] Grodzinsky AJ, Levenston ME, Jin M, Frank EH. Cartilage tissue remodeling in response to mechanical forces. Ann Rev Biomed Eng 2000;2:691–713.

[6] Ingber D. Mechanobiology and diseases of mechanotransduction. Ann Med 2003;35(8):564–77.

[7] Lammerding J, Kamm RD, Lee RT. Mechanotransduction in cardiac myocytes. Ann N Y Acad Sci 2004;1015:53–70.

[8] Giorgi M, Verbruggen SW, Lacroix D. In silico bone mechanobiology: modeling a multifaceted biological system. Wiley Interdiscip Rev Syst Biol Med 2016;8(6):485–505.

[9] Wang N, Tytell JD, Ingber DE. Mechanotransduction at a distance: mechanically coupling the extracellular matrix with the nucleus. Nat Rev Mol Cell Biol 2009;10(1):75.

[10] Eyckmans J, Boudou T, Yu X, Chen Christopher S. A hitchhiker's guide to mechanobiology. Dev Cell 2011;21(1):35–47.

[11] Harris A, Wild P, Stopak D. Silicone rubber substrata: a new wrinkle in the study of cell locomotion. Science (New York, NY) 1980;208(4440):177–9.

[12] Kung C. A possible unifying principle for mechanosensation. Nature 2005;436(7051):647–54.

[13] Civelekoglu-Scholey G, Scholey J. Mitotic force generators and chromosome segregation. Cell Mol Life Sci 2010;67(13):2231–50.

[14] Renkawitz J, Sixt M. Mechanisms of force generation and force transmission during interstitial leukocyte migration. EMBO Rep 2010;11(10):744–50.

[15] Robling AG, Hinant FM, Burr DB, Turner CH. Improved bone structure and strength after long-term mechanical loading is greatest if loading is separated into short bouts. J Bone Miner Res 2002;17(8):1545–54.

[16] Stamatialis DF, Papenburg BJ, Gironés M, Saiful S, Bettahalli SNM, Schmitmeier S, et al. Medical applications of membranes: drug delivery, artificial organs and tissue engineering. J Membr Sci 2008;308(1):1–34.

[17] Tondon A, Kaunas R. The direction of stretch-induced cell and stress fiber orientation depends on collagen matrix stress. PLoS One 2014;9(2):e89592.

# References

[18] Castillo AB, Blundo JT, Chen JC, Lee KL, Yereddi NR, Jang E, et al. Focal adhesion kinase plays a role in osteoblast mechanotransduction in vitro but does not affect load-induced bone formation in vivo. PLoS One 2012;7(9):e43291.

[19] Merzouk S, Deuve JL, Dubois A, Navarro P, Avner P, Morey C. Lineage-specific regulation of imprinted X inactivation in extraembryonic endoderm stem cells. Epigenetics Chromatin 2014;7(1):11.

[20] Adachi T, Aonuma Y, S-i I, Tanaka M, Hojo M, Takano-Yamamoto T, et al. Osteocyte calcium signaling response to bone matrix deformation. J Biomech 2009;42(15):2507–12.

[21] Sugiyama T, Saxon LK, Zaman G, Moustafa A, Sunters A, Price JS, et al. Mechanical loading enhances the anabolic effects of intermittent parathyroid hormone (1–34) on trabecular and cortical bone in mice. Bone 2008;43(2):238–48.

[22] McNulty AL, Guilak F. Mechanobiology of the meniscus. J Biomech 2015;48(8):1469–78.

[23] Verbruggen SW, Vaughan TJ, McNamara LM. Fluid flow in the osteocyte mechanical environment: a fluid–structure interaction approach. Biomech Model Mechanobiol 2014;13(1):85–97.

[24] Ofek G, Athanasiou K. Micromechanical properties of chondrocytes and chondrons: relevance to articular cartilage tissue engineering. J Mech Mater Struct 2007;2(6):1059–86.

[25] Reynolds N. Experimental and computational investigation of the active force generation of cells subjected to static and dynamic loading (Doctoral dissertation). 2016.

[26] Rodriguez ML, McGarry PJ, Sniadecki NJ. Review on cell mechanics: experimental and modeling approaches. Appl Mech Rev 2013;65(6):060801.

[27] Hochmuth RM. Micropipette aspiration of living cells. J Biomech 2000;33(1):15–22.

[28] Ashkin A. Acceleration and trapping of particles by radiation pressure. Phys Rev Lett 1970;24(4):156.

[29] Ashkin A, Dziedzic JM. Optical trapping and manipulation of viruses and bacteria. Science (New York, NY) 1987;235:1517–21.

[30] Neuman KC, Nagy A. Single-molecule force spectroscopy: optical tweezers, magnetic tweezers and atomic force microscopy. Nat Methods 2008;5(6):491–505.

[31] Rotsch C, Jacobson K, Radmacher M. Dimensional and mechanical dynamics of active and stable edges in motile fibroblasts investigated by using atomic force microscopy. Proc Natl Acad Sci U S A 1999;96(3):921–6.

[32] Rotsch C, Radmacher M. Drug-induced changes of cytoskeletal structure and mechanics in fibroblasts: an atomic force microscopy study. Biophys J 2000;78(1):520–35.

[33] Stolz M, Raiteri R, Daniels A, VanLandingham MR, Baschong W, Aebi U. Dynamic elastic modulus of porcine articular cartilage determined at two different levels of tissue organization by indentation-type atomic force microscopy. Biophys J 2004;86(5):3269–83.

[34] Pelling AE, Dawson DW, Carreon DM, Christiansen JJ, Shen RR, Teitell MA, et al. Distinct contributions of microtubule subtypes to cell membrane shape and stability. Nanomed: Nanotechnol Biol Med 2007;3(1):43–52.

[35] Rico F, Roca-Cusachs P, Gavara N, Farré R, Rotger M, Navajas D. Probing mechanical properties of living cells by atomic force microscopy with blunted pyramidal cantilever tips. Phys Rev E 2005;72(2):021914.

[36] Mitchison J, Swann M. The mechanical properties of the cell surface. J Exp Biol 1954;31(3):461–72.

[37] Van Vliet K, Bao G, Suresh S. The biomechanics toolbox: experimental approaches for living cells and biomolecules. Acta Mater 2003;51(19):5881–905.

[38] Jones WR, Ping Ting-Beall H, Lee GM, Kelley SS, Hochmuth RM, Guilak F. Alterations in the Young's modulus and volumetric properties of chondrocytes isolated from normal and osteoarthritic human cartilage. J Biomech 1999;32(2):119–27.

[39] Sato M, Levesque MJ, Nerem RM. Micropipette aspiration of cultured bovine aortic endothelial cells exposed to shear stress. Arterioscl Thromb Vasc Biol 1987;7(3):276–86.

[40] Sato M, Ohshima N, Nerem RM. Viscoelastic properties of cultured porcine aortic endothelial cells exposed to shear stress. J Biomech 1996;29(4):461–7.

[41] Guilak F, Tedrow JR, Burgkart R. Viscoelastic properties of the cell nucleus. Biochem Biophys Res Commun 2000;269(3):781–6.

[42] Ribeiro AJS, Tottey S, Taylor RWE, Bise R, Kanade T, Badylak SF, et al. Mechanical characterization of adult stem cells from bone marrow and perivascular niches. J Biomech 2012;45(7):1280–7.

[43] Pravincumar P, Bader DL, Knight MM. Viscoelastic cell mechanics and actin remodelling are dependent on the rate of applied pressure. PLoS One 2012;7(9):e43938.

[44] Reynolds NH, Ronan W, Dowling EP, Owens P, McMeeking RM, McGarry JP. On the role of the actin cytoskeleton and nucleus in the biomechanical response of spread cells. Biomaterials 2014;35(13):4015–25.

[45] Reynolds NH, McGarry JP. Single cell active force generation under dynamic loading—part II: active modelling insights. Acta Biomater 2015;27:251–63.

[46] Deguchi S, Maeda K, Ohashi T, Sato M. Flow-induced hardening of endothelial nucleus as an intracellular stress-bearing organelle. J Biomech 2005;38(9):1751–9.

[46a] Kurth F, Eyer K, Franco-Obregón A, Dittrich PS. A new mechanobiological era: microfluidic pathways to apply and sense forces at the cellular level. Curr Opin Chem Biol 2012;16(3–4):400–8.

[46b] Bhatia SN, Ingber DE. Microfluidic organs-on-chips. Nat Biotechnol 2014;32(8):760.

[46c] Young EW. Cells, tissues, and organs on chips: challenges and opportunities for the cancer tumor microenvironment. Integrat Biol 2013;5(9):1096–109.

[46d] Hou HW, Lee WC, Leong MC, Sonam S, Vedula SRK, Lim CT. Microfluidics for applications in cell mechanics and mechanobiology. Cell Mol Bioeng 2011;4(4):591–602.

[47] Polacheck WJ, Chen CS. Measuring cell-generated forces: a guide to the available tools. Nat Methods 2016;13(5):415–23.

[48] Dembo M, Wang Y-L. Stresses at the cell-to-substrate interface during locomotion of fibroblasts. Biophys J 1999;76(4):2307–16.

[49] Hall MS, Long R, Feng X, Huang Y, Hui C-Y, Wu M. Toward single cell traction microscopy within 3D collagen matrices. Exp Cell Res 2013;319(16):2396–408.

[50] Style RW, Boltyanskiy R, German GK, Hyland C, MacMinn CW, Mertz AF, et al. Traction force microscopy in physics and biology. Soft Matter 2014;10(23):4047–55.

[51] Wang JH, Lin J-S. Cell traction force and measurement methods. Biomech Model Mechanobiol 2007;6(6):361–71.

[52] Balaban NQ, Schwarz US, Riveline D, Goichberg P, Tzur G, Sabanay I, et al. Force and focal adhesion assembly: a close relationship studied using elastic micropatterned substrates. Nat Cell Biol 2001;3(5):466.

[53] Beningo KA, Dembo M, Kaverina I, Small JV, Y-l W. Nascent focal adhesions are responsible for the generation of strong propulsive forces in migrating fibroblasts. J Cell Biol 2001;153(4):881–8.

[54] Dembo M, Oliver T, Ishihara A, Jacobson K. Imaging the traction stresses exerted by locomoting cells with the elastic substratum method. Biophys J 1996;70(4):2008–22.

[55] Jannat RA, Dembo M, Hammer DA. Traction forces of neutrophils migrating on compliant substrates. Biophys J 2011;101(3):575–84.

[56] Plotnikov SV, Pasapera AM, Sabass B, Waterman CM. Force fluctuations within focal adhesions mediate ECM-rigidity sensing to guide directed cell migration. Cell 2012;151(7):1513–27.

[57] Engler AJ, Sen S, Sweeney HL, Discher DE. Matrix elasticity directs stem cell lineage specification. Cell 2006;126(4):677–89.

[58] Kraning-Rush CM, Califano JP, Reinhart-King CA. Cellular traction stresses increase with increasing metastatic potential. PLoS One 2012;7(2):e32572.

[59] Lemmon CA, Sniadecki NJ, Ruiz SA, Tan JL, Romer LH, Chen CS. Shear force at the cell-matrix interface: enhanced analysis for microfabricated post array detectors. Mech Chem Biosyst 2005;2(1):1.

[60] Tan JL, Tien J, Pirone DM, Gray DS, Bhadriraju K, Chen CS. Cells lying on a bed of microneedles: an approach to isolate mechanical force. Proc Natl Acad Sci U S A 2003;100(4):1484–9.

[61] Du Roure O, Saez A, Buguin A, Austin RH, Chavrier P, Siberzan P, et al. Force mapping in epithelial cell migration. Proc Natl Acad Sci U S A 2005;102(7):2390–5.

[62] Sochol R, Higa A, Janairo R, Li S, Lin L. Effects of micropost spacing and stiffness on cell motility. Micro Nano Lett 2011;6(5):323–6.

[63] Ricart BG, Yang MT, Hunter CA, Chen CS, Hammer DA. Measuring traction forces of motile dendritic cells on micropost arrays. Biophys J 2011;101(11):2620–8.

[64] Rodriguez AG, Han SJ, Regnier M, Sniadecki NJ. Substrate stiffness increases twitch power of neonatal cardiomyocytes in correlation with changes in myofibril structure and intracellular calcium. Biophys J 2011;101(10):2455–64.

[65] Kural MH, Billiar KL. Mechanoregulation of valvular interstitial cell phenotype in the third dimension. Biomaterials 2014;35(4):1128–37.

[66] Fu J, Wang Y-K, Yang MT, Desai RA, Yu X, Liu Z, et al. Mechanical regulation of cell function with geometrically modulated elastomeric substrates. Nat Methods 2010;7(9):733–6.

[67] Liu Z, Tan JL, Cohen DM, Yang MT, Sniadecki NJ, Ruiz SA, et al. Mechanical tugging force regulates the size of cell–cell junctions. Proc Natl Acad Sci U S A 2010;107(22):9944–9.

[68] Ganz A, Lambert M, Saez A, Silberzan P, Buguin A, Mège RM, et al. Traction forces exerted through N-cadherin contacts. Biol Cell 2006;98(12):721–30.

[69] Pörtner R, Nagel-Heyer S, Goepfert C, Adamietz P, Meenen NM. Bioreactor design for tissue engineering. J Biosci Bioeng 2005;100(3):235–45.

[70] Darling EM, Athanasiou KA. Biomechanical strategies for articular cartilage regeneration. Ann Biomed Eng 2003;31(9):1114–24.

[71] Partap S, Plunkett N, O'brien F. Bioreactors in tissue engineering, In: Tissue engineering. InTechOpen; 2010, https://www.intechopen.com/books/tissue-engineering/bioreactors-in-tissue-engineering.

[72] Vunjak-Novakovic G, Obradovic B, Martin I, Bursac PM, Langer R, Freed LE. Dynamic cell seeding of polymer scaffolds for cartilage tissue engineering. Biotechnol Prog 1998;14(2):193–202.

[73] Bancroft GN, Sikavitsas VI, Mikos AG. Design of a flow perfusion bioreactor system for bone tissue-engineering applications. Tissue Eng 2003;9(3):549–54.

[74] Youngstrom DW, Barrett JG. Engineering tendon: scaffolds, bioreactors, and models of regeneration. Stem Cells Int 2015;2016:11 pages.

[75] Youngstrom DW, Rajpar I, Kaplan DL, Barrett JG. A bioreactor system for in vitro tendon differentiation and tendon tissue engineering. J Orthop Res 2015;33(6):911–8.

[76] Youngstrom DW, LaDow JE, Barrett JG. Tenogenesis of bone marrow-, adipose-, and tendon-derived stem cells in a dynamic bioreactor. Connect Tissue Res 2016;57(6):454–65.

[77] Burk J, Plenge A, Brehm W, Heller S, Pfeiffer B, Kasper C. Induction of tenogenic differentiation mediated by extracellular tendon matrix and short-term cyclic stretching. Stem Cells Int 2016;2016:11 pages.

[78] Mauck RL, Soltz MA, Wang CC, Wong DD, Chao P-HG, Valhmu WB, et al. Functional tissue engineering of articular cartilage through dynamic loading of chondrocyte-seeded agarose gels. J Biomech Eng 2000;122(3):252–60.

[79] Mauck RL, Nicoll SB, Seyhan SL, Ateshian GA, Hung CT. Synergistic action of growth factors and dynamic loading for articular cartilage tissue engineering. Tissue Eng 2003;9(4):597–611.

[80] Mauck RL, Seyhan SL, Ateshian GA, Hung CT. Influence of seeding density and dynamic deformational loading on the developing structure/function relationships of chondrocyte-seeded agarose hydrogels. Ann Biomed Eng 2002;30(8):1046–56.

[81] Mauck R, Byers B, Yuan X, Tuan R. Regulation of cartilaginous ECM gene transcription by chondrocytes and MSCs in 3D culture in response to dynamic loading. Biomech Model Mechanobiol 2007;6(1):113–25.

[82] McCoy RJ, O'Brien FJ. Influence of shear stress in perfusion bioreactor cultures for the development of three-dimensional bone tissue constructs: a review. Tissue Eng Part B: Rev 2010;16(6):587–601.

[83] Birmingham E, Kreipke T, Dolan E, Coughlin T, Owens P, McNamara LM, et al. Mechanical stimulation of bone marrow in situ induces bone formation in trabecular explants. Ann Biomed Eng 2015;43(4):1036–50.

[84] David V, Guignandon A, Martin A, Malaval L, Lafage-Proust M-H, Rattner A, et al. Ex vivo bone formation in bovine trabecular bone cultured in a dynamic 3D bioreactor is enhanced by compressive mechanical strain. Tissue Eng Part A 2008;14(1):117–26.

[85] Huang C, Charles Y, Hagar KL, Frost LE, Sun Y, Cheung HS. Effects of cyclic compressive loading on chondrogenesis of rabbit bone-marrow derived mesenchymal stem cells. Stem Cells 2004;22(3):313–23.

[86] Viens M, Chauvette G, Langelier È. A roadmap for the design of bioreactors in mechanobiological research and engineering of load-bearing tissues. J Med Devices 2011;5(4):041006–11.

[87] Henstock JR, El Haj AJ. Bioreactors. In: Mechanobiology: exploitation for medical benefit. Wiley Online Library; 2017. p. 275–96.

[88] Lanyon L, Hampson W, Goodship A, Shah J. Bone deformation recorded in vivo from strain gauges attached to the human tibial shaft. Acta Orthop Scand 1975;46(2):256–68.

[89] Usui T, Maki K, Toki Y, Shibasaki Y, Takanobu H, Takanishi A, et al. Measurement of mechanical strain on mandibular surface with mastication robot: influence of muscle loading direction and magnitude. Orthod Craniofac Res 2003;6(s1):163–7.
[90] Klein-Nulend J, Bacabac R, Bakker A. Mechanical loading and how it affects bone cells: the role of the osteocyte cytoskeleton in maintaining our skeleton. Eur Cell Mater 2012;24:278–91.
[91] Robling AG, Burr DB, Turner CH. Skeletal loading in animals. J Musculoskel Neuron Interact 2001;1(3):249–62.
[92] Turner C, Akhter M, Raab D, Kimmel D, Recker R. A noninvasive, in vivo model for studying strain adaptive bone modeling. Bone 1991;12(2):73–9.
[93] Torrance A, Mosley J, Suswillo R, Lanyon L. Noninvasive loading of the rat ulna in vivo induces a strain-related modeling response uncomplicated by trauma or periosteal pressure. Calcif Tissue Int 1994;54(3):241–7.
[94] Robling AG, Burr DB, Turner CH. Recovery periods restore mechanosensitivity to dynamically loaded bone. J Exp Biol 2001;204(19):3389–99.
[95] Akhter M, Raab D, Turner C, Kimmel D, Recker R. Characterization of in vivo strain in the rat tibia during external application of a four-point bending load. J Biomech 1992;25(10):1241–6.
[96] Akhter M, Cullen D, Pedersen E, Kimmel D, Recker R. Bone response to in vivo mechanical loading in two breeds of mice. Calcif Tissue Int 1998;63(5):442–9.
[97] Lee K, Maxwell A, Lanyon L. Validation of a technique for studying functional adaptation of the mouse ulna in response to mechanical loading. Bone 2002;31(3):407–12.
[98] Baumann AP, Aref MW, Turnbull TL, Robling AG, Niebur GL, Allen MR, et al. Development of an in vivo rabbit ulnar loading model. Bone 2015;75:55–61.
[99] van der Meulen MCH, Huiskes R. Why mechanobiology?: a survey article. J Biomech 2002;35(4):401–14.
[100] Rolfe R, Roddy K, Murphy P. Mechanical regulation of skeletal development. Curr Osteoporos Rep 2013;11(2):107–16.
[101] Stern CD. The chick; a great model system becomes even greater. Dev Cell 2005;8(1):9–17.
[102] King ED, Munger BL. Myotome and early neurogenesis in chick embryos. Anat Rec 1990;228(2):191–210.
[103] Hamburger V. Some aspects of the embryology of behavior. Q Rev Biol 1963;38:342–65.
[104] Nowlan NC, Prendergast PJ, Murphy P. Identification of mechanosensitive genes during embryonic bone formation. PLoS Comput Biol 2008;4(12):e1000250.
[105] Osborne AC, Lamb KJ, Lewthwaite JC, Dowthwaite GP, Pitsillides AA. Short-term rigid and flaccid paralyses diminish growth of embryonic chick limbs and abrogate joint cavity formation but differentially preserve pre-cavitated joints. J Musculoskelet Neuronal Interact 2002;2(5):448–56.
[106] Rolfe RA, Bezer JH, Kim T, Zaidon AZ, Oyen ML, Iatridis JC, et al. Abnormal fetal muscle forces result in defects in spinal curvature and alterations in vertebral segmentation and shape. J Orthop Res 2017;35(10):2135–44.
[107] Murray PD, Drachman DB. The role of movement in the development of joints and related structures: the head and neck in the chick embryo. J Embryol Exp Morphol 1969;22(3):349–71.

[108] Nowlan NC, Chandaria V, Sharpe J. Immobilized chicks as a model system for early-onset developmental dysplasia of the hip. J Orthop Res 2014;32(6):777–85.
[109] Persson M. The role of movements in the development of sutural and diarthrodial joints tested by long-term paralysis of chick embryos. J Anat 1983;137(Pt 3):591–9.
[110] Roddy KA, Prendergast PJ, Murphy P. Mechanical influences on morphogenesis of the knee joint revealed through morphological, molecular and computational analysis of immobilised embryos. PLoS One 2011;6(2):e17526.
[111] Suzue T. Movements of mouse fetuses in early stages of neural development studied in vitro. Neurosci Lett 1996;218(2):131–4.
[112] Rudnicki MA, Schnegelsberg PN, Stead RH, Braun T, Arnold HH, Jaenisch R. MyoD or Myf-5 is required for the formation of skeletal muscle. Cell 1993;75(7):1351–9.
[113] Pai AC. Developmental genetics of a lethal mutation, muscular dysgenesis (mdg), in the mouse. II. Developmental analysis. Dev Biol 1965;11:93–109.
[114] Franz T, Kothary R, Surani MA, Halata Z, Grim M. The Splotch mutation interferes with muscle development in the limbs. Anat Embryol (Berl) 1993;187(2):153–60.
[115] Tajbakhsh S, Rocancourt D, Cossu G, Buckingham M. Redefining the genetic hierarchies controlling skeletal myogenesis: Pax-3 and Myf-5 act upstream of MyoD. Cell 1997;89(1):127–38.
[116] Franz T. The Splotch (Sp1H) and Splotch-delayed (Spd) alleles: differential phenotypic effects on neural crest and limb musculature. Anat Embryol (Berl) 1993;187(4):371–7.
[117] Vogan KJ, Epstein DJ, Trasler DG, Gros P. The splotch-delayed (Spd) mouse mutant carries a point mutation within the paired box of the Pax-3 gene. Genomics 1993;17 (2):364–9.
[118] Kassar-Duchossoy L, Gayraud-Morel B, Gomes D, Rocancourt D, Buckingham M, Shinin V, et al. Mrf4 determines skeletal muscle identity in Myf5:Myod double-mutant mice. Nature 2004;431(7007):466–71.
[119] Kablar B, Krastel K, Tajbakhsh S, Rudnicki MA. Myf5 and MyoD activation define independent myogenic compartments during embryonic development. Dev Biol 2003;258(2):307–18.
[120] Grifone R, Demignon J, Houbron C, Souil E, Niro C, Seller MJ, et al. Six1 and Six4 homeoproteins are required for Pax3 and Mrf expression during myogenesis in the mouse embryo. Development 2005;132(9):2235–49.
[121] Nowlan NC, Bourdon C, Dumas G, Tajbakhsh S, Prendergast PJ, Murphy P. Developing bones are differentially affected by compromised skeletal muscle formation. Bone 2010;46(5):1275–85.
[122] Hammond CL, Moro E. Using transgenic reporters to visualize bone and cartilage signaling during development in vivo. Front Endocrinol 2012;3:91.
[123] Brunt LH, Norton JL, Bright JA, Rayfield EJ, Hammond CL. Finite element modelling predicts changes in joint shape and cell behaviour due to loss of muscle strain in jaw development. J Biomech 2015;48(12):3112–22.
[124] Shwartz Y, Farkas Z, Stern T, Aszodi A, Zelzer E. Muscle contraction controls skeletal morphogenesis through regulation of chondrocyte convergent extension. Dev Biol 2012;370(1):154–63.
[125] Brunt LH, Skinner RE, Roddy KA, Araujo NM, Rayfield EJ, Hammond CL. Differential effects of altered patterns of movement and strain on joint cell behaviour and skeletal morphogenesis. Osteoarthritis Cartilage 2016;24(11):1940–50.
[126] Brunt LH, Begg K, Kague E, Cross S, Hammond CL. Wnt signalling controls the response to mechanical loading during Zebrafish joint development. Development 2017;144(15):2798–809.

[127] Shimomura O, Johnson FH, Saiga Y. Extraction, purification and properties of aequorin, a bioluminescent protein from the luminous hydromedusan, aequorea. J Cell Comp Physiol 1962;59(3):223–39.
[128] Lippincott-Schwartz J, Patterson GH. Development and use of fluorescent protein markers in living cells. Science (New York, NY) 2003;300(5616):87–91.
[129] Wang Y, Shyy JY, Chien S. Fluorescence proteins, live-cell imaging, and mechanobiology: seeing is believing. Annu Rev Biomed Eng 2008;10:1–38.
[130] Tsien RY. The green fluorescent protein. USA: Annual Reviews; 1998.
[131] Rowat A, Lammerding J, Ipsen JH. Mechanical properties of the cell nucleus and the effect of emerin deficiency. Biophys J 2006;91(12):4649–64.
[132] Wang N, Naruse K, Stamenović D, Fredberg JJ, Mijailovich SM, Tolić-Nørrelykke IM, et al. Mechanical behavior in living cells consistent with the tensegrity model. Proc Natl Acad Sci U S A 2001;98(14):7765–70.
[133] Hu S, Chen J, Fabry B, Numaguchi Y, Gouldstone A, Ingber DE, et al. Intracellular stress tomography reveals stress focusing and structural anisotropy in cytoskeleton of living cells. Am J Physiol Cell Physiol 2003;285(5):C1082–90.
[134] Kumar S, Maxwell IZ, Heisterkamp A, Polte TR, Lele TP, Salanga M, et al. Viscoelastic retraction of single living stress fibers and its impact on cell shape, cytoskeletal organization, and extracellular matrix mechanics. Biophys J 2006;90(10):3762–73.
[135] Hu K, Ji L, Applegate KT, Danuser G, Waterman-Storer CM. Differential transmission of actin motion within focal adhesions. Science (New York, NY) 2007;315 (5808):111–5.
[136] Lele TP, Pendse J, Kumar S, Salanga M, Karavitis J, Ingber DE. Mechanical forces alter zyxin unbinding kinetics within focal adhesions of living cells. J Cell Physiol 2006;207 (1):187–94.
[137] Kirchner J, Kam Z, Tzur G, Bershadsky AD, Geiger B. Live-cell monitoring of tyrosine phosphorylation in focal adhesions following microtubule disruption. J Cell Sci 2003;116(6):975–86.
[138] Giannone G, Dubin-Thaler BJ, Rossier O, Cai Y, Chaga O, Jiang G, et al. Lamellipodial actin mechanically links myosin activity with adhesion-site formation. Cell 2007;128 (3):561–75.
[139] Gupton SL, Waterman-Storer CM. Spatiotemporal feedback between actomyosin and focal-adhesion systems optimizes rapid cell migration. Cell 2006;125(7):1361–74.
[140] Zamir E, Katz M, Posen Y, Erez N, Yamada KM, Katz B-Z, et al. Dynamics and segregation of cell-matrix adhesions in cultured fibroblasts. Nat Cell Biol 2000; 2(4):191.
[141] Galbraith CG, Yamada KM, Sheetz MP. The relationship between force and focal complex development. J Cell Biol 2002;159(4):695–705.
[142] Mack PJ, Kaazempur-Mofrad MR, Karcher H, Lee RT, Kamm RD. Force-induced focal adhesion translocation: effects of force amplitude and frequency. Am J Physiol Cell Physiol 2004;287(4):C954–62.
[143] Sawada Y, Tamada M, Dubin-Thaler BJ, Cherniavskaya O, Sakai R, Tanaka S, et al. Force sensing by mechanical extension of the Src family kinase substrate p130Cas. Cell 2006;127(5):1015–26.
[144] Guo W-h, Y-l W. Retrograde fluxes of focal adhesion proteins in response to cell migration and mechanical signals. Mol Biol Cell 2007;18(11):4519–27.
[145] Wang Y, Chang J, Li Y-C, Li Y-S, Shyy JY-J, Chien S. Shear stress and VEGF activate IKK via the Flk-1/Cbl/Akt signaling pathway. Am J Physiol Heart Circ Physiol 2004;286(2):H685–92.

[146] Ji JY, Jing H, Diamond SL. Shear stress causes nuclear localization of endothelial glucocorticoid receptor and expression from the GRE promoter. Circ Res 2003;92(3):279–85.

[147] Ganguli A, Persson L, Palmer IR, Evans I, Yang L, Smallwood R, et al. Distinct NF-κB regulation by shear stress through Ras-dependent IκBα oscillations. Circ Res 2005;96(6):626–34.

[148] Oancea E, Wolfe JT, Clapham DE. Functional TRPM7 channels accumulate at the plasma membrane in response to fluid flow. Circ Res 2006;98(2):245–53.

[149] Ettinger A, Wittmann T. Fluorescence live cell imaging. Methods Cell Biol 2014;123:77–94.

[150] Chachisvilis M, Zhang Y-L, Frangos JA. G protein-coupled receptors sense fluid shear stress in endothelial cells. Proc Natl Acad Sci U S A 2006;103(42):15463–8.

[151] Zhao M, Wan X, Li Y, Zhou W, Peng L. Multiplexed 3D FRET imaging in deep tissue of live embryos. Sci Rep 2015;5:13991.

[152] Gomes P, Malfait M, Himpens B, Vereecke J. Intercellular Ca2+-transient propagation in normal and high glucose solutions in rat retinal epithelial (RPE-J) cells during mechanical stimulation. Cell Calcium 2003;34(2):185–92.

[153] Himpens B, Stalmans P, Gomez P, Malfait M, Vereecke J. Intra-and intercellular Ca2+ signaling in retinal pigment epithelial cells during mechanical stimulation. FASEB J 1999;13(9001):S63–8.

[154] Wolf K, Friedl P. Molecular mechanisms of cancer cell invasion and plasticity. Br J Dermatol 2006;154(s1):11–5.

[155] Pedersen JA, Swartz MA. Mechanobiology in the third dimension. Ann Biomed Eng 2005;33(11):1469–90.

[156] Even-Ram S, Yamada KM. Cell migration in 3D matrix. Curr Opin Cell Biol 2005;17(5):524–32.

[157] Goggin PM, Zygalakis KC, Oreffo RO, Schneider P. High-resolution 3D imaging of osteocytes and computational modelling in mechanobiology: insights on bone development, ageing, health and disease. Eur Cell Mater 2016;31:264–95.

[158] Periasamy A, Skoglund P, Noakes C, Keller R. An evaluation of two-photon excitation versus confocal and digital deconvolution fluorescence microscopy imaging in *Xenopus morphogenesis*. Microsc Res Tech 1999;47(3):172–81.

[159] Shotton DM. Confocal scanning optical microscopy and its applications for biological specimens. J Cell Sci 1989;94(2):175–206.

[160] Denk W, Strickler JH, Webb WW. Two-photon laser scanning fluorescence microscopy. Science (New York, NY) 1990;248(4951):73–6.

[161] Helmchen F, Denk W. Deep tissue two-photon microscopy. Nat Methods 2005;2(12):932–40.

[162] Rubart M. Two-photon microscopy of cells and tissue. Circ Res 2004;95(12):1154–66.

[163] Koutalos Y. Intracellular spreading of second messengers. J Physiol 1999;519(3):629.

[164] Weiss S. Fluorescence spectroscopy of single biomolecules. Science (New York, NY) 1999;283(5408):1676–83.

[165] Centonze VE, White JG. Multiphoton excitation provides optical sections from deeper within scattering specimens than confocal imaging. Biophys J 1998;75(4):2015–24.

[165a] Curley CJ, Dolan EB, Cavanagh B, O'Sullivan J, Duffy GP, Murphy BP. An in vitro investigation to assess procedure parameters for injecting therapeutic hydrogels into the myocardium. J Biomed Mater Res B 2017;105(8):2618–29.

[165b] Payne C, Dolan EB, O'Sullivan J, Cryan SA, Kelly H. A methylcellulose and collagen based temperature responsive hydrogel promotes encapsulated stem cell viability and proliferation in vitro. Drug Deliv Translat Res 2017;7:132.
[166] Steward AJ, Kelly DJ. Mechanical regulation of mesenchymal stem cell differentiation. J Anat 2014;.
[167] Potter CM, Lao KH, Zeng L, Xu Q. Role of biomechanical forces in stem cell vascular lineage differentiation. Arterioscler Thromb Vasc Biol 2014;34(10):2184–90.
[168] Pirrung MC, Southern EM. The genesis of microarrays. Biochem Mol Biol Educ 2014;42(2):106–13.
[169] Patel MJ, Liu W, Sykes MC, Ward NE, Risin SA, Risin D, et al. Identification of mechanosensitive genes in osteoblasts by comparative microarray studies using the rotating wall vessel and the random positioning machine. J Cell Biochem 2007;101 (3):587–99.
[170] Scholtes S, Kramer E, Weisser M, Roth W, Luginbuhl R, Grossner T, et al. Global chondrocyte gene expression after a single anabolic loading period: time evolution and re-inducibility of mechano-responses. J Cell Physiol 2018;233(1):699–711.
[171] Bougault C, Aubert-Foucher E, Paumier A, Perrier-Groult E, Huot L, Hot D, et al. Dynamic compression of chondrocyte-agarose constructs reveals new candidate mechanosensitive genes. PLoS One 2012;7(5):e36964.
[172] Sironen RK, Karjalainen HM, Elo MA, Kaarniranta K, Torronen K, Takigawa M, et al. cDNA array reveals mechanosensitive genes in chondrocytic cells under hydrostatic pressure. Biochim Biophys Acta 2002;1591(1–3):45–54.
[173] Rolfe RA, Nowlan NC, Kenny EM, Cormican P, Morris DW, Prendergast PJ, et al. Identification of mechanosensitive genes during skeletal development: alteration of genes associated with cytoskeletal rearrangement and cell signalling pathways. BMC Genomics 2014;15(1):48.
[174] Mortazavi A, Williams BA, McCue K, Schaeffer L, Wold B. Mapping and quantifying mammalian transcriptomes by RNA-Seq. Nat Methods 2008;5(7):621–8.
[175] Ozsolak F. Third-generation sequencing techniques and applications to drug discovery. Expert Opin Drug Discov 2012;7(3):231–43.
[176] Wilhelm BT, Landry JR. RNA-Seq-quantitative measurement of expression through massively parallel RNA-sequencing. Methods 2009;48(3):249–57.
[177] Bottomly D, Walter NA, Hunter JE, Darakjian P, Kawane S, Buck KJ, et al. Evaluating gene expression in C57BL/6J and DBA/2J mouse striatum using RNA-Seq and microarrays. PLoS One 2011;6(3):e17820.
[178] Lahiry P, Lee LJ, Frey BJ, Rupar CA, Siu VM, Blencowe BJ, et al. Transcriptional profiling of endocrine cerebro-osteodysplasia using microarray and next-generation sequencing. PLoS One 2011;6(9):e25400.
[179] Marioni JC, Mason CE, Mane SM, Stephens M, Gilad Y. RNA-seq: an assessment of technical reproducibility and comparison with gene expression arrays. Genome Res 2008;18(9):1509–17.
[180] Wang Z, Gerstein M, Snyder M. RNA-Seq: a revolutionary tool for transcriptomics. Nat Rev Genet 2009;10(1):57–63.
[181] Le Bleu HK, Kamal FA, Kelly M, Ketz JP, Zuscik MJ, Elbarbary RA. Extraction of high-quality RNA from human articular cartilage. Anal Biochem 2017;518:134–8.
[182] Poirion OB, Zhu X, Ching T, Garmire L. Single-cell transcriptomics bioinformatics and computational challenges. Front Genet 2016;7:163.

[183] Wu AR, Neff NF, Kalisky T, Dalerba P, Treutlein B, Rothenberg ME, et al. Quantitative assessment of single-cell RNA-sequencing methods. Nat Methods 2014;11(1):41–6.
[184] Livak KJ, Schmittgen TD. Analysis of relative gene expression data using real-time quantitative PCR and the 2(-Delta Delta C(T)) method. Methods 2001;25(4):402–8.
[185] Saha A, Rolfe R, Carroll S, Kelly DJ, Murphy P. Chondrogenesis of embryonic limb bud cells in micromass culture progresses rapidly to hypertrophy and is modulated by hydrostatic pressure. Cell Tissue Res 2016;.
[186] Juhasz T, Matta C, Somogyi C, Katona E, Takacs R, Soha RF, et al. Mechanical loading stimulates chondrogenesis via the PKA/CREB-Sox9 and PP2A pathways in chicken micromass cultures. Cell Signal 2014;26(3):468–82.
[187] Vinardell T, Rolfe RA, Buckley CT, Meyer EG, Ahearne M, Murphy P, et al. Hydrostatic pressure acts to stabilise a chondrogenic phenotype in porcine joint tissue derived stem cells. Eur Cell Mater 2012;23:121–32 [discussion 33-4].
[188] Derderian CA, Bastidas N, Lerman OZ, Bhatt KA, Lin S-E, Voss J, et al. Mechanical strain alters gene expression in an in vitro model of hypertrophic scarring. Ann Plast Surg 2005;55(1):69–75.
[189] Overby DR, Zhou EH, Vargas-Pinto R, Pedrigi RM, Fuchshofer R, Braakman ST, et al. Altered mechanobiology of Schlemm's canal endothelial cells in glaucoma. Proc Natl Acad Sci U S A 2014;111(38):13876–81.
[190] Dolan EB, Haugh MG, Voisin MC, Tallon D, McNamara LM. Thermally induced osteocyte damage initiates a remodelling signaling cascade. PLoS One 2015;10(3): e0119652.
[191] Dolan EB, Tallon D, Cheung W-Y, Schaffler MB, Kennedy OD, McNamara LM. Thermally induced osteocyte damage initiates pro-osteoclastogenic gene expression in vivo. J R Soc Interface 2016;13(119):20160337.
[192] Pardue ML, Gall JG. Molecular hybridization of radioactive DNA to the DNA of cytological preparations. Proc Natl Acad Sci U S A 1969;64(2):600–4.
[193] Jensen E. Technical review: in situ hybridization. Anat Rec (Hoboken, NJ) 2014;297 (8):1349–53.
[194] Kahn J, Shwartz Y, Blitz E, Krief S, Sharir A, Breitel DA, et al. Muscle contraction is necessary to maintain joint progenitor cell fate. Dev Cell 2009;16(5):734–43.
[195] Barreto S, Gonzalez-Vazquez A, Cameron AR, Cavanagh B, Murray DJ, O'Brien FJ. Identification of the mechanisms by which age alters the mechanosensitivity of mesenchymal stromal cells on substrates of differing stiffness: Implications for osteogenesis and angiogenesis. Acta Biomater 2017;53:59–69.
[196] Nowlan NC, Dumas G, Tajbakhsh S, Prendergast PJ, Murphy P. Biophysical stimuli induced by passive movements compensate for lack of skeletal muscle during embryonic skeletogenesis. Biomech Model Mechanobiol 2012;11(1-2):207–19.
[197] Singh PN, Ray A, Azad K, Bandyopadhyay A. A comprehensive mRNA expression analysis of developing chicken articular cartilage. Gene Expr Patterns 2016;20 (1):22–31.
[198] Kavanagh E, Church VL, Osborne AC, Lamb KJ, Archer CW, Francis-West PH, et al. Differential regulation of GDF-5 and FGF-2/4 by immobilisation in ovo exposes distinct roles in joint formation. Dev Dyn 2006;235(3):826–34.
[199] Roddy KA, Kelly GM, van Es MH, Murphy P, Prendergast PJ. Dynamic patterns of mechanical stimulation co-localise with growth and cell proliferation during morphogenesis in the avian embryonic knee joint. J Biomech 2011;.
[200] Wasik AA, Schiller HB. Functional proteomics of cellular mechanosensing mechanisms. Semin Cell Dev Biol 2017;71:118–28.

[201] Mahmood T, Yang PC. Western blot: technique, theory, and trouble shooting. N Am J Med Sci 2012;4(9):429–34.
[202] Carroll SF, Buckley CT, Kelly DJ. Cyclic hydrostatic pressure promotes a stable cartilage phenotype and enhances the functional development of cartilaginous grafts engineered using multipotent stromal cells isolated from bone marrow and infrapatellar fat pad. J Biomech 2014;47(9):2115–21.
[203] Thorpe SD, Buckley CT, Vinardell T, O'Brien FJ, Campbell VA, Kelly DJ. The response of bone marrow-derived mesenchymal stem cells to dynamic compression following TGF-beta3 induced chondrogenic differentiation. Ann Biomed Eng 2010;38(9):2896–909.
[204] Coughlin TR, Schiavi J, Alyssa Varsanik M, Voisin M, Birmingham E, Haugh MG, et al. Primary cilia expression in bone marrow in response to mechanical stimulation in explant bioreactor culture. Eur Cell Mater 2016;32:111–22.
[205] Espinha LC, Hoey DA, Fernandes PR, Rodrigues HC, Jacobs CR. Oscillatory fluid flow influences primary cilia and microtubule mechanics. Cytoskeleton (Hoboken, NJ) 2014;71(7):435–45.
[206] Upadhyay VS, Muntean BS, Kathem SH, Hwang JJ, AbouAlaiwi WA, Nauli SM. Roles of dopamine receptor on chemosensory and mechanosensory primary cilia in renal epithelial cells. Front Physiol 2014;5:72–8.
[207] Dupont S, Morsut L, Aragona M, Enzo E, Giulitti S, Cordenonsi M, et al. Role of YAP/TAZ in mechanotransduction. Nature 2011;474(7350):179–83.
[208] Engvall E, Perlmann P. Enzyme-linked immunosorbent assay (ELISA) quantitative assay of immunoglobulin G. Immunochemistry 1971;8(9):871–4.
[209] Kragstrup TW, Vorup-Jensen T, Deleuran B, Hvid M. A simple set of validation steps identifies and removes false results in a sandwich enzyme-linked immunosorbent assay caused by anti-animal IgG antibodies in plasma from arthritis patients. Springerplus 2013;2(1):263.
[210] Nagatomi J. Mechanobiology handbook. Florida, United States: CRC Press; 2011.
[211] Hartman RA, Bell KM, Debski RE, Kang JD, Sowa GA. Novel ex-vivo mechanobiological intervertebral disc culture system. J Biomech 2012;45(2):382–5.
[212] Merryman WD, Youn I, Lukoff HD, Krueger PM, Guilak F, Hopkins RA, et al. Correlation between heart valve interstitial cell stiffness and transvalvular pressure: implications for collagen biosynthesis. Am J Physiol Heart Circ Physiol 2006;290(1):H224–31.
[213] Polacheck WJ, Li R, Uzel SG, Kamm RD. Microfluidic platforms for mechanobiology. Lab Chip 2013;13(12):2252–67.
[214] Mei Y, Wang Y, Chen H, Sun ZS, Ju XD. Recent progress in CRISPR/Cas9 technology. J Genet Genomics 2016;43(2):63–75.
[215] Wade M. High-throughput silencing using the CRISPR-Cas9 system: a review of the benefits and challenges. J Biomol Screen 2015;20(8):1027–39.
[216] Jinek M, Chylinski K, Fonfara I, Hauer M, Doudna JA, Charpentier E. A programmable dual-RNA-guided DNA endonuclease in adaptive bacterial immunity. Science (New York, NY) 2012;337(6096):816–21.
[217] Suzuki H, Ito Y, Shinohara M, Yamashita S, Ichinose S, Kishida A, et al. Gene targeting of the transcription factor Mohawk in rats causes heterotopic ossification of Achilles tendon via failed tenogenesis. Proc Natl Acad Sci U S A 2016;113(28):7840–5.
[218] Gordon CT, Rodda FA, Farlie PG. The RCAS retroviral expression system in the study of skeletal development. Dev Dyn 2009;238(4):797–811.

[219] Muramatsu T, Mizutani Y, Ohmori Y, Okumura J. Comparison of three nonviral transfection methods for foreign gene expression in early chicken embryos in ovo. Biochem Biophys Res Commun 1997;230(2):376–80.
[220] Muramatsu TMY, Okumura J. Live detection of the firefly luciferase gene expression by bioluminescence in incubating chicken embryos. Anim Sci Technol (Jpn) 1996;67(10):906–9.
[221] Potter H, Weir L, Leder P. Enhancer-dependent expression of human kappa immunoglobulin genes introduced into mouse pre-B lymphocytes by electroporation. Proc Natl Acad Sci U S A 1984;81(22):7161–5.
[222] Miller-Delaney SF, Lieberam I, Murphy P, Mitchell KJ. Plxdc2 is a mitogen for neural progenitors. PLoS One 2011;6(1):e14565.
[223] Scaal M, Gros J, Lesbros C, Marcelle C. In ovo electroporation of avian somites. Dev Dyn 2004;229(3):643–50.
[224] Durand C, Bangs F, Signolet J, Decker E, Tickle C, Rappold G. Enhancer elements upstream of the SHOX gene are active in the developing limb. Eur J Hum Genet 2010;18(5):527–32.
[225] Datar S, Bhonde RR. Shell-less chick embryo culture as an alternative in vitro model to investigate glucose-induced malformations in mammalian embryos. Rev Diabet Stud 2005;2(4):221–7.
[226] El-Ghali N, Rabadi M, Ezin AM, De Bellard ME. New methods for chicken embryo manipulations. Microsc Res Tech 2010;73(1):58–66.
[227] Luo J, Yan X, Lin J, Rolfs A. Gene transfer into older chicken embryos by ex ovo electroporation. J Vis Exp 2012;(65):4078–81.
[228] Schomann T, Qunneis F, Widera D, Kaltschmidt C, Kaltschmidt B. Improved method for ex ovo-cultivation of developing chicken embryos for human stem cell xenografts. Stem Cells Int 2013;2013:960958.
[229] Yalcin HC, Shekhar A, Rane AA, Butcher JT. An ex-ovo chicken embryo culture system suitable for imaging and microsurgery applications. J Vis Exp 2010;(44):2154–8.
[230] Giorgi M, Carriero A, Shefelbine SJ, Nowlan NC. Effects of normal and abnormal loading conditions on morphogenesis of the prenatal hip joint: application to hip dysplasia. J Biomech 2015;48(12):3390–7.
[231] Verbruggen SW, Loo JHW, Hayat TTA, Hajnal JV, Rutherford MA, Phillips ATM, et al. Modeling the biomechanics of fetal movements. Biomech Model Mechanobiol 2016;15(4):995–1004.
[231a] Verbruggen SW, Kainz B, Shelmerdine SC, Hajnal JV, Rutherford MA, Arthurs OJ, Phillips AT, Nowlan NC. Stresses and strains on the human fetal skeleton during development. J R Soc Interface 2018;15(138):20170593.
[231b] Verbruggen SW, Oyen ML, Phillips AT, Nowlan NC. Function and failure of the fetal membrane: modelling the mechanics of the chorion and amnion. PLoS ONE 2017;12(3):e0171588.
[232] Wong SY, Chiam K-H, Lim CT, Matsudaira P. Computational model of cell positioning: directed and collective migration in the intestinal crypt epithelium. J R Soc Interface 2010;7(Suppl. 3):S351–63.
[233] Humphrey JD, Holzapfel GA. Mechanics, mechanobiology, and modeling of human abdominal aorta and aneurysms. J Biomech 2012;45(5):805–14.
[234] Gonzalez CF, Cho YI, Ortega HV, Moret J. Intracranial aneurysms: flow analysis of their origin and progression. Am J Neuroradiol 1992;13(1):181–8.
[235] Robertson AM, Watton PN. Computational fluid dynamics in aneurysm research: critical reflections, future directions. Am J Neuroradiol 2012;33(6):992–5.

[236] Wong GK, Poon W. Current status of computational fluid dynamics for cerebral aneurysms: the clinician's perspective. J Clin Neurosci 2011;18(10):1285–8.
[237] Banerjee RK, Back LH, Back MR, Cho YI. Physiological flow simulation in residual human stenoses after coronary angioplasty. J Biomech Eng 2000;122(4):310–20.
[238] Sazonov I, Yeo SY, Bevan RLT, Xie X, van Loon R, Nithiarasu P. Modelling pipeline for subject-specific arterial blood flow—a review. Int J Numer Methods Biomed Eng 2011;27(12):1868–910.
[239] Zhang J-M, Zhong L, Su B, Wan M, Yap JS, Tham JPL, et al. Perspective on CFD studies of coronary artery disease lesions and hemodynamics: a review. Int J Numer Methods Biomed Eng 2014;30(6):659–80.
[240] Dahl KN, Kalinowski A, Pekkan K. Mechanobiology and the microcirculation: cellular, nuclear and fluid mechanics. Microcirculation (New York, NY) 2010;17(3):179–91.
[241] AlMomani T, Udaykumar HS, Marshall JS, Chandran KB. Micro-scale dynamic simulation of erythrocyte–platelet interaction in blood flow. Ann Biomed Eng 2008;36(6):905–20.
[242] Bagchi P, Johnson PC, Popel AS. Computational fluid dynamic simulation of aggregation of deformable cells in a shear flow. J Biomech Eng 2005;127(7):1070–80.
[243] Jung J, Lyczkowski RW, Panchal CB, Hassanein A. Multiphase hemodynamic simulation of pulsatile flow in a coronary artery. J Biomech 2006;39(11):2064–73.
[244] Abouali O, Modareszadeh A, Ghaffariyeh A, Tu J. Numerical simulation of the fluid dynamics in vitreous cavity due to saccadic eye movement. Med Eng Phys 2012;34(6):681–92.
[245] Steck R, Niederer P, Knothe Tate ML. A finite element analysis for the prediction of load-induced fluid flow and mechanochemical transduction in bone. J Theor Biol 2003;220(2):249–59.
[246] Mak AFT, Huang DT, Zhang JD, Tong P. Deformation-induced hierarchical flows and drag forces in bone canaliculi and matrix microporosity. J Biomech 1997;30(1):11–8.
[247] Anderson E, Kaliyamoorthy S, Alexander J, Tate M. Nano-microscale models of periosteocytic flow show differences in stresses imparted to cell body and processes. Ann Biomed Eng 2005;33(1):52–62.
[248] Anderson EJ, Knothe Tate ML. Idealization of pericellular fluid space geometry and dimension results in a profound underprediction of nano-microscale stresses imparted by fluid drag on osteocytes. J Biomech 2008;41(8):1736–46.
[249] Kamioka H, Kameo Y, Imai Y, Bakker AD, Bacabac RG, Yamada N, et al. Microscale fluid flow analysis in a human osteocyte canaliculus using a realistic high-resolution image-based three-dimensional model. Integr Biol 2012;4(10):1198–206.
[250] Anderson E, Kreuzer S, Small O, Knothe Tate M. Pairing computational and scaled physical models to determine permeability as a measure of cellular communication in micro- and nano-scale pericellular spaces. Microfluid Nanofluid 2008;4(3):193–204.
[251] Lemaire T, Lemonnier S, Naili S. On the paradoxical determinations of the lacuno-canalicular permeability of bone. Biomech Model Mechanobiol 2012;11(7):933–46.
[252] Lemonnier S, Naili S, Oddou C, Lemaire T. Numerical determination of the lacuno-canalicular permeability of bone. Comput Methods Biomech Biomed Eng 2011;14(sup1):133–5.
[253] Lemaire T, Naïli S, Rémond A. Multiscale analysis of the coupled effects governing the movement of interstitial fluid in cortical bone. Biomech Model Mechanobiol 2006;5(1):39–52.

[254] Sansalone V, Kaiser J, Naili S, Lemaire T. Interstitial fluid flow within bone canaliculi and electro-chemo-mechanical features of the canalicular milieu. Biomech Model Mechanobiol 2012;12(3):1–21.

[255] Lemaire T, Naïli S, Rémond A. Study of the influence of fibrous pericellular matrix in the cortical interstitial fluid movement with hydroelectrochemical effects. J Biomech Eng 2008;130(1):011001.

[256] Birmingham E, Grogan JA, Niebur GL, McNamara LM, McHugh PE. Computational modelling of the mechanics of trabecular bone and marrow using fluid structure interaction techniques. Ann Biomed Eng 2013;41(4):814–26.

[257] Teo JCM, Teoh SH. Permeability study of vertebral cancellous bone using microcomputational fluid dynamics. Comput Methods Biomech Biomed Eng 2012;15(4):417–23.

[258] Porter B, Zauel R, Stockman H, Guldberg R, Fyhrie D. 3-D computational modeling of media flow through scaffolds in a perfusion bioreactor. J Biomech 2005;38(3):543–9.

[259] Huiskes R, Chao E. A survey of finite element analysis in orthopedic biomechanics: the first decade. J Biomech 1983;16(6):385–409.

[260] Prendergast PJ. Finite element models in tissue mechanics and orthopaedic implant design. Clin Biomech 1997;12(6):343–66.

[261] Doblaré M, García J, Gómez M. Modelling bone tissue fracture and healing: a review. Eng Fract Mech 2004;71(13):1809–40.

[262] Giorgi M, Carriero A, Shefelbine SJ, Nowlan NC. Mechanobiological simulations of prenatal joint morphogenesis. J Biomech 2014;47(5):989–95.

[263] Isaksson H. Recent advances in mechanobiological modeling of bone regeneration. Mech Res Commun 2012;42:22–31.

[264] Müller R, Rüegsegger P. Three-dimensional finite element modelling of non-invasively assessed trabecular bone structures. Med Eng Phys 1995;17(2):126–33.

[265] van Rietbergen B, Weinans H, Huiskes R, Odgaard A. A new method to determine trabecular bone elastic properties and loading using micromechanical finite-element models. J Biomech 1995;28(1):69–81.

[266] Coughlin TR, Niebur GL. Fluid shear stress in trabecular bone marrow due to low-magnitude high-frequency vibration. J Biomech 2012;45(13):2222–9.

[267] Metzger TA, Kreipke TC, Vaughan TJ, McNamara LM, Niebur GL. The in situ mechanics of trabecular bone marrow: the potential for mechanobiological response. J Biomech Eng 2015;137(1):011006–7.

[268] Metzger TA, Niebur GL. Comparison of solid and fluid constitutive models of bone marrow during trabecular bone compression. J Biomech 2016;49(14):3596–601.

[269] Dolan EB, Vaughan TJ, Niebur GL, Casey C, Tallon D, McNamara LM. How bone tissue and cells experience elevated temperatures during orthopaedic cutting: an experimental and computational investigation. J Biomech Eng 2014;136(2):021019.

[270] Rath Bonivtch A, Bonewald LF, Nicolella DP. Tissue strain amplification at the osteocyte lacuna: a microstructural finite element analysis. J Biomech 2007;40(10):2199–206.

[271] Varga P, Hesse B, Langer M, Schrof S, Männicke N, Suhonen H, et al. Synchrotron X-ray phase nano-tomography-based analysis of the lacunar–canalicular network morphology and its relation to the strains experienced by osteocytes in situ as predicted by case-specific finite element analysis. Biomech Model Mechanobiol 2015;14(2):267–82.

[272] Verbruggen SW, Vaughan TJ, McNamara LM. Strain amplification in bone mechanobiology: a computational investigation of the in vivo mechanics of osteocytes. J R Soc Interface 2012;9(75):2735–44.
[273] Mullen CA, Vaughan TJ, Voisin MC, Brennan MA, Layrolle P, McNamara LM. Cell morphology and focal adhesion location alters internal cell stress. J R Soc Interface 2014;11(101):20140885.
[274] Mullen Conleth A, Vaughan Ted J, Billiar Kristen L, McNamara Laoise M. The effect of substrate stiffness, thickness, and cross-linking density on osteogenic cell behavior. Biophys J 2015;108(7):1604–12.
[275] Barreto S, Perrault CM, Lacroix D. Structural finite element analysis to explain cell mechanics variability. J Mech Behav Biomed Mater 2014;38:219–31.
[276] Cataloglu A, Clark RE, Gould P. Refined stress analysis of human aortic heart valves. J Eng Mech Div 1976;102(1):135–50.
[277] Cataloglu A, Clark RE, Gould PL. Stress analysis of aortic valve leaflets with smoothed geometrical data. J Biomech 1977;10(3):153–8.
[278] Chong M, Eng M, Missirlis Y. Aortic valve mechanics part II: a stress analysis of the porcine aortic valve leaflets in diastole. Biomater Med Devices Artif Organs 1978;6(3):225–44.
[279] Hamid M, Sabbah HN, Stein PD. Comparison of finite element stress analysis of aortic valve leaflet using either membrane elements or solid elements. Comput Struct 1985;20(6):955–61.
[280] Hamid MS, Sabbah HN, Stein PD. Large-deformation analysis of aortic valve leaflets during diastole. Eng Fract Mech 1985;22(5):773–85.
[281] Grande KJ, Cochran RP, Reinhall PG, Kunzelman KS. Stress variations in the human aortic root and valve: the role of anatomic asymmetry. Ann Biomed Eng 1998;26(4):534–45.
[282] Grande KJ, Cochran RP, Reinhall PG, Kunzelman KS. Mechanisms of aortic valve incompetence in aging: a finite element model. J Heart Valve Dis 1999;8(2):149–56.
[283] Grande-Allen KJ, Cochran RP, Reinhall PG, Kunzelman KS. Finite-element analysis of aortic valve-sparing: influence of graft shape and stiffness. IEEE Trans Biomed Eng 2001;48(6):647–59.
[284] Kunzelman K, Cochran R, Chuong C, Ring W, Verrier ED, Eberhart R. Finite element analysis of the mitral valve. J Heart Valve Dis 1993;2(3):326–40.
[285] Black M, Howard I, Huang X, Patterson E. A three-dimensional analysis of a bioprosthetic heart valve. J Biomech 1991;24(9):793797–5801.
[286] Howard I, Patterson E, Yoxall A. On the opening mechanism of the aortic valve: some observations from simulations. J Med Eng Technol 2003;27(6):259–66.
[287] Sacks MS, Yoganathan AP. Heart valve function: a biomechanical perspective. Philos Trans R Soc Lond B: Biol Sci 2007;362(1484):1369–91.
[288] Lim C, Zhou E, Quek S. Mechanical models for living cells—a review. J Biomech 2006;39(2):195–216.
[289] Mofrad MR, Kamm RD. Cytoskeletal mechanics: models and measurements in cell mechanics. Cambridge, United Kingdom: Cambridge University Press; 2006.
[290] Carter DR, Wong M. Modelling cartilage mechanobiology. Philos Trans R Soc Lond B: Biol Sci 2003;358(1437):1461–71.
[291] Buganza Tepole A, Joseph Ploch C, Wong J, Gosain AK, Kuhl E. Growing skin: a computational model for skin expansion in reconstructive surgery. J Mech Phys Solids 2011;59(10):2177–90.

[292] Socci L, Pennati G, Gervaso F, Vena P. An axisymmetric computational model of skin expansion and growth. Biomech Model Mechanobiol 2007;6(3):177–88.
[293] Zöllner AM, Buganza Tepole A, Kuhl E. On the biomechanics and mechanobiology of growing skin. J Theor Biol 2012;297:166–75.
[294] Zöllner AM, Holland MA, Honda KS, Gosain AK, Kuhl E. Growth on demand: reviewing the mechanobiology of stretched skin. J Mech Behav Biomed Mater 2013;28:495–509.
[295] Li NY, Heris HK, Mongeau L. Current understanding and future directions for vocal fold mechanobiology. J Cytol Mol Biol 2013;1(1):001.
[296] Kelleher JE, Zhang K, Siegmund T, Chan RW. Spatially varying properties of the vocal ligament contribute to its eigenfrequency response. J Mech Behav Biomed Mater 2010;3(8):600–9.
[297] Vaughan TJ, Verbruggen SW, McNamara LM. Are all osteocytes equal? Multiscale modelling of cortical bone to characterise the mechanical stimulation of osteocytes. Int J Numer Methods Biomed Eng 2013;29(12):1361–72.
[298] Vaughan T, Voisin M, Niebur G, McNamara L. Multiscale modeling of trabecular bone marrow: understanding the micromechanical environment of mesenchymal stem cells during osteoporosis. ASME J Biomech Eng 2014;.
[299] Nishii K, Reese G, Moran EC, Sparks JL. Multiscale computational model of fluid flow and matrix deformation in decellularized liver. J Mech Behav Biomed Mater 2016;57:201–14.
[300] Wyczalkowski MA, Chen Z, Filas BA, Varner VD, Taber LA. Computational models for mechanics of morphogenesis. Birth Defects Res C: Embryo Today 2012;96(2):132–52.
[301] Tang D, Yang C, Geva T, Rathod R, Yamauchi H, Gooty V, et al. A multiphysics modeling approach to develop right ventricle pulmonary valve replacement surgical procedures with a contracting band to improve ventricle ejection fraction. Comput Struct 2013;122(Suppl. C):78–87.
[302] Vaughan TJ, Haugh MG, McNamara LM. A fluid–structure interaction model to characterize bone cell stimulation in parallel-plate flow chamber systems. J R Soc Interface 2013;10(81).
[303] Zhao F, Vaughan T, McNamara L. Multiscale fluid–structure interaction modelling to determine the mechanical stimulation of bone cells in a tissue engineered scaffold. Biomech Model Mechanobiol 2015;14(2):231–43.
[304] Zhao F, Vaughan T, McNamara L. Quantification of fluid shear stress in bone tissue engineering scaffolds with spherical and cubical pore architectures. Biomech Model Mechanobiol 2015;1–17.
[305] Khayyeri H, Barreto S, Lacroix D. Primary cilia mechanics affects cell mechanosensation: a computational study. J Theor Biol 2015;379:38–46.
[306] Verbruggen SW, Vaughan TJ, McNamara LM. Mechanisms of osteocyte stimulation in osteoporosis. J Mech Behav Biomed Mater 2016;62:158–68.
[307] Vaughan TJ, Mullen CA, Verbruggen SW, McNamara LM. Bone cell mechanosensation of fluid flow stimulation: a fluid–structure interaction model characterising the role integrin attachments and primary cilia. Biomech Model Mechanobiol 2015;14(4):703–18.
[308] Boyle C, Lennon AB, Early M, Kelly D, Lally C, Prendergast P. Computational simulation methodologies for mechanobiological modelling: a cell-centred approach to neointima development in stents. Philos Trans R Soc Lond A: Math Phys Eng Sci 2010;368(1921):2919–35.

[309] Garbey M, Rahman M, Berceli S. A multiscale computational framework to understand vascular adaptation. J Comput Sci 2015;8:32–47.

[310] Hoekstra AG, Alowayyed S, Lorenz E, Melnikova N, Mountrakis L, van Rooij B, et al. Towards the virtual artery: a multiscale model for vascular physiology at the physics–chemistry–biology interface. Philos Trans R Soc A 2016;374(2080): 20160146.

[311] Nolan DR, Lally C. An investigation of damage mechanisms in mechanobiological models of in-stent restenosis. J Comput Sci 2017;24:132–42.

[312] Zahedmanesh H, Lally C. A multiscale mechanobiological modelling framework using agent-based models and finite element analysis: application to vascular tissue engineering. Biomech Model Mechanobiol 2012;11(3):363–77.

[313] Zahedmanesh H, Van Oosterwyck H, Lally C. A multi-scale mechanobiological model of in-stent restenosis: deciphering the role of matrix metalloproteinase and extracellular matrix changes. Comput Meth Biomech Biomed Eng 2014;17(8):813–28.

[314] Caiazzo A, Evans D, Falcone J-L, Hegewald J, Lorenz E, Stahl B, et al. A complex automata approach for in-stent restenosis: two-dimensional multiscale modelling and simulations. J Comput Sci 2011;2(1):9–17.

[315] Marom G. Numerical methods for fluid–structure interaction models of aortic valves. Arch Comput Meth Eng 2015;22(4):595–620.

[316] Keating T, Wolf P, Scarpace F. An improved method of digital image correlation. Photogramm Eng Remote Sens 1975;41(8).

[317] Pan B, Qian K, Xie H, Asundi A. Two-dimensional digital image correlation for in-plane displacement and strain measurement: a review. Meas Sci Technol 2009;20 (6):062001.

[318] Berfield TA, Patel JK, Shimmin RG, Braun PV, Lambros J, Sottos NR. Fluorescent image correlation for nanoscale deformation measurements. Small 2006;2(5):631–5.

[319] Evans SL, Holt CA. Measuring the mechanical properties of human skin in vivo using digital image correlation and finite element modelling. J Strain Anal Eng Des 2009;44 (5):337–45.

[320] Staloff IA, Guan E, Katz S, Rafailovitch M, Sokolov A, Sokolov S. An in vivo study of the mechanical properties of facial skin and influence of aging using digital image speckle correlation. Skin Res Technol 2008;14(2):127–34.

[321] Genovese K, Casaletto L, Humphrey JD, Lu J, editors. Digital image correlation-based point-wise inverse characterization of heterogeneous material properties of gallbladder in vitro. Proc R Soc A 2014;470(2167):20140152.

[322] Genovese K, Lee Y, Lee A, Humphrey J. An improved panoramic digital image correlation method for vascular strain analysis and material characterization. J Mech Behav Biomed Mater 2013;27:132–42.

[323] Zhang D, Eggleton CD, Arola DD. Evaluating the mechanical behavior of arterial tissue using digital image correlation. Exp Mech 2002;42(4):409–16.

[324] Zhou B, Ravindran S, Ferdous J, Kidane A, Sutton MA, Shazly T. Using digital image correlation to characterize local strains on vascular tissue specimens. J Vis Exp 2016; (107):e53625.

[325] Cheng T, Dai C, Gan RZ. Viscoelastic properties of human tympanic membrane. Ann Biomed Eng 2007;35(2):305–14.

[326] Verhulp E, Bv R, Huiskes R. A three-dimensional digital image correlation technique for strain measurements in microstructures. J Biomech 2004;37(9):1313–20.

[327] Krehbiel JD, Lambros J, Viator J, Sottos N. Digital image correlation for improved detection of basal cell carcinoma. Exp Mech 2010;50(6):813–24.

[328] Huang J, Pan X, Peng X, Zhu T, Qin L, Xiong C, et al. High-efficiency cell-substrate displacement acquisition via digital image correlation method using basis functions. Opt Lasers Eng 2010;48(11):1058–66.

[329] Verbruggen SW, Mc Garrigle MJ, Haugh MG, Voisin MC, McNamara LM. Altered mechanical environment of bone cells in an animal model of short- and long-term osteoporosis. Biophys J 2015;108(7):1587–98.

[330] Wan LQ, Ronaldson K, Park M, Taylor G, Zhang Y, Gimble JM, et al. Micropatterned mammalian cells exhibit phenotype-specific left-right asymmetry. Proc Natl Acad Sci U S A 2011;108(30):12295–300.

[331] Ahola A, Kiviaho AL, Larsson K, Honkanen M, Aalto-Setälä K, Hyttinen J. Video image-based analysis of single human induced pluripotent stem cell derived cardiomyocyte beating dynamics using digital image correlation. Biomed Eng Online 2014;13(1):39.

[332] Westerweel J, Elsinga GE, Adrian RJ. Particle image velocimetry for complex and turbulent flows. Annu Rev Fluid Mech 2013;45:409–36.

[333] Adrian RJ. Particle-imaging techniques for experimental fluid mechanics. Annu Rev Fluid Mech 1991;23(1):261–304.

[334] Willert CE, Gharib M. Digital particle image velocimetry. Exp Fluids 1991;10(4):181–93.

[335] Arjunon S, Rathan S, Jo H, Yoganathan AP. Aortic valve: mechanical environment and mechanobiology. Ann Biomed Eng 2013;41(7):1331–46.

[336] Gunning PS, Saikrishnan N, McNamara LM, Yoganathan AP. An in vitro evaluation of the impact of eccentric deployment on transcatheter aortic valve hemodynamics. Ann Biomed Eng 2014;42(6):1195–206.

[337] Kini V, Bachmann C, Fontaine A, Deutsch S, Tarbell JM. Integrating particle image velocimetry and laser doppler velocimetry measurements of the regurgitant flow field past mechanical heart valves. Artif Organs 2001;25(2):136–45.

[338] Saikrishnan N, Yap C-H, Milligan NC, Vasilyev NV, Yoganathan AP. In vitro characterization of bicuspid aortic valve hemodynamics using particle image velocimetry. Ann Biomed Eng 2012;40(8):1760–75.

[339] Campos Marin A, Grossi T, Bianchi E, Dubini G, Lacroix D. 2D μ-particle image velocimetry and computational fluid dynamics study within a 3D porous scaffold. Ann Biomed Eng 2017;45(5):1341–51.

[340] De Boodt S, Truscello S, Özcan SE, Leroy T, Van Oosterwyck H, Berckmans D, et al. Bi-modular flow characterization in tissue engineering scaffolds using computational fluid dynamics and particle imaging velocimetry. Tissue Eng Part C: Methods 2010;16(6):1553–64.

[341] Sucosky P, Osorio DF, Brown JB, Neitzel GP. Fluid mechanics of a spinner-flask bioreactor. Biotechnol Bioeng 2004;85(1):34–46.

[342] Engelmayr GC, Soletti L, Vigmostad SC, Budilarto SG, Federspiel WJ, Chandran KB, et al. A novel flex-stretch-flow bioreactor for the study of engineered heart valve tissue mechanobiology. Ann Biomed Eng 2008;36(5):700–12.

[343] Rui L, Shigeo W, Ken-ichi T, Takami Y. Confocal micro-PIV measurements of three-dimensional profiles of cell suspension flow in a square microchannel. Meas Sci Technol 2006;17(4):797.

[344] Patrick MJ, Chen C-Y, Frakes DH, Dur O, Pekkan K. Cellular-level near-wall unsteadiness of high-hematocrit erythrocyte flow using confocal μPIV. Exp Fluids 2011;50(4):887–904.

[345] Fan Z, Sun Y, Di C, Tay D, Chen W, Deng CX, et al. Acoustic tweezing cytometry for live-cell subcellular modulation of intracellular cytoskeleton contractility. Sci Rep 2013;3:2176.

[346] Okeyo KO, Adachi T, Sunaga J, Hojo M. Actomyosin contractility spatiotemporally regulates actin network dynamics in migrating cells. J Biomech 2009;42(15):2540–8.

[347] Doxzen K, Vedula SRK, Leong MC, Hirata H, Gov NS, Kabla AJ, et al. Guidance of collective cell migration by substrate geometry. Integr Biol 2013;5(8):1026–35.

[348] Zamir EA, Czirók A, Rongish BJ, Little CD. A digital image-based method for computational tissue fate mapping during early avian morphogenesis. Ann Biomed Eng 2005;33(6):854–65.

[349] Serra-Picamal X, Conte V, Vincent R, Anon E, Tambe DT, Bazellieres E, et al. Mechanical waves during tissue expansion. Nat Phys 2012;8(8):628–34.

# CHAPTER 2

# Cell geometric control of nuclear dynamics and its implications

Abhishek Kumar*, Ekta Makhija[†,‡], A.V. Radhakrishnan[†,§,¶],
Doorgesh Sharma Jokhun[†], G.V. Shivashankar[†,||]

*Yale Cardiovascular Research Center (YCVRC), Department of Internal Medicine, Yale School of Medicine, Yale University, New Haven, CT, United States* Mechanobiology Institute (MBI), National University of Singapore, Singapore[†] BioSystems and Micromechanics Group, Singapore-MIT Alliance for Research & Technology, CREATE, Singapore[‡] Raman Research Institute, Bangalore, India[§] Interdisciplinary Institute for Neuroscience, UMR 5297, CNRS, Bordeaux, France[¶] Institute of Molecular Oncology (IFOM), Italian Foundation for Cancer Research, Milan, Italy[||]*

## ABBREVIATIONS

| | |
|---|---|
| **ASFs** | actin stress fibers |
| **ATP** | adenosine triphosphate |
| **CSI** | cell shape index |
| **DNA** | deoxyribonucleic acid |
| **ECM** | extracellular matrix |
| **EGFP** | enhanced green fluorescent protein |
| **ER** | endoplasmic reticulum |
| **KASH** | Klarsicht, ANC-1, Syne Homology |
| **LINC** | linker of nucleoskeleton and cytoskeleton |
| **MSD** | mean squared displacement |
| **MTOC** | microtubule organizing center |
| **SUN** | Sad1p, UNC-84 |

## 1 INTRODUCTION

Nuclear dynamics is the study of translational and rotational motion of the nucleus as a rigid body and the fluctuations of the nuclear envelope [1–3]. In a living cell, the nucleus can be treated as a viscoelastic body residing in an active gel [1]. It moves within a confined region defined by the cell boundary as a result of various forces acting on it [1,3]. Some of these forces, such as actin and microtubule polymerization and actomyosin contractility, are active in nature, while others, such as thermal fluctuations, are passive. The combination of all forces that have varying magnitudes, acting at different length and timescales, ultimately dictates the resultant motion of the nucleus. A systematic analysis of the motion of the nucleus gives information about the forces and torques experienced [1,4,5].

The position and dynamics of the nucleus in the cellular environment has been shown to be critical for many physiological functions such as migration, mitosis, polarization, wound healing, fertilization, and cell differentiation [6,7]. Various studies have shown that the nucleus undergoes translational and rotational dynamics during cell migration and wound healing [8–11]. The altered nuclear position and dynamics during these processes has been attributed to various cytoplasmic factors.

Additionally, just as in particle tracking microrheological studies, a tracer particle is used for measuring the rheological properties of a medium; the nuclear translational dynamics can provide insights about the physical architecture of the cytoplasm in eukaryotic cells. A time-series analysis of the trajectory and shape changes of the nucleus gives a measure of an effective mesh size and the viscoelastic response of the microenvironment. Since cellular processes depend critically on the coordination between the chemical and physical realms, such analyses hold tremendous potential for a new generation of diagnostic tools for various pathological conditions including cancer and laminopathies [12].

## 1.1 PHYSICAL LINK BETWEEN NUCLEUS AND CYTOSKELETON

The eukaryotic cell nucleus is coupled to a dynamic cytoskeleton [13,14]. The cytoskeletal filaments provide a scaffold to the cell; these include contractile actomyosin complexes, microtubule filaments that constantly undergo dynamic instability, and load-bearing intermediate filaments [14–25]. These cytoplasmic-to-nuclear connections, which emerge through differentiation and development, maintain the nucleus in a prestressed state [19,20,25–34]. Actomyosin complexes apply a contractile force on the nucleus via links with the outer nuclear membrane, while microtubules apply a compressive load pushing the nucleus inward [14,33]. Apical actin filaments have been shown to form a "perinuclear actin cap" shaping the nucleus [35–37]. Importantly, these actin stress fibers (ASFs) act like a bridge for force transmission to the nucleus via SUN (Sad1p, UNC-84) and KASH (Klarsicht, ANC-1, Syne Homology) domain linker proteins forming the LINC complex (linker of nucleoskeleton and cytoskeleton). Similarly, dorsal actin filaments have also been found to be part of transmembrane actin-associated nuclear (TAN) lines and drive nuclear movement [38]. Collectively, the prestressed nuclear morphology and dynamics are tightly controlled by cellular mechanical constraints that modulate the cytoskeletal architecture [39–43].

Cytoskeletal forces from outside the nucleus balance the entropic forces of chromatin polymer from inside to maintain the size and shape of the nucleus [30]. Spatiotemporal changes in either cytoskeletal or entropic forces can result in fluctuations of size and shape of the nucleus. Since fluctuations in size of a soft body involve either compressibility of its constituents or a flux of material transporting through its surface, they are expected to occur at longer timescales compared with fluctuations in shape at constant size [28]. Investigating the nuclear dynamics (translation, rotation, and envelope fluctuations) can help us understand the forces and torques continuously being relayed to the nucleus by the cytoskeleton. This chapter discusses these in detail.

## 1.2 MICRORHEOLOGY OF THE NUCLEUS

There are two types of forces acting on the nucleus: (1) passive force that is ATP (adenosine triphosphate) hydrolysis-independent and (2) active force that is ATP hydrolysis-dependent. Passive forces come from thermal fluctuations of molecules/proteins in the cytoplasmic milieu. For example, thermal kick from a water molecule, which is 0.275 nm in size and has $K_B T$ energy ($K_B = 1.38 \times 10^{-23}$ J K$^{-1}$ is Boltzmann constant and $T$ is temperature in kelvin), would generate a force of ~pN over a nanometer distance. If the nucleus is treated as a freely diffusing particle of radius 10 μm in the cytoplasmic medium, which has a viscosity of ~$10^{-2}$ Pa s, in accordance with the Stokes-Einstein equation, its diffusion constant at 37°C is approximately $2 \times 10^{-3}$ μm$^2$/s. Hence, in the absence of confinement caused by cytoskeletal meshwork, the cell nucleus is expected to move ~45 nm per second due to the thermal forces. However, the nucleus, just like many other cellular organelles and vesicles, can also have directional motion, rather than random and diffusive, when observed at timescales longer than that associated with collision of small molecules, which is ~1 ps. Such directional or super diffusive motion of the nucleus can arise from active forces, of which there are two main sources in the cytoplasm—pushing force from polymerizing cytoskeletal filaments [44] and pulling force from cytoskeletal motors [32,44–50]. The force arising from single polymerizing filament of actin or microtubule is ~1 pN, while a single myosin or kinesin motor generates ~1–5 pN and moves in steps of ~8 nm [51,52]. In addition, the microtubules and intermediate filaments form a mesh or cagelike structure around the nucleus, which can transmit active forces to the nucleus while also serving as 3D confinement for nuclear motion. The cytoplasm, a viscoelastic medium, also applies a drag force on the nucleus. Thus, the nucleus can exhibit random and free-diffusive motion, directional and superdiffusive motion, and confined and subdiffusive motion, depending on the net forces and confinement from its microenvironment at various timescales.

## 1.3 BOUNDARY CONDITIONS

Most studies in cell biology have been carried out in freely migrating cells in tissue culture dishes. In physiology, however, the cells in tissues are surrounded by neighboring cells and the extracellular matrix (ECM). The chemical and physical surroundings of the cells are referred to as the boundary conditions, and their effects on cellular and nuclear architecture and dynamics are only recently being appreciated [53]. Various in vitro systems have been developed using microfluidics and micropatterning to provide geometric confinement to cells. Confinement of stem cells or progenitor cells on various geometries has shown that their fate is governed not only by chemical signals from the differentiation medium but also by these physical boundary conditions [54]. The nuclear morphology and orientation have also been shown to depend on the cell geometry. While nuclear dynamics is attributed to forces from actomyosin and microtubule motors, the organization and dynamics of the cytoskeleton filaments themselves depend on the cell geometry.

The study of nuclear morphology and dynamics in confined cells and the microrheological parameters of the cytoplasmic microenvironment derived from these can, in principle, be used as potential biophysical markers for disease diagnostics. Analogous to the development and propagation of internal cracks in a wall, when deforming forces act on it, we can figure out the defects in a cell, when subjected to different conditions, and observe how it reacts to each of them in terms of the biophysical markers [12]. One way of achieving this is by tuning the geometric shape of the cell using polydimethylsiloxane (PDMS) stamped fibronectin (or any ECM protein)-coated micropatterned substrates. These patterns mimic the physiological and pathological conditions in various tumors where the boundary conditions of cells are significantly altered.

In the following sections, we discuss various studies that describe nuclear (1) translation, (2) rotation, and (3) fluctuation, and deformation in geometrically confined cells. In each section, we first explain the image processing and computational methods for quantifying the particular type of nuclear motion. Next, we discuss which cell geometries are most likely to elicit the particular type of motion of the nucleus and the differential force contribution from various cytoskeletal structures in these geometries. Lastly, we discuss the implications of the particular type of nuclear motion for various cellular processes.

## 2 NUCLEAR TRANSLATIONAL MOTION
### 2.1 IMAGE PROCESSING AND COMPUTATIONAL METHODS

To be able to capture nuclear translational motion arising from thermal fluctuations and from active forces, the imaging frame rate must be chosen as the estimated time required for the nucleus to displace by one pixel. The spatial ($XY$) resolution of a fluorescence confocal microscope is around $\sim 200$ nm (for a $100\times$, 1.4 NA oil immersion objective). Using the known value of diffusion constant of a nuclear-sized particle ($10^{-3}$ $\mu m^2/s$) in a viscous medium with a viscosity comparable with cytoplasm ($10^{-2}$ Pa s) and accounting for directed motions, the relevant frame rate for imaging is $\sim 1$ frame/s to capture nuclear translational motion arising from thermal fluctuations and from active forces [3]. Using this frame rate, time-lapse (up to 30 min) imaging was carried out for cells, with the nucleus labeled with H2B tagged enhanced green fluorescent protein (EGFP) (or any deoxyribonucleic acid (DNA) binding fluorescent dyes such as Hoechst and DRAQ5 may be used [55]), plated on fibronectin-coated substrate (Fig. 1A).

To calculate parameters characterizing the nuclear motion, nuclear displacement was calculated for each time point. For this, each gray scale image of the nucleus was thresholded to create a binary image. The centroid of the binary image represents the nuclear centroid. Fig. 1B shows trajectories of the nuclear centroid for cells plated on circular and rectangular geometries. Nuclear displacement can then be calculated as distance between centroid positions in consecutive time frames.

Two parameters that can be computed from the nuclear displacement time series are speed and area explored by the nucleus. Measurement of speed provides an estimate of the net forces on the nucleus, while the measurement of the area explored

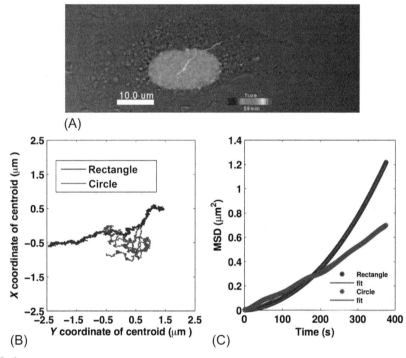

**FIG. 1**

Geometric regulation of nuclear translations. (A) Translational motion of the nucleus in NIH 3T3 cells stably expressing H2B-EGFP, recorded for an hour at 1 frame/s. The instantaneous direction of motion of the centroid of the nucleus is indicated by the path advancing in left-down direction with time. (B) Trajectories of nuclear centroid recorded for 1500 s at 1 frame/s, in two well-defined geometric constraints (more directed along the horizontal axis in 1800 $\mu m^2$ rectangle compared to 500 $\mu m^2$ circle). (C) MSD curve for nucleus in rectangular geometry starting with lower values and further crossing over the values of MSD in circular geometry at later time scales. The MSD fit to moving corral model is also shown for each geometry.

*Adapted from Radhakrishnan AV, Jokhun DS, Venkatachalapathy S, Shivashankar GV. Nuclear positioning and its translational dynamics are regulated by cell geometry. Biophys J 2017;112(9):1920–8.*

conveys how well the nucleus is anchored to its preferred site. Area explored is computed as the width of the distribution function or the histogram of the nuclear displacements. Cells on rectangular geometry have larger width compared with that of circular geometry. While these quantities are quite straightforward and intuitive, they do not provide complete information about the mechanisms underlying the motion or the ways in which the nucleus interacts with its microenvironment. They also do not distinguish between the different types of motion that might be at play as a function of the timescale of the observations. This prompts one to calculate the mean squared displacement (MSD), computed using the following equation:

$$\langle r^2(\tau) \rangle = \frac{1}{N-n} \sum_{k=1}^{N-n} [r((k-1)\Delta t + n\Delta t) - r((k-1)\Delta t)]^2$$

Here, $r$ is the position vector of the nucleus at each time point, and $N$ is the total number of measured points. For normal diffusion, the MSD depends linearly on $\tau$, $\langle r^2(\tau) \rangle = 2dD\tau$, where $d$ is the spatial dimension (for $d = 1$, 2, and 3 for one-, two-, and three-dimensional motions, respectively) and $D$ is the diffusion coefficient. MSD quantifies how much space a particle explores as a function of observational time intervals ($\tau$). A linear change of MSD with time indicates a normal diffusive behavior, while a nonlinear change where $\langle r^2(\tau) \rangle = 2dD\tau^\alpha$ indicates subdiffusive ($\alpha < 1$) or superdiffusive ($\alpha > 1$) behavior.

## 2.2 GEOMETRIC CONTROL OF NUCLEAR TRANSLATION

Recent studies of nuclear dynamics with high spatial and temporal resolution in elongated and small circular geometry revealed different diffusive behavior with multiple regimes of diffusion in a cell-geometry-dependent manner [3]. Typical MSD curve for the nucleus on rectangular and circular geometry is shown in Fig. 1C. These curves have distinct timescale regimes, called crossover timescale, governed by different scaling (with $\alpha = 1$ or $< 1$ or $> 1$) behavior. A crossover of the nuclear centroid motion from subdiffusion to superdiffusion happens at $\sim$18 s in rectangular geometry, whereas at $\sim$3 s in circular geometry and $\sim$36 s, the motion becomes diffusive (Fig. 1C) [3].

## 2.3 ROLE OF CYTOSKELETON IN NUCLEAR TRANSLATION

While polarized cells have long stress fibers and apical actin, rounded cells have short actin filaments. The percentage of cells showing apical ASFs increases with increasing aspect ratio (12% in 1:1, 79% in 1:3, and 92% in 1:5) [36,37]. ASFs are shown to stabilize nuclear positioning [36], and disruption of ASFs leads to instability and amplified diffusive motion of the nucleus. In NIH3T3 cells with specific geometric shape, actin fibers play a dominant role in positioning the nucleus [3]. In a different study, the nucleus in a premitotic tobacco BY2 cell moves from the periphery and settles in the central region of the cell, and this translation is not prevented by actin disruption, suggesting a role for microtubules in premitotic migration of the nucleus in these cells [56]. Cells confined on elongated geometry, that is, rectangles or stripes, have been reported to exhibit translational dynamics [1]. Nuclear translational speed is inversely related to cell shape index (CSI), while rotational speed is directly proportional to CSI. This is again arising from the reduced number of ASFs that results in the loss of directionality of nuclear motion [1,3,36].

Several studies have quantified nuclear speed in different cell types from different organisms and have attributed the driving force to one or more cytoskeletal components [57]. Translational motion depends on the cell type and their intracellular structural details [58]. The actin cap is found to have a major role in nuclear translation [59]. Though the molecular motor dynein is shown not to affect translation speed [60], however, kinesin-1 and dynein at the nuclear envelope are shown to mediate the bidirectional migration of nuclei. Opposing microtubule motors drive robust nuclear dynamics in developing muscle cells [47]. The robustness in several cases arises from microtubules, which undergo buckling deformation and maintain the average position upon the relaxation of applied stress, as revealed by both in vitro and in vivo experiments. Microtubule asters are positioned in a geometry-dependent manner arising from the pushing forces. Protein Op18 is shown to biochemically promote microtubule catastrophes and is sensitive to aster localization dynamics [61]. Coordinated activities of the protein molecules Unc-83, kinesin-1, and dynein at the nuclear envelope during nuclear migration have also been reported [46]. LINC complex mediates the transfer of these forces to the nuclear envelope. In addition to LINC complex, toca-1 is required to move nuclei in *Caenorhabditis elegans* [62]. A balance of cytoskeletal forces generated by microtubule motors that shear the nuclear surface, actomyosin forces that pull, push, and shear the nucleus, results in the apparent nuclear position and shape. Intermediate filaments passively resist the nuclear decentering and deformation [4]. In fibroblasts, the nucleus positions itself at the center point of the traction-force balance, known as the point of maximum tension. In migrating cells, this point of force balance keeps changing, and the nucleus must undergo translational motion to reposition itself [4]. Lamin-deficient cells show higher nuclear translation speed, and in immobilized lamin-deficient cells, the nuclear diffusion gets amplified in different geometries [3].

These results show the diversity and complexity of the nuclear positioning dynamics. Further experiments in well-controlled geometry will help to decipher these complex dynamics. Numerous diseases resulting from genetic alterations in the proteins involved in nuclear movement confirm the significance of proper nuclear positioning. In a well-defined cell geometry that mimics tissue environments, nuclear positioning and its translational dynamics in single cells need careful examination, and these sensitive and unique biophysical markers may find applications in single-cell assays for early disease diagnosis. In the next section, we discuss another kind of motion exhibited by the cell nucleus—rotational motion.

## 3 NUCLEAR ROTATIONAL MOTION
### 3.1 IMAGE PROCESSING AND COMPUTATIONAL METHODS

The change in the angular position of the cell nucleus along the axis perpendicular to the plane of cell-ECM attachment is referred as nuclear rotation. Nuclear rotation and translation occur simultaneously; hence, it is necessary to extract the rotation parameters from the total movement. Fig. 2A shows a cell plated on a triangular

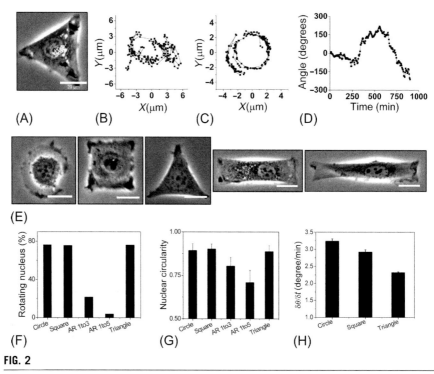

**FIG. 2**

Geometric regulation of nuclear rotations. (A) Phase-contrast image of NIH3T3 fibroblast with overlaid tracks of two nucleoli on a triangular geometry of area 1600 μm$^2$. Scale bar = 20 μm. (B) Translational and (C) rotational coordinates of the nucleus computed from the two tracks. (D) Plot of the nucleus rotation angle with time. (E) Modulation of cell shape using fibronectin-coated micropatterns. Phase-contrast image of single cells grown on different patterns—circles, squares, triangles, and rectangles of aspect ratio (AR) 1:3 and 1:5. Pattern area = 1600 μm$^2$. Scale bar = 20 μm. (F) Fraction of rotating nuclei on circle, square, triangle, and rectangle patterns. (G) Nuclear circularity calculated from phase-contrast image of cells on the above mentioned patterns. (H) Instantaneous angular velocity of nucleus on rotationally symmetrical patterns: circle, square, and triangle.

*Adapted from Kumar A, Maitra A, Sumit M, Ramaswamy S, Shivashankar GV. Actomyosin contractility rotates the cell nucleus. Sci Rep 2014;4:3781.*

micropattern and overlaid is the trajectory of two diagonally opposite nucleoli present in the nucleus.

If $(x_1, y_1)$ and $(x_2, y_2)$ are the tracked coordinates of two nucleoli located at the diagonally opposite ends, then nuclear rotation is characterized by determining the nuclear angle from the slope of the line formed between this mean position and one of the nucleoli and is given by $\theta = \tan^{-1}(y_r/x_r)$ where $(x_r = x_1 - x_t)$ and $(y_r = y_1 - y_t)$ are rotational coordinates of one of the nucleoli and $(x_t, y_t) = \left(\frac{x_1+x_2}{2}, \frac{y_1+y_2}{2}\right)$ are translational coordinates. The extracted translational and rotation trajectory of the nucleus is plotted in Fig. 2B and C. Fig. 2D shows change in angle of the nucleus over time. It is evident that the nucleus shows rotation, stalls/pauses, and reversal.

## 3.2 CELL GEOMETRIC REGULATION OF NUCLEAR ROTATION

Cell geometry can be precisely engineered by plating them on micropatterns of various shapes (Fig. 2E). Percentage of nuclei that exhibit rotation are the highest on circular, then square, and then triangular geometries. Rectangles of high aspect ratio have very few cells with a rotating nucleus (Fig. 2F). On 1:5 aspect ratio rectangles, some nuclei exhibit rotation in z-axis (perpendicular to cell-ECM attachment) (data not shown). Nuclear circularity, calculated as $\left(4 \cdot \pi \cdot Area / Perimeter^2\right)$, decreases significantly on rectangles of high aspect ratio (Fig. 2G). Speed of rotation is highly dependent on the cell geometry. Instantaneous nuclear rotation speed (change in $\theta$ per unit time) is the largest on circular, compared with square and triangular geometries.

## 3.3 ROLE OF CYTOSKELETON IN NUCLEAR ROTATION

Nuclear rotation has been observed during cell migration and wound healing [5,8–10]. Different cytoplasmic and nuclear proteins, both cytoskeletal and linker proteins, have been implicated to control nuclear dynamics during such processes. Microtubule, along with microtubule organizing center (MTOC), forms an asymmetrical cagelike structure around the nucleus. During cell migration, proper positioning and orientation of the nucleus and MTOC is inevitable. MTOC is positioned between the nucleus and the leading edge of the migrating cell. The nucleus is clasped by the dynamic microtubule network and thus experiences a fluctuating torque due to microtubule-associated motor proteins dynein and kinesin [10,46]. In C. elegans, kinesin-1 and dynein are required to move the nucleus during nuclear migration via adapter protein UNC-83 [46]. Forward movement by kinesin-1 and backward movement by dynein are required to maintain the nuclear position; however, the nucleus also rolls/rotates at a speed of 30 degrees/min while migrating. This could be a mechanism for it to pass the accumulation of cytoplasmic granules during migration through narrow cells.

Knockdown of dynein heavy chain or p150-glued, a subunit of dynactin decreases leading edge protrusion and nuclear speed during wound healing [10]. Concomitantly, there is also a decrease in the speed of nuclear rotation.

Further, microtubule depolymerization due to high dose of nocodazole also decreased nuclear angular speed, while decreasing actomyosin contractility through the application of blebbistatin did not have any effect on nuclear angular speed in migrating cells. These experiments suggest a role for dynein motor in nuclear rotation [10].

Unlike the microtubule network, intermediate filaments vimentin form a symmetrical network around the nucleus, connecting via linker protein plectin to it, and it has been shown to act like a shock absorber [13]. Expression of full-length vimentin or domains of vimentin that bind plectin, a LINC protein, in MFT-16 cells that lack vimentin, decreased nuclear rotation suggesting that vimentin acts like a brake for nuclear rotation [63].

Intact cytoplasm to nuclear connection via SUN and KASH domain adapter proteins is necessary to maintain the mechanical properties of the cell. The loss of lamin A/C leads to a decrease in emerin and nesprin3 protein from the nuclear envelope, which causes delayed nuclear orientation and MTOC positioning during cell migration [64]. Nesprin3 via plectin links intermediate filaments to the nucleus, thus compromising the nuclear mechanical integrity upon lamin A/C loss. Rotational speeds are drastically decreased in lamin A/C knockdown cells, and these have impaired wound healing. On the other hand, the loss of other nuclear-intermediate filament—lamin B1—has been shown to increase nuclear rotation [65]. The loss of lamin A/C makes the nucleus deformable, while in the absence of lamin B1, the nucleus maintains its circularity and spins like a rigid particle, without any apparent change in the nucleoli position. During rotation, the nucleus temporally delinks from the surrounding cytoplasm, and hence, there is no change in the position of the cytoplasmic endoplasmic reticulum (ER) and mitochondria [65]. Since cytoplasmic organelles like the ER are not observed to rotate with the nucleus, cytoplasmic connections to the nucleus must be very dynamic and must be linking and delinking at timescales smaller than the rotation speed so as to maintain the mechanical rigidity. Nuclear envelope KASH domain protein nesprin-2G has been shown to have a dynamic connection with the actin cytoskeleton [25]; however, other nuclear connections at these timescales need to be explored.

The prestressed state of the nucleus, which depends on the organization of the actin cytoskeleton [33] and its contractility, was shown to govern both the translational and rotational movements of the nucleus [1,36]. There are physical connections between the cytoskeleton and the nucleus through nuclear outer membrane proteins [13,21,66–71], the alteration of nuclear prestress by plating cells on different geometries of varied shapes and aspect ratio results in modulations in nuclear dynamics. While a number of components including cytoskeleton and motor proteins have been implicated in driving nuclear dynamics in migrating cells, as discussed previously, our results on single cells confined to specific geometric constraints suggest a role for actomyosin contractility in nuclear dynamics [1]. Geometrically isotropic cells on geometries like circles and squares have sustained and faster rotating nuclei compared with higher-aspect-ratio rectangles [1,36,59]. Decreasing actomyosin contractility, using blebbistatin, decreased the tangential actin flow without affecting the radial actin flow [1], thus leading to a decrease in nuclear angular velocity. Increase in nuclear rotation on isotropic geometries could be due to compromised actin to nucleus links as seen by diminished number of apical ASFs [36].

In migrating cells, nuclear rotation is driven by microtubule-associated dynein motors, which maintain the centrality of the MTOC and the nucleus by tethering the plus end to the cortex and asserting torsional force directly on the nucleus [10]. The nucleus rotates due to the torque generated by the dynein motor. This rotation persists for tens of minutes due to the asymmetrical organization of microtubule around the nucleus; however, when this asymmetry changes due to dynamic instability of microtubule, the direction of rotation changes, and hence at longer timescales, nuclear rotation is random. The prediction from a model based on these is that the speed of angular rotation is dependent on the distance between the MTOC and the nucleus. This was true when angular speeds of cells on micropatterns were compared with that of unpatterned where the above distance is larger [5].

Work from our laboratory showed that actomyosin contractility is an important regulator of nuclear rotational dynamics on confined geometries like circle, square, and triangle. We rationalized these observations through a hydrodynamic approach in which the nucleus was treated as a rigid tracer particle residing in a fluid of orientable filaments with active and viscous stresses [1]. The two predictions based on this simple model are that actomyosin is crucial for nuclear rotation and that the angle- and time-averaged angular velocity of the flow will be the maximum away from the nucleus. The first prediction was validated by decreasing actomyosin contractility using blebbistatin that reduced angular rotation speed. The second prediction was validated by measuring actin flow speed as a function of radial distance from the nucleus. Both the predictions were found to be true, thus showing that the actomyosin contractility rotates the cell nucleus on isotropic geometries.

## 3.4 IMPLICATIONS OF NUCLEAR ROTATION IN CELLULAR FUNCTIONS

Nuclear rotation is observed in various cell types including nasal mucosa cells, HeLa, fibroblasts, and skeletal muscle cells [72–74]. During cell migration, nuclear orientation is important for persistent movement. MTOC positions between the leading edge of the cell and nucleus present at the rear end. Thus, nuclear rotation is necessary for the establishment of front-back polarity of cell during wound healing and cell migration [64,75]. Recent studies on patterned circular geometries have shown a chiral pattern of actomyosin emerges when the cell is plated on a confined area [76]. On these geometries, alpha-actinin-rich radial fibers and myosin-IIA-rich transverse fibers show a transition from radial to chiral patterns. Earlier, we demonstrated that on similar geometries, the nucleus also rotates, driven by this actin-myosin system [1]. This self-organization of actomyosin could be the basic mechanism underlying the development of left-right asymmetry in tissues in developing organisms. In the next section, we will focus on shape fluctuations of the nucleus and the role of cell geometry and cytoskeletal forces in such fluctuations.

## 4 NUCLEAR ENVELOPE FLUCTUATIONS
### 4.1 IMAGE PROCESSING AND COMPUTATIONAL METHODS

Ideally, the shape fluctuations must be calculated from temporal measurement of surface area, which requires time-lapse confocal z-stack imaging of the nucleus. However, to shorten imaging time per nucleus (to decrease bleaching and phototoxicity), shape fluctuations can be calculated from temporal measurement of projected area, which requires time-lapse wide-field imaging of the nucleus (Fig. 3A).

Next, to compute the amplitude of the projected area fluctuations, the time-series data of projected area is first detrended by fitting to a third-order polynomial (Fig. 3B). Then, the residual fluctuations are normalized by the respective value of the projected area at the corresponding time point (Fig. 3C). Standard deviation of the distribution of such normalized residual fluctuations can be used as a measure of the amplitude of the projected area fluctuations (Fig. 3D and E).

### 4.2 CELL GEOMETRIC REGULATION OF NUCLEAR FLUCTUATIONS

NIH3T3 fibroblasts in culture dishes adopt an average spreading area of $\sim 1800\,\mu m^2$ and an elongated and polarized shape. When these cells are confined on small isotropic fibronectin micropatterned islands of $500\,\mu m^2$ area, they exhibit threefold higher nuclear area fluctuations (5.3%) compared with cells confined on large polarized fibronectin islands of $1800\,\mu m^2$ area (1.65%) (Fig. 3A–E). The time period of the fluctuations was observed to be 5–8 min, which is close to that of stem cell nuclei [31]. Interestingly, the nuclei of large polarized cells start exhibiting such fluctuations upon perturbation of actin or lamin A/C, which will be discussed later.

### 4.3 ROLE OF CYTOSKELETON IN NUCLEAR FLUCTUATIONS

Systematic perturbation and stabilization of actin polymers in the small isotropic and large polarized geometries showed a nonmonotonous dependence of nuclear fluctuations on actin polymerization, that is, cells with intermediate state of actin polymerization showed highest fluctuations (Fig. 3F). Further, myosin and formin were also shown to be necessary for nuclear envelope fluctuations, suggesting that in small isotropic cells, small units of actin, myosin, and formin apply dynamic forces on the nucleus (Fig. 3G). On the other hand, in cells with large and polarized geometry, actin is present as long and stable stress fibers that keep the nucleus under prestress and prohibit fluctuations [2]. Another source of external force contributing to nuclear fluctuations arises from the polymerization of microtubules, as observed in *Drosophila* embryo during cellularization [77]. Such microtubule-induced fluctuations were repressed in the absence of the *Drosophila* nuclear envelope protein—kugelkern.

In addition to enhanced forces, reduced nuclear stiffness could also lead to enhanced nuclear fluctuations. Fibroblasts constrained on small isotropic geometry showed transcriptional downregulation of lamin A/C, a protein that confers stiffness

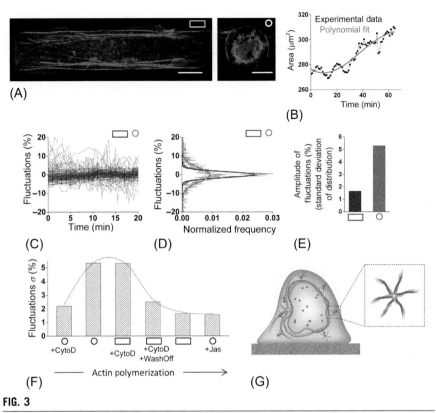

**FIG. 3**

Geometric regulation of nuclear fluctuations. (A) Typical fibroblast cells cultured on large polarized and small isotropic fibronectin micropatterns. Actin is labeled with phalloidin and nucleus is labeled with DAPI. Scale bar = 10 μm. (B) Detrending of a typical nuclear projected area versus time trace for large polarized cell nucleus using third-order polynomial fit.
(C) Residual nuclear projected area fluctuations in the two cell geometries after detrending, normalized with respect to the value of the fit at corresponding time point; circular geometry has larger residual fluctuations compared to rectangle (D) Normalized histogram of fluctuations shown in (C) showing wider distribution for circular geometry. (E) Amplitude of nuclear projected area fluctuations in the two cell geometries, calculated as standard deviation of the fluctuations shown in (C). (F) Amplitude of nuclear projected area fluctuations as a function of actin polymerization states obtained using polymerization perturbing and stabilizing pharmacological reagents cytochalasin D and jasplakinolide, respectively.
(G) Schematic of a small isotropic cell showing floppy nuclear envelope at different time points and uncorrelated displacements of heterochromatin foci inside the nucleus at two different time points. Microtubule is shown indenting on the nucleus, and small actin-myosin-formin asters are shown surrounding the nucleus.

*Adapted from Makhija E, Jokhun DS, Shivashankar GV. Nuclear deformability and telomere dynamics are regulated by cell geometric constraints. Proc Natl Acad Sci U S A 2016;113(1):E32–40.*

to the nucleus [2]. Also, fibroblasts from a lamin A mutation patient and Hutchinson-Gilford progeria syndrome patient show enhanced nuclear fluctuations [78]. In summary, fluctuations of the nuclear envelope can arise from temporal changes in cytoskeletal forces. Alternatively, nuclear fluctuations can also become more evident if the nucleus becomes more compliant and therefore more sensitive to force changes, due to the loss of lamin A/C.

Fluctuations of the nucleus, measured as projected area fluctuations, and dynamics of the chromatin, measured as recovery timescales in photobleaching experiments, both are higher in stem cells as compared with differentiated cells [26,31,79]. Interestingly, the source of these fluctuations in stem cells is also same as those in fibroblasts constrained on small isotropic geometry, that is, short actin polymerized structures and lower levels of lamin A/C [31]. Thus, enhanced nuclear and chromatin fluctuations in small isotropic fibroblast cells could be a mark of initiation of reverse transcriptome programming in these cells. Microarray studies have shown that the gene expression profile of small isotropic cells is considerably distinct from large polarized cells, after 3 h of plating on these geometries [40]. Further experiments are required to confirm whether mechanical signals in the form of constrained cell geometry can lead to reverse transcriptome programming.

## 5 DISCUSSION

### 5.1 SUMMARY OF NUCLEAR TRANSLATION, ROTATION, AND ENVELOPE FLUCTUATIONS

Nuclear shape and dynamics (translation, rotation, and shape fluctuations) depend considerably on the cell geometry. In a polarized elongated cell, the nucleus is elongated, and translation is the predominant type of motion. However, in isotropic cells, the nucleus is circular, and both translational and rotational motion is observed [1]. Moreover, nuclear envelope fluctuations are suppressed in geometries that enhance the formation of ASFs [2].

Nuclear lamina is dynamically coupled to the cytoskeleton via LINC complexes present on the nuclear envelope [25]. These physical connections transmit and react to forces from focal adhesions and cell-cell junctions to the nucleus [80,81]. Small fluctuations in these forces cause the nucleus to move around its mean position, and thus, at short timescales (~seconds), the nucleus displays a constrained diffusive motion. At longer timescales (seconds to minutes), the net force in a particular direction causes the nucleus to undergo a directed motion (superdiffusion). Since the cell itself constitutes a physical boundary, the net force eventually shifts direction, and the nucleus moves in the opposite direction. This oscillatory motion will result in constrained diffusion being observed at even larger timescales (hours to days). Since the cell geometry regulates the cytoskeletal organization, it also defines the timescales at which the various types of motions are observed. For instance, in a small isotropic NIH3T3 fibroblast, the transition from constrained diffusion to directed motion is observed at timescale of ~5 s, which is half (~10 s) of the time required in the case of large elongated cells [3]. Moreover, actin being a major component of the active gel surrounding the nucleus, differential actin organization also results in differential effective viscosity experienced by the nucleus. Thus, changes in actin density affect nuclear diffusion.

Microtubules emanating from the MTOC form a cage and link to the nucleus via transmembrane proteins, adapter proteins, and motor proteins dynein and kinesin. The asymmetrical distribution of motors together with the dynamic nature of the microtubule network subjects the nucleus to a fluctuating torque, thus regulating its rotation [10,46]. Rotation of the nucleus, without any apparent rotation of the ER, suggests that the cytoplasmic-to-nuclear links are very dynamic [25,65]. Nuclear lamins dictate the structural integrity and stiffness of the nucleus and therefore regulate how the physical forces are experienced by the nucleus. Nuclei lacking lamin A/C are softer and therefore more prone to local deformations rather than global rotations. Additionally, changes in lamin A/C have been shown to alter levels of emerin and nesprin3 at the nuclear envelope, further influencing nuclear rotation by altering the physical connections between the nucleoskeleton and the cytoskeleton [64]. Since many of these are directly or indirectly regulated by cell geometry, hence, the latter has a significant bearing on nuclear rotation. Isotropic geometries such as circles and squares have been shown to elicit faster and persistent nuclear rotations. Cell geometry-dependent hydrodynamic actin flow has been shown to be an important player in mediating nuclear rotation under these conditions [1,36,59].

While the net force and torque determine the translational and rotational motions, respectively, tension sets the overall morphology of the nucleus. Thus, conditions favoring contractile apical actomyosin bundles (large elongated geometries) result in flatter prestressed nuclei, and conditions hindering contractile apical actomyosin bundles (small geometries) result in more spherical nuclei [36]. While the decreased tension on the nucleus renders it more susceptible to deformation under the action of other external forces, small isotropic geometries have also been shown to downregulate the expression of lamin A/C. Since lamin A/C provides stiffness to the nucleus, nuclei on small isotropic geometries become even more compliant to external forces. Moreover, under these conditions, small actin-myosin-formin asters are formed, and these directly and/or indirectly exert forces on the nucleus. As a result, increased nuclear area fluctuations are observed [2].

## 5.2 IMPLICATIONS IN GENE EXPRESSION

Confining about a meter length of genetic material within a ~10–20 μm interphase nucleus while maintaining its functional integrity requires very intricate and highly organized packing of the DNA. For instance, regions of the genome that do not have to be transcribed are often tightly packed into distinct heterochromatin domains. Active genes, on the other hand, need to be easily accessible by transcription factors and other components of the transcription machinery. Therefore, they form the more loosely packed euchromatin regions. Proper cell function requires very precise spatiotemporal regulation of chromatin inside the nucleus, with certain processes requiring the coclustering of specific loci from distant parts of the genome [82]. The cytoskeleton-mediated nuclear and chromatin dynamics play an important role in spatiotemporally regulating the genome as a function of cell shape. This is reflected in changes in chromatin dynamics and interchromosome intermingling that occur when the cell geometry is altered [2]. This mechanical regulation of the genome contributes to changes in gene expression profiles, when cells are confined to different geometries [40]. At the molecular level, dynamic compaction and decompaction of chromatin rely on various histone proteins binding and unbinding to and from the DNA strands. Fluorescence recovery after photobleaching (FRAP) analysis of core histones has revealed that perturbing the actin network by ribonucleic acid interference (RNAi) results in changes in the turnover rate of core H2B histones on the chromatin [83,84]. These strongly suggest that cell geometry also influences the chromatin compaction/decompaction or the heterochromatin-to-euchromatin ratios.

While the cell geometry regulates the shuttling of G-actin binding MAL (transcription factor) in and out of the nucleus by regulating the F-actin/G-actin ratio, cell geometry can also influence shuttling of factors by altering nuclear dynamics [40]. If the nucleus is considered as a sponge, one can imagine how having a fluctuating shape can enhance the flux of soluble materials to and from the cytoplasm. Additionally, increased dynamics inside the nucleus results in increased mobility of soluble factors throughout the nucleus. This could play an important role in conditions where most of the genome is highly active (e.g., stem cells).

### 5.3 IMPLICATIONS IN PHYSIOLOGY

During development, the geometry of cells is altered in a very precise manner [27], concomitant with precise alterations in cell function. Embryonic stem cells are circular in shape with decompact genome and lack apical ASFs [31]. These, together with the absence of lamin A/C, result in the stem cell nucleus and chromatin being highly plastic and dynamic [31], ideal for supporting high levels of genomic activity. As the cell differentiates, it starts to adopt a defined morphology, and the emerging actin cytoskeletal network starts to apply forces on the nucleus [85]. Specific parts of the genome are deactivated, and heterochromatin domains are formed. Simultaneously, there is upregulation of lamin A/C; the nucleus becomes less dynamic and stiffens into particular shapes that favor the chromosomal configuration (intermingling) required for proper functioning of the differentiated cell. The importance of cell geometry during stem cell differentiation is further highlighted by experiments that show that the fate of the differentiated cell is greatly influenced by its physical confinement [54,86]. Additionally, studies with geometrically confined cells have shown that chiral patterns of actin flow can emerge from circular cells [76]. This could potentially be a way of setting up polarity during early developmental processes.

Another process whereby cell geometry and nuclear dynamics are relevant is cell migration. Rotation of the nucleus is enhanced, while the cell migrates, and this rotation has been shown to rely on the microtubule network and its associated motors [10,46]. Interestingly, vimentin, which forms a symmetrical meshwork of intermediate filaments around the nucleus, has been shown to dampen nuclear rotation [63]. From the nuclear side, lamins have been shown to play important roles in mediating nuclear rotation during migration [64]. The coordination between these different cytoskeletal and nucleoskeletal systems is critical in regulating nuclear rotation. Moreover, during migration, cells often have to pass through physical constraints that require deformation and squeezing of the nucleus. Both actin and microtubules have been shown to play an important role in dragging and deforming the nucleus. Disruptions of the LINC complexes, which mechanically couple the cytoskeleton to the nuclear envelope, have been shown to affect cell migration. This shows that proper regulation/coordination of physical forces on the nucleus is critical during this process [57]. After passing through the constraint, the nucleus is translated back to the cell center as the cell adopts once again its typical morphology. The regular force balance on the nucleus is finally restored. Such transient deformations, however, are not expected to cause long-term changes in cell function. There could, therefore, exist minimum threshold time and length scales beyond which proper cell function is permanently disrupted. This could potentially provide a way for mechanically reprograming cells.

## 5.4 CONCLUSIONS

Cell geometry sets the mechanical homeostasis of the nucleus by defining the balance of forces between the cytoskeleton and nucleoskeleton. It is therefore not surprising that abnormal gene expression profiles are observed in several diseases wherein the cytoskeleton or nucleoskeleton is impaired. From a research perspective, tuning the geometry of cells can serve as a powerful means for stimulating distinct cytoskeletal configurations and studying the role of nuclear mechanics on genome regulation. On the other hand, from an application point of view, investigating nuclear dynamics offers a plethora of potential new tools for diagnosing diseases.

## REFERENCES

[1] Kumar A, Maitra A, Sumit M, Ramaswamy S, Shivashankar GV. Actomyosin contractility rotates the cell nucleus. Sci Rep 2014;4:3781.

[2] Makhija E, Jokhun DS, Shivashankar GV. Nuclear deformability and telomere dynamics are regulated by cell geometric constraints. Proc Natl Acad Sci U S A 2016;113(1): E32–40.

[3] Radhakrishnan AV, Jokhun DS, Venkatachalapathy S, Shivashankar GV. Nuclear positioning and its translational dynamics are regulated by cell geometry. Biophys J 2017;112(9):1920–8.

[4] Neelam S, Chancellor TJ, Li Y, Nickerson JA, Roux KJ, Dickinson RB, et al. Direct force probe reveals the mechanics of nuclear homeostasis in the mammalian cell. Proc Natl Acad Sci U S A 2015;112(18):5720–5.
[5] Wu J, Lee KC, Dickinson RB, Lele TP. How dynein and microtubules rotate the nucleus. J Cell Physiol 2011;226(10):2666–74.
[6] Hagan I, Yanagida M. Evidence for cell cycle-specific, spindle pole body-mediated, nuclear positioning in the fission yeast Schizosaccharomyces pombe. J Cell Sci 1997;110(Pt 16):1851–66.
[7] Starr DA. Communication between the cytoskeleton and the nuclear envelope to position the nucleus. Mol Biosyst 2007;3(9):583–9.
[8] Brosig M, Ferralli J, Gelman L, Chiquet M, Chiquet-Ehrismann R. Interfering with the connection between the nucleus and the cytoskeleton affects nuclear rotation, mechanotransduction and myogenesis. Int J Biochem Cell Biol 2010;42(10):1717–28.
[9] Lee JS, Chang MI, Tseng Y, Wirtz D. Cdc42 mediates nucleus movement and MTOC polarization in Swiss 3T3 fibroblasts under mechanical shear stress. Mol Biol Cell 2005;16(2):871–80.
[10] Levy JR, Holzbaur EL. Dynein drives nuclear rotation during forward progression of motile fibroblasts. J Cell Sci 2008;121(Pt 19):3187–95.
[11] Wu CY, Rolfe PA, Gifford DK, Fink GR. Control of transcription by cell size. PLoS Biol 2010;8(11):e1000523.
[12] Radhakrishnan A, Damodaran K, Soylemezoglu AC, Uhler C, Shivashankar GV. Machine learning for nuclear mechano-morphometric biomarkers in cancer diagnosis. Sci Rep 2017;7(1):17946.
[13] Dahl KN, Ribeiro AJ, Lammerding J. Nuclear shape, mechanics, and mechanotransduction. Circ Res 2008;102(11):1307–18.
[14] Tzur YB, Wilson KL, Gruenbaum Y. SUN-domain proteins: 'Velcro' that links the nucleoskeleton to the cytoskeleton. Nat Rev Mol Cell Biol 2006;7(10):782–8.
[15] Crisp M, Liu Q, Roux K, Rattner JB, Shanahan C, Burke B, et al. Coupling of the nucleus and cytoplasm: role of the LINC complex. J Cell Biol 2006;172(1):41–53.
[16] Dasanayake NL, Michalski PJ, Carlsson AE. General mechanism of actomyosin contractility. Phys Rev Lett 2011;107(11):118101.
[17] Haque F, Lloyd DJ, Smallwood DT, Dent CL, Shanahan CM, Fry AM, et al. SUN1 interacts with nuclear lamin A and cytoplasmic nesprins to provide a physical connection between the nuclear lamina and the cytoskeleton. Mol Cell Biol 2006;26(10):3738–51.
[18] Houben F, Ramaekers FC, Snoeckx LH, Broers JL. Role of nuclear lamina-cytoskeleton interactions in the maintenance of cellular strength. Biochim Biophys Acta 2006;1773(5):675–86.
[19] Ingber DE, Tensegrity I. Cell structure and hierarchical systems biology. J Cell Sci 2003;116(Pt 7):1157–73.
[20] Ingber DE, Tensegrity II. How structural networks influence cellular information processing networks. J Cell Sci 2003;116(Pt 8):1397–408.
[21] King MC, Drivas TG, Blobel G. A network of nuclear envelope membrane proteins linking centromeres to microtubules. Cell 2008;134(3):427–38.
[22] Mackintosh FC, Schmidt CF. Active cellular materials. Curr Opin Cell Biol 2010;22(1):29–35.
[23] Theriot JA. The polymerization motor. Traffic 2000;1(1):19–28.
[24] Zhang Q, Ragnauth CD, Skepper JN, Worth NF, Warren DT, Roberts RG, et al. Nesprin-2 is a multi-isomeric protein that binds lamin and emerin at the nuclear envelope and forms a subcellular network in skeletal muscle. J Cell Sci 2005;118(Pt 4):673–87.

# References

[25] Kumar A, Shivashankar GV. Dynamic interaction between actin and nesprin2 maintain the cell nucleus in a prestressed state. Methods Appl Fluoresc 2016;4(4):044008.

[26] Bhattacharya D, Talwar S, Mazumder A, Shivashankar GV. Spatio-temporal plasticity in chromatin organization in mouse cell differentiation and during Drosophila embryogenesis. Biophys J 2009;96(9):3832–9.

[27] Kumar A, Shivashankar GV. Mechanical force alters morphogenetic movements and segmental gene expression patterns during Drosophila embryogenesis. PLoS One 2012;7(3):e33089.

[28] Mazumder A, Roopa T, Basu A, Mahadevan L, Shivashankar GV. Dynamics of chromatin decondensation reveals the structural integrity of a mechanically prestressed nucleus. Biophys J 2008;95(6):3028–35.

[29] Mazumder A, Roopa T, Kumar A, Iyer KV, Ramdas NM, Shivashankar GV. Prestressed nuclear organization in living cells. Methods Cell Biol 2010;98:221–39.

[30] Mazumder A, Shivashankar GV. Gold-nanoparticle-assisted laser perturbation of chromatin assembly reveals unusual aspects of nuclear architecture within living cells. Biophys J 2007;93(6):2209–16.

[31] Talwar S, Kumar A, Rao M, Menon GI, Shivashankar GV. Correlated spatio-temporal fluctuations in chromatin compaction states characterize stem cells. Biophys J 2013; 104(3):553–64.

[32] Wang N, Tytell JD, Ingber DE. Mechanotransduction at a distance: mechanically coupling the extracellular matrix with the nucleus. Nat Rev Mol Cell Biol 2009; 10(1):75–82.

[33] Mazumder A, Shivashankar GV. Emergence of a prestressed eukaryotic nucleus during cellular differentiation and development. J R Soc Interface 2010;7(Suppl. 3):S321–30.

[34] Ingber DE, Prusty D, Sun Z, Betensky H, Wang N. Cell shape, cytoskeletal mechanics, and cell cycle control in angiogenesis. J Biomech 1995;28(12):1471–84.

[35] Khatau SB, Hale CM, Stewart-Hutchinson PJ, Patel MS, Stewart CL, Searson PC, et al. A perinuclear actin cap regulates nuclear shape. Proc Natl Acad Sci U S A 2009;106 (45):19017–22.

[36] Li Q, Kumar A, Makhija E, Shivashankar GV. The regulation of dynamic mechanical coupling between actin cytoskeleton and nucleus by matrix geometry. Biomaterials 2014;35(3):961–9.

[37] Li QS, Lee GY, Ong CN, Lim CT. AFM indentation study of breast cancer cells. Biochem Biophys Res Commun 2008;374(4):609–13.

[38] Luxton GWG, Gomes ER, Folker ES, Vintinner E, Gundersen GG. Linear arrays of nuclear envelope proteins harness retrograde actin flow for nuclear movement. Science 2010;329(5994):956–9.

[39] Folker ES, Östlund C, Luxton GWG, Worman HJ, Gundersen GG. Lamin A variants that cause striated muscle disease are defective in anchoring transmembrane actin-associated nuclear lines for nuclear movement. Proc Natl Acad Sci U S A 2011;108(1):131–6.

[40] Jain N, Iyer KV, Kumar A, Shivashankar GV. Cell geometric constraints induce modular gene-expression patterns via redistribution of HDAC3 regulated by actomyosin contractility. Proc Natl Acad Sci U S A 2013;110(28):11349–54.

[41] Roca-Cusachs P, Alcaraz J, Sunyer R, Samitier J, Farre R, Navajas D. Micropatterning of single endothelial cell shape reveals a tight coupling between nuclear volume in G1 and proliferation. Biophys J 2008;94(12):4984–95.

[42] Thomas CH, Collier JH, Sfeir CS, Healy KE. Engineering gene expression and protein synthesis by modulation of nuclear shape. Proc Natl Acad Sci U S A 2002; 99(4):1972–7.

[43] Vergani L, Grattarola M, Nicolini C. Modifications of chromatin structure and gene expression following induced alterations of cellular shape. Int J Biochem Cell Biol 2004;36(8):1447–61.
[44] Friedl P, Wolf K, Lammerding J. Nuclear mechanics during cell migration. Curr Opin Cell Biol 2011;23(1):55–64.
[45] Sims JR, Karp S, Ingber DE. Altering the cellular mechanical force balance results in integrated changes in cell, cytoskeletal and nuclear shape. J Cell Sci 1992;103(Pt 4):1215–22.
[46] Fridolfsson HN, Starr DA. Kinesin-1 and dynein at the nuclear envelope mediate the bidirectional migrations of nuclei. J Cell Biol 2010;191(1):115–28.
[47] Adames NR, Cooper JA. Microtubule interactions with the cell cortex causing nuclear movements in *Saccharomyces cerevisiae*. J Cell Biol 2000;149(4):863–74.
[48] Meyerzon M, Fridolfsson HN, Ly N, McNally FJ, Starr DA. UNC-83 is a nuclear-specific cargo adaptor for kinesin-1-mediated nuclear migration. Development 2009;136(16):2725–33.
[49] Chancellor TJ, Lee J, Thodeti CK, Lele T. Actomyosin tension exerted on the nucleus through nesprin-1 connections influences endothelial cell adhesion, migration, and cyclic strain-induced reorientation. Biophys J 2010;99(1):115–23.
[50] Wu D, Flannery AR, Cai H, Ko E, Cao K. Nuclear localization signal deletion mutants of lamin A and progerin reveal insights into lamin A processing and emerin targeting. Nucleus 2014;5(1):66–74.
[51] Finer JT, Simmons RM, Spudich JA. Single myosin molecule mechanics: piconewton forces and nanometre steps. Nature 1994;368(6467):113–9.
[52] Coy DL, Wagenbach M, Howard J. Kinesin takes one 8-nm step for each ATP that it hydrolyzes. J Biol Chem 1999;274(6):3667–71.
[53] Vahey MD, Fletcher DA. The biology of boundary conditions: cellular reconstitution in one, two, and three dimensions. Curr Opin Cell Biol 2014;26:60–8.
[54] Kilian KA, Bugarija B, Lahn BT, Mrksich M. Geometric cues for directing the differentiation of mesenchymal stem cells. Proc Natl Acad Sci U S A 2010;107(11):4872–7.
[55] Martin RM, Leonhardt H, Cardoso MC. DNA labeling in living cells. Cytometry A 2005;67(1):45–52.
[56] Katsuta J, Hashiguchi Y, Shibaoka H. The role of the cytoskeleton in positioning of the nucleus in premitotic tobacco BY-2 cells. J Cell Sci 1990;95(3):413–22.
[57] Gundersen GG, Worman HJ. Nuclear positioning. Cell 2013;152(6):1376–89.
[58] Kiss A, Horvath P, Rothballer A, Kutay U, Csucs G. Nuclear motility in glioma cells reveals a cell-line dependent role of various cytoskeletal components. PLoS One 2014;9(4):e93431.
[59] Kim DH, Cho S, Wirtz D. Tight coupling between nucleus and cell migration through the perinuclear actin cap. J Cell Sci 2014;127(Pt 11):2528–41.
[60] Cadot B, Gache V, Gomes ER. Moving and positioning the nucleus in skeletal muscle—one step at a time. Nucleus 2015;6(5):373–81.
[61] Faivre-Moskalenko C, Dogterom M. Dynamics of microtubule asters in microfabricated chambers: the role of catastrophes. Proc Natl Acad Sci U S A 2002;99(26):16788–93.
[62] Chang YT, Dranow D, Kuhn J, Meyerzon M, Ngo M, Ratner D, et al. toca-1 is in a novel pathway that functions in parallel with a SUN-KASH nuclear envelope bridge to move nuclei in *Caenorhabditis elegans*. Genetics 2013;193(1):187–200.
[63] Gerashchenko MV, Chernoivanenko IS, Moldaver MV, Minin AA. Dynein is a motor for nuclear rotation while vimentin IFs is a "brake". Cell Biol Int 2009;33(10):1057–64.

[64] Houben F, Willems CH, Declercq IL, Hochstenbach K, Kamps MA, Snoeckx LH, et al. Disturbed nuclear orientation and cellular migration in A-type lamin deficient cells. Biochim Biophys Acta 2009;1793(2):312–24.

[65] Ji JY, Lee RT, Vergnes L, Fong LG, Stewart CL, Reue K, et al. Cell nuclei spin in the absence of lamin b1. J Biol Chem 2007;282(27):20015–26.

[66] Shivashankar GV. Mechanosignaling to the cell nucleus and gene regulation. Annu Rev Biophys 2011;40:361–78.

[67] Rowat AC, Lammerding J, Ipsen JH. Mechanical properties of the cell nucleus and the effect of emerin deficiency. Biophys J 2006;91(12):4649–64.

[68] Morgan JT, Pfeiffer ER, Thirkill TL, Kumar P, Peng G, Fridolfsson HN, et al. Nesprin-3 regulates endothelial cell morphology, perinuclear cytoskeletal architecture, and flow-induced polarization. Mol Biol Cell 2011;22(22):4324–34.

[69] Lombardi ML, Jaalouk DE, Shanahan CM, Burke B, Roux KJ, Lammerding J. The interaction between nesprins and sun proteins at the nuclear envelope is critical for force transmission between the nucleus and cytoskeleton. J Biol Chem 2011;286(30):26743–53.

[70] Lombardi ML, Lammerding J. Keeping the LINC: the importance of nucleocytoskeletal coupling in intracellular force transmission and cellular function. Biochem Soc Trans 2011;39(6):1729–34.

[71] Jaalouk DE, Lammerding J. Mechanotransduction gone awry. Nat Rev Mol Cell Biol 2009;10(1):63–73.

[72] Pomerat CM. Rotating nuclei in tissue cultures of adult human nasal mucosa. Exp Cell Res 1953;5(1):191–6.

[73] Capers CR. Multinucleation of skeletal muscle in vitro. J Biophys Biochem Cytol 1960;7:559–66.

[74] Leone V, Hsu TC, Pomerat CM. Cytological studies on HeLa, a strain of human cervical carcinoma. II. On rotatory movements of the nuclei. Z Zellforsch Mikrosk Anat 1955;41(5):481–92.

[75] Maninova M, Klimova Z, Parsons JT, Weber MJ, Iwanicki MP, Vomastek T. The reorientation of cell nucleus promotes the establishment of front-rear polarity in migrating fibroblasts. J Mol Biol 2013;425(11):2039–55.

[76] Tee YH, Shemesh T, Thiagarajan V, Hariadi RF, Anderson KL, Page C, et al. Cellular chirality arising from the self-organization of the actin cytoskeleton. Nat Cell Biol 2015;17(4):445–57.

[77] Hampoelz B, Azou-Gros Y, Fabre R, Markova O, Puech PH, Lecuit T. Microtubule-induced nuclear envelope fluctuations control chromatin dynamics in Drosophila embryos. Development 2011;138(16):3377–86.

[78] De Vos WH, Houben F, Hoebe RA, Hennekam R, van Engelen B, Manders EM, et al. Increased plasticity of the nuclear envelope and hypermobility of telomeres due to the loss of A-type lamins. Biochim Biophys Acta 2010;1800(4):448–58.

[79] Pajerowski JD, Dahl KN, Zhong FL, Sammak PJ, Discher DE. Physical plasticity of the nucleus in stem cell differentiation. Proc Natl Acad Sci U S A 2007;104(40):15619–24.

[80] Guilluy C, Osborne LD, Van Landeghem L, Sharek L, Superfine R, Garcia-Mata R, et al. Isolated nuclei adapt to force and reveal a mechanotransduction pathway in the nucleus. Nat Cell Biol 2014;16(4):376–81.

[81] Kumar A, Ouyang M, Van den Dries K, McGhee EJ, Tanaka K, Anderson MD, et al. Talin tension sensor reveals novel features of focal adhesion force transmission and mechanosensitivity. J Cell Biol 2016;213(3):371–83.

[82] Fanucchi S, Shibayama Y, Burd S, Weinberg MS, Mhlanga MM. Chromosomal contact permits transcription between coregulated genes. Cell 2013;155(3):606–20.
[83] Ramdas NM, Shivashankar GV. Cytoskeletal control of nuclear morphology and chromatin organization. J Mol Biol 2015;427(3):695–706.
[84] Toh KC, Ramdas NM, Shivashankar GV. Actin cytoskeleton differentially alters the dynamics of lamin A, HP1alpha and H2B core histone proteins to remodel chromatin condensation state in living cells. Integr Biol (Camb) 2015;7(10):1309–17.
[85] Olson EN, Nordheim A. Linking actin dynamics and gene transcription to drive cellular motile functions. Nat Rev Mol Cell Biol 2010;11(5):353–65.
[86] Ruiz SA, Chen CS. Emergence of patterned stem cell differentiation within multicellular structures. Stem Cells 2008;26(11):2921–7.

# CHAPTER

# Mechanobiology throughout development

# 3

**Jason P. Gleghorn, Megan L. Killian**
*Department of Biomedical Engineering, University of Delaware, Newark, DE, United States*

## 1 INTRODUCTION

In our day-to-day life, we can appreciate growth through observations, whether a child grows too big to fit into their favorite shirt or the seeds we planted in our garden break through the soil as budding plants. We can visualize, albeit abstractly, the process of cell division, an essential part of growth and morphogenesis, if we slice a tomato or a ball of dough in half and then into quarters. Unlike the tomato and dough on our kitchen counter, as cells divide, they not only change their size but also change their shape and their interactions with neighboring cells. A divided cell becomes two, which then becomes four, and so on, at an exponential rate. This rapid expansion is constrained, in part, by the microenvironment of the cell and the biological and mechanical responses of surrounding cells. At this point, the process of differentiation plays a key role in integrating information of the spatial microenvironment in order to assign an undifferentiated cell to a particular functional role in the embryo. Differentiation is initiated upon the establishment of a cellular, morphological axis, pushing cells to function in a specific, oriented direction. These initial stages of morphogenesis allow an organ to form from individual cells and layers of tissues in an orchestrated fashion, with fine rhythms of movement and forces guiding the way.

Morphogenesis is a physical process that integrates cellular behaviors and interactions with each other and the surrounding microenvironment, at multiple time and length scales to generate structure that is intimately tied to tissue and organ function. The identity and spatial composition of extracellular matrix proteins guide local and global tissue geometry from which emergent collective cellular behaviors arise. The cells, in turn, remodel their extracellular microenvironment, changing the chemical identity (e.g., the presence, absence, or concentration of collagen, laminin, and fibronectin) and physical properties (e.g., stiffness or deformation) to temporally evolve the architecture. Over time, the established architecture feeds back into the mechanical environment experienced by the cells. As such, it is important to recognize the spatial and temporal nature of this interplay of individuals and populations of cells with their environment. In this chapter, we highlight some of the mechanical forces and mechanobiology underlying the physical mechanisms that drive tissue and organ morphogenesis and review methods that have been developed to quantify and understand the intricate dance cells perform to build functional tissues.

## 2 MECHANOBIOLOGY IN EARLY DEVELOPMENT

The division of cells is the first step in understanding developmental biology. The symmetry of cell division is governed by the positioning of mitotic spindles, and whether cells undergo symmetrical or asymmetrical cell division depends on how the cytoskeletal microtubules interact with contractile actin microfilaments [1–5]. The division of cells and their subsequent role in modeling and remodeling of embryonic structures play a crucial role in how the embryo is shaped [6]. From the very beginning, there are thought to be specific mechanically regulated forces required for sperm to fertilize an egg, whether it is the spring forces required for the fertilization of horseshoe crab eggs during spawning [6] or osmotic pressure gradients that alter cell shape and membrane tension during the migration of *Drosophila* eggs from the ovary to the uterus [6,7]. In the human embryo, primarily intrinsic forces drive early morphogenesis, and extrinsic, environmental forces drive fetal growth and maturation (Fig. 1).

The size, shape, and interactions of cells during development are influenced by both internal and external forces. Growth of the cell results from an increase in the volume of the cell and is associated with cell proliferation and its production of

**FIG. 1**

Morphogenesis of the embryo and maturation of the fetus experience a wide range of mechanical forces. During early morphogenesis, the primary forces are intrinsic, such as cell proliferation, migration, and rearrangements. During maturation, the formation of active tissues like skeletal and smooth muscle shifts the mechanical environment to more extrinsic, applied forces. Intrauterine pressure and neighboring-tissue constraints also play a role in the maturation of the fetus.

*Reproduced with permissions from Creative Commons.*

extracellular matrix. This individual cell's growth exerts pressure on its surrounding cells. Cell shape is derived in part from cell-generated traction on its neighboring cells and surrounding environment. Cell hypertrophy, which is a factor of cell growth, also contributes to cell shape. The interactions between size and shape are key during morphogenesis, as there are several ways in which cells interact and build tissues (Fig. 2). Cell growth and movements during gastrulation induce dramatic changes in the morphology of the early embryo and are the primary driver of cell-generated pressure. For example, when a cell increases in volume (i.e.,

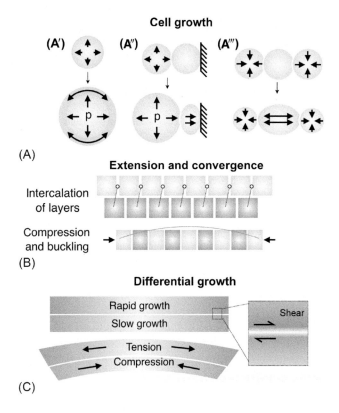

**FIG. 2**

(A) Cell growth results in localized forces, such as (A′) intracellular pressure (p) and, upon expansion, membrane tension. As a cell grows, its neighboring cell may be (A″) constrained due to its location within tissue, and its extracellular matrix interactions or (A‴) stretched due to cell-cell adhesions. (B) The process of extension and convergence results in intercalation of two or more cell layers, causing a thin cell sheet to compress and buckle due to space constraints and its new geometry. (C) The differential growth rates of tissue layers or in buckling tissues impart localized tissue loads, such as tension on the stretched surface and compression on the inner surface. This also leads to shear loading between facing tissue layers.

grows), it is constrained by its plasma membrane, inducing a passive stretch across the surface of the cell. The local forces observed by neighboring cells are also influenced by growing cells. Epiboly is the first event of coordinated cell movement that drives tissue morphology in the embryo, and it is also one of the most extensively studied mechanisms of cell movements. During epiboly, the cell layers of the early embryo undergo thinning and spreading. This process requires both intracellular adhesion and cortical tension to keep the cells attached and able to spread. Following spreading, elongation of the longitudinal direction occurs under intercalation, whereby two or more rows of cells move between one another and generate longer, thinner arrays of cells [8]. Convergent extension, a specialized form of intercalation, is highly directional, and convergence of cells occurs perpendicular to the axis of extension. This results in an overall extension of the tissue in a preferred direction of growth, forming the epithelium. The formation of the thin epithelium, which is a two-dimensional tissue, undergoes buckling, folding, and wrinkling due to its thin, elongated geometry. Buckling of the epithelia during growth can result from constrained expansion in the same direction as elongation, imparting high compressive loads and causing the epithelium to fold. Differential growth rates of tissue layers can also induce buckling, leading to increased tension on the surface undergoing rapid growth and increased compression on the regions of slower growth.

In several developing structures, cellular gradients play a critical role in the patterning and formation of tissues. During tissue folding in the *Drosophila* mesoderm, multicellular myosin gradients have been shown to promote this tissue's three-dimensional form [9]. This gradient in myosin contraction drives apical contractility that is required for proper tissue folding [9]. Regulators of contraction, such as permanently attached membrane proteins (e.g., transmembrane protein in Drosophila, T48), ion channels, and gap junctions (e.g., connexin-43), may regulate patterning with their spatial control of the contraction. Gradients in cell density and cell adhesion likely act inversely during morphogenesis of the embryo, with cells preferentially migrating toward regions of lower cell density and toward concentrated adhesion [10]. Other gradients form in the developing embryo, such as anterior/posterior gradients [11–13] and proximodistal gradients [14–20], driving morphogenesis of the lung and limb.

We will focus primarily on the mechanical and mechanobiological aspects of neural tube, lung, and limb formation here as examples of important structures with varying levels of complexity, in order to highlight a range of physical mechanisms the embryo uses to sculpt organs. However, readers are encouraged to pursue works in the field of developmental biology to get a more thorough understanding of the signaling pathways involved in each of these unique patterning events.

## 2.1 NEURAL TUBE MORPHOGENESIS

In a series of cellular migrations, rearrangements, contractions, and differentiation events, the neural tube is formed [21]. This closed tube filled with cerebrospinal fluid is the precursor structure that eventually becomes the various components of the central nervous system. In early development, partitioning of the neural tube is driven and maintained by circumferential actomyosin contractions to generate local constrictions called sulci [22]. Lined by the neural ectoderm, a polarized epithelium, the neural tube maintains a positive internal luminal pressure and begins to enlarge at the onset of closure [23]. The positive pressures in the neural tube are critical for proper morphogenesis as dysregulated pressures lead to a variety of neural tube defects [23,24]. The necessary fluid pressures are generated and regulated by controlling lumen fluid composition. The ectoderm actively transports $Na^+$ into the lumen through the action of ion pumps [25,25a] and secretes chondroitin sulfate proteoglycans to retain water within the lumen [26]. Experiments increasing or decreasing the presence of these osmolytes dramatically changed the neural tube swelling, thereby confirming the role of an osmotic gradient [26].

In addition to the need for a positive pressure in the neural tube, data suggest that flow-induced pressure is also a necessary factor for the original differentiation and enlargement of the brain in early stages of embryogenesis [27]. Cilia in the neural tube and embryonic brain beat in a synchronized motion to regulate fluidic pressure by pushing cerebrospinal fluid through the neural tube and across the ventricles of the brain [23,28]. In fact, several neural tube defects including hydrocephaly are linked to altered ciliary fluid flows secondary to malformed or immotile cilia [29,30]. New evidence demonstrates that mutations affecting proteins necessary for ciliogenesis or cilia motility, like cadherin EGF LAG seven-pass G-type receptor 2 and 3 (Celsr2, Celsr3), and axonemal dynein heavy chain (MDnah-5), decrease flow velocity and persistence, leading to neural malformations [28,30]. Indeed, during the earliest stages of embryogenesis, the controlled modulation of localized pressure plays a critical role in the formation of the neural tube. These mechanisms that modulate pressure during neural tube morphogenesis are not exclusive to this structure. In fact, pressure plays a key role during lung morphogenesis, which is one of the last mammalian organs to fully develop.

## 2.2 LUNG MORPHOGENESIS

The embryonic lung is a prime example of larger-scale mechanical forces coupled with cellular mechanotransduction to induce proper organ formation. The epithelial airways undergo branching morphogenesis into a surrounding mesenchyme to form an airway architecture that is stereotyped and similar within individuals of the same species [31–33]. A tonic positive pressure within the developing airways is created due to fluid secretion from the epithelium. Initially, chloride ($Cl^-$) is actively pumped into the luminal space by cystic fibrosis transmembrane conductance regulator (CFTR) and other channels [34–37]. The $Cl^-$ gradient draws sodium and water through leaky cell junctions into the lumen [38,39]. The resulting positive airway luminal pressure is directly correlated to fetal development [25a]. Chemical disruption of the $Cl^-$ flux [35,40,41] or physically draining the luminal fluid [42,43] decreases lung pressure and tissue distension and causes lung hypoplasia. Conversely, artificially increasing luminal pressures has shown rescue of lung hypoplasia in animal models [44,45]. In addition to tonic fluid pressure exerting forces on the newly formed airways, contractile smooth muscle wraps circumferentially around the proximal airways [46,47]. Peristaltic-like waves of contraction can be observed in explant culture that moves the luminal lung fluid through the airways toward the closed distal tips [48–50]. These contractions are driven by calcium flux [51,52], and ex vivo lung culture has shown that alteration of contraction is directly correlated to growth [25a,53,54]. Airway smooth muscle, once termed vestigial, is now thought responsible for triggering local growth events by mechanically stretching the distal airway epithelium to spatially activate cellular mechanotransduction pathways and associated growth factor signaling to further airway elaboration [55–58].

The specific events underlying airway morphogenesis are still an active area of investigation. In the developing chick lung, apical constriction of airway epithelial cells was shown experimentally and computationally to be the underlying mechanism of initial branch lateral formation of the main bronchus [59]. These cell shape changes—constricting the apical surface to produce a wedge-shaped geometry—locally drive the folding of the epithelium into the surrounding lung mesenchyme. In the developing mouse lung, new airway initiation events are less clear. Differentiating smooth muscle cells appear at the bifurcating airway tips, the site at which one branch splits into two daughter branches. Blocking smooth muscle differentiation or surgically removing the smooth muscle disrupts this bifurcation process [60]. The epithelial folding at these sites may be driven by an active contraction of the newly formed smooth muscle [25a,60] or postnatally by mechanical buckling [61–63] that has been identified in the morphogenesis of other tissues [64,65]. Whereas spatial epithelial proliferation is not necessary to drive new airway morphogenesis, oriented cell division has been demonstrated to be responsible and necessary for branch elongation [66]. Quantitative methods to determine the physical mechanisms that drive branching morphogenesis are an emergent need to better understand this complex developmental process.

## 2.3 APPENDICULAR LIMB DEVELOPMENT

The developmental processes involved in constructing the vertebrate limb bud into its complex, fully formed structure are complicated and well coordinated. The formation of the appendicular limb is initiated by budding, subsequent three-dimensional cellular and tissue polarization, and outgrowth in the proximodistal direction. The patterned structures that form during the outgrowth process require a commitment of progenitor cells that will eventually become the bone template, joints, tendons and ligaments, and muscle. While the mechanical cues required for establishing the early vertebrate limb are the primary focus of this section, the induction and orchestrated patterning of elements in the limb are unique and worth noting. The limb bud contains mesoderm, surrounded by ectoderm, and expands in a proximal-to-distal progression to form the skeletal elements known as the stylopod (single element), zeugopod (double element), and autopod (digit elements), in that sequence. In addition, the coordinated outgrowth of the bud and the anterior/posterior and dorsal/ventral axes of patterning in three-dimensions are regulated by cell polarization, proliferative events, and growth-related strains throughout the length of the limb. These specified patterning events are driven by tightly controlled molecular cues, via homeobox family of genes (HOX) [67–69], transforming growth fact-beta (TGF-β) [70,71], bone morphogenetic protein (BMP) [71–75], fibroblast growth factors (FGF) [72,73,76–80], Indian/Sonic hedgehog (IHH/SHH) [72,76], parathyroid hormone related protein (PTHrP) [72], and wingless-type family (WNT) [79,81] signaling. Interestingly, the biological patterning that drives appendicular limb development closely parallels that of few other organs but has striking similarities in many (but not all) ways to the formation of the genitalia [82].

The initiation of limb patterning occurs in the absence of muscle loading, yet it is clear that the induction and elongation of limb elements undergo mechanobiological processes similar to other morphogenetic processes described for the neural tube and lung. In the developing limb, patterns of differential growth of neighboring tissues play a role in the generation of compression and tension along surfaces, such as during the initial mesenchymal condensation [83,84]. The induction of mesenchymal progenitors to undergo chondrogenic differentiation, thus establishing the cartilage template, is encouraged by compressive forces and their interaction with surrounding matrix [85–89]. The morphogenesis of the cartilage template from mesenchymal condensation in vivo results in the separation of the progenitor condensation into distinct cell pools, such as chondroblasts and osteoblasts. These progenitor pools are encapsulated with a thin layer of mesenchymal progenitors that establish the perichondrium. The template eventually becomes vascularized, initiating the formation of growth plates and elongation of bone. At the same time, skeletal muscles migrate into the limb bud from the dermomyotome and attach to tendons dispersed along the perichondral layer. Although the limb does not require the migration of muscle or innervation to form the skeletal elements, it does require muscle loading to maintain its progenitor populations for limb maturation.

The response of limb progenitors during morphogenesis has been interrogated using in vivo models, as well as isolated stem cell culture and in vitro explant models. Long and Linsenmayer used chick limb culture models to remove regions of the perichondrium and determine differential growth of cartilage [90]. In the developing limb, the perichondrium acts as a fiber-like constraint that prevents outward growth and promotes directionality of long bones. The groove of Ranvier, which is a circumferential groove peripheral to the growth plate [91–93], also plays a role in promoting longitudinal growth, although its exact function is debated. The perichondrium and groove and eventually the cortical bone and periosteum may provide the growth plate and diaphysis with reactive circumferential/hoop stress that increase the compressive forces on the growth plate and diaphyseal progenitor cells [94,95]. However, the direct mechanical role of the periosteum to growing bone has been contested [96], suggesting a more biological role in its regulation of longitudinal bone growth.

Once the cartilage template has its generalized form, it requires intermittent embryonic movement, driven primarily by muscle contraction, for maintaining its proper shape and progenitor fate [97–102]. The lack of muscle contraction in the developing embryo leads to striking abnormalities in the shape and formation of articulating joints [97,101,103–112]. Interestingly, muscle loading is not required for the initiation of bone shape or the initial stage of articulating joint development (i.e., formation of the interzone) [103,105,106]. However, the prolonged lack of muscle loads can lead to abnormal bone curvature [113] and shallow/shorter bone landmarks, like the intercondyloid fossa [113], acetabulum [114], and the deltoid tuberosity [103]. The role that skeletal muscle loading plays in joint formation is primarily through the commitment of progenitor cells to maintain the joint following the formation of the interzone. However, applied contractile loading from skeletal muscle is required throughout embryonic growth in order to drive bone and joint shape. These forces act primarily through the tendons [115], which transfer loads to the skeleton and drive joint movement, stabilization, and posture. Not surprisingly, impaired tendon formation also regulates impaired bone shape [116], primarily through the failure to initialize the formation of bone ridges [103]. The timing of these embryonic events is synchronized during morphogenesis; whereas the initiation of skeletal muscle development is independent of its innervation [117], the maintenance of skeletal muscle depends on its subsequent innervation. The migration of myogenic progenitors from the dermomyotome during early limb bud growth is initiated around embryonic day 9.5 (E9.5) in mice, around the same time that the mesenchymal condensation of the cartilage template has formed. The establishment of skeletal muscle bodies in the limb, their attachment to limb tendons [118], and their innervation via neuromuscular synapse formation all coalesce around E13 [119]. This leads to the active loading from skeletal muscle through tendons to the developing bone template, giving rise to the shape, elongation, and alignment of bones [97,99,100,103,111,120,121]. Clearly, muscle loading is required throughout development, including during postnatal growth [122–131], for proper maturation of the tendon, cartilage, and bone, even after the establishment of these structures.

Identifying the biophysical cues that drive joint maturation in the limb has been elucidated in recent years through computational modeling and advanced imaging techniques (Fig. 3) [107,132,133,133a]. Intermittent and continuous compressive forces can lead to increased calcification of growth plate cartilage in vitro [134]. In addition, applied bending forces from skeletal muscle contraction on long bones have been shown to be withstood through rapid adaptation of the long bone's circumference via asymmetrical mineral deposition and cortical thickening [99]. Muscle paralysis during embryonic growth has been shown to lead to cartilaginous fusion in the spine [135] and synovial joints [97,110,136]. Mechanical loading, primarily driven from contracting skeletal muscle, has been shown repeatedly to play an essential role in joint formation [97,104–106,108–110,137–139]. Decreased movement during fetal development has also been shown to lead to asymmetry in growth plate shape, the loss of ridges [103,140], and sesamoid bones [106] that provide articulating joints with mechanical leverage. Muscle loading is also required for establishing proper shape of long bones [141,142], and the loss of muscle contraction leads to shorter and more cylindrically shaped long bones [99,100]. Computational and experimental models of skeletal development that focused primarily on muscle contraction and strain have been used to reveal potential mechanisms of musculoskeletal and joint disorders [101,102,107,133,143–146]. Delays or defects in how the bones and joints form during fetal development have been linked to disorders and disease

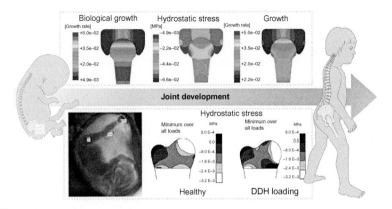

**FIG. 3**

The proper formation of the hip joint, shown by the top inset *dashed box*, requires (A) symmetrical movements guided by the mechanobiological growth rate and hydrostatic stresses estimated during physiological motion (top images). (B) Fetal joint displacements in utero, measured using magnetic resonance imaging, can be applied to kinematic models for estimating joint loads and properties during fetal development. (C) Modeling of joint development using computational tools complements morphological observations, as shown in (D) with the growth of the interphalangeal joint.

*Modified from Giorgi M, Verbruggen SW, Lacroix D. In silico bone mechanobiology: modeling a multifaceted biological system. Wiley Interdiscip Rev Syst Biol Med 2016;8(6):485–505; Verbruggen SW, Loo JHW, Hayat TTA, Hajnal JV, Rutherford MA, Phillips ATM, et al. Modeling the biomechanics of fetal movements. Biomech Model Mechanobiol 2016;15(4):995–1004; Giorgi M, Carriero A, Shefelbine SJ, Nowlan NC. Effects of normal and abnormal loading conditions on morphogenesis of the prenatal hip joint: application to hip dysplasia. J Biomech 2015;48(12):3390–7 with permissions from Creative Commons.*

that are diagnosed during postnatal stages [112,147], such as congenital dysplasia of the hip, clubfoot, and arthrogryposis.

## 3 ELUCIDATING THE MECHANICS OF DEVELOPING TISSUES

Whereas the mechanical forces and physical mechanisms are critically important in understanding morphogenesis and pathogenesis, it is often difficult to determine the microenvironmental cues that cells are experiencing in the developing embryo. Several tools have been developed to attempt to quantify the mechanics in development.

### 3.1 TRACKING CELL FATE

What drives cell fate? Changes in the local stiffness of extracellular matrix, for example, along with the locally applied forces by the cells themselves, can result in changes in gene transcription that can then drive cell fate switching. Mechanical factors such as tissue surface tension, regulated by cell adhesion and cortical tension, were thought to be crucial for determining cell sorting and tissue layering during development [148]. However, recent findings suggest that directed cell migration also plays a key role in driving progenitor cell segregation [149]. Visualization of cell fate is critical for understanding how tissues develop and form during embryogenesis. Fate mapping was developed by Walter Vogt in the 1920s and is now commonly used to trace the development of specific regions [150]. Fate mapping techniques leverage the use of vital dyes to stain and image individual or clusters of cells in living tissue. This technique allowed for Dr. Vogt and others to generate accurate fate maps of living embryos and study morphogenesis. Advances in time-lapse imaging have accelerated accuracy and visualization of cells using fate mapping techniques. Commonly used species for fate mapping techniques include *Xenopus* and *Drosophila*, and isolated organs from larger-order species, such as mouse and chick, can also be used with optimized culture conditions. Nonetheless, visualizing the fate of cells remains difficult. While the surface of developing tissues is relatively easy to image, deep-tissue fate mapping requires three-dimensional visualization with confocal, deconvolution, or structured illumination microscopy. Imaging of developing tissues for cell shape and orientation also depends on ultrastructural morphological traits, and scanning electron microscopy is commonly utilized at discrete time points of growth and morphogenesis.

Visualizing cell migration/invagination requires dynamic visualization, which limits some of the techniques described above, such as electron microscopy. For example, bottle cells of the dorsal marginal zone create a local invagination during gastrulation and then push inward, forcing deep material and cells upward. Apical constriction may also be visualized with SEM or confocal imaging. However, mechanical cutting of the cells from their surrounding matrix/neighboring cells can lead to cell condensation, and lateral cuts cause the bottle cells to pull together (generating a "wound healing" response). Taking into account how the processing of tissues prior to fate mapping and cell/tissue visualization might influence the interpretation of results is critical.

### 3.2 MECHANICAL BEHAVIOR OF DEVELOPING TISSUES

The material and structural properties of tissues change from early developmental stages to adulthood as tissues form, adapt, and remodel [151,152]. Defining how tissue materials in the embryo behave in response to an applied load can provide insight into the mechanical environment of the embryo. The hierarchical organization that many embryonic structures develop is not formed instantaneously or spontaneously but rather through an orchestrated process. Therefore, temporal and spatial mapping of mechanical properties are important for understanding morphogenesis.

Feasibly, it is difficult to characterize the mechanical properties of developing tissues in the embryo due to their size and constraints of material test stands. Development of material test systems specifically designed for testing small-scale structures has alleviated some of this burden. An extended, albeit not exhaustive, list of approaches for characterizing embryonic tissues with visual and/or mechanical testing is described in Table 1. Small-scale devices have been custom-built for these

**Table 1** Approaches for Mechanical Testing and Characterization of Embryonic Tissue During Development

| Developmental Stage | Tissue Type | Testing Modality | Property | Reference |
|---|---|---|---|---|
| Gastrulation | Developing epithelia, neurulation | Fiducial markers | Deformation | [153,154] |
| | | Laser wounding (ablation/drilling) | Duration of cell recoil (ms) | [155] |
| | | Microsurgical cuts | Residual stresses, deformations | [156] |
| | | Microindentation/ creep test | Elasticity (stiffness), viscoelasticity | [157] |
| | | Magnetic repulsion | Force | [158] |
| | | Agar gel deformation | Contraction forces (convergent extension) | [159] |
| | | Uniaxial tensile test | Stiffness modulus | [159–161] |
| Invagination | Eye | Microindentation | Stiffness; opening angles; deformation | [162] |
| Involution | Gut tube/mesentery | Uniaxial tensile test | Stiffness | [64,163] |
| Organ morphogenesis | Cardiac looping and cardiovascular development | Optical coherence tomography | Deformation | [164] |
| | | Plethysmography | Pressure | [165] |
| | | Microsurgical cuts | Residual stress/strain | [166–168] |
| | | Microindentation/ stress relaxation | Hysteresis, viscoelasticity | [169] |
| | Brain | Microindentation | Stiffness | [170] |
| | | Microsurgical cuts | Opening angles, residual stress | [170] |
| | Developing bone and cartilage | Indentation | Stiffness, shear modulus, stress relaxation | [171,172] [173] |
| | | Four-point bending | Young's modulus | [174] |
| | | Static and dynamic compression | Compressive (confined or free swelling) modulus, instantaneous and equilibrium moduli (viscoelasticity) | [172,175–177] |

purposes and specifically tailored for characterizing tissues, such as the nano Newton force measurement device (nNFMD) [157] and the Nanstron [163]. In addition to uniaxial tensile test configurations, characterization of tissue stiffness and viscoelasticity can be performed using indentation testing, and the scalability of indentation test configurations ranges widely [178]. Nanoindentation, using industrial nanoindentation equipment, can be correlated using image-processing techniques to surface characteristics determined through Raman spectroscopy or scanning electron microscopy. Additionally, pairing Raman spectroscopy with atomic force nanoindenters can also provide complimentary structure and function relationships of surfaces of developing and mature tissues.

Use of small-scale mechanical testing devices and interest in developmental biology have pushed forward empirical measurements of tissue stiffness during morphogenesis. In addition, characterization of stiffness in living tissues has bolstered our understanding of how the ECM and cells both contribute to the local mechanical environment of the developing embryo. For example, vertebrate embryo undergoes increased stiffening from gastrula to neurula stages in the *Xenopus*, and this stiffness is regulated primarily by actomyosin contractility [157]. Pairing of mechanical testing devices with imaging-based modalities will shift the paradigm of approaches in developmental biology.

## 4 CONCLUSION

To say that morphogenesis of the developing embryo is complicated is an understatement. Great strides have been made in identifying the intricacies of the mechanobiological cues at different stages of morphogenesis, yet the field is wide open for exploration. The mechanics of morphogenesis is a burgeoning field that will continue to expand with the advancements of new technologies, such as imaging modalities and micromechanical sensing. As such, the field of morphogenesis also requires multidisciplinary approaches. By leveraging tools from physical sciences and biological sciences, especially developmental biology, our understanding of the most fundamental physiological process will continue to grow.

## ACKNOWLEDGMENTS

Special thanks to the members of the Killian and Gleghorn laboratories for the great discussion, especially Connor Leek and Nicholas Ruggiero. Work in musculoskeletal development in the Killian laboratory is supported by grants from the National Institutes of Health (K12HD073945 and P30GM103333) and the University of Delaware Research Foundation (16A01396). Work in morphogenesis in the Gleghorn laboratory is supported by grants from the National Institutes of Health (R01HL133163, R21ES027962, and P20GM103446), National Science Foundation (1537256), and the March of Dimes Foundation Basil O'Connor Award (5-FY16-33).

# REFERENCES

[1] Mammoto T, Mammoto A, Ingber DE. Mechanobiology and developmental control. Annu Rev Cell Dev Biol 2013;29(1):27–61.

[2] Wühr M, Dumont S, Groen AC, Needleman DJ, Mitchison TJ. How does a millimeter-sized cell find its center? Cell Cycle 2009;8(8):1115–21.

[3] Reinsch S, Gonczy P. Mechanisms of nuclear positioning. J Cell Sci 1998;111(16):2283–95.

[4] Kunda P, Baum B. The actin cytoskeleton in spindle assembly and positioning. Trends Cell Biol 2009;19(4):174–9.

[5] Grill SW, Hyman AA. Spindle positioning by cortical pulling forces. Dev Cell 2005;8(4):461–5.

[6] Mammoto T, Ingber DE. Mechanical control of tissue and organ development. Development 2010;137:1407–20.

[7] Horner VL, Wolfner MF. Transitioning from egg to embryo: triggers and mechanisms of egg activation. Dev Dyn 2008;237(3):527–44.

[8] Levayer R, Lecuit T. Oscillation and polarity of E-cadherin asymmetries control actomyosin flow patterns during morphogenesis. Dev Cell 2013;26(2):162–75.

[9] Heer NC, Miller PW, Chanet S, Stoop N, Dunkel J, Martin AC. Actomyosin-based tissue folding requires a multicellular myosin gradient. Development 2017;144:1876–86.

[10] Oster GF, Murray JD, Harris AK. Mechanical aspects of mesenchymal morphogenesis. J Embryol Exp Morphol 1983;78:83–125.

[11] Tickle C, Summerbell D, Wolpert L. Positional signalling and specification of digits in chick limb morphogenesis. Nature 1975;254(5497):199–202.

[12] Aono H, Ide H. A gradient of responsiveness to the growth-promoting activity of ZPA (zone of polarizing activity) in the chick limb bud. Dev Biol 1988;128(1):136–41.

[13] Wang B, Fallon JF, Beachy PA. Hedgehog-regulated processing of Gli3 produces an anterior/posterior repressor gradient in the developing vertebrate limb. Cell 2000;100(4):423–34.

[14] Kwasigroch TE, Kochhar DM. Production of congenital limb defects with retinoic acid: phenomenological evidence of progressive differentiation during limb morphogenesis. Anat Embryol 1980;161(1):105–13.

[15] Tickle C. Morphogen gradients in vertebrate limb development. Semin Cell Dev Biol 1999;10(3):345–51.

[16] Raspopovic J, Marcon L, Russo L, Sharpe J. Digit patterning is controlled by a Bmp-Sox9-Wnt Turing network modulated by morphogen gradients. Science 2014;345:566–70.

[17] Tabin C, Wolpert L. Rethinking the proximodistal axis of the vertebrate limb in the molecular era. Genes Dev 2007;21(12):1433–42.

[18] Towers M, Wolpert L, Tickle C. Gradients of signalling in the developing limb. Curr Opin Cell Biol 2012;24(2):181–7.

[19] Rutz R, Haney C, Hauschka S. Spatial analysis of limb bud myogenesis: a proximodistal gradient of muscle colony-forming cells in chick embryo leg buds. Dev Biol 1982;90(2):399–411.

[20] Dollé P, Ruberte E, Kastner P, Petkovich M, Stoner CM, Gudas LJ, et al. Differential expression of genes encoding α, β and γ retinoic acid receptors and CRABP in the developing limbs of the mouse. Nature 1989;342(6250):702–5.

[21] Galea GL, Cho Y-J, Galea G, Molè MA, Rolo A, Savery D, et al. Biomechanical coupling facilitates spinal neural tube closure in mouse embryos. Proc Natl Acad Sci U S A 2017;114(26):E5177–86.

[22] Garcia KE, Okamoto RJ, Bayly PV, Taber LA. Contraction and stress-dependent growth shape the forebrain of the early chicken embryo. J Mech Behav Biomed Mater 2017;65(Suppl. C):383–97.

[23] Desmond ME, Jacobson AG. Embryonic brain enlargement requires cerebrospinal fluid pressure. Dev Biol 1977;57(1):188–98.

[24] Gato A, Alonso MI, Martín C, Carnicero E, Moro JA, De la Mano A, et al. Embryonic cerebrospinal fluid in brain development: neural progenitor control. Croat Med J 2014;55(4):299–305.

[25] Li X, Desmond M, editors. Modulation of Na+/K+ ATPase pumps in the heart of the chick embryo influences brain expansion. Soc Neurosci Abstr 1991.

[25a] Nelson CM, Gleghorn JP, Pang M-F, Jaslove JM, Goodwin K, Varner VD, Miller E, Radisky DC, Stone HA. Microfluidic chest cavities reveal that transmural pressure controls the rate of lung development. Development 2017;144:4328–35.

[26] Gato A, Moro J, Alonso M, Pastor J, Represa J, Barbosa E. Chondroitin sulphate proteoglycan and embryonic brain enlargement in the chick. Anat Embryol 1993;188(1):101–6.

[27] Gilbert RM, Morgan JT, Marcin ES, Gleghorn JP. Fluid mechanics as a driver of tissue-scale mechanical signaling in organogenesis. Curr Pathobiol Rep 2016;4(4):199–208.

[28] Tissir F, Qu Y, Montcouquiol M, Zhou L, Komatsu K, Shi D, et al. Lack of cadherins Celsr2 and Celsr3 impairs ependymal ciliogenesis, leading to fatal hydrocephalus. Nat Neurosci 2010;13(6):700–7.

[29] Rao Damerla R, Gabriel GC, Li Y, Klena NT, Liu X, Chen Y, et al. Role of cilia in structural birth defects: insights from ciliopathy mutant mouse models. Birth Defects Res C Embryo Today 2014;102(2):115–25.

[30] Ibañez-Tallon I, Pagenstecher A, Fliegauf M, Olbrich H, Kispert A, Ketelsen U-P, et al. Dysfunction of axonemal dynein heavy chain Mdnah5 inhibits ependymal flow and reveals a novel mechanism for hydrocephalus formation. Hum Mol Genet 2004;13(18):2133–41.

[31] Tzou D, Spurlin JW, Pavlovich AL, Stewart CR, Gleghorn JP, Nelson CM. Morphogenesis and morphometric scaling of lung airway development follows phylogeny in chicken, quail, and duck embryos. EvoDevo 2016;7:12.

[32] Metzger RJ, Klein OD, Martin GR, Krasnow MA. The branching programme of mouse lung development. Nature 2008;453(7196):745–50.

[33] Gleghorn JP, Kwak J, Pavlovich AL, Nelson CM. Inhibitory morphogens and monopodial branching of the embryonic chicken lung. Dev Dyn 2012;241(5):852–62.

[34] Larson JE, Delcarpio JB, Farberman MM, Morrow SL, Cohen JC. CFTR modulates lung secretory cell proliferation and differentiation. Am J Physiol Lung Cell Mol Physiol 2000;279(2):L333–41.

[35] Olver R, Strang L. Ion fluxes across the pulmonary epithelium and the secretion of lung liquid in the foetal lamb. J Physiol 1974;241(2):327–57.

[36] Olver RE, Walters DV, Wilson SM. Developmental regulation of lung liquid transport. Annu Rev Physiol 2004;66:77–101.

[37] Brennan SC, Wilkinson WJ, Tseng H-E, Finney B, Monk B, Dibble H, et al. The extracellular calcium-sensing receptor regulates human fetal lung development via CFTR. Sci Rep 2016;6:21975.

[38] Folkesson HG, Norlin A, Baines DL. Salt and water transport across the alveolar epithelium in the developing lung: correlations between function and recent molecular biology advances. Int J Mol Med 1998;2(5):515–46.

[39] Harding R, Hooper S. Regulation of lung expansion and lung growth before birth. J Appl Physiol 1996;81(1):209–24.

[40] Cassin S, Gause G, Perks A. The effects of bumetanide and furosemide on lung liquid secretion in fetal sheep. Proc Soc Exp Biol Med 1986;181(3):427–31.

[41] Carlton D, Cummings J, Chapman D, Poulain F, Bland R. Ion transport regulation of lung liquid secretion in foetal lambs. J Dev Physiol 1992;17(2):99–107.

[42] Fewell JE, Hislop AA, Kitterman JA, Johnson P. Effect of tracheostomy on lung development in fetal lambs. J Appl Physiol 1983;55(4):1103–8.

[43] Alcorn D, Adamson T, Lambert T, Maloney J, Ritchie B, Robinson P. Morphological effects of chronic tracheal ligation and drainage in the fetal lamb lung. J Anat 1977;123(Pt 3):649.

[44] Kotecha S, Barbato A, Bush A, Claus F, Davenport M, Delacourt C, et al. Congenital diaphragmatic hernia. Eur Respir Soc 2012;39:820–9. https://doi.org/10.1183/09031936.00066511.

[45] Larson JE, Cohen JC. Improvement of pulmonary hypoplasia associated with congenital diaphragmatic hernia by in utero CFTR gene therapy. Am J Physiol Lung Cell Mol Physiol 2006;291(1):L4–L10.

[46] Mailleux AA, Kelly R, Veltmaat JM, De Langhe SP, Zaffran S, Thiery JP, et al. Fgf10 expression identifies parabronchial smooth muscle cell progenitors and is required for their entry into the smooth muscle cell lineage. Development 2005;132(9):2157–66.

[47] Sparrow MP, Lamb JP. Ontogeny of airway smooth muscle: structure, innervation, myogenesis and function in the fetal lung. Respir Physiol Neurobiol 2003;137(2):361–72.

[48] Pandya HC, Innes J, Hodge R, Bustani P, Silverman M, Kotecha S. Spontaneous contraction of pseudoglandular-stage human airspaces is associated with the presence of smooth muscle-α-actin and smooth muscle-specific myosin heavy chain in recently differentiated fetal human airway smooth muscle. Neonatology 2006;89(4):211–9.

[49] PB Jr MC. Spontaneous contractility of human fetal airway smooth muscle. Am J Respir Cell Mol Biol 1993;8(5):573–80.

[50] Lewis MR. Spontaneous rhythmical contraction of the muscles of the bronchial tubes and air sacs of the chick embryo. Am J Physiol Legacy Content 1924;68(2):385–8.

[51] Schittny JC, Miserocchi G, Sparrow MP. Spontaneous peristaltic airway contractions propel lung liquid through the bronchial tree of intact and fetal lung explants. Am J Respir Cell Mol Biol 2000;23(1):11–8.

[52] Jesudason EC, Smith NP, Connell MG, Spiller DG, White MR, Fernig DG, et al. Developing rat lung has a sided pacemaker region for morphogenesis-related airway peristalsis. Am J Respir Cell Mol Biol 2005;32(2):118–27.

[53] Jesudason EC, Smith NP, Connell MG, Spiller DG, White M, Fernig DG, et al. Peristalsis of airway smooth muscle is developmentally regulated and uncoupled from hypoplastic lung growth. Am J Physiol Lung Cell Mol Physiol 2006;291(4):L559–65.

[54] Yang Y, Beqaj S, Kemp P, Ariel I, Schuger L. Stretch-induced alternative splicing of serum response factor promotes bronchial myogenesis and is defective in lung hypoplasia. J Clin Investig 2000;106(11):1321.

[55] Jesudason E. Airway smooth muscle: an architect of the lung? Thorax 2009;64(6):541–5.

[56] Warburton D, El-Hashash A, Carraro G, Tiozzo C, Sala F, Rogers O, et al. Chapter three-lung organogenesis. Curr Top Dev Biol 2010;90:73–158.
[57] Nelson CM, Gleghorn JP. Sculpting organs: mechanical regulation of tissue development. Annu Rev Biomed Eng 2012;14:129–54.
[58] George UZ, Bokka KK, Warburton D, Lubkin SR. Quantifying stretch and secretion in the embryonic lung: implications for morphogenesis. Mech Dev 2015;138:356–63.
[59] Kim HY, Varner VD, Nelson CM. Apical constriction initiates new bud formation during monopodial branching of the embryonic chicken lung. Development 2013;140(15):3146–55.
[60] Kim HY, Pang M-F, Varner VD, Kojima L, Miller E, Radisky DC, et al. Localized smooth muscle differentiation is essential for epithelial bifurcation during branching morphogenesis of the mammalian lung. Dev Cell 2015;34(6):719–26.
[61] Varner VD, Gleghorn JP, Miller E, Radisky DC, Nelson CM. Mechanically patterning the embryonic airway epithelium. Proc Natl Acad Sci U S A 2015;112(30):9230–5.
[62] Schnatwinkel C, Niswander L. Multiparametric image analysis of lung-branching morphogenesis. Dev Dyn 2013;242(6):622–37.
[63] Taber LA. Biomechanics of growth, remodeling, and morphogenesis. Evolution 1995;490:6.
[64] Savin T, Kurpios NA, Shyer AE, Florescu P, Liang H, Mahadevan L, et al. On the growth and form of the gut. Nature 2011;476(7358):57.
[65] Shyer AE, Tallinen T, Nerurkar NL, Wei Z, Gil ES, Kaplan DL, et al. Villification: how the gut gets its villi. Science 2013;342(6155):212–8.
[66] Tang N, Marshall WF, McMahon M, Metzger RJ, Martin GR. Control of mitotic spindle angle by the RAS-regulated ERK1/2 pathway determines lung tube shape. Science 2011;333(6040):342–5.
[67] Zákány J, Kmita M, Duboule D. A dual role for Hox genes in limb anterior-posterior asymmetry. Science 2004;304(5677):1669–72.
[68] Coates MI, Cohn MJ. Fins, limbs, and tails: outgrowths and axial patterning in vertebrate evolution. Bioessays 1998;20(5):371–81.
[69] Burke AC, Nelson CE, Morgan BA, Tabin C. Hox genes and the evolution of vertebrate axial morphology. Development 1995;121(2):333–46.
[70] Capdevila J, Belmonte JCI. Patterning mechanisms controlling vertebrate limb development. Annu Rev Cell Dev Biol 2001;17(1):87–132.
[71] Ganan Y, Macias D, Duterque-Coquillaud M, Ros M, Hurle J. Role of TGF beta s and BMPs as signals controlling the position of the digits and the areas of interdigital cell death in the developing chick limb autopod. Development 1996;122(8):2349–57.
[72] Long F, Ornitz DM. Development of the endochondral skeleton. Cold Spring Harb Perspect Biol 2013;5(1):a008334.
[73] Merino R, Gañan Y, Macias D, Economides AN, Sampath KT, Hurle JM. Morphogenesis of digits in the avian limb is controlled by FGFs, TGFβs, and noggin through BMP signaling. Dev Biol 1998;200(1):35–45.
[74] Merino R, Rodriguez-Leon J, Macias D, Ganan Y, Economides A, Hurle J. The BMP antagonist Gremlin regulates outgrowth, chondrogenesis and programmed cell death in the developing limb. Development 1999;126(23):5515–22.
[75] Hogan B. Bone morphogenetic proteins: multifunctional regulators of vertebrate development. Genes Dev 1996;10(13):1580–94.
[76] Laufer E, Nelson CE, Johnson RL, Morgan BA, Tabin C. Sonic hedgehog and Fgf-4 act through a signaling cascade and feedback loop to integrate growth and patterning of the developing limb bud. Cell 1994;79(6):993–1003.

[77] Xu X, Weinstein M, Li C, Naski M, Cohen RI, Ornitz DM, et al. Fibroblast growth factor receptor 2 (FGFR2)-mediated reciprocal regulation loop between FGF8 and FGF10 is essential for limb induction. Development 1998;125:753–65.

[78] Crossley PH, Minowada G, MacArthur CA, Martin GR. Roles for FGF8 in the induction, initiation, and maintenance of chick limb development. Cell 1996;84(1): 127–36.

[79] ten Berge D, Brugmann SA, Helms JA, Nusse R. Wnt and FGF signals interact to coordinate growth with cell fate specification during limb development. Development 2008;135(19):3247–57.

[80] Fallon JF, Lopez A, Ros MA, Savage MP, Olwin BB, Simandl BK. FGF-2: apical ectodermal ridge growth signal for chick limb development. Science 1994;264 (5155):104–8.

[81] Brunt LH, Begg K, Kague E, Cross S, Hammond CL. Wnt signalling controls the response to mechanical loading during zebrafish joint development. Development 2017;144:2798–809.

[82] Cohn MJ. Development of the external genitalia: conserved and divergent mechanisms of appendage patterning. Dev Dyn 2011;240(5):1108–15.

[83] Henderson JH, Carter DR. Mechanical induction in limb morphogenesis: the role of growth-generated strains and pressures. Bone 2002;31(6):645–53.

[84] Hall BK, Miyake T. All for one and one for all: condensations and the initiation of skeletal development. Bioessays 2000;22(2):138–47.

[85] Thorpe SD, Buckley CT, Vinardell T, O'Brien FJ, Campbell VA, Kelly DJ. The response of bone marrow-derived mesenchymal stem cells to dynamic compression following TGF-β3 induced chondrogenic differentiation. Ann Biomed Eng 2010;38 (9):2896–909.

[86] Huang AH, Farrell MJ, Mauck RL. Mechanics and mechanobiology of mesenchymal stem cell-based engineered cartilage. J Biomech 2010;43(1):128–36.

[87] Burdick JA, Vunjak-Novakovic G. Engineered microenvironments for controlled stem cell differentiation. Tissue Eng Part A 2008;15(2):205–19.

[88] Guilak F, Cohen DM, Estes BT, Gimble JM, Liedtke W, Chen CS. Control of stem cell fate by physical interactions with the extracellular matrix. Cell Stem Cell 2009;5 (1):17–26.

[89] Takahashi I, Nuckolls GH, Takahashi K, Tanaka O, Semba I, Dashner R, et al. Compressive force promotes sox9, type II collagen and aggrecan and inhibits IL-1beta expression resulting in chondrogenesis in mouse embryonic limb bud mesenchymal cells. J Cell Sci 1998;111:2067–76.

[90] Long F, Linsenmayer TF. Regulation of growth region cartilage proliferation and differentiation by perichondrium. Development 1998;125:1067–73.

[91] Karuppaiah K, Yu K, Lim J, Chen J, Smith C, Long F, et al. FGF signaling in the osteoprogenitor lineage non-autonomously regulates postnatal chondrocyte proliferation and skeletal growth. Development 2016;143:1811–22.

[92] Yang W, Wang J, Moore DC, Liang H, Dooner M, Wu Q, et al. Ptpn11 deletion in a novel cartilage cell causes metachondromatosis by activating hedgehog signaling. Nature 2013;499(7459):491–5.

[93] Shapiro F, Holtrop M, Glimcher M. Organization and cellular biology of the perichondrial ossification groove of ranvier: a morphological study in rabbits. JBJS 1977;59 (6):703–23.

[94] Warrell E, Taylor JF. The role of periosteal tension in the growth of long bones. J Anat 1979;128(Pt 1):179–84.

[95] Tanck E, Hannink G, Ruimerman R, Buma P, Burger EH, Huiskes R. Cortical bone development under the growth plate is regulated by mechanical load transfer. J Anat 2006;208(1):73–9.
[96] Foolen J, van Donkelaar CC, Murphy P, Huiskes R, Ito K. Residual periosteum tension is insufficient to directly modulate bone growth. J Biomech 2009;42(2):152–7.
[97] Kahn J, Shwartz Y, Blitz E, Krief S, Sharir A, Breitel DA, et al. Muscle contraction is necessary to maintain joint progenitor cell fate. Dev Cell 2009;16(5):734–43.
[98] Rolfe RA, Bezer JH, Kim T, Zaidon AZ, Oyen ML, Iatridis JC, et al. Abnormal fetal muscle forces result in defects in spinal curvature and alterations in vertebral segmentation and shape. J Orthop Res 2017;35(10):2135–44.
[99] Sharir A, Stern T, Rot C, Shahar R, Zelzer E. Muscle force regulates bone shaping for optimal load-bearing capacity during embryogenesis. Development 2011;138 (15):3247–59.
[100] Shwartz Y, Farkas Z, Stern T, Aszodi A, Zelzer E. Muscle contraction controls skeletal morphogenesis through regulation of chondrocyte convergent extension. Dev Biol 2012;370(1):154–63.
[101] Nowlan N, Chandaria V, Sharpe J. Immobilized chicks as a model system for early-onset developmental dysplasia of the hip. J Orthop Res 2014;23(6):777–85.
[102] Brunt LH, Norton JL, Bright JA, Rayfield EJ, Hammond CL. Finite element modelling predicts changes in joint shape and cell behaviour due to loss of muscle strain in jaw development. J Biomech 2015;48(12):3112–22.
[103] Blitz E, Viukov S, Sharir A, Shwartz Y, Galloway JL, Pryce BA, et al. Bone ridge patterning during musculoskeletal assembly is mediated through SCX regulation of Bmp4 at the tendon-skeleton junction. Dev Cell 2009;17(6):861–73.
[104] Dowthwaite GP, Flannery CR, Flannelly J, Lewthwaite JC, Archer CW, Pitsillides AA. A mechanism underlying the movement requirement for synovial joint cavitation. Matrix Biol 2003;22(4):311–22.
[105] Drachman DB, Sokoloff L. The role of movement in embryonic joint development. Dev Biol 1966;14(3):401–20.
[106] Eyal S, Blitz E, Shwartz Y, Akiyama H, Schweitzer R, Zelzer E. On the development of the patella. Development 2015;142(10):1831–9.
[107] Giorgi M, Carriero A, Shefelbine SJ, Nowlan NC. Effects of normal and abnormal loading conditions on morphogenesis of the prenatal hip joint: application to hip dysplasia. J Biomech 2015;48(12):3390–7.
[108] Lelkes G. Experiments in vitro on the role of movement in the development of joints. J Embryol Exp Morphol 1958;6(2):183–6.
[109] Mitrovic D. Development of the articular cavity in paralyzed chick embryos and in chick embryo limb buds cultured on chorioallantoic membranes. Acta Anat 1982;113(4):313–24.
[110] Osborne AC, Lamb KJ, Lewthwaite JC, Dowthwaite GP, Pitsillides AA. Short-term rigid and flaccid paralyses diminish growth of embryonic chick limbs and abrogate joint cavity formation but differentially preserve pre-cavitated joints. J Musculoskelet Neuronal Interact 2002;2(5):448–56.
[111] Shwartz Y, Blitz E, Zelzer E. One load to rule them all: mechanical control of the musculoskeletal system in development and aging. Differ Res Biol Divers 2013;86 (3):104–11.
[112] Smeeton J, Askary A, Crump JG. Building and maintaining joints by exquisite local control of cell fate. Wiley Interdiscip Rev Dev Biol 2017;6(1):e245.

[113] Murray PDF, Selby D. Intrinsic and extrinsic factors in the primary development of the skeleton. Wilhelm Roux Arch Entwickl Org 1930;122(3):629–62.

[114] Ford CA, Nowlan NC, Thomopoulos S, Killian ML. Effects of imbalanced muscle loading on hip joint development and maturation. J Orthop Res 2017;35(5):1128–36.

[115] Killian ML, Cavinatto L, Galatz LM, Thomopoulos S. The role of mechanobiology in tendon healing. J Shoulder Elbow Surg 2012;21(2):228–37.

[116] Murchison ND, Price BA, Conner DA, Keene DR, Olson EN, Tabin CJ, et al. Regulation of tendon differentiation by scleraxis distinguishes force-transmitting tendons from muscle-anchoring tendons. Development 2007;134(14):2697–708.

[117] Broadie K, Bate M. Muscle development is independent of innervation during Drosophila embryogenesis. Development 1993;119(2):533–43.

[118] Huang AH, Riordan TJ, Wang L, Eyal S, Zelzer E, Brigande JV, et al. Repositioning forelimb superficialis muscles: tendon attachment and muscle activity enable active relocation of functional myofibers. Dev Cell 2013;26(5):544–51.

[119] Escher P, Lacazette E, Courtet M, Blindenbacher A, Landmann L, Bezakova G, et al. Synapses form in skeletal muscles lacking neuregulin receptors. Science 2005;308(5730):1920–3.

[120] Stern T, Aviram R, Rot C, Galili T, Sharir A, Kalish Achrai N, et al. Isometric scaling in developing long bones is achieved by an optimal epiphyseal growth balance. PLoS Biol 2015;13(8):e1002212.

[121] Blecher R, Krief S, Galili T, Biton IE, Stern T, Assaraf E, et al. The proprioceptive system masterminds spinal alignment: insight into the mechanism of scoliosis. Dev Cell 2017;42(4):388–99. e3.

[122] Kim HM, Galatz LM, Das R, Patel N, Thomopoulos S. Musculoskeletal deformities secondary to neurotomy of the superior trunk of the brachial plexus in neonatal mice. J Orthop Res 2010;28(10):1391–8.

[123] Thomopoulos S, Genin GM, Galatz LM. The development and morphogenesis of the tendon-to-bone insertion—what development can teach us about healing. J Musculoskelet Neuronal Interact 2010;10(1):35–45.

[124] Thomopoulos S. The role of mechanobiology in the attachment of tendon to bone. IBMS BoneKey 2011;8:271–85.

[125] Schwartz A, Lipner J, Pasteris J, Genin G, Thomopoulos S. Muscle loading is necessary for the formation of a functional tendon enthesis. Bone 2013;55(1):44–51.

[126] Schwartz AG, Pasteris JD, Genin GM, Daulton TL, Thomopoulos S. Mineral distributions at the developing tendon enthesis. PLoS One 2012;7.

[127] Liu Y, Schwartz AG, Birman V, Thomopoulos S, Genin GM. Stress amplification during development of the tendon-to-bone attachment. Biomech Model Mechanobiol 2013;13:973–83.

[128] Zelzer E, Blitz E, Killian M, Thomopoulos S. Tendon-to-bone attachment: from development to maturity. Birth Defects Res C Embryo Today 2014;102(1):101–12.

[129] Schwartz AG, Long F, Thomopoulos S. Enthesis fibrocartilage cells originate from a population of Hedgehog-responsive cells modulated by the loading environment. Development 2015;142(1):196–206.

[130] Ford C, Nowlan N, Thomopoulos S, Killian M. The effects of imbalanced muscle loading on hip joint development and maturation. J Orthop Res 2016;.

[131] Killian M, Thomopoulos S. Scleraxis is required for the development of a functional tendon enthesis. FASEB J 2016;30(1):301–11.

[132] Giorgi M, Verbruggen SW, Lacroix D. In silico bone mechanobiology: modeling a multifaceted biological system. Wiley Interdiscip Rev Syst Biol Med 2016;8(6): 485–505.

[133] Verbruggen SW, Loo JHW, Hayat TTA, Hajnal JV, Rutherford MA, Phillips ATM, et al. Modeling the biomechanics of fetal movements. Biomech Model Mechanobiol 2016;15(4):995–1004.

[133a] Verbruggen SW, Kainz B, Shelmerdine SC, Hajnal JV, Rutherford MA, Arthurs OJ, Phillips AT, Nowlan NC. Stresses and strains on the human fetal skeleton during development. J R Soc Interface 2018;15(138):20170593.

[134] Klein-Nulend J, Veldhuijzen JP, Burger EH. Increased calcification of growth plate cartilage as a result of compressive force in vitro. Arthritis Rheum 1986;29(8): 1002–9.

[135] Sullivan G. Prolonged paralysis of the chick embryo, with special reference to effects on the vertebral column. Aust J Zool 1966;14(1):1–17.

[136] Brunt LH, Skinner RE, Roddy KA, Araujo NM, Rayfield EJ, Hammond CL. Differential effects of altered patterns of movement and strain on joint cell behaviour and skeletal morphogenesis. Osteoarthritis Cartilage 2016;24(11):1940–50.

[137] Ogawa H, Kozhemyakina E, Hung H-H, Grodzinsky AJ, Lassar AB. Mechanical motion promotes expression of Prg4 in articular cartilage via multiple CREB-dependent, fluid flow shear stress-induced signaling pathways. Genes Dev 2014;28 (2):127–39.

[138] Ruano-Gil D, Nardi-Vilardaga J, Tejedo-Mateu A. Influence of extrinsic factors on the development of the articular system. Acta Anat 1978;101(1):36–44.

[139] Hamburger V, Waugh M. The primary development of the skeleton in nerveless and poorly innervated limb transplants of chick embryos. Physiol Zool 1940;13(4): 367–82.

[140] Blitz E, Sharir A, Akiyama H, Zelzer E. Tendon-bone attachment unit is formed modularly by a distinct pool of Scx- and Sox9-positive progenitors. Development 2013;140 (13):2680–90.

[141] Galea GL, Meakin LB. Modulating skeletal responses to mechanical loading by targeting estrogen receptor signaling. In: Mechanobiology. Hoboken, NJ: John Wiley & Sons, Inc.; 2017. p. 115–29.

[142] Pollard AS, Charlton BG, Hutchinson JR, Gustafsson T, McGonnell IM, Timmons JA, et al. Limb proportions show developmental plasticity in response to embryo movement. Sci Rep 2017;7:41926. https://doi.org/10.1038/srep41926.

[143] Nowlan NC, Murphy P, Prendergast PJ. A dynamic pattern of mechanical stimulation promotes ossification in avian embryonic long bones. J Biomech 2008;41(2):249–58.

[144] Nowlan NC, Dumas G, Tajbakhsh S, Prendergast PJ, Murphy P. Biophysical stimuli induced by passive movements compensate for lack of skeletal muscle during embryonic skeletogenesis. Biomech Model Mechanobiol 2012;11(1–2):207–19.

[145] Rolfe RA, Nowlan NC, Kenny EM, Cormican P, Morris DW, Prendergast PJ, et al. Identification of mechanosensitive genes during skeletal development: alteration of genes associated with cytoskeletal rearrangement and cell signalling pathways. BMC Genomics 2014;15(1):48.

[146] Giorgi M, Carriero A, Shefelbine SJ, Nowlan NC. Mechanobiological simulations of prenatal joint morphogenesis. J Biomech 2014;47(5):989–95.

[147] Baker-LePain JC, Lane NE. Relationship between joint shape and the development of osteoarthritis. Curr Opin Rheumatol 2010;22(5):538–43.

[148] Krens S, Heisenberg C-P. Cell sorting in development. Curr Top Dev Biol 2011;95:189–213.

[149] Krens SFG, Veldhuis JH, Barone V, Čapek D, Maître J-L, Brodland GW, et al. Interstitial fluid osmolarity modulates the action of differential tissue surface tension in progenitor cell segregation during gastrulation. Development 2017;144:1798–806.

[150] Stern CD, Fraser SE. Tracing the lineage of tracing cell lineages. Nat Cell Biol 2001;3(9):E216–8.

[151] Davidson L, von Dassow M, Zhou J. Multi-scale mechanics from molecules to morphogenesis. Int J Biochem Cell Biol 2009;41(11):2147–62.

[152] Heisenberg C-P, Bellaïche Y. Forces in tissue morphogenesis and patterning. Cell 2013;153(5):948–62.

[153] Filas BA, Knutsen AK, Bayly PV, Taber LA. A new method for measuring deformation of folding surfaces during morphogenesis. J Biomech Eng 2008;130(6):061010. https://doi.org/10.1115/1.2979866.

[154] Rauzi M, Krzic U, Saunders TE, Krajnc M, Ziherl P, Hufnagel L, et al. Embryo-scale tissue mechanics during Drosophila gastrulation movements. Nat Commun 2015;6:8677.

[155] Ma X, Lynch HE, Scully PC, Hutson MS. Probing embryonic tissue mechanics with laser hole drilling. Phys Biol 2009;6(3):036004.

[156] Varner VD, Voronov D, Taber LA. Mechanics of head fold formation: investigating tissue-level forces during early development. Development 2010;137(22):3801–11.

[157] Zhou J, Kim HY, Davidson LA. Actomyosin stiffens the vertebrate embryo during crucial stages of elongation and neural tube closure. Development 2009;136(4):677–88.

[158] Selman GG. The forces producing neural closure in amphibia. J Embryol Exp Morphol 1958;6(3):448–65.

[159] Vijayraghavan DS, Davidson LA. Mechanics of neurulation: from classical to current perspectives on the physical mechanics that shape, fold, and form the neural tube. Birth Defects Res 2017;109(2):153–68.

[160] Wiebe C, Brodland GW. Tensile properties of embryonic epithelia measured using a novel instrument. J Biomech 2005;38(10):2087–94.

[161] Benko R, Brodland GW. Measurement of in vivo stress resultants in neurulation-stage amphibian embryos. Ann Biomed Eng 2007;35(4):672–81.

[162] Oltean A, Huang J, Beebe DC, Taber LA. Tissue growth constrained by extracellular matrix drives invagination during optic cup morphogenesis. Biomech Model Mechanobiol 2016;15(6):1405–21.

[163] Nerurkar NL, Mahadevan L, Tabin CJ. BMP signaling controls buckling forces to modulate looping morphogenesis of the gut. Proc Natl Acad Sci U S A 2017;114(9):2277–82.

[164] Filas BA, Efimov IR, Taber LA. Optical coherence tomography as a tool for measuring morphogenetic deformation of the looping heart. Anat Rec 2007;290(9):1057–68.

[165] Girard H. Arterial pressure in the chick embryo. Am J Physiol 1973;224(2):454–60.

[166] Shi Y, VD V, Taber LA. Why is cytoskeletal contraction required for cardiac fusion before but not after looping begins? Phys Biol 2015;12(1):016012.

[167] Varner VD, Taber LA. Not just inductive: a crucial mechanical role for the endoderm during heart tube assembly. Development 2012;139(9):1680–90.

[168] Voronov DA, Alford P, Xu G, Taber LA. The role of mechanical forces in dextral rotation during cardiac looping in the chick embryo. Dev Biol 2004;272(2):339–50.

[169] Yao J, VD V, Brilli LL, Young JM, Taber LA, Perucchio R. Viscoelastic material properties of the myocardium and cardiac jelly in the looping chick heart. J Biomech Eng 2012;134(2):024502.

[170] Xu G, Kemp PS, Hwu JA, Beagley AM, Bayly PV, Taber LA. Opening angles and material properties of the early embryonic chick brain. J Biomech Eng 2009;132(1) 011005. https://doi.org/10.1115/1.4000169.

[171] Mulder L, Koolstra JH, den Toonder JM, van Eijden TM. Relationship between tissue stiffness and degree of mineralization of developing trabecular bone. J Biomed Mater Res A 2008;84(2):508–15.

[172] Mikic B, Isenstein AL, Chhabra A. Mechanical modulation of cartilage structure and function during embryogenesis in the chick. Ann Biomed Eng 2004;32(1):18–25.

[173] Berteau JP, Oyen M, Shefelbine SJ. Permeability and shear modulus of articular cartilage in growing mice. Biomech Model Mechanobiol 2016;15(1):205–12.

[174] Tanck E, Van Donkelaar CC, Jepsen KJ, Goldstein SA, Weinans H, Burger EH, et al. The mechanical consequences of mineralization in embryonic bone. Bone 2004;35(1):186–90.

[175] Williamson AK, Chen AC, Sah RL. Compressive properties and function-composition relationships of developing bovine articular cartilage. J Orthop Res 2001;19(6):1113–21.

[176] Brown TD, Singerman RJ. Experimental determination of the linear biphasic constitutive coefficients of human fetal proximal femoral chondroepiphysis. J Biomech 1986;19(8):597–605.

[177] Nowlan NC, Murphy P, Prendergast PJ. Mechanobiology of embryonic limb development. Ann N Y Acad Sci 2007;1101(1):389–411.

[178] Chevalier NR, Gazguez E, Dufour S, Fleury V. Measuring the micromechanical properties of embryonic tissues. Methods (San Diego, CA) 2016;94:120–8.

# CHAPTER 4

# Cartilage mechanobiology: How chondrocytes respond to mechanical load

**Sophie J. Gilbert, Emma J. Blain**

*Arthritis Research UK Biomechanics and Bioengineering Centre, Biomedicine Division, School of Biosciences, Cardiff University, Cardiff, United Kingdom*

## ABBREVIATIONS

| | |
|---|---|
| **CITED2** | Cbp/p300-interacting transactivator with ED-rich tail 2 |
| **COMP** | cartilage oligomeric matrix protein |
| **ECM** | extracellular matrix |
| **ERK** | extracellular-regulated kinase |
| **FAK** | focal adhesion kinase |
| **FGF** | fibroblast growth factor |
| **FGFR** | FGF receptor |
| **GAG** | glycosaminoglycan |
| **Ihh** | Indian hedgehog |
| **JNK** | Jun N-terminal kinase |
| **lincRNA** | large-intergenic noncoding RNA |
| **lncRNA** | long, noncoding RNA |
| **MAPK** | mitogen-activated protein kinases |
| **MATN** | matrilin |
| **MMP** | matrix metalloproteinase |
| **NO** | nitric oxide |
| **ncRNA** | noncoding RNA |
| **PCM** | pericellular matrix |
| **PGE$_2$** | prostaglandin E$_2$ |
| **piRNA** | piwi-interacting RNA |
| **PKC** | protein kinase C |
| **snoRNA** | small nucleolar RNA |
| **TM** | territorial matrix |
| **TRPV4** | transient receptor potential vanilloid 4 |

# 1 INTRODUCTION

Articular cartilage is a highly specialized connective tissue that covers the long bones of load-bearing diarthrodial joints such as the knee, hip, shoulder, and ankle. It is responsible for providing a smooth, lubricated surface to allow low-friction joint movement and promoting load transmission. Its ability to deform under dynamic compressive loading is critical for reducing stress to the underlying bone [1]. The organization and interactions of the extracellular matrix (ECM) of articular cartilage are critical for providing the tissue with its unique mechanical properties [2–4] and its ability to withstand a lifetime of joint loading [5].

# 2 STRUCTURAL OVERVIEW

Articular cartilage is composed of 70%–80% water, chondrocytes, the only cellular component, and the ECM that is composed of largely collagen and proteoglycans; other noncollagenous proteins and glycoproteins are present in lesser amounts [6,7]. Collagen fibrils within articular cartilage are composed largely of type II collagen with the minor collagens, types IX, XI, XII, XIV, XVI, and XXII helping to provide cartilage with tensile strength and its unique properties that contribute to the physical nature of the mature matrix [8]. Articular cartilage is avascular, aneural, and alymphatic, resulting in a tissue that is intrinsically difficult to repair if damaged (reviewed in Ref. [9]), and as such, the immediate pericellular environment of the chondrocytes appears to play a critical role in regulating cell activity [10]. Articular chondrocytes are largely postmitotic with their metabolism being stimulated by physiological magnitudes of mechanical load, inducing biosynthesis of molecules to preserve the tissue integrity and maintain a balance of catabolic and anabolic processes (Fig. 1) [11–14]. The nature of this response is dependent on strain rate, frequency, loading history, and loading amplitude; superphysiological magnitudes fail to enhance matrix synthesis and static or low-frequency loads inhibit matrix synthesis, whereas dynamic compression can markedly stimulate matrix production [11,15–19]. The cell's response to load occurs through the process of "mechanotransduction" whereby compression of the tissue results in matrix and cellular deformation; the development of hydrostatic and osmotic pressures; fluid flow; and changes in ion concentrations, fixed charge density, and water content [20–23]. Primary cilia detect these changes that, along with other mechanosensors on the cell surface such as integrins and ion channels, initiate intracellular signaling pathways resulting in tissue remodeling processes.

## 2.1 THE ECM

The ECM is composed of a pericellular, territorial, and interterritorial region (Fig. 2) (reviewed in Ref. [8]). The pericellular matrix (PCM) and the chondrocyte make up a structural unit known as the "chondron"; it has a critical biomechanical role

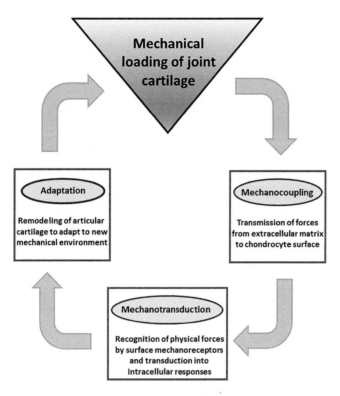

**FIG. 1**

Schematic indicating how articular cartilage responds to and adapts to mechanical loads to ensure tissue functionality.

absorbing, redistributing, and transmitting physiological compressive and shearing forces [24]. The PCM consists of high concentrations of hyaluronan-aggrecan-link protein complexes and small leucine-rich proteoglycans, biglycan, and decorin [24] and is associated with rapid rates of proteoglycan deposition and the greatest sensitivity to static and dynamic loads [25]. Fine heterotypic fibrils of collagen types II, IX, and XI and the glycoprotein, fibronectin, are also located within the pericellular environment [24]. Small amounts of type III collagen have also been found to colocalize to type II collagen around the cells throughout the depth of the cartilage and play a role in regulating fibril diameter [26,27]. In addition, type VI collagen, which preferentially localizes to the PCM [28,29], anchors the chondrocyte to the ECM, mediates cell-matrix interactions, and acts as a transducer for biomechanical signals from the cartilage ECM (reviewed in Refs. [30–33]). Various molecules including nidogen [34], laminin [35,36], and type IV collagen [35,37,38], the functional equivalent to a basement membrane, are also located within the PCM and play a role in regulating biological functions, such as maintaining chondrocyte phenotype and viability and modifying existing fibril networks [37, 39] (reviewed in Ref. [40]).

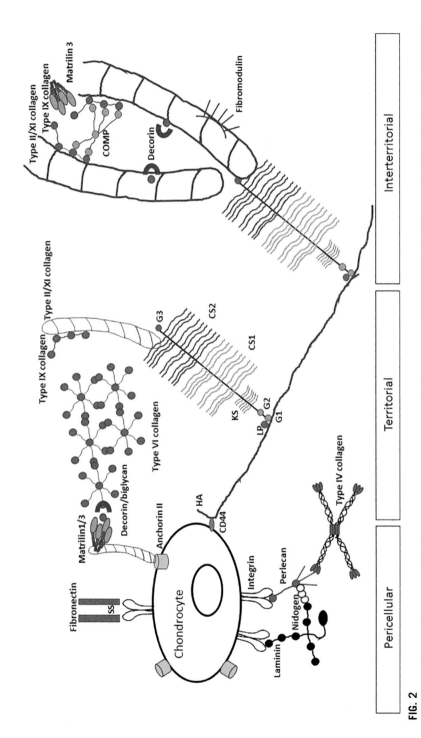

**FIG. 2**

The ECM of articular cartilage. Chondrocytes of the articular cartilage are embedded within the ECM composed of large collagen and proteoglycans, with other noncollagenous proteins and glycoproteins present in lesser amounts. Cartilage proteoglycans contain chondroitin sulfate (CS) and keratan sulfate (KS), covalently linked to hyaluronic acid (HA) by link protein (LP) to form large aggregates such as aggrecan that interact with collagens. Perifibrillar adapter proteins such as matrilin-1, matrilin-3, and matrilin-4; decorin; COMP; and type IX collagen link the type II/XI collagen fibrils and mediate interactions between the collagen network and the aggrecan matrix. G globular and SS disulfide bond.

*Adapted from Heinegård D, Saxne T. The role of the cartilage matrix in osteoarthritis. Nat Rev Rheumatol 2011;7(1):50–6.*

The chondrons are encased in a basketlike weave of thin collagen fibrils, collectively termed the territorial matrix (TM) [41]. This matrix, dense in chondroitin sulfate-rich proteoglycans, may protect the chondrocytes against mechanical stresses contributing to the resilience of the articular cartilage structure and its ability to withstand substantial loads [24,42]. The TM is situated within an interterritorial matrix, a complex network of posttranslationally modified proteins including fibrillar and nonfibrillar collagens, proteoglycans, laminins, and fibronectin, which contributes most to the biomechanical properties of the articular cartilage [43]. The PCM and TM act as an integrated mechanotransducer and are critical for transmitting biomechanical and biochemical signals between the ECM and the cells [25,41,44]. Perifibrillar adapter proteins such as matrilin-1, matrilin-3, and matrilin-4; decorin; cartilage oligomeric matrix protein (COMP); and type IX collagen link the type II/XI collagen fibrils and mediate interactions between the collagen network and the aggrecan matrix [45]. These adapter proteins are critical for the mechanical stabilization of the ECM network in adult cartilage and are important for the regulation of fibril diameter during fibrillogenesis in immature cartilage [45]. Fibronectin binds to ECM components and interacts with the NC4 domain of type IX collagen in the PCM, thus playing an important role in the organization of the ECM network and in transmitting "outside-in" mechanical signals through integrins [2,46]. The anisotropic mechanical properties of articular cartilage and the diverse nature of collagen fibrils modify the response of the tissue under compression [47].

### 2.1.1 Zonal characteristics

Articular cartilage is divided into distinct zones—the superficial zone, the middle zone, the deep zone, and the calcified zone (Fig. 3) [8]. The three-dimensional orientation of the collagen fibrils vary with cartilage depth such that they are largely parallel to the surface in the superficial zone, randomly orientated in the mid zones, and perpendicular to the surface in the deep zone next to the subchondral bone (reviewed in Ref. [48]) [49]. The flattened superficial zone chondrocytes, which are subjected to fluid flow and high hydrostatic pressures, synthesize and maintain higher amounts of collagen compared to proteoglycans; type I collagen may be synthesized in addition to type II collagen [50,51]. The collagen fibrils within the superficial zone, which are largely composed of type II, XI, and IX collagens [50,52], are densely packed and orientated parallel to the cartilage surface to allow them to resist the shear, tensile, and compressive forces imposed by articulation [53]. Collagen fibrils within the superficial zone are thinner, and while there is a lower concentration of aggrecan, levels of decorin and COMP are higher [54]. Effective lubrication at the articular surface is reliant on the presence of lubricin, a highly O-glycosylated protein that has low-friction and nonadherent properties [55]. It is complexed with COMP, a homopentameric glycoprotein [56], fibronectin, and type II collagen at the articular surface of cartilage to provide effective wear against shear stress [57,58].

Within the mid zone, the fibrils are thicker and resist swelling of the tissue caused by the water binding capacity of the proteoglycans [59,60]. The major proteoglycan is aggrecan that comprises a core protein attached with glycosaminoglycans (GAGs)

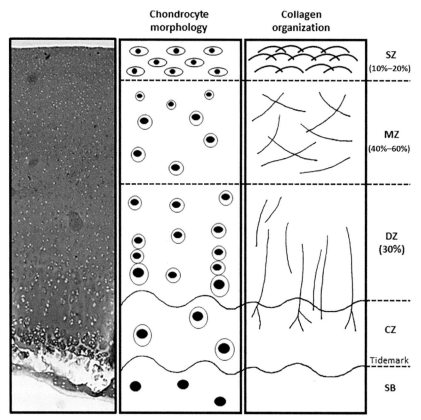

**FIG. 3**

Safranin O stained section and schematic representation of articular chondrocyte morphology and cartilage collagen organization. *SZ*, superficial zone; *MZ*, middle zone; *DZ*, deep zone; *CZ*, calcified zone; *SB*, subchondral bone.

and oligosaccharide chains [59]; the GAGs attract cations and water, and the resultant swelling pressure enables the cartilage to be resistant to compression [45]. The collagen fibrils within the deep zone have the largest diameter and lie perpendicular to the articular surface, thus providing the greatest resistance to compressive forces [61]. The tidemark separates the deep zone from the underlying calcified cartilage, an area populated by hypertrophic chondrocytes at low cell density, which serves to anchor the articular cartilage to the subchondral bone beneath [62]. Chondrocytes situated in the mid and deep zones are loaded under primarily hydrostatic pressures and thus synthesize high amounts of GAGs and type II collagen [63]. During joint loading, the compression of the proteoglycan aggregates within the mid zone reduces the pressure on the rest of the articular cartilage by distributing forces across the joint surface [20].

## 2.2 AGING OF ARTICULAR CARTILAGE

As cartilage ages, the ECM undergoes several changes including alterations in structure, total amount, proteolysis, and other posttranslational modifications [64,65]. The proteoglycans reduce in size due to proteolytic modifications resulting in a reduction in the amount of GAG bound to aggrecan; this not only decreases the hydration of articular cartilage [66] but also lowers its ability to respond to mechanical loading [66,67]. In addition, intermittent loading over time changes the sulfation pattern of the chondroitin sulfate chains and stimulates chondrocytes to produce elongated chondroitin sulfate chains, resulting in changes in the physiochemical properties of the proteoglycans in articular cartilage with age [68]. Type III collagen levels are also reduced with age, and type IX collagen becomes almost undetectable after approximately 40 years of age [26,29,69].

Advanced glycation end products increase over time with an increase in the formation of pentosidine cross-links between collagen fibrils [66,70], which has been associated with increased stiffness and brittleness of the articular cartilage with aging [71]. Deposition of calcium-containing crystals also occurs in aging cartilage [72] due to increased pyrophosphate production by the chondrocytes [73], resulting in a change of tissue stiffness that is likely to impact on the tissues' ability to respond to mechanical load.

In addition to changes in the matrix, the activity of the chondrocytes declines with age reducing their ability to maintain the ECM [64]. Taken together, these changes result in a tissue that has a reduced ability to withstand mechanical stress making it more susceptible to degenerative changes [66,67]. That said, even in old age, there is evidence to suggest that the chondrocyte attempts to maintain its pericellular environment and hence its mechanical role [69]. Recently, Madej et al. have reported that there is a reduced, or loss of, ability for mechanical regulation of various ECM components and essential growth factors in aged cartilage [74]. Collectively, these studies have revealed that there are age-related changes in articular cartilage that significantly impact on the tissues' ability to respond to mechanical signals.

## 2.3 THE RESPONSE OF ARTICULAR CARTILAGE TO PHYSIOLOGICAL MECHANICAL LOADING

Mechanical loading of articular cartilage is important for the development and maintenance of the tissue. During development, the geometry of the joint surface and the topological variation in cartilage thickness is determined by the mechanical environment of the joint [75]. In adulthood, moderate physiological mechanical loading is required for cartilage maintenance and health; immobilization of the joint results in a loss of proteoglycans and cartilage atrophy [76–79], while excessive loading ultimately leads to joint degeneration [80] (reviewed in Ref. [81]). During loading, articular cartilage is subjected to a wide range of static and dynamic mechanical stresses and strains [82,83], which in vivo are dependent on the anatomical location within the joint and the specific activity being undertaken [84,85]. In addition, the compressive strains received by the cartilage are nonuniform throughout its depth with >50% at the joint surface, 10%–20% in the transitional zone, and 0%–5% in the lower zones [86].

The biomechanical properties of articular cartilage are down to its biphasic nature; it contains a fluid phase and a solid phase. As discussed above, the fluid phase is largely made up of water, contributing up to 80% of the wet weight of the tissue [6] with a smaller component composed of inorganic ions such as sodium, calcium, chloride, and potassium. The solid phase is characterized by the ECM, which is porous and permeable [87,88]. The relationship between proteoglycan aggregates and interstitial fluid provides compressive resilience to cartilage through negative electrostatic repulsion forces [89,90]. The fixed negative charge of the proteoglycans dictates the extracellular ion composition; the ion concentration of the tissue is higher than the concentration of the surrounding joint fluid, resulting in increased pressure within the tissue [18]. This concentration difference results in fluid intake into the matrix and tension in the collagen fibers; this resultant hydrostatic pressure or Donnan osmotic pressure thus results in cartilage swelling [18,91]. In common daily activities, such as walking or running, the generation of very high cyclic hydrostatic pressures occurs by the formation of a thin fluid layer that forms between contact areas of the two opposing joint surfaces [75]. This increase in fluid pressure results in the squeezing out of fluid from the solid matrix; the low fluid permeability of the matrix limits this [92] so that the fluid flow is largely restricted to out of the superficial zone (reviewed in Ref. [75]). As fluid is squeezed out from the matrix, the chondrocytes and cellular components, such as the nucleus, experience compressive deformations that can be substantial due to the high water content of the ECM [75,93]. With continued cyclic loading, the loss of water and consolidation of the matrix leads to temporary, albeit to a small extent, thinning of the cartilage; upon cessation of load, fluid hydrostatic tension is created within the cartilage leading to the absorption of water back in and the restoration of normal cartilage thickness [94].

In summary, tissue pressure, deformation, and fluid flow are all important determinants in regulating the histomorphology of articular cartilage with depth-dependent variations in mechanical properties; the ECM is essential for resisting tensile and shear strains and the fluid component for resisting the high hydrostatic compressive stresses.

## 3 CARTILAGE LOADING AT CELLULAR LEVEL
### 3.1 IMPORTANCE OF MICROMECHANICAL ENVIRONMENT

As described above, articular cartilage is organized into discrete regions including the PCM in which the chondrocytes are embedded. The "chondron"—the thin layer of PCM and chondrocyte encapsulated within it [24]—is often thought of as a "mechanical unit" because it provides a physical link between the extensive ECM of the tissue and the chondrocytes, facilitating transmission of mechanical signals [44]. One of the principal components of the PCM is the matrilin (MATNs) group of oligomeric matrix proteins that form filamentous networks [95]. MATNs associate with specific integrin receptors at the cell surface and interact with other PCM constituents including type VI collagen and other ECM components including types II and IX collagen, aggrecan, and COMP (reviewed in Ref. [10]).

Of the MATNs expressed in chondrocytes, MATN1 is sensitive to mechanical load (5% elongation, 1 Hz, 15 min/h for 2 days); MATN1-deficient chondrocytes exhibit a diminished response to mechanical stimulation [96]. Furthermore, in engineered chondrocytes lacking MATN networks, mechanical signals are not transduced as evidenced by the inactivation of the reported downstream effector, that is, Indian hedgehog (Ihh) signaling [96]. Surprisingly, overexpression of MATNs also adversely affects the mechanosignaling response in chondrocytes; this is hypothesized to result from either saturation of these ECM binding sites by their ligands or a direct impact on the mechanical properties of the PCM. Thus, a threshold of chondrocyte mechanosensitivity may be decreased by mechanical loading through modification of the PCM, which could ultimately provide an adaptational response (via a negative feedback loop) to mechanical stimulation [96]. More recently, an investigation into the specific mechanotransducing role of MATN1, using MATN1 knockout mice, has demonstrated that the response of chondrocytes to a mechanical stimulus is impaired and that the mechanical properties (elastic modulus) of the cartilage matrix is altered [97]. Application of a loading regime (5% elongation, 1 Hz, 15 min/h for 24 h), previously shown to elicit an anabolic response in chondrocytes in vitro, is abolished in the absence of MATN1; furthermore, modifying the mechanical environment of the MATN1 knockout mice in vivo, by inducing joint instability via destabilization of the medial meniscus, results in a more progressive degeneration of the cartilage [97]. These in vitro and in vivo findings are suggestive that MATN1 is essential for effective mechanosignaling, enabling chondrocytes to adapt to and withstand applied mechanical stresses.

In addition to its molecular composition, the PCM also acts as a repository for several growth factors and regulatory molecules that have been associated, directly or indirectly, with transducing mechanical signals. The most widely reported is the release of fibroblast growth factor 2 (FGF-2) in response to mechanical trauma through injurious wounding of cartilage [98] or cyclic compression (0.01 MPa, 0.5 Hz, 2 min) [99]. FGF-2, sequestered on the heparan sulfate chains of perlecan, is released in response to mechanical perturbation and effects different intracellular responses depending on which FGF receptor (FGFR) the ligand binds to (reviewed in Ref. [100]); if FGF-2 mediated mechanoresponses are driven via FGFR1 ligation, the effect is catabolic, whereas if FGF-2 binds to FGFR3, a chondroprotective response is elicited [101]. Downstream pathways are subsequently activated in response to mechanical load, for example, with FGF-2 stimulating mitogen-activated protein kinase (MAPK) signaling [101], hence transmitting a physical load through the ECM/PCM to effect cellular responses.

## 3.2 CHONDROCYTE MECHANORECEPTORS

The mechanosignaling process is initiated in articular cartilage by stimulation of mechanoreceptors including ion channels, integrin receptors, and primary cilia located on the chondrocyte membrane, which are then able to effect intracellular responses (Fig. 4).

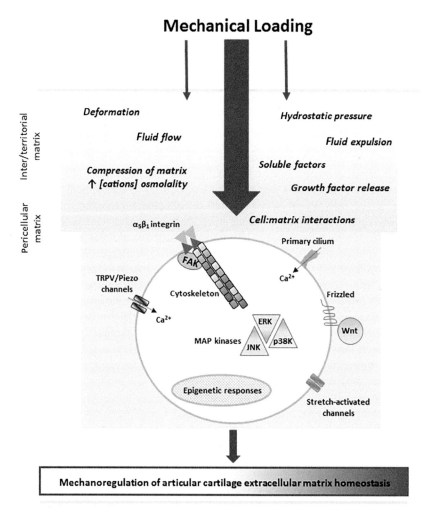

**FIG. 4**

Schematic representation of how mechanical load alters both the intra- and extracellular environment of articular cartilage chondrocytes. Upon application of a mechanical load, interactions between the pericellular environment and the chondrocyte facilitate transmission of the signal to the cell surface where mechanoreceptors become activated and initiate intracellular signaling cascades—the net effect being mechanoregulation of transcription, translation, and/or molecule secretion.

*Adapted from Blain EJ. Involvement of the cytoskeletal elements in articular cartilage mechanotransduction. In: Kamkin A, Kiseleva I, editors. Mechanosensitivity and mechanotransduction, vol. 4. Springer; 2011. p. 77–108.*

### 3.2.1 Ion channels

Mechanical stimulation regulates the activity of ion channels localized on the plasma membrane due to perturbations in membrane tension and lipid bilayer distortion. One of the earliest responses of chondrocytes to mechanical stimuli is the activation of $Ca^{2+}$ channels, thereby facilitating $Ca^{2+}$ influx into the chondrocyte and initiation of intracellular $Ca^{2+}$-dependent signaling pathways [102,103]. Under controlled mechanical displacement using a micropipette (25% deformation of the cell), an immediate and transient increase in intracellular $Ca^{2+}$ concentration $[Ca^{2+}]_i$ is evoked in isolated chondrocytes [102]. However, chondrocytes embedded in agarose and subjected to a 20% compressive strain (strain rate, 5%/s) demonstrate a slight delay (~200s) in $[Ca^{2+}]_i$ release preceded by localized areas of elevated $[Ca^{2+}]_i$ around the cell periphery, indicative that the localized $Ca^{2+}$ events are upstream of the global $Ca^{2+}$ mechanoresponse. A very recent study has attempted to elucidate which $[Ca^{2+}]_i$ signaling pathways are evoked by compressive loading in chondrocytes; of seven $[Ca^{2+}]_i$ pathways analyzed, all necessitate extracellular $Ca^{2+}$ stores to mediate $[Ca^{2+}]_i$ responses in situ [104]. Furthermore, three types of ion channels including transient receptor potential vanilloid 4 (TRPV4), T-type voltage-gated $Ca^{2+}$ channels (VGCCs), and mechanosensitive ion channels facilitate critical roles in controlling load-induced $[Ca^{2+}]_i$ responses in cartilage chondrocytes.

TRPV4 is a $Ca^{2+}$-permeable channel highly expressed in chondrocytes and activated in response to several stimuli including osmotic stress [105]. TRPV4 transduces compressive loading signals (10%, 1 Hz, 3 h/day) in chondrocytes, promoting an anabolic response by enhancing matrix accumulation [106], in keeping with its known role of regulating chondrogenesis [107]. TRPV4-mediated $Ca^{2+}$ signaling is thus believed to play a critical role in providing a physiological link between mechanical loading and regulation of cartilage ECM biosynthesis [106]. The PCM also impacts on TRPV4-mediated mechanosensing; although type VI collagen is not required directly for TRPV4-mediated $Ca^{2+}$ signaling, the loss of expression alters the mechanical properties of the pericellular environment resulting in cell swelling and osmotically induced TRPV4 activation, highlighting the role of the PCM in mechanotransduction [32].

More recently, Piezo channels (Piezo 1 and Piezo 2 in vertebrates), cation-permeable ion channels that are rapidly activated within 5 ms of mechanical stimulation, have been characterized in chondrocytes [108]. Both Piezo channels are expressed in chondrocytes and involved in transducing hyperphysiological, that is, injurious compressive loading signals ($> 45\%$ strain) by increasing $Ca^{2+}$ uptake [108]; knockdown inhibition of Piezo 1 or 2 channels prevents mechanically induced $Ca^{2+}$ transients from being evoked, suggestive that either of the Piezo channels can mediate mechanically induced $Ca^{2+}$ signaling in chondrocytes [108]. It is hypothesized that a synergy exists between Piezo 1 and 2 in transducing mechanical signals, particularly in relaying injurious loads, although the potential mechanism of action involved in cooperative transduction of these injurious signals requires elucidation [109]. Interestingly, the actin cytoskeleton is critical to the transduction of high strain-induced $Ca^{2+}$ transients mediated by Piezo 1 or 2 channel activation, as disruption of actin network organization inhibits the response in chondrocytes [109].

### 3.2.2 Primary cilia

The primary cilium, a microtubule-based structure extending from the cell surface and protruding into the PCM, is a principal sensor involved in cartilage chondrocyte mechanosignaling [110]. The primary cilium, positioned on the cell surface, is ideally situated to sense matrix deformation; ultrastructural studies have identified a direct connection between the extracellular collagen fibrils and the proteins that decorate the ciliary microtubules in chondrocytes, implying a matrix-cilium-Golgi continuum [111]. Furthermore, mechanical perturbation reversibly modifies chondrocyte cilium length; in chondrocytes embedded in agarose, extended periods of compression (0%–15%, 1 Hz, 48 h) significantly shorten cilia length [112]. In contrast, shorter loading intervals (<24 h) do not influence cilia length, suggestive of an adaptive mechanism to protect against prolonged loading episodes. However, it has been demonstrated that the cilia are themselves not the initial mechanoreceptors, but instead orchestrate downstream signaling events in response to the load stimulus [113]. In chondrocytes subjected to compression (15%, 1 Hz, 24 h), cilia are involved in promoting proteoglycan synthesis, through intracellular release of $Ca^{2+}$ (as described above) mediated by ATP release [113]; in the absence of cilia, chondrocytes are incapable of effecting such a mechanoresponse and exhibit reduced mechanical properties [114]. One downstream pathway reported to be activated by the cilia, $Ca^{2+}$ signaling, and ATP cascade involves the transcription regulator CITED2 (Cbp/p300-interacting transactivator with ED-rich tail 2). Compressive strain (5%, 1 Hz, 1 h) induces the transactivation of CITED2, leading to suppression of matrix metalloproteinase (MMP)-1 and MMP-13 expression, an event that is abolished in chondrocytes lacking cilia [115]. Interestingly, cilia serve as a receptacle for many of the mechanosignaling cascades in chondrocytes, for example, integrins, ion channels, $Ca^{2+}$, and the associated downstream pathways (as reviewed in Ref. [116]). For example, β1 integrin localizes to the chondrocyte cilium and mediates contact with matrix proteins in the pericellular environment [117].

### 3.2.3 Integrins

Integrins are a family of heterodimeric transmembrane proteins comprising α and β subunits that determine ligand specificity. Integrin subunits contain large extracellular domains that facilitate binding to PCM ligands and short cytoplasmic domains, devoid of intrinsic kinase activity, that interact with proteins that propagate intracellular kinase-mediated events [118]. These integrin cytoplasmic domains also bind to and orchestrate cytoskeletal actin organization [119], thereby serving as transmembrane mechanical links between extracellular molecules and the cell cytoskeleton. Chondrocytes express several members of the integrin family including α1β1, α10β1, and αvβ5 in addition to α5β1 [120]. Early studies demonstrate the role of α5β1 integrin as a primary mechanoreceptor in chondrocytes [121–123], mediating membrane hyperpolarization after cyclical pressure-induced strain (16 kPa, 0.33 Hz, 20 min) through the involvement of the actin cytoskeleton, the phospholipase C calmodulin pathway, and both tyrosine protein kinase and protein kinase C (PKC) activities [122]. Follow-on studies demonstrate that mechanical strain-induced α5β1-mediated membrane hyperpolarization stimulates transcription of aggrecan and inhibition of MMP-3 via an IL-4-dependent mechanism [124]. Furthermore, mechanical strain-induced membrane hyperpolarization, via α5β1 integrin activation, results in rapid tyrosine phosphorylation of focal adhesion kinase (pp125FAK), β-catenin, and paxillin [125], in addition to the rapid translocation of PKC to the chondrocyte cell membrane where it associates with RACK1 and β1 integrin subunits to mediate a mechanoresponse [126]. However, mechanical stimulation not only activates the integrin but also increases its expression in cartilage with compressive loading (1 MPa, 0.5 Hz, 6 or 24 h) inducing an increase in α5 integrin subunit expression [127]. More recently, chondrocyte cell death resulting from an injurious impact load has been attributed to integrin-mediated signaling through tyrosine phosphorylation of FAK and Src [128], further providing evidence for the involvement of α5β1 integrin as a conduit in transducing mechanical signals between the chondrocyte and the surrounding matrix to effect responses.

## 3.3 CHONDROCYTE MECHANOTRANSDUCTION

Following mechanostimulation of cell surface receptors and mechanically induced deformation of the ECM, biochemical effects are subsequently mediated via a series of intracellular signaling pathways including the cytoskeletal elements, Wnt molecules, and more recently miRNAs. Activation and/or suppression of these intracellular mechanosignaling cascades ultimately regulate cellular functions. However, the signaling intermediaries described here are by no means an exhaustive list, as many more signaling molecules and/or pathways have been implicated in cartilage chondrocyte mechanotransduction.

### *3.3.1 Cytoskeleton*

Involvement of the cytoskeleton in transducing mechanical signals within the cell is widely documented; the cytoskeletal proteins not only act as structural supports to withstand mechanical load but also orchestrate intracellular mechanosignaling cascades. The chondrocyte cytoskeleton comprises actin microfilaments, tubulin microtubules, and vimentin intermediate filaments [129].

A critical function of the actin cytoskeleton is to provide the chondrocyte with mechanical integrity to withstand compressive forces [130]. Studies in which the cytoskeletal actin network is disrupted using cytochalasin D demonstrate a significant 90% reduction in chondrocyte stiffness and 80% reduction in viscosity [131]; in addition, disruption of the actin cytoskeleton affects the relationship between matrix deformation and changes in nuclear height and shape, compromising downstream mechanoresponses [130]. The actin cytoskeleton is highly dynamic in nature, responding to mechanical perturbation by reversibly assembling and disassembling, as evidenced in several studies (reviewed in Refs. [132,133]). Exposure of chondrocytes to increasing hydrostatic pressure (15–30 MPa) results in a near complete disappearance of the actin network, which is reversible upon the removal of the stimulus [134]. In comparison, application of dynamic (10% or 15%, 1 Hz) or static compression (15%) results in cytoskeletal actin remodeling in chondrocytes embedded in agarose [135]; the classical cortical distribution of actin is replaced with a predominantly punctate organization. Exposure to alterations in the osmotic environment, an effect that can be propagated via application and then subsequent release of a compressive force, remodels actin in chondrocytes [136,137]. Load-induced rearrangement of the actin cytoskeleton in cartilage chondrocytes involves key actin accessory proteins including thymosin $\beta_4$ [138,139], cofilin, and destrin [135,140]. Differential responses to the loads applied could arise as a consequence of the activation of different signaling pathways [135,141].

There is little evidence to support a direct mechanical role for the microtubular network in chondrocytes, as tubulin disruption using colchicine does not alter cell stiffness or viscosity [131] and microtubular organization does not alter in response to hydrostatic pressure (15 MPa, 20 h) [141a]. As discussed above, microtubules do contribute to a mechanical role via their presence in the primary cilia of chondrocytes.

The vimentin network also provides mechanical integrity, as vimentin filament disruption using acrylamide decreases chondrocyte stiffness and viscosity [131]. Furthermore, it has recently been shown, using optical magnetic twisting cytometry and traction force microscopy, that following the application of a compressive force, cellular stiffness is compromised in vimentin-disrupted chondrocytes [142]. Vimentin-disrupted chondrocytes also exhibit a reduced response of fluidization-resolidification after compressive strain, suggestive of a principal role of vimentin in withstanding compressive forces and a minor role in regulating cytoskeletal tension [142]. A mechanical role for vimentin is not surprising since the vimentin cytoskeleton is more prominent in weight-bearing areas of immature rabbit articular cartilage in vivo [143]. In vitro studies have also evidenced the mechanosensitivity of the vimentin cytoskeleton in chondrocytes. Vimentin filament reorganization is observed in cartilage explants subjected to increasing levels of static compression (0–4 MPa, 1 h) [144] and in human osteoarthritic cartilage subjected to cyclic strain (7% elongation, 0.5 Hz) for 24 h [145]. Interestingly, vimentin network disassembly is observed in free swelling, that is, unloaded explant chondrocytes [144]. Vimentin disassembly is also evidenced when nonphysiological loads are applied to cartilage, with a characteristic juxtanuclear collapse of the vimentin network in response to a single impact load [146]. Regulation of the highly dynamic vimentin network, that is, its assembly and disassembly, is governed by phosphorylation of specific serine and threonine residues, with phosphorylation resulting in the disassembly of the vimentin filament and the generation of a pool of soluble vimentin subunits. Following dephosphorylation, polymerization is initiated, and the vimentin cytoskeleton undergoes further rearrangement [147]. Vimentin is reported to be one of the predominant phosphoproteins in the cytoplasm in vivo; furthermore, there is in vitro evidence that vimentin acts as a substrate for second messenger protein kinases, for example, the MAPK [148,149].

### 3.3.2 Mitogen-activated protein kinases

The MAPK family comprises extracellular-regulated kinase (ERK1/2), p38, and Jun N-terminal kinase (JNK); phosphorylation and activation of the MAPKs relay intracellular signals to facilitate many cellular processes including cell division, differentiation, apoptosis, and transcription [150]. Mechanoregulation of MAPK activities has been previously reported in chondrocytes. ERK1/2 and p38K activation is observed in cartilage explants subjected to either static compression (~50%, 2 or 24 h) or dynamic shear stress (3%, 0.1 Hz, 0.5–24 h) [150]; interestingly, inhibition of ERK1/2 or p38K abolishes mechanically induced transcription of aggrecan, type II collagen, and select MMPs [150]. Activation of p38K is observed in cartilage explants subjected to compression (0.1 MPa, 0.5 Hz, 24 h) concomitant with the production of the pro-inflammatory molecules nitric oxide (NO) and prostaglandin $E_2$ ($PGE_2$), suggestive that mechanical load induces an inflammatory phenotype in cartilage [151]. This is not completely unsurprising since ERK1/2 activation is reported to regulate the expression of pro-inflammatory cytokines, for example, IL-1β, IL-6, and TNF-α [152], thereby identifying a mechanism by which mechanical load elicits inflammatory and/or degenerative effects in cartilage. Cyclic pressurization (26.7 kPa, 0.33 Hz, 30 min–3 h) is reported to induce integrin-dependent JNK activation in isolated chondrocytes concomitant with increased proteoglycan synthesis [153], indicating a role for JNK in regulating load-induced matrix anabolism. These opposing effects of mechanical load on the downstream consequences of MAPK activation might reflect inherent differences in temporal dynamics and magnitudes of MAPK activation [150], conferring load-dependent activation of downstream transcription factors including AP-1 (*c-Fos/c-Jun*), RUNX-2, and HIF-2α to effect responses in articular cartilage [152]. However, it may well be that activation of other mechanically regulated signaling pathways requires phosphorylation of the MAPKs to initiate transcriptional responses due to high levels of interplay between many of these pathways.

### 3.3.3 Wnts

The Wnts, a large family of secreted glycoproteins, are fundamental to most cellular processes; furthermore, Wnt signaling is imperative during chondrogenesis and essential for chondrocyte function [154,155]. Wnt ligands utilize different intracellular signaling pathways to elicit responses, with divergence into three downstream pathways, namely, (i) β-catenin-dependent (canonical Wnt), noncanonical (ii) Wnt/$Ca^{2+}$, and (iii) planar cell polarity pathways. Downstream effects are varied but involve translocation of β-catenin to the nucleus that binds to TCF/LEF transcription factors to effect transcription, regulation of $Ca^{2+}$ release, and reorganization of the cytoskeleton, respectively. Interestingly, activation of the Wnt signaling pathway is another reported mechanism of mechanotransduction in cartilage chondrocytes. Expression of FrzB, a Wnt signaling inhibitor, is significantly downregulated in mechanically injured cartilage explants within 24 h [156]; it is hypothesized that a reduction in FrzB derepresses Wnt signaling. Mechanical injury activates canonical Wnt signaling in cartilage explants as evidenced by increased expression of two key

Wnt target genes: *Axin-2* and *c-jun* [156]. In this ex vivo mechanical injury model, in addition to *FrzB* downregulation, another Wnt component, specifically *Wnt16*, is upregulated at 24 h postinjury as assessed using microarray analysis [157]; translocation of β-catenin to the nucleus in injured cartilage is also observed, indicative of mechanically induced canonical Wnt signaling. Mechanically induced β-catenin nuclear localization is also observed in isolated chondrocytes (7.5% elongation, 1 Hz, 30 min), concomitant with induction of *c-jun* transcription as reported in our previous study [158]. Our subsequent studies conducted in cartilage explants not only have replicated these above findings but also have implicated noncanonical Wnt signaling in effecting mechanoresponses, particularly involving molecules associated with the Wnt/$Ca^{2+}$ noncanonical pathway (Al-Sabah et al., unpublished findings). Prolonged Wnt signaling (induced by exogenous Wnt3A stimulation of chondrocytes) inhibits markers of the chondrocyte phenotype including *SOX9*, *col2a1*, and *aggrecan* mRNAs [159]; thus, mechanoregulation of Wnt signaling molecules has considerable implications for maintaining cartilage integrity and functionality.

### 3.3.4 miRNAs

An emerging field over the last 8 years has been the involvement of miRNAs in chondrocyte mechanotransduction. miRNAs are very small (20–23 bp) cytoplasmic RNAs that control the posttranscriptional regulation of a third of all genes [160] and hence their involvement in many fundamental cellular processes. miRNAs instigate epigenetic silencing of transcription by binding to the 3'-UTR within the target mRNA followed by association with the RNA-induced silencing complex (RISC). Hence, target mRNA is degraded if the complementarity of miRNA with mRNA is exact, or if the match is not perfect, then inhibition of mRNA translation ensues. Expression patterns of most miRNAs depend on the tissue, the developmental stage, and the cellular environment [160,161]; furthermore, the functions of many miRNAs are reported to be induced by certain "stress" factors, including aging, inflammation, and mechanical load [160]. To date, there are several studies showing that mechanical stress has an impact on chondrocyte responses through the regulation of miRNAs [161–165].

The first study implicating miRNA expression in chondrocyte mechanotransduction was reported in 2009 following a comparison of miRNA profiles from regions of cartilage subjected to different amounts of weight-bearing stresses [161]. miRNAs found to be differentially regulated in the condylar anterior weight-bearing area exposed to maximum contact pressure included miRNA-221 and miRNA-222 (upregulated) and miRNA-148 (downregulated) when compared with the posterior nonweight-bearing area with minimum contact pressure [161]. Although the functions of these miRNAs are still to be completely elucidated, miRNA-221 is reported to be involved in chondrogenesis [166]; miRNA-222 is believed to drive chondrocyte proliferation [167]; and miRNA-148 has roles in cell proliferation, differentiation, and apoptosis [168], suggestive of a homeostatic response to mechanical load.

Cheleschi et al. (2017) have recently reported that application of hydrostatic pressure (1–5 MPa, 0.25 Hz, 3 h) on chondrocytes derived from human osteoarthritic cartilage significantly induces expression of miRNA-27a/b, miRNA-140, and miRNA-146a/b with a concomitant reduction in miRNA-365 [165]; interestingly, miRNA levels in chondrocytes obtained from normal cartilage were unaffected by this loading regime. Sustained upregulation of miRNA-27a, miRNA-140, and miRNA-146a is observed at 48 h postcessation of load and correlates with regulation of their known target genes including MMP-13, ADAMTS-5, and ADAMTS-4. HDACs modulate cell proliferation by regulating chromatin structure and repressing the activity of specific transcription factors. HDAC4, a class II HDAC expressed in prehypertrophic chondrocytes, acts to inhibit chondrocyte hypertrophy and endochondral bone formation by suppressing Runx2 activity [169]. The mechanoresponsiveness of HDAC-4 in chondrocytes is attributed to its regulation by miRNA-365. Yang et al. (2016) demonstrated the induction of miRNA-365 in chondrocytes obtained from macroscopically normal regions of OA cartilage exposed to 10% elongation for 24 h (1 Hz, 15 min/h); concomitant with this was an increase in type X collagen and MMP-13 transcription attributed to a reduction in HDAC-4 levels [164]. miRNA-365 levels are also significantly but transiently elevated in chick chondrocytes cultured in 3-D-sponge scaffolds and exposed to 5% elongation (1 Hz, 15 min/h) for 24 h [162]. This miRNA regulates chondrocyte proliferation and differentiation as ascertained by increased levels of type X collagen and Ihh expression and decreased HDAC-4 levels.

Mechanoregulation of miRNA-146a has also been demonstrated in normal human chondrocytes subjected to mechanical injury (10 MPa, 60 min), concomitant with increased VEGF and decreased Smad4 protein levels in a time-dependent manner following cessation of load (48 h postload) [163]. Mechanoregulation of miRNA-146a is hypothesized to drive the apoptosis observed in these chondrocytes following a mechanical pressure injury and thus has implications for injury-induced cartilage degeneration.

To date, the involvement of mechanical load on noncoding RNAs (ncRNAs) in cartilage chondrocytes has focused exclusively on the regulation of miRNA expression. However, other ncRNAs exist including small nucleolar RNAs (snoRNAs); piwi-interacting RNAs (piRNAs); and long, noncoding RNAs (lncRNAs) including large-intergenic noncoding RNA (lincRNAs). The importance of epitranscriptomics [170] and the role and function of snoRNAs, lncRNAs, and other ncRNAs in cartilage chondrocyte mechanobiology remain largely unexplored and merit further investigation.

## 4 CONCLUDING SUMMARY

Since the identification of the $\alpha_5\beta_1$ integrin as a chondrocyte mechanoreceptor in 1997 [122] and the subsequent studies performed over the last 20 years (some of which have been described here), our understanding of articular cartilage mechanobiology has increased exponentially. The importance of matrix composition, PCM interactions with the chondrocyte, mechanosensing at the cell membrane, and activation of key mechanoreceptors to induce $Ca^{2+}$ influx, cytoskeletal network reorganization, and transcriptional events is now well documented. Importantly, the influence of abnormal mechanical load on chondrocyte behavior, the mechanisms involved, and how this impacts on the ECM to induce a degenerative phenotype are also becoming better characterized; this has been largely attributed to the use of a complementary "toolbox" of in vitro (cell and explant) and in vivo loading models. The emergence of new signaling molecules/pathways, for example, the ncRNAs and their potential importance in cartilage mechanotransduction, is a largely unexplored field and warrants further investigation.

Collectively, our continually expanding knowledge will undoubtedly shed further light on the cartilage chondrocyte mechanobiology field to provide a more complete picture on (i) the mechanisms involved in mechanoregulation of the downstream processes of transcription, translation, and secretion; (ii) defining mechanical "thresholds" above which loads become deleterious; and (iii) characterizing the mechanisms that mediate this to develop therapeutic strategies for load-induced joint degeneration such as osteoarthritis. There is still much in the field of cartilage mechanobiology to be uncovered.

## ACKNOWLEDGMENT

The authors are funded by Arthritis Research UK (Arthritis Research UK Biomechanics and Bioengineering Centre, 510390).

## REFERENCES

[1] Korhonen RK, Laasanen MS, Töyräs J, Rieppo J, Hirvonen J, Helminen HJ, et al. Comparison of the equilibrium response of articular cartilage in unconfined compression, confined compression and indentation. J Biomech 2002;35(7):903–9.
[2] Parsons P, Gilbert SJ, Vaughan-Thomas A, Sorrell DA, Notman R, Bishop M, et al. Type IX collagen interacts with fibronectin providing an important molecular bridge in articular cartilage. J Biol Chem 2011;286(40):34986–97.
[3] Buckwalter JA, Mankin HJ, Grodzinsky AJ. Articular cartilage and osteoarthritis. Instr Course Lect 2005;54:465–80.
[4] Grodzinsky AJ, Levenston ME, Jin M, Frank EH. Cartilage tissue remodeling in response to mechanical forces. Annu Rev Biomed Eng 2000;2:691–713.
[5] Mow VC, Proctor CS, Kelly MA. Biomechanics of articular cartilage. In: Nordin M, Frankel V, editors. Basic biomechanics of the locomotor system. 2nd ed. Philadelphia: Lea and Febiger; 1989. p. 31–58.
[6] Buckwalter JA. Articular cartilage. Instr Course Lect 1983;32:349–70.
[7] Mow VC, Ratcliffe A, Poole AR. Cartilage and diarthrodial joints as paradigms for hierarchical materials and structures. Biomaterials 1992;13(2):67–97.
[8] Heinegård D, Saxne T. The role of the cartilage matrix in osteoarthritis. Nat Rev Rheumatol 2011;7(1):50–6.

[9] Khan IM, Gilbert SJ, Singhrao SK, Duance VC, Archer CW. Cartilage integration: evaluation of the reasons for failure of integration during cartilage repair. A review. Eur Cell Mater 2008;16:26–39.

[10] Wilusz RE, Sanchez-Adams J, Guilak F. The structure and function of the pericellular matrix of articular cartilage. Matrix Biol 2014;39:25–32.

[11] Wong M, Siegrist M, Cao X. Cyclic compression of articular cartilage explants is associated with progressive consolidation and altered expression pattern of extracellular matrix proteins. Matrix Biol 1999;18(4):391–9.

[12] Urban JP. The chondrocyte: a cell under pressure. Br J Rheumatol 1994;33(10):901–8.

[13] Grodzinsky AJ, Kim YJ, Buschmann MD, Garcia AM, Quinn TM, Hunziker EB. Response of the chondrocyte to mechanical stimuli. In: Brandt KD, Doherty M, Lohmander LS, editors. Osteoarthritis. Oxford: Oxford Univ. Press; 1998. p. 123–36.

[14] Bader DL, Salter DM, Chowdhury TT. Biomechanical influence of cartilage homeostasis in health and disease. Arthritis 2011;2011:979032.

[15] Guilak F, Meyer BC, Ratcliffe A, Mow VC. The effects of matrix compression on proteoglycan metabolism in articular cartilage explants. Osteoarthritis Cartilage 1994;2(2):91–101.

[16] Sah RL, Kim YJ, Doong JY, Grodzinsky AJ, Plaas AH, Sandy JD. Biosynthetic response of cartilage explants to dynamic compression. J Orthop Res 1989;7(5):619–36.

[17] Hall AC, Urban JP, Gehl KA. The effects of hydrostatic pressure on matrix synthesis in articular cartilage. J Orthop Res 1991;9(1):1–10.

[18] Urban JP, Hall AC, Gehl KA. Regulation of matrix synthesis rates by the ionic and osmotic environment of articular chondrocytes. J Cell Physiol 1993;154(2):262–70.

[19] Palmoski MJ, Brandt KD. Effects of static and cyclic compressive loading on articular cartilage plugs in vitro. Arthritis Rheum 1984;27(6):675–81.

[20] Eckstein F, Hudelmaier M, Putz R. The effects of exercise on human articular cartilage. J Anat 2006;208(4):491–512.

[21] Räsänen LP, Tanska P, Zbyň Š, van Donkelaar CC, Trattnig S, Nieminen MT, et al. The effect of fixed charge density and cartilage swelling on mechanics of knee joint cartilage during simulated gait. J Biomech 2017;61:34–44.

[22] Martel-Pelletier J. Pathophysiology of osteoarthritis. Osteoarthritis Cartilage 2004;12(Suppl. A):S31–3.

[23] Musumeci G. The effect of mechanical load on articular cartilage. J Funct Morphol Kinesiol 2016;1:154–61.

[24] Poole CA. Articular cartilage chondrons: form, function and failure. J Anat 1997;191(Pt 1):1–13.

[25] Quinn TM, Grodzinsky AJ, Buschmann MD, Kim YJ, Hunziker EB. Mechanical compression alters proteoglycan deposition and matrix deformation around individual cells in cartilage explants. J Cell Sci 1998;111(Pt 5):573–83.

[26] Wotton SF, Duance VC. Type III collagen in normal human articular cartilage. Histochem J 1994;26(5):412–6.

[27] Young RD, Lawrence PA, Duance VC, Aigner T, Monaghan P. Immunolocalization of collagen types II and III in single fibrils of human articular cartilage. J Histochem Cytochem 2000;48(3):423–32.

[28] Poole CA, Wotton SF, Duance VC. Localization of type IX collagen in chondrons isolated from porcine articular cartilage and rat chondrosarcoma. Histochem J 1988;20(10):567–74.

[29] Duance VC, Wotton SF. Changes in the distribution of mammalian cartilage collagens with age. Biochem Soc Trans 1991;19(4):376S.
[30] Luo Y, Sinkeviciute D, He Y, Karsdal M, Henrotin Y, Mobasheri A, et al. The minor collagens in articular cartilage. Protein Cell 2017;8(8):560–72.
[31] Pfaff M, Aumailley M, Specks U, Knolle J, Zerwes HG, Timpl R. Integrin and Arg-Gly-Asp dependence of cell adhesion to the native and unfolded triple helix of collagen type VI. Exp Cell Res 1993;206(1):167–76.
[32] Zelenski NA, Leddy HA, Sanchez-Adams J, Zhang J, Bonaldo P, Liedtke W, et al. Type VI collagen regulates pericellular matrix properties, chondrocyte swelling, and Mechanotransduction in mouse articular cartilage. Arthritis Rheumatol 2015;67(5):1286–94.
[33] Buckwalter JA, Mankin HJ. Articular cartilage: tissue design and chondrocyte-matrix interactions. Instr Course Lect 1998;47:477–86.
[34] Kruegel J, Sadowski B, Miosge N. Nidogen-1 and nidogen-2 in healthy human cartilage and in late-stage osteoarthritis cartilage. Arthritis Rheum 2008;58(5):1422–32.
[35] Foldager CB, Toh WS, Gomoll AH, Olsen BR, Spector M. Distribution of basement membrane molecules, laminin and collagen type IV, in normal and degenerated cartilage tissues. Cartilage 2014;5(2):123–32.
[36] Dürr J, Lammi P, Goodman SL, Aigner T, von der Mark K. Identification and immunolocalization of laminin in cartilage. Exp Cell Res 1996;222(1):225–33.
[37] Kvist AJ, Nyström A, Hultenby K, Sasaki T, Talts JF, Aspberg A. The major basement membrane components localize to the chondrocyte pericellular matrix—a cartilage basement membrane equivalent? Matrix Biol 2008;27(1):22–33.
[38] Eyre D. Collagen of articular cartilage. Arthritis Res 2002;4(1):30–5.
[39] Wu JJ, Weis MA, Kim LS, Eyre DR. Type III collagen, a fibril network modifier in articular cartilage. J Biol Chem 2010;285(24):18537–44.
[40] Sun Y, Wang TL, Toh WS, Pei M. The role of laminins in cartilaginous tissues: from development to regeneration. Eur Cell Mater 2017;34:40–54.
[41] Xia Y, Darling EM, Herzog W. Functional properties of chondrocytes and articular cartilage using optical imaging to scanning probe microscopy. J Orthop Res 2018;36(2):620–31.
[42] Szirmai JA. Structure of cartilage. In: Engel A, Larsson T, editors. Aging of connective and skeletal tissue. Stockholm, Sweden: Nordiska; 1969. p. 163–200.
[43] Mow VC, Guo XE. Mechano-electrochemical properties of articular cartilage: their inhomogeneities and anisotropies. Annu Rev Biomed Eng 2002;4:175–209.
[44] Guilak F, Alexopoulos LG, Upton ML, Youn I, Choi JB, Cao L, et al. The pericellular matrix as a transducer of biomechanical and biochemical signals in articular cartilage. Ann N Y Acad Sci 2006;1068:498–512.
[45] Firner S, Zaucke F, Michael J, Dargel J, Schiwy-Bochat KH, Heilig J, et al. Extracellular distribution of collagen II and perifibrillar adapter proteins in healthy and osteoarthritic human knee joint cartilage. J Histochem Cytochem 2017;65(10):593–606.
[46] Kadler KE, Hill A, Canty-Laird EG. Collagen fibrillogenesis: fibronectin, integrins, and minor collagens as organizers and nucleators. Curr Opin Cell Biol 2008;20(5):495–501.
[47] Korhonen RK, Laasanen MS, Töyräs J, Lappalainen R, Helminen HJ, Jurvelin JS. Fibril reinforced poroelastic model predicts specifically mechanical behavior of normal, proteoglycan depleted and collagen degraded articular cartilage. J Biomech 2003;36(9):1373–9.

[48] Eyre DR, Wu JJ, Fernandes RJ, Pietka TA, Weis MA. Recent developments in cartilage research: matrix biology of the collagen II/IX/XI heterofibril network. Biochem Soc Trans 2002;30(Pt 6):893–9.
[49] Inamdar SR, Knight DP, Terrill NJ, Karunaratne A, Cacho-Nerin F, Knight MM, et al. The secret life of collagen: temporal changes in nanoscale fibrillar pre-strain and molecular organization during physiological loading of cartilage. ACS Nano 2017;11 (10):9728–37.
[50] Duance VC. Surface of articular cartilage: immunohistological studies. Cell Biochem Funct 1983;1(3):143–4.
[51] Muir H, Bullough P, Maroudas A. The distribution of collagen in human articular cartilage with some of its physiological implications. J Bone Joint Surg Br 1970; 52(3):554–63.
[52] Eyre DR. The collagens of articular cartilage. Semin Arthritis Rheum 1991; 21(3 Suppl. 2):2–11.
[53] Askew MJ, Mow VC. The biomechanical function of the collagen fibril ultrastructure of articular cartilage. J Biomech Eng 1978;100(3):105–15.
[54] Lorenzo P, Bayliss MT, Heinegard D. A novel cartilage protein (CILP) present in the mid-zone of human articular cartilage increases with age. J Biol Chem 1998; 273(36):23463–8.
[55] Jones AR, Gleghorn JP, Hughes CE, Fitz LJ, Zollner R, Wainwright SD, et al. Binding and localization of recombinant lubricin to articular cartilage surfaces. J Orthop Res 2007;25(3):283–92.
[56] Flowers SA, Kalamajski S, Ali L, Björkman LI, Raj JR, Aspberg A, et al. Cartilage oligomeric matrix protein forms protein complexes with synovial lubricin via non-covalent and covalent interactions. Osteoarthritis Cartilage 2017;25(9):1496–504.
[57] Flowers SA, Zieba A, Örnros J, Jin C, Rolfson O, Björkman LI, et al. Lubricin binds cartilage proteins, cartilage oligomeric matrix protein, fibronectin and collagen II at the cartilage surface. Sci Rep 2017;7(1):13149.
[58] Andresen Eguiluz RC, Cook SG, Brown CN, Wu F, Pacifici NJ, Bonassar LJ, et al. Fibronectin mediates enhanced wear protection of lubricin during shear. Biomacromolecules 2015;16(9):2884–94.
[59] Caterson B, Flannery CR, Hughes CE, Little CB. Mechanisms involved in cartilage proteoglycan catabolism. Matrix Biol 2000;19(4):333–44.
[60] Hardingham TE, Fosang AJ. Proteoglycans: many forms and many functions. FASEB J 1992;6(3):861–70.
[61] Shirazi R, Shirazi-Adl A. Deep vertical collagen fibrils play a significant role in mechanics of articular cartilage. J Orthop Res 2008;26(5):608–15.
[62] Huber M, Trattnig S, Lintner F. Anatomy, biochemistry, and physiology of articular cartilage. Invest Radiol 2000;35(10):573–80.
[63] Wong M, Wuethrich P, Eggli P, Hunziker E. Zone-specific cell biosynthetic activity in mature bovine articular cartilage: a new method using confocal microscopic stereology and quantitative autoradiography. J Orthop Res 1996;14(3):424–32.
[64] Roughley PJ. Age-associated changes in cartilage matrix: implications for tissue repair. Clin Orthop Relat Res 2001;(391 Suppl.):S153–60.
[65] Maroudas A. Different ways of expressing concentration of cartilage constituents with special reference to the tissues organization and functional properties. In: Maroudas A, Kuettner KE, editors. Methods in cartilage research. New York: Academic Press; 1990. p. 211–9.

[66] Loeser RF. Aging and the etiopathogenesis and treatment of osteoarthritis. Rheum Dis Clin North Am 2000;26(3):547–67.
[67] Hudelmaier M, Glaser C, Hohe J, Englmeier KH, Reiser M, Putz R, et al. Age-related changes in the morphology and deformational behavior of knee joint cartilage. Arthritis Rheum 2001;44(11):2556–61.
[68] Sauerland K, Plaas AH, Raiss RX, Steinmeyer J. The sulfation pattern of chondroitin sulfate from articular cartilage explants in response to mechanical loading. Biochim Biophys Acta 2003;1638(3):241–8.
[69] Vaughan-Thomas A, Dudhia J, Bayliss MT, Kadler KE, Duance VC. Modification of the composition of articular cartilage collagen fibrils with increasing age. Connect Tissue Res 2008;49(5):374–82.
[70] Verzijl N, Bank RA, TeKoppele JM, DeGroot J. AGEing and osteoarthritis: a different perspective. Curr Opin Rheumatol 2003;15(5):616–22.
[71] Bank RA, Bayliss MT, Lafeber FP, Maroudas A, Tekoppele JM. Ageing and zonal variation in post-translational modification of collagen in normal human articular cartilage. The age-related increase in non-enzymatic glycation affects biomechanical properties of cartilage. Biochem J 1998;330(Pt 1):345–51.
[72] Mitsuyama H, Healey RM, Terkeltaub RA, Coutts RD, Amiel D. Calcification of human articular knee cartilage is primarily an effect of aging rather than osteoarthritis. Osteoarthritis Cartilage 2007;15(5):559–65.
[73] Rosen F, McCabe G, Quach J, Solan J, Terkeltaub R, Seegmiller JE, et al. Differential effects of aging on human chondrocyte responses to transforming growth factor beta: increased pyrophosphate production and decreased cell proliferation. Arthritis Rheum 1997;40(7):1275–81.
[74] Madej W, van Caam A, Blaney Davidson E, Buma P, van der Kraan PM. Unloading results in rapid loss of TGFβ signaling in articular cartilage: role of loading-induced TGFβ signaling in maintenance of articular chondrocyte phenotype? Osteoarthritis Cartilage 2016;24(10):1807–15.
[75] Wong M, Carter DR. Articular cartilage functional histomorphology and mechanobiology: a research perspective. Bone 2003;33(1):1–13.
[76] Haapala J, Arokoski JP, Hyttinen MM, Lammi M, Tammi M, Kovanen V, et al. Remobilization does not fully restore immobilization induced articular cartilage atrophy. Clin Orthop Relat Res 1999;362:218–29.
[77] Buckwalter JA, Mankin HJ. Articular cartilage: degeneration and osteoarthritis, repair, regeneration, and transplantation. Instr Course Lect 1998;47:487–504.
[78] Palmoski M, Perricone E, Brandt KD. Development and reversal of a proteoglycan aggregation defect in normal canine knee cartilage after immobilization. Arthritis Rheum 1979;22(5):508–17.
[79] Palmoski MJ, Colyer RA, Brandt KD. Joint motion in the absence of normal loading does not maintain normal articular cartilage. Arthritis Rheum 1980;23(3):325–34.
[80] Torzilli PA, Grigiene R, Borrelli J, Helfet DL. Effect of impact load on articular cartilage: cell metabolism and viability, and matrix water content. J Biomech Eng 1999;121(5):433–41.
[81] Buckwalter JA, Anderson DD, Brown TD, Tochigi Y, Martin JA. The roles of mechanical stresses in the pathogenesis of osteoarthritis: implications for treatment of joint injuries. Cartilage 2013;4(4):286–94.
[82] Hodge WA, Fijan RS, Carlson KL, Burgess RG, Harris WH, Mann RW. Contact pressures in the human hip joint measured in vivo. Proc Natl Acad Sci U S A 1986;83(9):2879–83.

[83] Harris MD, Anderson AE, Henak CR, Ellis BJ, Peters CL, Weiss JA. Finite element prediction of cartilage contact stresses in normal human hips. J Orthop Res 2012; 30(7):1133–9.

[84] Sanchez-Adams J, Leddy HA, McNulty AL, O'Conor CJ, Guilak F. The mechanobiology of articular cartilage: bearing the burden of osteoarthritis. Curr Rheumatol Rep 2014;16(10):451.

[85] Arokoski JP, Jurvelin JS, Väätäinen U, Helminen HJ. Normal and pathological adaptations of articular cartilage to joint loading. Scand J Med Sci Sports 2000;10(4): 186–98.

[86] Eckstein F, Tieschky M, Faber S, Englmeier KH, Reiser M. Functional analysis of articular cartilage deformation, recovery, and fluid flow following dynamic exercise in vivo. Anat Embryol (Berl) 1999;200(4):419–24.

[87] Ateshian GA, Warden WH, Kim JJ, Grelsamer RP, Mow VC. Finite deformation biphasic material properties of bovine articular cartilage from confined compression experiments. J Biomech 1997;30(11–12):1157–64.

[88] Mow VC, Holmes MH, Lai WM. Fluid transport and mechanical properties of articular cartilage: a review. J Biomech 1984;17(5):377–94.

[89] Mow VC, Ratcliffe A. Structure and function of articular cartilage and meniscus. 2nd ed. Philadelphia, PA: Lippincott-Raven; 1997.

[90] Maroudas A. Physiochemical properties of articular cartilage. In: Freeman MAR, editor. Adult articular cartilage. Kent, UK: Cambridge University Press; 1979. p. 215–90.

[91] Lai WM, Hou JS, Mow VC. A triphasic theory for the swelling and deformation behaviors of articular cartilage. J Biomech Eng 1991;113(3):245–58.

[92] Maroudas A, Bullough P. Permeability of articular cartilage. Nature 1968;219(5160): 1260–1.

[93] Mow VC, Setton LA, Bachrach NM, Guilak F. Stress, strain, pressure and flow fields in articular cartilage and chondrocytes. In: Mow VC, Tran-Son-Tay R, Guilak F, Hochmuth RM, editors. Cell mechanics and cellular engineering. New York: Springer Verlag; 1994. p. 345–79.

[94] Suh JK, Li Z, Woo SL. Dynamic behavior of a biphasic cartilage model under cyclic compressive loading. J Biomech 1995;28(4):357–64.

[95] Deak F, Wagener R, Kiss I, Paulsson M. The matrilins: a novel family of oligomeric extracellular matrix proteins. Matrix Biol 1999;18(1):55–64.

[96] Kanbe K, Yang X, Wei L, Sun C, Chen Q. Pericellular matrilins regulate activation of chondrocytes by cyclic load-induced matrix deformation. J Bone Miner Res 2007; 22(2):318–28.

[97] Chen Y, Cossman J, Jayasuriya CT, Li X, Guan Y, Fonseca V, et al. Deficient mechanical activation of anabolic transcripts and post-traumatic cartilage degeneration in matrilin-1 knockout mice. PLoS One 2016;11(6):e0156676.

[98] Vincent T, Hermansson M, Bolton M, Wait R, Saklatvala J. Basic FGF mediates an immediate response of articular cartilage to mechanical injury. Proc Natl Acad Sci U S A 2002;99(12):8259–64.

[99] Vincent TL, McLean CJ, Full LE, Peston D, Saklatvala J. FGF-2 is bound to perlecan in the pericellular matrix of articular cartilage, where it acts as a chondrocyte mechanotransducer. Osteoarthritis Cartilage 2007;15(7):752–63.

[100] Vincent TL. Targeting mechanotransduction pathways in osteoarthritis: a focus on the pericellular matrix. Curr Opin Pharmacol 2013;13(3):449–54.

[101] Chia SL, Sawaji Y, Burleigh A, McLean C, Inglis J, Saklatvala J, et al. Fibroblast growth factor 2 is an intrinsic chondroprotective agent that suppresses ADAMTS-5

and delays cartilage degradation in murine osteoarthritis. Arthritis Rheum 2009; 60(7):2019–27.

[102] Guilak F, Zell RA, Erickson GR, Grande DA, Rubin CT, McLeod KJ, et al. Mechanically induced calcium waves in articular chondrocytes are inhibited by gadolinium and amiloride. J Orthop Res 1999;17(3):421–9.

[103] Pingguan-Murphy B, El-Azzeh M, Bader DL, Knight MM. Cyclic compression of chondrocytes modulates a purinergic calcium signalling pathway in a strain rate- and frequency-dependent manner. J Cell Physiol 2006;209(2):389–97.

[104] Lv M, Zhou Y, Chen X, Han L, Wang L, Lu XL. Calcium signaling of in situ chondrocytes in articular cartilage under compressive loading: roles of calcium sources and cell membrane ion channels. J Orthop Res 2018;36(2):730–8.

[105] Phan MN, Leddy HA, Votta BJ, Kumar S, Levy DS, Lipshutz DB, et al. Functional characterization of TRPV4 as an osmotically sensitive ion channel in porcine articular chondrocytes. Arthritis Rheum 2009;60(10):3028–37.

[106] O'Conor CJ, Leddy HA, Benefield HC, Liedtke WB, Guilak F. TRPV4-mediated mechanotransduction regulates the metabolic response of chondrocytes to dynamic loading. Proc Natl Acad Sci U S A 2014;111(4):1316–21.

[107] Muramatsu S, Wakabayashi M, Ohno T, Amano K, Ooishi R, Sugahara T, et al. Functional gene screening system identified TRPV4 as a regulator of chondrogenic differentiation. J Biol Chem 2007;282(44):32158–67.

[108] Lee W, Leddy HA, Chen Y, Lee SH, Zelenski NA, McNulty AL, et al. Synergy between Piezo1 and Piezo2 channels confers high-strain mechanosensitivity to articular cartilage. Proc Natl Acad Sci U S A 2014;111(47):E5114–22.

[109] Lee W, Guilak F, Liedtke W. Role of piezo channels in joint health and injury. Curr Top Membr 2017;79:263–73.

[110] Jensen CG, Poole CA, McGlashan SR, Marko M, Issa ZI, Vujcich KV, et al. Ultrastructural, tomographic and confocal imaging of the chondrocyte primary cilium in situ. Cell Biol Int 2004;28(2):101–10.

[111] Poole CA, Zhang ZJ, Ross JM. The differential distribution of acetylated and detyrosinated alpha-tubulin in the microtubular cytoskeleton and primary cilia of hyaline cartilage chondrocytes. J Anat 2001;199(Pt 4):393–405.

[112] McGlashan SR, Knight MM, Chowdhury TT, Joshi P, Jensen CG, Kennedy S, et al. Mechanical loading modulates chondrocyte primary cilia incidence and length. Cell Biol Int 2010;34(5):441–6.

[113] Wann AK, Zuo N, Haycraft CJ, Jensen CG, Poole CA, McGlashan SR, et al. Primary cilia mediate mechanotransduction through control of ATP-induced Ca2+ signaling in compressed chondrocytes. FASEB J 2012;26(4):1663–71.

[114] Irianto J, Ramaswamy G, Serra R, Knight MM. Depletion of chondrocyte primary cilia reduces the compressive modulus of articular cartilage. J Biomech 2014;47(2):579–82.

[115] He Z, Leong DJ, Zhuo Z, Majeska RJ, Cardoso L, Spray DC, et al. Strain-induced mechanotransduction through primary cilia, extracellular ATP, purinergic calcium signaling, and ERK1/2 transactivates CITED2 and downregulates MMP-1 and MMP-13 gene expression in chondrocytes. Osteoarthritis Cartilage 2016;24(5):892–901.

[116] Ruhlen R, Marberry K. The chondrocyte primary cilium. Osteoarthritis Cartilage 2014;22(8):1071–6.

[117] McGlashan SR, Jensen CG, Poole CA. Localization of extracellular matrix receptors on the chondrocyte primary cilium. J Histochem Cytochem 2006;54(9):1005–14.

[118] Hynes RO. Integrins: bidirectional, allosteric signaling machines. Cell 2002;110(6):673–87.
[119] Wolfenson H, Lavelin I, Geiger B. Dynamic regulation of the structure and functions of integrin adhesions. Dev Cell 2013;24(5):447–58.
[120] Loeser RF. Integrins and cell signaling in chondrocytes. Biorheology 2002;39(1–2):119–24.
[121] Salter DM, Hughes DE, Simpson R, Gardner DL. Integrin expression by human articular chondrocytes. Br J Rheumatol 1992;31(4):231–4.
[122] Wright MO, Nishida K, Bavington C, Godolphin JL, Dunne E, Walmsley S, et al. Hyperpolarisation of cultured human chondrocytes following cyclical pressure-induced strain: evidence of a role for alpha 5 beta 1 integrin as a chondrocyte mechanoreceptor. J Orthop Res 1997;15(5):742–7.
[123] Millward-Sadler SJ, Wright MO, Lee H, Caldwell H, Nuki G, Salter DM. Altered electrophysiological responses to mechanical stimulation and abnormal signalling through alpha5beta1 integrin in chondrocytes from osteoarthritic cartilage. Osteoarthritis Cartilage 2000;8(4):272–8.
[124] Millward-Sadler SJ, Wright MO, Lee H, Nishida K, Caldwell H, Nuki G, et al. Integrin-regulated secretion of interleukin 4: a novel pathway of mechanotransduction in human articular chondrocytes. J Cell Biol 1999;145(1):183–9.
[125] Lee HS, Millward-Sadler SJ, Wright MO, Nuki G, Salter DM. Integrin and mechanosensitive ion channel-dependent tyrosine phosphorylation of focal adhesion proteins and beta-catenin in human articular chondrocytes after mechanical stimulation. J Bone Miner Res 2000;15(8):1501–9.
[126] Lee HS, Millward-Sadler SJ, Wright MO, Nuki G, Al-Jamal R, Salter DM. Activation of integrin-RACK1/PKCalpha signalling in human articular chondrocyte mechanotransduction. Osteoarthritis Cartilage 2002;10(11):890–7.
[127] Lucchinetti E, Bhargava MM, Torzilli PA. The effect of mechanical load on integrin subunits alpha5 and beta1 in chondrocytes from mature and immature cartilage explants. Cell Tissue Res 2004;315(3):385–91.
[128] Jang KW, Buckwalter JA, Martin JA. Inhibition of cell-matrix adhesions prevents cartilage chondrocyte death following impact injury. J Orthop Res 2014;32(3):448–54.
[129] Benjamin M, Archer CW, Ralphs JR. Cytoskeleton of cartilage cells. Microsc Res Tech 1994;28(5):372–7.
[130] Guilak F. Compression-induced changes in the shape and volume of the chondrocyte nucleus. J Biomech 1995;28(12):1529–41.
[131] Trickey WR, Vail TP, Guilak F. The role of the cytoskeleton in the viscoelastic properties of human articular chondrocytes. J Orthop Res 2004;22(1):131–9.
[132] Blain EJ. Involvement of the cytoskeletal elements in articular cartilage homeostasis and pathology. Int J Exp Pathol 2009;90(1):1–15.
[133] Blain EJ. Involvement of the cytoskeletal elements in articular cartilage mechanotransduction. In: Kamkin A, Kiseleva I, editors. Mechanosensitivity and mechanotransduction. vol. 4. Dordrecht, Heidelberg, London, New York: Springer; 2011. p. 77–108.
[134] Parkkinen JJ, Lammi MJ, Inkinen R, Jortikka M, Tammi M, Virtanen I, et al. Influence of short-term hydrostatic pressure on organization of stress fibers in cultured chondrocytes. J Orthop Res 1995;13(4):495–502.
[135] Knight MM, Toyoda T, Lee DA, Bader DL. Mechanical compression and hydrostatic pressure induce reversible changes in actin cytoskeletal organisation in chondrocytes in agarose. J Biomech 2006;39(8):1547–51.

[136] Erickson GR, Northrup DL, Guilak F. Hypo-osmotic stress induces calcium-dependent actin reorganization in articular chondrocytes. Osteoarthritis Cartilage 2003;11(3): 187–97.

[137] Chao PH, West AC, Hung CT. Chondrocyte intracellular calcium, cytoskeletal organization, and gene expression responses to dynamic osmotic loading. Am J Physiol Cell Physiol 2006;291(4):C718–25.

[138] Blain EJ, Mason DJ, Duance VC. The effect of thymosin beta4 on articular cartilage chondrocyte matrix metalloproteinase expression. Biochem Soc Trans 2002;30(Pt 6): 879–82.

[139] Blain EJ, Mason DJ, Duance VC. The effect of cyclical compressive loading on gene expression in articular cartilage. Biorheology 2003;40(1–3):111–7.

[140] Campbell JJ, Blain EJ, Chowdhury TT, Knight MM. Loading alters actin dynamics and up-regulates cofilin gene expression in chondrocytes. Biochem Biophys Res Commun 2007;361(2):329–34.

[141] Knight MM, Bomzon Z, Kimmel E, Sharma AM, Lee DA, Bader DL. Chondrocyte deformation induces mitochondrial distortion and heterogeneous intracellular strain fields. Biomech Model Mechanobiol 2006;5(2–3):180–91.

[141a] Jortikka MO, Parkkinen JJ, Inkinen RI, Kärner J, Järveläinen HT, Nelimarkka LO, Tammi MI, Lammi MJ. The role of microtubules in the regulation of proteoglycan synthesis in chondrocytes under hydrostatic pressure. Arch Biochem Biophys 2000;374:172–80.

[142] Chen C, Yin L, Song X, Yang H, Ren X, Gong X, et al. Effects of vimentin disruption on the mechanoresponses of articular chondrocyte. Biochem Biophys Res Commun 2016;469(1):132–7.

[143] Eggli PS, Hunziker EB, Schenk RK. Quantitation of structural features characterizing weight- and less-weight-bearing regions in articular cartilage: a stereological analysis of medial femoral condyles in young adult rabbits. Anat Rec 1988;222(3):217–27.

[144] Durrant LA, Archer CW, Benjamin M, Ralphs JR. Organisation of the chondrocyte cytoskeleton and its response to changing mechanical conditions in organ culture. J Anat 1999;194(Pt 3):343–53.

[145] Lahiji K, Polotsky A, Hungerford DS, Frondoza CG. Cyclic strain stimulates proliferative capacity, alpha2 and alpha5 integrin, gene marker expression by human articular chondrocytes propagated on flexible silicone membranes. In Vitro Cell Dev Biol Anim 2004;40(5–6):138–42.

[146] Henson FM, Vincent TA. Alterations in the vimentin cytoskeleton in response to single impact load in an in vitro model of cartilage damage in the rat. BMC Musculoskelet Disord 2008;9:94.

[147] Herrmann H, Aebi U. Intermediate filaments and their associates: multi-talented structural elements specifying cytoarchitecture and cytodynamics. Curr Opin Cell Biol 2000;12(1):79–90.

[148] Perlson E, Michaelevski I, Kowalsman N, Ben-Yaakov K, Shaked M, Seger R, et al. Vimentin binding to phosphorylated Erk sterically hinders enzymatic dephosphorylation of the kinase. J Mol Biol 2006;364(5):938–44.

[149] Kim S, Coulombe PA. Intermediate filament scaffolds fulfill mechanical, organizational, and signaling functions in the cytoplasm. Genes Dev 2007;21(13):1581–97.

[150] Fitzgerald JB, Jin M, Chai DH, Siparsky P, Fanning P, Grodzinsky AJ. Shear- and compression-induced chondrocyte transcription requires MAPK activation in cartilage explants. J Biol Chem 2008;283(11):6735–43.

[151] Fermor B, Weinberg JB, Pisetsky DS, Misukonis MA, Fink C, Guilak F. Induction of cyclooxygenase-2 by mechanical stress through a nitric oxide-regulated pathway. Osteoarthritis Cartilage 2002;10(10):792–8.
[152] Mariani E, Pulsatelli L, Facchini A. Signaling pathways in cartilage repair. Int J Mol Sci 2014;15(5):8667–98.
[153] Zhou Y, Millward-Sadler SJ, Lin H, Robinson H, Goldring M, Salter DM, et al. Evidence for JNK-dependent up-regulation of proteoglycan synthesis and for activation of JNK1 following cyclical mechanical stimulation in a human chondrocyte culture model. Osteoarthritis Cartilage 2007;15(8):884–93.
[154] Yuasa T, Otani T, Koike T, Iwamoto M, Enomoto-Iwamoto M. Wnt/beta-catenin signaling stimulates matrix catabolic genes and activity in articular chondrocytes: its possible role in joint degeneration. Lab Invest 2008;88(3):264–74.
[155] Tamamura Y, Otani T, Kanatani N, Koyama E, Kitagaki J, Komori T, et al. Developmental regulation of Wnt/beta-catenin signals is required for growth plate assembly, cartilage integrity, and endochondral ossification. J Biol Chem 2005;280(19): 19185–95.
[156] Dell'Accio F, De Bari C, El Tawil NM, Barone F, Mitsiadis TA, O'Dowd J, et al. Activation of WNT and BMP signaling in adult human articular cartilage following mechanical injury. Arthritis Res Ther 2006;8(5):R139.
[157] Dell'accio F, De Bari C, Eltawil NM, Vanhummelen P, Pitzalis C. Identification of the molecular response of articular cartilage to injury, by microarray screening: Wnt-16 expression and signaling after injury and in osteoarthritis. Arthritis Rheum 2008;58 (5):1410–21.
[158] Thomas RS, Clarke AR, Duance VC, Blain EJ. Effects of Wnt3A and mechanical load on cartilage chondrocyte homeostasis. Arthritis Res Ther 2011;13(6):R203.
[159] Nalesso G, Sherwood J, Bertrand J, Pap T, Ramachandran M, De Bari C, et al. WNT-3A modulates articular chondrocyte phenotype by activating both canonical and noncanonical pathways. J Cell Biol 2011;193(3):551–64.
[160] Miyaki S, Asahara H. Macro view of microRNA function in osteoarthritis. Nat Rev Rheumatol 2012;8(9):543–52.
[161] Dunn W, DuRaine G, Reddi AH. Profiling microRNA expression in bovine articular cartilage and implications for mechanotransduction. Arthritis Rheum 2009;60(8): 2333–9.
[162] Guan YJ, Yang X, Wei L, Chen Q. MiR-365: a mechanosensitive microRNA stimulates chondrocyte differentiation through targeting histone deacetylase 4. FASEB J 2011; 25(12):4457–66.
[163] Jin L, Zhao J, Jing W, Yan S, Wang X, Xiao C, et al. Role of miR-146a in human chondrocyte apoptosis in response to mechanical pressure injury in vitro. Int J Mol Med 2014;34(2):451–63.
[164] Yang X, Guan Y, Tian S, Wang Y, Sun K, Chen Q. Mechanical and IL-1beta responsive miR-365 contributes to osteoarthritis development by targeting histone deacetylase 4. Int J Mol Sci 2016;17(4):436.
[165] Cheleschi S, De Palma A, Pecorelli A, Pascarelli NA, Valacchi G, Belmonte G, et al. Hydrostatic pressure regulates MicroRNA expression levels in osteoarthritic chondrocyte cultures via the Wnt/beta-catenin pathway. Int J Mol Sci 2017;18(1).
[166] Kim D, Song J, Jin EJ. MicroRNA-221 regulates chondrogenic differentiation through promoting proteosomal degradation of slug by targeting Mdm2. J Biol Chem 2010; 285(35):26900–7.

[167] Goldring MB, Marcu KB. Epigenomic and microRNA-mediated regulation in cartilage development, homeostasis, and osteoarthritis. Trends Mol Med 2012;18(2):109–18.
[168] Friedrich M, Pracht K, Mashreghi MF, Jack HM, Radbruch A, Seliger B. The role of the miR-148/-152 family in physiology and disease. Eur J Immunol 2017;47(12):2026–38.
[169] Vega RB, Matsuda K, Oh J, Barbosa AC, Yang X, Meadows E, et al. Histone deacetylase 4 controls chondrocyte hypertrophy during skeletogenesis. Cell 2004;119(4):555–66.
[170] Helm M, Motorin Y. Detecting RNA modifications in the epitranscriptome: predict and validate. Nat Rev Genet 2017;18(5):275–91.

# CHAPTER 5

# Advances in tendon mechanobiology

James H.-C. Wang, Bhavani P. Thampatty

*MechanoBiology Laboratory, Department of Orthopaedic Surgery, University of Pittsburgh School of Medicine, Pittsburgh, PA, United States*

## ABBREVIATIONS

| | |
|---|---|
| **AGEs** | advanced glycation end products |
| **BMP-2** | bone morphogenetic protein 2 |
| **CFTR** | cystic fibrosis transmembrane conductance regulator |
| **CITED2** | Cbp/p300-interacting transactivator 2 |
| **CREB** | cAMP response element binding |
| **ECM** | extracellular matrix |
| **Egr** | early growth response |
| **FM** | fascicular matrix |
| **GDF** | growth differentiation factor |
| **Gtf2ird1** | general transcription factor II-I repeat domain-containing protein 1 |
| **HARP** | heparin affin regulatory peptide |
| **IF** | intermediate filament |
| **IFM** | intrafascicular matrix |
| **ITR** | intensive treadmill running |
| **MAPK** | mitogen-activated protein kinase |
| **MGF** | mechanogrowth factor |
| **Mkx** | Mohawk |
| **MMP** | matrix metalloproteinases |
| **MSC** | mesenchymal stem cell |
| **MTR** | moderate treadmill running |
| **NS** | nucleostemin |
| **Oct-4** | octamer-binding transcription factor 4 |
| **PGE$_2$** | prostaglandin E$_2$ |
| **PI3K** | phosphoinositide 3-kinase |
| **PPARγ** | peroxisome proliferator-activated receptor γ |
| **Runx-2** | runt-related transcription factor 2 |
| **SA β-gal** | senescence-associated β-galactosidase |
| **SASP** | senescence-associated secretory phenotype |
| **Sox-6** | sex determining region Y (Sry)-box6 |

| | |
|---|---|
| **Sox-9** | sex determining region Y (Sry)-box9 |
| **SSEA** | stage-specific embryonic antigen |
| **TDSC** | tendon-derived stem cell |
| **TGF-β1** | transforming growth factor-β1 |
| **TSCs** | tendon stem/progenitor cells |

# 1 INTRODUCTION

Tendons are dense bands of connective tissues that link muscle to bone and as such transmit muscular forces to enable joint movements. Tendon has a hierarchical structure starting with collagen molecules, the smallest building blocks of tendon, which bind together to form fibrils [1]. The fibrils aggregate to form fibers, which aggregate again to form the largest tendon subunit, the fascicle. A connective tissue compartment called endotenon or intrafascicular matrix (IFM) surrounds each fascicle. Yet another connective tissue called epitenon covers the tendon, which is often surrounded by another layer of connective tissue, the paratenon (Fig. 1) [2].

Tendon mainly consists of collagen, proteoglycans, water, and cells. Type I collagen, which provides mechanical stability and elastic energy storage, is the

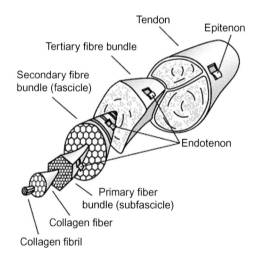

**FIG. 1**

The structure of tendon. The diagram illustrates the highly hierarchical structural units of tendon. Collagen molecules aggregate to form fibrils that assemble into fibers. The bundles of fibers form subfascicles and fascicles. A thin layer of connective tissue, endotenon, surrounds the fiber bundles. Another thin layer, epitenon, binds the whole tendon, and an outer layer, paratenon (not shown), surrounds most tendons.

*Adapted with permission from Fig. 1 in Riley G. Chronic tendon pathology: molecular basis and therapeutic implications. Expert Rev Mol Med 2005;7(5):1–25.*

most abundant extracellular matrix (ECM) component of tendon [3]. Tendon matrix also contains small amounts of type III and V collagens. The major cell types of tendons are a mixed population of fibroblast-like cells called tenocytes that reside between the collagen fibers within the fascicles, and tendon stem/progenitor cells (TSCs) sparsely distributed in ECM. Tenocytes are the predominant cell type (90%–95%), while TSCs are adult stem cells that constitute <5%, with typical characteristics of adult stem cells such as clonogenicity, self-renewal, and multidifferentiation potential [4–6]. Under normal physiological conditions, TSCs differentiate into tenocytes that are primarily responsible for the tendon maintenance, remodeling, and repair.

Tendons, such as patellar and Achilles tendons, are load-bearing connective tissues, and their development, homeostasis, and repair greatly depend on mechanical loading. Although tendons as live tissues have the unique ability to withstand high tensile force, they are susceptible to injuries, which are common clinical problems costing billions of health-care dollars every year in America [7,8]. Regular exercise promotes tendon anabolic response in young and old alike and enhances structural integrity and mechanical strength of tendons; however, excessive mechanical loading is considered as the major factor for the development of chronic tendon injuries or tendinopathy. Aging also brings degenerative changes in tendons that eventually cause tendinopathy, and in such cases, prolonged rehabilitation is required for tendon repair. Disorganized collagen fibers, increase in cellularity and proteoglycans, infiltration of blood vessels, and tissue calcification are some typical features of tendinopathic tendons [2,9]. Such injured tendons have slow and incomplete healing, which makes the tendons vulnerable to further injuries. Currently, there are no adequate effective treatment options for tendinopathy, and the disease poses a major health-care problem. Limited understanding of tendon mechanobiology and the lack of knowledge regarding pathogenesis of tendinopathy are hindrances to the development of novel, effective treatment methods. Therefore, tendon mechanobiological studies are essential for a better understanding of the etiology, pathology, and proper management of tendinopathy in young and old alike. The recent discovery of TSCs provides a promising opportunity for advancing our understanding of tendon mechanobiology, stem cell-based mechanisms of tendinopathy, and cell therapy for effective repair of injured tendons.

## 2 DISCOVERY OF TSCs

Until 2007, tendon cell population used to be considered primarily composed of tenocytes. A unique feature of tendon is its plasticity, so the possibility of the presence of stem cells in tendon has arisen. Stem cells are broadly classified into two categories based on their origin from either embryo or the adult. The usually quiescent adult stem cells retain their properties throughout life and enter cell cycle when stimulated [10]. Several lines of evidence for the presence of adult stem cells in tendons were suggested based on the previous observations that human and mouse tendons develop fibrocartilage and ossification in response to injury, and tendon-derived immortalized cell lines and human tendon-derived fibroblasts possess multidifferentiation capabilities in vitro [11–13]. Before long, a unique cell population that displays the characteristics of adult stem cells had been isolated and identified from human, mouse, rat, and rabbit tendons [4–6]. Different names are used in literature to describe these tendon-specific adult stem cells, namely, tendon stem/progenitor cells (TSCs/TSPCs) and tendon-derived stem cells (TDSCs). In vivo, TSCs are contained in FM and IFM; however, it seems that TSCs in IFM are rounder and metabolically more active compared with TSCs in FM [14].

While TSCs belong to the category of mesenchymal stem cells (MSCs), they are distinguished from MSCs in gene expression and response to diverse growth factors such as TGF-β and GDF-5 [4,5,15]. For example, TSCs express tissue-specific characteristics such as tendon-related genes including scleraxis and tenomodulin and hence functionally differ from MSCs. However, there are no specific markers that could uniquely identify TSCs in vitro and in vivo, and markers expressed in vivo could be altered by in vitro culture.

To isolate TSCs from tendon tissues, tendon samples are minced and digested with collagenase and dispase [6]. The isolated cells are then plated individually to allow cells to form individual colonies. The size and density of the colonies largely vary suggesting differences in proliferation among colonies [16]. Therefore, TSCs in vitro are most likely a heterogeneous population of stem cells and progenitor cells. A successful isolation and culture of TSCs depend on several factors such as optimal seeding cell density, appropriate culture conditions, and age of animals and individuals. The absence of uniformity in the appearance of TSCs is also common, because the cell shape may depend on the species, tissue origins, cell passages, and culture conditions in general. In addition, long-term culturing can result in the loss of multilineage differentiation potential of TSCs and even cellular senescence [17]. It should be noted that many studies, particularly those before 2007, did not use such isolation procedures and proper culture conditions, and consequently, the references to tendon cells or tendon fibroblasts may be in fact a mixture of tenocytes and TSCs or different progenitor populations.

## 3 THE PROPERTIES OF TSCs VS. TENOCYTES

TSCs differ from tenocytes on many aspects such as shape, proliferation, clonogenicity, and differentiation potential [16,18]. When reaching confluent conditions, TSCs are cobblestone-like with smaller cell bodies and larger nuclei compared with tenocytes, which are elongated (Fig. 2).

TSCs proliferate more quickly in culture than tenocytes, and unlike tenocytes, they form round colonies. In addition, TSCs in culture express stem cell markers, such as Oct-4, SSEA-4, and nucleostemin (NS) (Fig. 2), whereas tenocytes, which are considered to be fully differentiated fibroblast-like cells, express none of these markers [19]. Also, stem cells derived from the human patellar tendon express

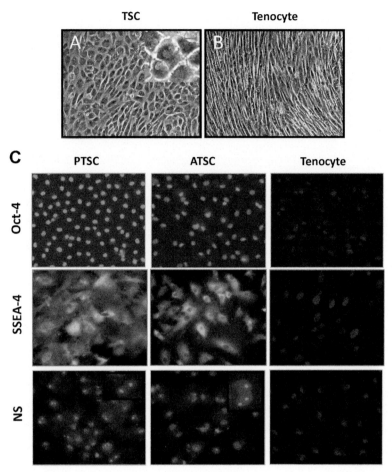

**FIG. 2**

The properties of tendon stem/progenitor cells (TSCs) and tenocytes in culture. (A) TSCs in culture. The cells are cobblestone-like in culture when grown to confluence. They also have large nuclei (inset). (B) Tenocytes in culture. These cells are fibroblast-like and assume a highly elongated shape in confluent conditions. (C) The expression of stem cell markers Oct-4, SSEA-1, and NS in patellar and Achilles tendon-derived TSCs. Both patellar tendon stem/progenitor cells (PTSCs) and Achilles tendon stem/progenitor cells (ATSCs) abundantly express the three stem cells markers, whereas tenocytes express none of these markers.

CD73, CD44, CD90, and CD105, but not CD34 and CD45, as shown by flow cytometry [20]. Moreover, TSCs have the capability to differentiate not only into tenocytes but also into adipocytes, chondrocytes, and osteocytes in vitro and form tendon-like, cartilage-like, and bone-like tissues in vivo [18,21]. The multidifferentiation capability of TSCs raises the possibility that TSCs may play a vital role in the development of degenerative tendinopathy, because at late stages, lipid deposition, proteoglycan accumulation, and calcification, either alone or in combination, are found inside the tendinopathic tendons [22].

## 4 THE FACTORS THAT INFLUENCE THE FATE OF TSCs

The source of TSCs could be both vascular and nonvascular. There is evidence that vasculature of tendon tissues and perivascular cells of tendon capillaries might harbor TSCs [23]. Also, the stem/progenitor cells in the peritenon might be more vascular in origin [24]. Different progenitor populations exist within discrete niches of the tendon proper versus peritenon in mature Achilles tendon [25]. Although both stem/progenitor cell populations from epitenon and paratenon are multipotent, only cells isolated from the tendon proper are able to produce a calcified matrix [25]. Moreover, stem cells isolated from tendon proper express greater levels of scleraxis and tenomodulin compared with those from peritenon. TSCs isolated from distinct regional location within the tendon tissue may exhibit differences in colony-forming potential and stem cell marker expression [26].

The specialized microenvironment or niche of TSCs maintains their "stemness" properties such as self-renewal and cell fate [23]. Several factors that define stem cell niche include mechanical stresses on TSCs, oxygen tension, growth factors, cytokines, and ECM proteins [10,27]. Among these factors, biglycan and fibromodulin are important ECM niche components of TSCs [4]. Biglycan and fibromodulin double knockout (KO) mouse has impaired patellar tendon formation, and TSCs isolated from this mouse KO model formed bone-like in addition to tendon-like tissues. The interplay between TSCs and their niche creates a "bidirectional system" where cells influence their surrounding microenvironment, which in turn influences the cells' property [27]. Under normal physiological conditions, TSCs differentiate into tenocytes by default, which is often induced and regulated by transcription factors including Mkx, Egr1, and Egr2, and growth factors such as TGF-$\beta$2 and GDF-5. Nestin is a type IV intermediate filament (IF) protein that is highly expressed in TSCs from Achilles tendon, and nestin$^+$ TSCs exhibit strong capacity for self-renewal and tenogenesis [28]. Nestin knockdown results in the loss of self-renewal, phenotypic maintenance, tenogenic capacity, and tendon repair and regeneration capacity of TSCs. BMP-2 is considered as a key molecule responsible for the aberrant differentiation TSCs into other lineages such as adipocytes, chondrocytes, osteocytes [21].

The typical features of mature TSCs in tendons, such as self-renewal and clonogenicity, however, change during aging of animals, which presumably causes cell division to cease completely or better-termed cell senescence [18]. TSCs from aging human tendons exhibit profound deficits in self-renewal and clonogenicity compared with those from young; these aging cells are also characterized by premature entry into cellular senescence [29]. Because of aging-induced cellular senescence, aging TSCs have reduced migratory ability, matrix production and integrin expression, and dysregulated cell-matrix interactions and actin dynamics [29]. Senescent TSCs however may produce abundant MMPs, causing aging-related tendinopathy [30,31].

## 5 THE STUDY OF TSC MECHANOBIOLOGY

Mechanical loads determine the physiological and pathological state of tendon cells. As living tissues, tendons respond to mechanical loading by changing their metabolism in turn bringing changes in their structural and mechanical properties [32]. The investigation into the role of mechanobiological response of TSCs in tendon homeostasis, repair, and healing is an emerging field ever since the identification of TSCs. In mechanobiological studies of tendon, cell stretching is the most popular way to apply mechanical stimulation to TSCs in vitro, and treadmill running of rodents is a common method to induce mechanical loading on tendons in vivo.

### 5.1 MECHANOBIOLOGICAL RESPONSES OF TSCs IN VITRO

In vitro, moderate level stretching is usually applied to TSCs at a low magnitude at physiological level for short durations, whereas excessive stretching is imposed to TSCs at higher magnitudes and longer durations. Several custom made cell stretching devices are in use for in vitro mechanical stretching studies that mimic in vivo mechanical loading on tendons. Cells are seeded onto silicone chambers for uniaxial stretching and harvested for molecular analysis. Such systems are used to identify stretch-activated genes and proteins as cellular responses. For example, Flexcell system is often used for cell stretching studies [33]. In this system, the cells are uniaxially loaded by placing loading rectangle posts beneath each well of the UniFlex culture plates in a gasketed base plate and applying vacuum to deform the flexible membranes downward. The flexible membrane deforms downward along the long sides of the loading posts thus applying uniaxial strain to cells attached to the membrane. Other popular commercial cell stretching system includes STB-140 system (STREX) [34,35]. Stretch chambers are flexible polydimethylsiloxane (PDMS) culture dishes developed for applying mechanical loading to cultured cells. An optically transparent, ultrathin (100–200 μm) membrane at the well bottom makes stretch chambers compatible with optical microscopy techniques but with fluorescence detection and microscopy as well. Stretch direction is uniaxial, which provides highly reproducible cyclic stretching over prolonged periods at stretching magnitudes of 1%–20%. Another cell stretching system controls cells in a manner that mimics cell shape and organization, as well as loading conditions in vivo. Moreover, in vivo-like, well-controlled multiple mechanical loading conditions can also be applied to cells plated in this system to examine TSCs' mechanobiological responses (Fig. 3).

Moderate mechanical stretching of tendon fibroblasts induces many anabolic effects such as increased cell proliferation, elevated collagen type I expression, and TGF-β release, as shown in previous studies [36,37]. Similar studies have been

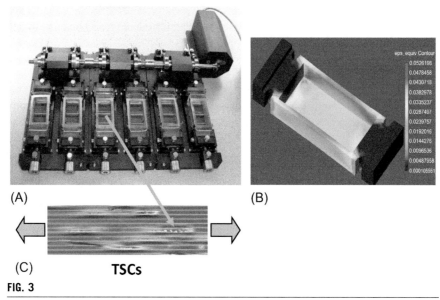

**FIG. 3**

A novel in vitro model system to study mechanobiology of TSCs. (A) The stretching apparatus. Elastic silicone dishes are used to support cell attachment and stretch tendon cells. (B) Strain distribution in the silicone dish determined by finite element method. It shows that the surface strains are uniform in the central region of the culture surface where cells are attached and stretched. (C) TSCs on microgrooved culture surfaces (an arrow points to tendon cell residing on microgrooves with a width of 10 μm). Tendon cell density, cell shape, and organization can be controlled in this model system to mimic in vivo conditions in the tendon.

extended to isolated TSCs recently. TSCs differentially respond to mechanical loading in vitro depending on the stretching magnitude and duration. The responses of TSCs strongly favor tenogenic differentiation under moderate stretching. Moderate stretching in rabbit patellar and Achilles tendon-derived TSCs suggests that it is beneficial in terms of promoting differentiation into tenocytes that are essential for tendon homeostasis via increased proliferation and enhanced expression of tenocyte-related genes, collagen type I, and tenomodulin [38,39]. Tenomodulin predominantly expressed in tendon is a regulator of tenocyte proliferation that is involved in collagen fibril maturation and organization and possibly in tendon vascularity [40]. Previous studies in embryonic stem cells and human bone marrow mesenchymal cells demonstrated mechanical stimulation that regulates stem cell proliferation and differentiation [41,42]. Moderate stretching of human TSCs can also upregulate matrix proteins (fibromodulin, lumican, and versican), collagen type I-binding integrins ($\alpha 1$, $\alpha 2$, and $\alpha 11$), and MMPs (MMP9, MMP13, and MMP14) that play an important role in TSC niche composition, cell survival, mechanosignaling, and matrix remodeling. These anabolic responses are recognized to be mediated through ERK1/2 and p38 signaling pathways that are also activated in parallel by mechanical stretch [43]. Upregulation of integrin expression is important to ensure a sufficient number of receptors for the conversion of mechanical signals to cytoplasmic signals through mechanotransduction, which in turn affects a number of cellular processes including proliferation, migration, differentiation, gene expression, and cell survival [44].

Apparently, the transcription factor Mohawk (Mkx), which regulates tendon-related gene expression, is required for mechanical loading-induced TSC differentiation into tenocytes [45]. This conclusion is based on the observations that moderate mechanical stretching enhances the expression of tendon-related genes Mkx, Col1a1, and Col3a1 only in Achilles TSCs of Mkx overexpressing ($Mkx^{+/+}$) mice, while the same mechanical stimulation enhances chondrogenic markers, Sox-6, Sox-9, and Acan in TSCs from $Mkx^{-/-}$ mice [45]. Moreover, moderate mechanical stretching induces Mkx and the downstream tendon-associated genes, Tnmd, Col1a1, and Col1a2 in rat tendon cells [46]. Evidently, an upstream transcription factor, general transcription factor II-I repeat domain-containing protein 1 (Gtf2ird1), is essential for Mkx transcription because Gtf2ird1 expressed in the cytoplasm of unstressed tendon cells translocates to nucleus upon mechanical stretching to activate Mkx promoter through chromatin regulation [46].

Cystic fibrosis transmembrane conductance regulator (CFTR), a stretch-mediated activation chloride ion channel, also regulates tenogenic differentiation of TSCs, since CFTR is also upregulated along with Mkx, Col1a1, and decorin during tenogenic differentiation under moderate mechanical stretching [47]. Tendon tissues in CFTR dysfunctional mice (DF508) exhibit irregular cell arrangement, uneven altered fibril diameter distribution, weak mechanical properties, and less matrix formation in a tendon defect model. Moreover, both tendon tissues and TSCs isolated from DF508 mice show significantly decreased levels of tendon markers, such as scleraxis, tenomodulin, Col1a1, and decorin. The role of CFTR in tenogenic differentiation and tendon regeneration is believed to be via inhibition of the β-catenin/pERK1/2 signaling pathway [47].

While mechanical stretching at physiological levels induces anabolic effects in young tendon cells, excessive stretching causes major catabolic effects. The detrimental effects of excessive mechanical stretching in primary human tendon cells, rat tail tendon cells, and patellar tendon fibroblasts are well documented by numerous studies [48–53]. These effects include actin depolymerization, influx of extracellular $Ca^{2+}$, cytoskeleton disruption, increase in collagenases and inflammatory mediators such as COX-2 and $PGE_2$, induction of autophagy, and cell death.

In patellar and Achilles TSCs, while moderate mechanical stretching favors tenogenic differentiation, excessive stretching drives TSCs toward adipogenic, chondrogenic, and osteogenic differentiation (Fig. 4).

Specifically, small stretching (4%) significantly enhanced only the expression of tenocyte-related gene, collagen type I, while none of the nontenocyte-related genes, PPARγ, collagen type II, Sox-9, and Runx-2 that are markers of adipocytes, chondrocytes, and osteocytes, respectively, were affected. However, the expression of both tenocyte-related and nontenocyte-related genes was significantly increased by large

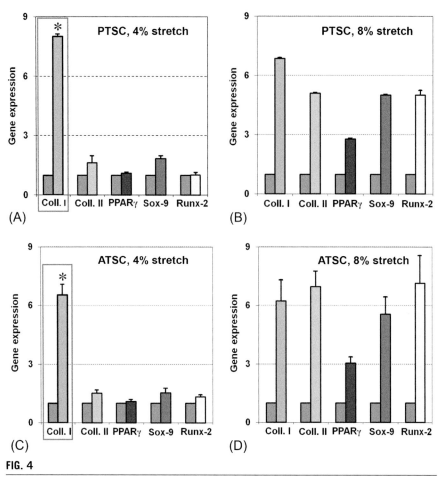

**FIG. 4**

Gene expression profiles in TSCs derived from mouse patellar and Achilles tendons depend on mechanical loading conditions in vitro. Patellar tendon stem cells (PTSCs) and Achilles tendon stem cells (ATSCs) stretched to 4% show specific upregulation of collagen type I (Coll. I), which is a tenocyte-related gene, but not the nontenocyte-related genes, collagen type II (Coll. II), Sox-9, PPARγ, and Runx-2 (A, C). However, 8% stretching upregulates both tenocyte and nontenocyte genes to varying degrees in both PTSCs and ATSCs (B, D). Gray bars are controls without loading. *$P<.05$ with respect to control.

mechanical stretching at 8%. Upregulated expression of BMP-2, adipocyte-related gene (PPARγ), chondrocyte-related genes (collagen type II and Sox-9), osteocyte-related gene (Runx-2), and Wnt proteins are evident by excessive mechanical stretching from numerous studies [38,39,54–56]. In brief, along with contribution of upregulated tendon-related genes that are regulated by transcription factors such as Mkx and CFTR, moderate stretching favors the differentiation of TSCs into tenocytes.

In addition, moderate stretching upregulates the gene expression of many molecules that are essential for mechanosignaling and matrix remodeling.

## 5.2 MECHANOBIOLOGICAL RESPONSES OF TSCs IN VIVO

To study tendon mechanobiology in vivo, treadmill running of rodents at short or long durations is commonly used to induce moderate, excessive, and one bout mechanical loading on tendons, respectively (Fig. 5).

Moderate loading is achieved by runs that last for a short duration of less than an hour/day and do not exceed 4–6 weeks. However, to induce excessive mechanical loading on the tendon, intensive running regimens with longer duration, at least 1 h/day, over 10–12 weeks, are widely applied. One bout of rigorous treadmill running often uses a single run until the rodents are exhausted [57].

Increase in glucose uptake, oxygen demand, and blood flow in tendon during physical exertion suggests that tendon is a metabolically active tissue [58]. Physiological loading of tendons leads to increases in tendon mass, cross-sectional area, cell density, and collagen type I content [59]. Exercise induces anabolic effects in tendons with increased rates of collagen synthesis and levels of growth factors such

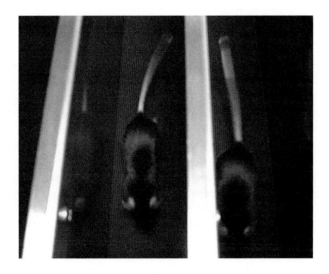

**FIG. 5**

Mouse treadmill running used to study tendon mechanobiology in vivo. Typically, mice run on the treadmill with six running lanes according to a specified running regimen, such as moderate treadmill running (MTR) or intensive treadmill running (ITR). MTR regimen is defined as follows: a training period of 1 week (13 m/min, 15 min/day for 5 days/week) followed by 3 weeks of running (same speed, 50 min/day, 5 days/week). On the other hand, ITR regimen can also be applied to mice. In this regimen, the training program is the same as that in MTR. After training, mice then run at the same speed as in MTR (13 m/min) for 3, 4, and 5 h/day for 5 days in the second, third, and fourth weeks [39].

as IGF-1 and TGF-β [59,60]. Moderate mechanical loading of tendon in animal models of treadmill running induces many anabolic effects potentially via a TSC-based mechanism. For instance, moderate treadmill running (MTR) exerts anabolic effects in young rodents by increasing TSC proliferation, TSC-related collagen production, and elevated expression of mechanogrowth factor (MGF) that promote tissue growth, collagen type I, and tenomodulin; hence, tendons may be able to enhance mechanical properties via stem-cell-based mechanism [39,61]. In response to the mechanical load placed on tendons, more tendon cells, in particular TSCs, are generated for the repair and/or remodeling of tendons. Higher numbers of quickly growing TSCs are present in mechanically loaded tendons compared with control tendons [39]. Under moderate loading conditions, the fast growing TSC population provides progenitors of tenocytes that enhance the remodeling of tendons. Such biological responses due to appropriate mechanical loading may explain the beneficial anabolic changes in tendon structure due to exercise.

Additionally, MTR exerts anabolic effects by increasing scleraxis, tenomodulin, and collagen type I gene expression in adult mice that express GFP under the control of the scleraxis promoter (ScxGFP) when subjected to a 6-week treadmill training program designed to induce adaptive growth in Achilles tendons [62]. Following treadmill training, the cell density increases with the presence of scleraxis-expressing tendon cells that appear to be migrating into tendon fascicles from the epitenon. The results suggest that a population of tendon progenitor cells exists in the epitenon. Moreover, this study indicates that scleraxis expression is important in the early stages of activating these cells in response to physiological loading. Moderate mechanical loading restores the tendon properties with a decreased number of chondroid cells and GAG in TGF-β-induced murine tendinopathy model [63]. Various tendon-associated genes, such as Scx and Egr1/Egr2 (Egr1/2), tenascin C, and collagen type I and III are responsive to mechanical stimulation in bioartificial tendons and hBMSCs [64,65]. Mesenchymal stem cell-to-tenocyte differentiation is strongly associated with cumulative tensile loads on the cells [65]. As observed in mechanical stretching studies in vitro [45,46], moderate mechanical loading by treadmill running also induces Mkx and tendon-associated genes (Tnmd, Col1a1, and Col1a2) in wild-type mice, whereas in $Mkx^{-/-}$ mice, tendons fail to respond in a similar manner to the same mechanical stimulation [46].

While moderate mechanical loading brings about many anabolic effects, excessive mechanical loading can induce many detrimental effects even in young tendons. Intensive treadmill running (ITR) changes the gross appearance of mouse patellar and Achilles tendons from a normal white glistening to a more vascularized appearance [39]. Intensive treadmill running upregulates both tenocyte-related genes (collagen type I and tenomodulin) and nontenocyte-related genes (LPL, Sox-9, Runx-2, and Osterix) in the major load-bearing patellar and Achilles tendons (Fig. 6) [39].

In addition, tendon matrix changes such as decreased collagen fiber organization and collagen bundle disintegration, increased expression of collagen type III and GAG, increased cellularity and cell nuclei, and tenocyte shape change to a more

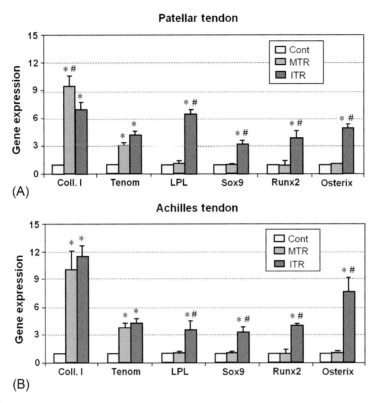

**FIG. 6**

Mechanical loading regulates tenocyte- and nontenocyte-related gene expression in tendons. Moderate treadmill running (MTR) only increases the expression of tenocyte-related genes, but intensive treadmill running (ITR) upregulates both tenocyte-related genes, collagen type I (Coll. I) and tenomodulin (Tenom), and nontenocyte-related genes, LPL (adipocyte marker), Sox-9 (chondrocyte marker), Runx-2, and Osterix (osteocyte markers) in mouse patellar tendons (A) and Achilles tendons (B). *$P<.05$ with respect to cage control; #$P<.05$ with respect to MTR. In (A), however, # in Coll. I represents $P<.05$ with respect to ITR.

round form are some detrimental changes that take effect after excessive mechanical loading in the form of intense treadmill running of animal models [66–69]. Early molecular events such as increased mRNA levels of GAG, Sox-9, decreased collagen type I transcript, elevated protein levels of heparin affin regulatory protein (HARP), enhanced metalloproteinase activity, and significant upregulation of chondrogenic markers Acan and Sox-9 are observed in overused rotator cuff tendons [70–72]. Such pathological changes may occur due to aberrant differentiation of TSCs as a result of mechanical overloading.

Tendons under excessively mechanical loading conditions exhibit higher expression of tenocyte and nontenocyte-related genes. The higher expression of nontenocyte-related genes suggests the presence of TSCs that are responding to mechanical loading in such a way that they undergo aberrant differentiation. In addition, one bout of treadmill running increases the production of potent inflammatory mediator, $PGE_2$, in mouse tendons [57]. In isolated TSCs in culture, high levels of $PGE_2$ decrease cell proliferation and induce adipogenic and osteogenic differentiation in an apparent dose-dependent fashion [57]. In contrast, at low levels, $PGE_2$ maintains the stemness of TSCs [73]. Taken together, moderate mechanical loading of tendons for short durations induces anabolic effects in terms of cell proliferation, collagen production, increased growth factors, and tendon-related gene expression possibly via a TSC-based mechanism. In contrast, excessive loading may cause at least some TSCs to undergo aberrant differentiation into nontenocytes.

## 6 THE ROLE OF TSCs IN THE DEVELOPMENT OF LOADING-INDUCED TENDINOPATHY

The broad spectrum of pathological conditions of tendinopathy includes hypercellularity, hypervascularity, matrix disturbance with an increase in proteoglycan deposition, ECM degradation, rounding of cell nuclei and acquisition of chondrocyte phenotypes, occasional adipose, and bony metaplasia [11,74–76]. Excessive mechanical loading placed on rat tendons by forced treadmill running is known to induce proliferation of so-called round tenocytes, which are most likely chondrocytes differentiated from TSCs, as well as glycosaminoglycan accumulation and collagen fragmentation [77,78]. Specifically, chronic mechanical loading of tendons in animal treadmill running models produces many histopathologic features similar to those observed in human tendon disease. These features include decreased collagen fiber organization, more intense collagen staining, and increased cell nuclei numbers [66]. Abnormal tenocyte morphology, typified by the appearance of tenocytes with prominent cytoplasm and round morphology, is one of the responses of mechanical loading in tendons [67]. Biomechanical changes such as decrease in stiffness and ultimate tensile strength, and histological changes including tenocyte proliferation, change in tenocytes appearance, and collagen bundle disintegration are common in the mechanically overloaded tendons [68,79]. These changes are suggestive of pathological changes similar to those in tendinopathic tendons. The molecular alterations, such as increased gene expressions of GAG proteins (aggrecan, versican, biglycan, and decorin), Sox-9, and decreased collagen type I expression that are characteristics of phenotypic shift of tendon cells toward chondrogenic phenotype, are evident in overuse rotator cuff model of rats [70]. Additionally, enhanced protein levels of HARP, a regulatory cytokine in developmental chondrocyte formation in this overuse model, suggest altered phenotype of tendon cells. Increased metalloproteinase activity such as MMP-2 upregulation associated with an increase of CD147 may lead to enhanced tendon degeneration by mechanical overloading [71]. Altered transcriptional regulation of chondrogenic genes Acan and Sox-9 in overused tendons may explain the biological mechanism of rotator cuff injuries [72].

The current evidence supports the idea that the erroneous differentiation of TSCs under altered mechanical loading may induce pathological changes in the tendon [38,80]. Moreover, both in vitro and in vivo studies of mechanical loading support the notion that TSCs play a major pathogenic role in the development of tendinopathy. This is due to the capacity of TSCs to differentiate into multiple cell lineages, namely, adipocytes, chondrocytes, and osteoblasts [5,39]. These cells produce fats, cartilage-like tissues, and bony tissues that are often seen in tendinopathic tendons at the late stages [22,81].

Mechanobiological responses of TSCs induced by excessive mechanical loading accelerate tendon degeneration and propel the development of typical tendinopathic features in tendons. According to recent studies, aberrant differentiation of TSC into nontenocyte phenotype is the possible mode for the abnormal production of matrix components such as GAG accumulation, calcification, and acquisition of chondrocyte phenotype by excessive mechanical stimulation [38,39]. Loading-induced differentiation is not reversible by rest; in vitro and in vivo studies show that once TSCs begin differentiation, the process becomes irreversible, even after removal of the mechanical load [38,82]. Such an abnormal differentiation might explain the formation of chondrocytes and accumulation of proteoglycans in rat tendons after long-term treadmill running [67,78].

## 7 THE MECHANOBIOLOGICAL STUDIES IN AGING TENDONS
### 7.1 FEATURES OF AGING TENDONS

Aging accelerates tendon degeneration and subsequent injuries. Cellular senescence is an important contributor to age-related diseases in tissues and organs [31]. Currently, senescence is considered to inflict permanent cell cycle arrest in response to various stressors; such cellular senescence contributes to impairment in tissue renewal and function. Senescent cells produce proinflammatory and matrix degrading molecules called senescence-associated secretory phenotypes (SASPs) [83]. Senescence-associated β-galactosidase (SA β-gal) and p16 are two of the key markers of senescent cells.

With advancing age, distinct cellular and molecular changes become evident in TSCs such as premature entry into cellular senescence accompanied by deficit in self-renewal, decrease in number, decreased proliferation, reduced colony formation, decreased tendon lineage gene expression, and tendency of TSCs toward adipocytic differentiation [84–87]. Aging reduces the stemness of TSCs [88]. Concomitant with increasing age, the number of TSCs expressing stem cell markers (Oct-4, SSEA-1, Sca-1, and NS) markedly decreases (Fig. 7).

The expression of senescence markers, such as SA β-gal and p16$^{INK4a}$, markedly increases with aging [89]. Extrinsic and intrinsic factors including inflammatory cytokines, Wnt pathway activators, and reactive oxygen species influence stem cell aging [86]. Increased expression of CD44 in aged TSCs compared with young cells

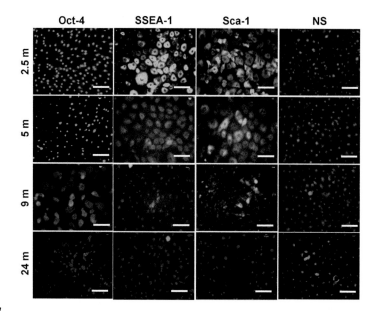

**FIG. 7**

The effects of aging on TSCs. Expression of the stem cell markers, Oct-4, SSEA-1, Sca-1, and nucleostemin (NS) in TSCs from 2.5-, 5-, 9-, and 24-month-old mice are shown. Increase in the mouse age decreases the number of TSCs expressing the stem cell markers, which are abundant in TSCs from 2.5-month-old mice but scant in TSCs from 24-month-old mice. Bar 100 μm.

contributes to reduced repair capacity of TSCs as CD44 knockdown improves the mechanical properties of healing tendon and increases matrix and cytokine expression in a mouse patellar tendon injury model [84].

Senescence is not limited to aging cells; premature senescence may occur in young cells in stress and disease conditions [87,90]. However, moderate exercise maintains structural and functional integrity of aging tendons. MTR shows many beneficial effects in aging animal tendons including increased stem cell number and marker expression, enhanced TSC function, decreased lipid and proteoglycan accumulation, calcification, and AGE adducts [88,91]. Moderate mechanical loading rejuvenates aging tendons as evidenced by both in vitro and in vivo studies [88]. Moderate stretching of aging TSCs significantly increases stem cell and tenocyte-related genes such as Nanog, collagen type I, and tenomodulin without having much effect on the expression of nontenocyte-related genes such as LPL, Sox-9, and Runx-2 (Fig. 8).

However, excessive stretching can supersede the beneficial effects of moderate mechanical stretching in aging tendons; it can in fact upregulate the expression of nontenocyte-related genes in aging TSCs (Fig. 8). in vivo study also highlights

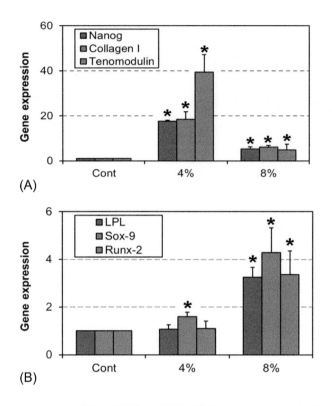

**FIG. 8**

Moderate mechanical loading rejuvenates aging tendons. (A) Stem cell marker gene (Nanog) and tenocyte-related genes (collagen type I and tenomodulin) remain low in control mouse PTSCs, but their expression is significantly enhanced after small stretching (4%); in contrast, excessive stretching at 8% has a smaller effect on the expression of the same genes. (B) Small stretching at 4% slightly upregulates the expression of nontenocyte-related gene Sox-9 (chondrocyte) without altering the expression of nontenocyte-related genes, LPL (adipocyte) and Runx-2 (osteocyte). However, 8% stretching significantly increases the expression of all three nontenocyte-related genes. *$P<.05$ with respect to control.

the positive effects of moderate loading in aging tendons [88]. Extensive lipid accumulation in aging tendons significantly decreases after moderate mechanical loading in the form of treadmill exercise in aging tendons (Fig. 9).

Moderate loading also accelerates wound healing in aging tendon. Lowered vascularization, increased expression of tenocyte-related genes and stem cell markers, and decreased nontenocyte-related gene expression are evident in moderately loaded aging tendon suggesting that preexercise may enhance healing process through TSC-based mechanism [92]. This conclusion is reinforced by the finding that transplanted TSCs could promote earlier and better recovery after tendon injuries [93].

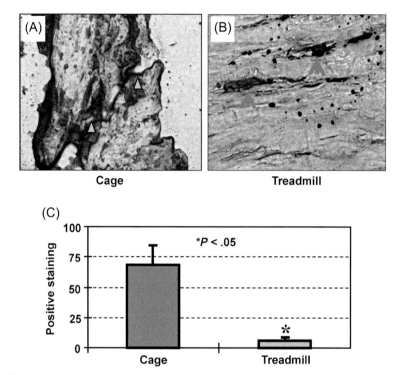

**FIG. 9**

Mechanical loading in the form of moderate treadmill running (MTR) reduces nontendinous tissue formation in aging tendons. (A) Extensive lipid deposition (*green arrows* point to accumulated lipids) is observed in cage control mouse tendon (*Oil Red O staining*). (B) After MTR, much less lipid tissues remain; only a few residual lipids *(red arrows)* are present. (C) Semiquantification results of histochemical staining for lipids in aging tendons.

## 7.2 THE ROLE OF TSCs IN THE DEVELOPMENT OF AGING-INDUCED TENDINOPATHY

Aging is considered as a primary risk factor for tendinopathy in old individuals. The role of TSCs in the development of aging-induced tendinopathy is not well investigated yet, although animal studies suggest plethora of changes in TSCs due to aging. These include decreased proliferation and differentiation potential, delay in cell cycle progression, altered cell fate, increase in adipocytic differentiation, and downregulation of a transcriptional modulator involved in cell growth and senescence [18,94]. Such altered properties of TSCs can lead to tendon pathology. Cellular senescence is implicated in age-related decline of matrix turnover rate in tendons. Senescent cells are not just quiescent cells; they exert harmful effects on the tissue microenvironments by acquisition of secretory phenotypes that increase the secretion of various inflammatory cytokines (e.g., IL-1, IL-6, and IL-8) and MMPs [95]. Increased levels of MMPs can accelerate tendon degeneration.

Loss of tenomodulin that maintains tendon homeostasis can lead to reduced self-renewal and augmented senescence of TSCs [96]. Diverse cellular processes including senescence are mediated via miRNAs, and miRNA-135a modulates TSC senescence via suppressing ROCK1 [97]. Pin1, a highly conserved protein that participates in diverse cellular processes including cell cycle progression, cell survival, and lineage commitment, is also implicated to have an important role in TSC aging since prolonged in vitro culture of human TSCs decreases Pin1 mRNA and protein expression levels [98]. Overexpression of Pin1 delays the progression of cellular senescence as confirmed by the downregulation of SA β-gal and increased telomerase activity, both of which promote cellular senescence.

## 8 MECHANOSIGNALING IN TSC DIFFERENTIATION

Mechanically responsive tissues like tendons detect and process mechanical signals that are transduced by cells to stimulate biochemical pathways and to initiate various cellular processes including migration, proliferation, and differentiation [32,99]. This process, termed mechanotransduction, refers to the initiation of an intracellular response by forces that generate deformation of tendon matrix and cells that in turn ensues increased transcription of genes and protein synthesis. Externally applied strains are directly transduced to the cells via sensors like integrins that link ECM to the cytoskeleton and nucleus. Tenocytes alter the expression of integrins in response to tensile stress [100]. A number of mechanisms and signaling pathways are involved in mechanotransduction that respond by rapid changes in milliseconds and longer changes in minutes-hours-days [99]. The rapid responses may include activation of various ion channels, second messengers (e.g., $PGE_2$), and kinases [100]. Subsequent responses include focal adhesion rearrangement, cytoskeletal changes, kinase signaling, and activation of transcription and translation factors. The long-term changes influence various cell processes such as cell division, migration, differentiation, and apoptosis.

It becomes increasingly clear that mechanical loading influences TSC proliferation, self-renewal, and differentiation. However, the exact mechanisms by which the mechanical cues are translated to biochemical signals are yet unclear. A broad spectrum of sensors, signaling molecules, and pathways is suggested that are capable of regulating TSC differentiation in tendon. As an example, the primary cilium, a single sensory cellular extension that is present in most cells, including tenocytes, has a mechanosensory role in many tissues, including kidney and liver [101]. Primary cilia are now known to function as a hub for various signaling pathways and many pathways whose receptors are localized in the cilium such as Wnt and Hedgehog signaling, and it is believed to be involved in mechanotransduction and MSC lineage commitment [102]. The elongation of the primary cilia has been shown to increase the mechanosensitivity of cells to mechanical signals and has been identified as a biomarker of alterations in cellular homeostasis. Primary cilia with a wide variety of lengths were mainly found not only on tenocytes but possibly also on TSCs since a mixture of cells were used in previous studies [14,103]. Tendon cell-cilia elongate in the absence of mechanical load and adapt their length in response to various mechanical loading conditions, and cilia length in tendon cells is affected by mechanical signals such as force and deformation from the extracellular matrix [14,104]. In situ deflection of tendon cell-cilia by tensile loading supports the concept that cilia are involved in mechanoresponse of tendon cells [105]. Primary cilia in the tendon fascicular matrix (FM) and endotenon or interfascicular matrix (IFM) respond differently to stress deprivation with elongation being greater in IFM that is associated with greater loss of biomechanical integrity in this region [104].

Various signaling molecules such as transcription and growth factors induce and regulate TSC differentiation into tenocytes. Mechanical loading and gene expression, growth factor production, and collagen synthesis in tendon are tightly linked, although the causal relationship is yet to be defined. Growth factors such as the members of the TGF-β superfamily and GDF-5 and certain crucial transcription factors including Mkx and Egr1/2 are identified as the major regulators of TSC differentiation to tenogenic lineage in tendons [15]. Two signaling pathways, TGFβ-SMAD2/3 and FGF-ERK/MAPK are believed to be mainly responsible for mechanotransduction in tendon [2,106]. Potential anabolic effects of mechanical stretching at physiological level in human tenocytes is elicited via decrease in two main collagenases (MMP-1 and MMP-13) accompanied by an increase in Col1a1 possibly via TGF-β signaling mechanism [107].

Mohawk promotes the tenogenesis of mesenchymal stem cells through activation of the TGF-β signaling pathway [108,109]. A similar pathway may also control tenogenesis in TSCs. In fact, TNF-α and TGF-β facilitate the differentiation and proliferation of TSCs in vitro [110]. The expression of tenogenic/osteogenic-related markers (Scx, Tnmd, Col1a1, Runx-2, and Alpl) and proliferation significantly enhance after simultaneous or sequential treatment of TSCs with TGF-β1 and TNF-α. TGF-β and BMP signaling pathways are highly activated since Smad2/3 and Smad1/5/8 are highly phosphorylated after treatment [110]. Additional studies may define the precise molecular mechanisms regarding how Mkx coordinates expression of diverse set of genes in TSCs and the role of TGF-β and BMP signaling pathways in TSC proliferation and/or differentiation, which may be helpful in the development of therapeutic targets for tendon repair.

Erroneous differentiation of TSCs under excessive mechanical loading may drive cell differentiation toward more adipocytic, chondrocytic, and osteogenic phenotype potentially through various signaling mechanisms. $PGE_2$ has a major role in inducing abnormal differentiation of TSCs as evidenced by both in vitro and in vivo findings. Intensive mechanical loading of tendons induces $PGE_2$, and treatment of TSCs with $PGE_2$ decreases cell proliferation and induces TSC differentiation into adipocytes and osteocytes [57]. $PGE_2$-treated TSCs when implanted into nude rats form lipid deposition and GAG accumulation. However, $PGE_2$ can exhibit biphasic effects in TSCs. While higher concentrations can decrease cell proliferation and induce nontenocyte differentiation, lower concentrations can enhance cell proliferation and expression of stem cell markers [73], that is, they stimulate maintenance of the stemness of TSCs. Nevertheless, the molecular mechanisms for the biphasic actions of $PGE_2$, particularly under mechanical loading conditions, remain to be elucidated.

BMPs, the group of signaling molecules that belong to TGF-β superfamily, are yet other candidate molecules that are suggested to be involved in TSC differentiation into nontenocyte phenotypes. BMPs signal via canonical Smad-dependent pathway and noncanonical MAPK pathways [111]. Activated Smads induce the expression of Runx-2, one of the major regulators of osteogenesis [112]. TSCs express a high level of BMP receptors [113], and mechanical stretching increases BMP-2 expression in TSCs, facilitating osteogenic differentiation [55]. Moreover, $PGE_2$ activates Akt and induces osteogenic differentiation in TSCs via BMP-2 production [57,114,115]. Phosphoinositide 3-kinase (PI3K) signaling cascade is essential for $PGE_2$-induced BMP-2 production and BMP-mediated osteogenic differentiation in TSCs [114]. IGF-1 and BMP-2 together mediate $PGE_2$-induced adipogenic differentiation of TSCs via a CREB and Smad-dependent mechanism [116]. $PGE_2$ increases the phosphorylation of CREB and Smad via IGF-1 and BMP-2. In addition, ERK1/2 pathway, which is a branch of MAPK signaling pathway, is also suggested to be involved in osteogenic differentiation of TSCs [117].

Wnt signaling that plays a vital role in the regulation of differentiation in MSCs [118] is also important in directing erroneous TSC differentiation leading toward the pathogenesis of tendinopathy [119]. Mechanical stretching-induced osteogenic differentiation of TSCs is mediated via Wnt5a-Rho A pathway and Wnt5a/5b/JNK pathway [54,56]. Wnt3a promotes osteogenic differentiation in TSCs. Moreover, Wnt mediators are increased in animal models and clinical samples of tendinopathy that might contribute to tissue metaplasia and failed healing in some cases of tendinopathy [119].

## 9 CONCLUDING REMARKS

Chronic tendon injuries or tendinopathies are frequent especially in athletes and in older populations. Clinically, the complete restoration of such injured tendons continues to be a challenge as the treatments are mostly palliative. Therefore, more basic scientific studies on tendons are needed so that new, effective treatment strategies can be devised. The identification of TSCs with self-renewal and multidifferentiation potential is particularly remarkable because it offers an opportunity for tendon researchers to explore the potential stem cell mechanisms involved in mechanical loading and aging-induced tendinopathy.

There are several future research directions to be explored in TSC mechanobiology. Currently, TSCs are characterized only in vitro, which are likely a mixture of adult stem cells from tendon core and peritenon that may contain different types of stem cells. The properties of stem cells may also differ depending on the anatomical source of isolation. The in vivo identities and niche of TSCs with respect to anatomical locations and regulatory factors in tendons remain to be investigated [120,121], both under normal and injury conditions. This requires lineage specific marking/tracing of TSCs. A recent study showed that α-SMA positive progenitor cells in the paratenon migrate into the tendon wound to participate in its healing [122]. Other potential sources of "TSCs" in the healing of tendon injury may include stem cells in the circulatory system; adipose tissue near tendons; and the bursa, a fluid-filled sac that acts as a cushion between a bone and other moving parts such as muscles, tendons, or skin [123,124].

In addition, the effects of aging on TSC mechanobiology should be thoroughly investigated. While pending definitive demonstration, aging is presumed to cause tendon cells to undergo the process of senescence, which refers to permanent cell cycle arrest that cannot be reactivated [89]. Tendon cell senescence may lead to tendon degeneration and impair tendon healing, which are often seen in aging patients. But the relationship between aging and reduced tendon potential to heal needs to be demonstrated. These detrimental effects caused by aging can be reversed or lessened by applying moderate physiological loads on tendons that induce anabolic changes by modulating TSCs, which are in senescent state [88]. Thus, two lines of research can be pursued in future research: the first is to test the notion that cell senescence in aging tendons may be at early stages, such that moderate mechanical loading in the form of exercise can reverse the process of tendon cell senescence. The second is to perform research on the possibility of using moderate exercise as a therapy to improve aging-associated impairment in healing tendons.

## ACKNOWLEDGMENTS

We gratefully acknowledge the funding support from National Institutes of Health grants AR061395, AR065949, and AR070340 (JHW).

## REFERENCES

[1] Thorpe CT, Screen HR. Tendon structure and composition. Adv Exp Med Biol 2016;920:3–10.
[2] Nourissat G, Berenbaum F, Duprez D. Tendon injury: from biology to tendon repair. Nat Rev Rheumatol 2015;11(4):223–33.
[3] Kannus P. Structure of the tendon connective tissue. Scand J Med Sci Sports 2000;10(6):312–20.
[4] Bi YM, Ehirchiou D, Kilts TM, Inkson CA, Embree MC, Sonoyama W, et al. Identification of tendon stem/progenitor cells and the role of the extracellular matrix in their niche. Nat Med 2007;13(10):1219–27.
[5] Rui YF, Lui PP, Li G, Fu SC, Lee YW, Chan KM. Isolation and characterization of multipotent rat tendon-derived stem cells. Tissue Eng Part A 2010;16(5):1549–58.
[6] Zhang J, Wang JH. Characterization of differential properties of rabbit tendon stem cells and tenocytes. BMC Musculoskelet Disord 2010;11:10.
[7] Maffulli N, Wong J, Almekinders LC. Types and epidemiology of tendinopathy. Clin Sports Med 2003;22(4):675–92.

[8] Andarawis-Puri N, Flatow EL, Soslowsky LJ. Tendon basic science: development, repair, regeneration, and healing. J Orthop Res 2015;33(6):780–4.
[9] Riley G. Chronic tendon pathology: molecular basis and therapeutic implications. Expert Rev Mol Med 2005;7(5):1–25.
[10] Moore KA, Lemischka IR. Stem cells and their niches. Science 2006;311 (5769):1880–5.
[11] Fenwick S, Harrall R, Hackney R, Bord S, Horner A, Hazleman B, et al. Endochondral ossification in Achilles and patella tendinopathy. Rheumatology (Oxford) 2002;41 (4):474–6.
[12] Salingcarnboriboon R, Yoshitake H, Tsuji K, Obinata M, Amagasa T, Nifuji A, et al. Establishment of tendon-derived cell lines exhibiting pluripotent mesenchymal stem cell-like property. Exp Cell Res 2003;287(2):289–300.
[13] de Mos M, Koevoet WJ, Jahr H, Verstegen MM, Heijboer MP, Kops N, et al. Intrinsic differentiation potential of adolescent human tendon tissue: an in-vitro cell differentiation study. BMC Musculoskelet Disord 2007;8:16.
[14] Gardner K, Arnoczky SP, Lavagnino M. Effect of in vitro stress-deprivation and cyclic loading on the length of tendon cell cilia in situ. J Orthop Res 2011;29(4):582–7.
[15] Costa-Almeida RG, Gershovich P, Rodrigues MT, Reis RL, Gomes ME. Tendon stem cell niche. Ottava, ON, Canada: Humana Press, Springer; 2015.
[16] Lui PP, Chan KM. Tendon-derived stem cells (TDSCs): from basic science to potential roles in tendon pathology and tissue engineering applications. Stem Cell Rev 2011;7(4):883–97.
[17] Tan Q, Lui PP, Rui YF. Effect of in vitro passaging on the stem cell-related properties of tendon-derived stem cells-implications in tissue engineering. Stem Cells Dev 2012;21 (5):790–800.
[18] Wang JH, Komatsu I. Tendon stem cells: mechanobiology and development of tendinopathy. Adv Exp Med Biol 2016;920:53–62.
[19] Zhang J, Wang JH. Human tendon stem cells better maintain their stemness in hypoxic culture conditions. PLoS ONE 2013;8(4):e61424.
[20] Lee WY, Lui PP, Rui YF. Hypoxia-mediated efficient expansion of human tendon-derived stem cells in vitro. Tissue Eng Part A 2012;18(5–6):484–98.
[21] Rui YF, Lui PP, Wong YM, Tan Q, Chan KM. BMP-2 stimulated non-tenogenic differentiation and promoted proteoglycan deposition of tendon-derived stem cells (TDSCs) in vitro. J Orthop Res 2013;31(5):746–53.
[22] Kannus P, Jozsa L. Histopathological changes preceding spontaneous rupture of a tendon. A controlled study of 891 patients. J Bone Joint Surg Am 1991;73(10):1507–25.
[23] Lui PP. Identity of tendon stem cells—how much do we know? J Cell Mol Med 2013;17(1):55–64.
[24] Tempfer H, Wagner A, Gehwolf R, Lehner C, Tauber M, Resch H, et al. Perivascular cells of the supraspinatus tendon express both tendon- and stem cell-related markers. Histochem Cell Biol 2009;131(6):733–41.
[25] Mienaltowski MJ, Adams SM, Birk DE. Regional differences in stem cell/progenitor cell populations from the mouse achilles tendon. Tissue Eng Part A 2013;19 (1–2):199–210.
[26] Mienaltowski MJ, Adams SM, Birk DE. Tendon proper- and peritenon-derived progenitor cells have unique tenogenic properties. Stem Cell Res Ther 2014;5(4):86.
[27] Spanoudes K, Gaspar D, Pandit A, Zeugolis DI. The biophysical, biochemical, and biological toolbox for tenogenic phenotype maintenance in vitro. Trends Biotechnol 2014;32(9):474–82.

[28] Yin Z, Hu JJ, Yang L, Zheng ZF, An CR, Wu BB, et al. Single-cell analysis reveals a nestin+ tendon stem/progenitor cell population with strong tenogenic potentiality. Sci Adv 2016;2(11):e1600874.
[29] Kohler J, Popov C, Klotz B, Alberton P, Prall WC, Haasters F, et al. Uncovering the cellular and molecular changes in tendon stem/progenitor cells attributed to tendon aging and degeneration. Aging Cell 2013;12(6):988–99.
[30] Coppe JP, Desprez PY, Krtolica A, Campisi J. The senescence-associated secretory phenotype: the dark side of tumor suppression. Annu Rev Pathol 2010;5:99–118.
[31] Childs BG, Durik M, Baker DJ, van Deursen JM. Cellular senescence in aging and age-related disease: from mechanisms to therapy. Nat Med 2015;21(12):1424–35.
[32] Wang JH. Mechanobiology of tendon. J Biomech 2006;39(9):1563–82.
[33] Sun L, Qu L, Zhu R, Li H, Xue Y, Liu X, et al. Effects of mechanical stretch on cell proliferation and matrix formation of mesenchymal stem cell and anterior cruciate ligament fibroblast. Stem Cells Int 2016;2016:9842075.
[34] Riehl BD, Park JH, Kwon IK, Lim JY. Mechanical stretching for tissue engineering: two-dimensional and three-dimensional constructs. Tissue Eng Part B Rev 2012;18(4):288–300.
[35] Hatta T, Sano H, Sakamoto N, Kishimoto KN, Sato M, Itoi E. Nicotine reduced MMP-9 expression in the primary porcine tenocytes exposed to cyclic stretch. J Orthop Res 2013;31(4):645–50.
[36] Skutek M, van Griensven M, Zeichen J, Brauer N, Bosch U. Cyclic mechanical stretching modulates secretion pattern of growth factors in human tendon fibroblasts. Eur J Appl Physiol 2001;86(1):48–52.
[37] Yang G, Crawford RC, Wang JH. Proliferation and collagen production of human patellar tendon fibroblasts in response to cyclic uniaxial stretching in serum-free conditions. J Biomech 2004;37(10):1543–50.
[38] Zhang J, Wang JH. Mechanobiological response of tendon stem cells: implications of tendon homeostasis and pathogenesis of tendinopathy. J Orthop Res 2010;28(5):639–43.
[39] Zhang J, Wang JH. The effects of mechanical loading on tendons—an in vivo and in vitro model study. PLoS ONE 2013;8(8):e71740.
[40] Docheva D, Hunziker EB, Fassler R, Brandau O. Tenomodulin is necessary for tenocyte proliferation and tendon maturation. Mol Cell Biol 2005;25(2):699–705.
[41] Yamamoto K, Sokabe T, Watabe T, Miyazono K, Yamashita JK, Obi S, et al. Fluid shear stress induces differentiation of Flk-1-positive embryonic stem cells into vascular endothelial cells in vitro. Am J Physiol Heart Circ Physiol 2005;288(4):H1915–24.
[42] Song G, Ju Y, Soyama H, Ohashi T, Sato M. Regulation of cyclic longitudinal mechanical stretch on proliferation of human bone marrow mesenchymal stem cells. Mol Cell Biomech 2007;4(4):201–10.
[43] Popov C, Burggraf M, Kreja L, Ignatius A, Schieker M, Docheva D. Mechanical stimulation of human tendon stem/progenitor cells results in upregulation of matrix proteins, integrins and MMPs, and activation of p38 and ERK1/2 kinases. BMC Mol Biol 2015;16:6.
[44] Mammoto A, Mammoto T, Ingber DE. Mechanosensitive mechanisms in transcriptional regulation. J Cell Sci 2012;125(Pt 13):3061–73.
[45] Suzuki H, Ito Y, Shinohara M, Yamashita S, Ichinose S, Kishida A, et al. Gene targeting of the transcription factor Mohawk in rats causes heterotopic ossification of Achilles tendon via failed tenogenesis. Proc Natl Acad Sci U S A 2016;113(28):7840–5.

[46] Kayama T, Mori M, Ito Y, Matsushima T, Nakamichi R, Suzuki H, et al. Gtf2ird1-dependent Mohawk expression regulates mechanosensing properties of the tendon. Mol Cell Biol 2016;36(8):1297–309.

[47] Liu Y, Xu J, Xu L, Wu T, Sun Y, Lee YW, et al. Cystic fibrosis transmembrane conductance regulator mediates tenogenic differentiation of tendon-derived stem cells and tendon repair: accelerating tendon injury healing by intervening in its downstream signaling. FASEB J 2017;31(9):3800–15.

[48] Mousavizadeh R, Khosravi S, Behzad H, McCormack RG, Duronio V, Scott A. Cyclic strain alters the expression and release of angiogenic factors by human tendon cells. PLoS ONE 2014;9(5):e97356.

[49] Mousavizadeh R, Scott A, Lu A, Ardekani GS, Behzad H, Lundgreen K, et al. Angiopoietin-like 4 promotes angiogenesis in the tendon and is increased in cyclically loaded tendon fibroblasts. J Physiol 2016;594(11):2971–83.

[50] Lavagnino M, Gardner KL, Arnoczky SP. High magnitude, in vitro, biaxial, cyclic tensile strain induces actin depolymerization in tendon cells. Muscles Ligaments Tendons J 2015;5(2):124–8.

[51] Chen H, Chen L, Cheng B, Jiang C. Cyclic mechanical stretching induces autophagic cell death in tenofibroblasts through activation of prostaglandin E2 production. Cell Physiol Biochem 2015;36(1):24–33.

[52] Chen W, Deng Y, Zhang J, Tang K. Uniaxial repetitive mechanical overloading induces influx of extracellular calcium and cytoskeleton disruption in human tenocytes. Cell Tissue Res 2015;359(2):577–87.

[53] Huisman E, Lu A, Jamil S, Mousavizadeh R, McCormack R, Roberts C, et al. Influence of repetitive mechanical loading on MMP2 activity in tendon fibroblasts. J Orthop Res 2016;34(11):1991–2000.

[54] Liu X, Chen W, Zhou Y, Tang K, Zhang J. Mechanical tension promotes the Osteogenic differentiation of rat tendon-derived stem cells through the Wnt5a/Wnt5b/JNK signaling pathway. Cell Physiol Biochem 2015;36(2):517–30.

[55] Rui YF, Lui PP, Ni M, Chan LS, Lee YW, Chan KM. Mechanical loading increased BMP-2 expression which promoted osteogenic differentiation of tendon-derived stem cells. J Orthop Res 2011;29(3):390–6.

[56] Shi Y, Fu Y, Tong W, Geng Y, Lui PP, Tang T, et al. Uniaxial mechanical tension promoted osteogenic differentiation of rat tendon-derived stem cells (rTDSCs) via the Wnt5a-RhoA pathway. J Cell Biochem 2012;113(10):3133–42.

[57] Zhang J, Wang JH. Production of PGE(2) increases in tendons subjected to repetitive mechanical loading and induces differentiation of tendon stem cells into non-tenocytes. J Orthop Res 2010;28(2):198–203.

[58] Kjaer M, Langberg H, Miller BF, Boushel R, Crameri R, Koskinen S, et al. Metabolic activity and collagen turnover in human tendon in response to physical activity. J Musculoskelet Neuronal Interact 2005;5(1):41–52.

[59] Kjaer M, Magnusson P, Krogsgaard M, Boysen Moller J, Olesen J, Heinemeier K, et al. Extracellular matrix adaptation of tendon and skeletal muscle to exercise. J Anat 2006;208(4):445–50.

[60] Langberg H, Ellingsgaard H, Madsen T, Jansson J, Magnusson SP, Aagaard P, et al. Eccentric rehabilitation exercise increases peritendinous type I collagen synthesis in humans with Achilles tendinosis. Scand J Med Sci Sports 2007;17(1):61–6.

[61] Zhang J, Pan T, Liu Y, Wang JH. Mouse treadmill running enhances tendons by expanding the pool of tendon stem cells (TSCs) and TSC-related cellular production of collagen. J Orthop Res 2010;28(9):1178–83.

[62] Mendias CL, Gumucio JP, Bakhurin KI, Lynch EB, Brooks SV. Physiological loading of tendons induces scleraxis expression in epitenon fibroblasts. J Orthop Res 2012;30(4):606–12.

[63] Bell R, Li J, Gorski DJ, Bartels AK, Shewman EF, Wysocki RW, et al. Controlled treadmill exercise eliminates chondroid deposits and restores tensile properties in a new murine tendinopathy model. J Biomech 2013;46(3):498–505.

[64] Scott A, Danielson P, Abraham T, Fong G, Sampaio AV, Underhill TM. Mechanical force modulates scleraxis expression in bioartificial tendons. J Musculoskelet Neuronal Interact 2011;11(2):124–32.

[65] Morita Y, Watanabe S, Ju Y, Xu B. Determination of optimal cyclic uniaxial stretches for stem cell-to-tenocyte differentiation under a wide range of mechanical stretch conditions by evaluating gene expression and protein synthesis levels. Acta Bioeng Biomech 2013;15(3):71–9.

[66] Glazebrook MA, Wright Jr. JR, Langman M, Stanish WD, Lee JM. Histological analysis of achilles tendons in an overuse rat model. J Orthop Res 2008;26(6):840–6.

[67] Abraham T, Fong G, Scott A. Second harmonic generation analysis of early Achilles tendinosis in response to in vivo mechanical loading. BMC Musculoskelet Disord 2011;12:26.

[68] Ng GY, Chung PY, Wang JS, Cheung RT. Enforced bipedal downhill running induces Achilles tendinosis in rats. Connect Tissue Res 2011;52(6):466–71.

[69] Jafari L, Vachon P, Beaudry F, Langelier E. Histopathological, biomechanical, and behavioral pain findings of Achilles tendinopathy using an animal model of overuse injury. Physiol Rep 2015;3(1).

[70] Attia M, Scott A, Duchesnay A, Carpentier G, Soslowsky LJ, Huynh MB, et al. Alterations of overused supraspinatus tendon: a possible role of glycosaminoglycans and HARP/pleiotrophin in early tendon pathology. J Orthop Res 2012;30(1):61–71.

[71] Attia M, Huet E, Gossard C, Menashi S, Tassoni MC, Martelly I. Early events of overused supraspinatus tendons involve matrix metalloproteinases and EMMPRIN/CD147 in the absence of inflammation. Am J Sports Med 2013;41(4):908–17.

[72] Reuther KE, Thomas SJ, Evans EF, Tucker JJ, Sarver JJ, Ilkhani-Pour S, et al. Returning to overuse activity following a supraspinatus and infraspinatus tear leads to joint damage in a rat model. J Biomech 2013;46(11):1818–24.

[73] Zhang J, Wang JH. Prostaglandin E2 (PGE2) exerts biphasic effects on human tendon stem cells. PLoS ONE 2014;9(2):e87706.

[74] Riley GP, Harrall RL, Constant CR, Cawston TE, Hazleman BL. Prevalence and possible pathological significance of calcium phosphate salt accumulation in tendon matrix degeneration. Ann Rheum Dis 1996;55(2):109–15.

[75] Maffulli N, Reaper J, Ewen SW, Waterston SW, Barrass V. Chondral metaplasia in calcific insertional tendinopathy of the Achilles tendon. Clin J Sport Med 2006;16(4):329–34.

[76] Fu SC, Chan KM, Rolf CG. Increased deposition of sulfated glycosaminoglycans in human patellar tendinopathy. Clin J Sport Med 2007;17(2):129–34.

[77] Archambault JM, Jelinsky SA, Lake SP, Hill AA, Glaser DL, Soslowsky LJ. Rat supraspinatus tendon expresses cartilage markers with overuse. J Orthop Res 2007;25(5):617–24.

[78] Scott A, Cook JL, Hart DA, Walker DC, Duronio V, Khan KM. Tenocyte responses to mechanical loading in vivo: a role for local insulin-like growth factor 1 signaling in early tendinosis in rats. Arthritis Rheum 2007;56(3):871–81.

[79] Silva RD, Glazebrook MA, Campos VC, Vasconcelos AC. Achilles tendinosis: a morphometrical study in a rat model. Int J Clin Exp Pathol 2011;4(7):683–91.

[80] Rui YF, Lui PP, Chan LS, Chan KM, Fu SC, Li G. Does erroneous differentiation of tendon-derived stem cells contribute to the pathogenesis of calcifying tendinopathy? Chin Med J 2011;124(4):606–10.

[81] Riley G. The pathogenesis of tendinopathy. A molecular perspective. Rheumatology (Oxford) 2004;43(2):131–42.

[82] Wang JZ, Nirmala J. X. Advancements in the treatment and repair of tendon injuries. Curr Tissue Eng 2014;3(2):71–81.

[83] Rodier F, Coppe JP, Patil CK, Hoeijmakers WA, Munoz DP, Raza SR, et al. Persistent DNA damage signalling triggers senescence-associated inflammatory cytokine secretion. Nat Cell Biol 2009;11(8):973–9.

[84] Zhou Z, Akinbiyi T, Xu L, Ramcharan M, Leong DJ, Ros SJ, et al. Tendon-derived stem/progenitor cell aging: defective self-renewal and altered fate. Aging Cell 2010;9(5):911–5.

[85] Alt EU, Senst C, Murthy SN, Slakey DP, Dupin CL, Chaffin AE, et al. Aging alters tissue resident mesenchymal stem cell properties. Stem Cell Res 2012;8(2):215–25.

[86] Liu L, Rando TA. Manifestations and mechanisms of stem cell aging. J Cell Biol 2011;193(2):257–66.

[87] Unterluggauer H, Hampel B, Zwerschke W, Jansen-Durr P. Senescence-associated cell death of human endothelial cells: the role of oxidative stress. Exp Gerontol 2003;38(10):1149–60.

[88] Zhang J, Wang JH. Moderate exercise mitigates the detrimental effects of aging on tendon stem cells. PLoS ONE 2015;10(6):e0130454.

[89] Campisi J, d'Adda di Fagagna F. Cellular senescence: when bad things happen to good cells. Nat Rev Mol Cell Biol 2007;8(9):729–40.

[90] Tsirpanlis G. Cellular senescence, cardiovascular risk, and CKD: a review of established and hypothetical interconnections. Am J Kidney Dis 2008;51(1):131–44.

[91] Wood LK, Brooks SV. Ten weeks of treadmill running decreases stiffness and increases collagen turnover in tendons of old mice. J Orthop Res 2016;34(2):346–53.

[92] Zhang J, Yuan T, Wang JH. Moderate treadmill running exercise prior to tendon injury enhances wound healing in aging rats. Oncotarget 2016;7(8):8498–512.

[93] Ni M, Lui PP, Rui YF, Lee YW, Tan Q, Wong YM, et al. Tendon-derived stem cells (TDSCs) promote tendon repair in a rat patellar tendon window defect model. J Orthop Res 2012;30(4):613–9.

[94] Hu C, Zhang Y, Tang K, Luo Y, Liu Y, Chen W. Downregulation of CITED2 contributes to TGFbeta-mediated senescence of tendon-derived stem cells. Cell Tissue Res 2017;368(1):93–104.

[95] Ren JL, Pan JS, Lu YP, Sun P, Han J. Inflammatory signaling and cellular senescence. Cell Signal 2009;21(3):378–83.

[96] Alberton P, Dex S, Popov C, Shukunami C, Schieker M, Docheva D. Loss of tenomodulin results in reduced self-renewal and augmented senescence of tendon stem/progenitor cells. Stem Cells Dev 2015;24(5):597–609.

[97] Chen L, Wang GD, Liu JP, Wang HS, Liu XM, Wang Q, et al. miR-135a modulates tendon stem/progenitor cell senescence via suppressing ROCK1. Bone 2015;71:210–6.

[98] Chen L, Liu J, Tao X, Wang G, Wang Q, Liu X. The role of Pin1 protein in aging of human tendon stem/progenitor cells. Biochem Biophys Res Commun 2015;464(2):487–92.

[99] Banes AJ, Tsuzaki M, Yamamoto J, Fischer T, Brigman B, Brown T, et al. Mechanoreception at the cellular level: the detection, interpretation, and diversity of responses to mechanical signals. Biochem Cell Biol = Biochimie et biologie cellulaire 1995;73(7–8):349–65.

[100] Lavagnino M, Wall ME, Little D, Banes AJ, Guilak F, Arnoczky SP. Tendon mechanobiology: current knowledge and future research opportunities. J Orthop Res 2015;33(6):813–22.

[101] Christensen ST, Pedersen LB, Schneider L, Satir P. Sensory cilia and integration of signal transduction in human health and disease. Traffic 2007;8(2):97–109.

[102] Hoey DA, Tormey S, Ramcharan S, O'Brien FJ, Jacobs CR. Primary cilia-mediated mechanotransduction in human mesenchymal stem cells. Stem Cells 2012;30(11):2561–70.

[103] Donnelly E, Ascenzi MG, Farnum C. Primary cilia are highly oriented with respect to collagen direction and long axis of extensor tendon. J Orthop Res 2010;28(1):77–82.

[104] Rowson D, Knight MM, Screen HR. Zonal variation in primary cilia elongation correlates with localized biomechanical degradation in stress deprived tendon. J Orthop Res 2016;34(12):2146–53.

[105] Lavagnino M, Arnoczky SP, Gardner K. In situ deflection of tendon cell-cilia in response to tensile loading: an in vitro study. J Orthop Res 2011;29(6):925–30.

[106] Havis E, Bonnin MA, Esteves de Lima J, Charvet B, Milet C, Duprez D. TGF beta and FGF promote tendon progenitor fate and act downstream of muscle contraction to regulate tendon differentiation during chick limb development. Development 2016;143(20):3839–51.

[107] Jones ER, Jones GC, Legerlotz K, Riley GP. Cyclical strain modulates metalloprotease and matrix gene expression in human tenocytes via activation of TGFbeta. Biochim Biophys Acta 2013;1833(12):2596–607.

[108] Liu H, Zhang C, Zhu S, Lu P, Zhu T, Gong X, et al. Mohawk promotes the tenogenesis of mesenchymal stem cells through activation of the TGFbeta signaling pathway. Stem Cells 2015;33(2):443–55.

[109] Otabe K, Nakahara H, Hasegawa A, Matsukawa T, Ayabe F, Onizuka N, et al. Transcription factor Mohawk controls tenogenic differentiation of bone marrow mesenchymal stem cells in vitro and in vivo. J Orthop Res 2015;33(1):1–8.

[110] Han P, Cui Q, Yang S, Wang H, Gao P, Li Z. Tumor necrosis factor-alpha and transforming growth factor-beta1 facilitate differentiation and proliferation of tendon-derived stem cells in vitro. Biotechnol Lett 2017;39(5):711–9.

[111] Wang RN, Green J, Wang Z, Deng Y, Qiao M, Peabody M, et al. Bone morphogenetic protein (BMP) signaling in development and human diseases. Genes Dis 2014;1(1):87–105.

[112] Lee KS, Hong SH, Bae SC. Both the Smad and p38 MAPK pathways play a crucial role in Runx2 expression following induction by transforming growth factor-beta and bone morphogenetic protein. Oncogene 2002;21(47):7156–63.

[113] Rui YF, Lui PP, Lee YW, Chan KM. Higher BMP receptor expression and BMP-2-induced osteogenic differentiation in tendon-derived stem cells compared with bone-marrow-derived mesenchymal stem cells. Int Orthop 2012;36(5):1099–107.

[114] Liu J, Chen L, Tao X, Tang K. Phosphoinositide 3-kinase/Akt signaling is essential for prostaglandin E2-induced osteogenic differentiation of rat tendon stem cells. Biochem Biophys Res Commun 2013;435(4):514–9.

[115] Zhang J, Wang JH. BMP-2 mediates PGE(2)-induced reduction of proliferation and osteogenic differentiation of human tendon stem cells. J Orthop Res 2012;30(1):47–52.

[116] Liu J, Chen L, zhou Y, Liu X, Tang K. Insulin-like growth factor-1 and bone morphogenetic protein-2 jointly mediate prostaglandin E2-induced adipogenic differentiation of rat tendon stem cells. PLoS ONE 2014;9(1):e85469.

[117] Li P, Xu Y, Gan Y, Song L, Zhang C, Wang L, et al. Role of the ERK1/2 signaling pathway in osteogenesis of rat tendon-derived stem cells in normoxic and hypoxic cultures. Int J Med Sci 2016;13(8):629–37.

[118] Ling L, Nurcombe V, Cool SM. Wnt signaling controls the fate of mesenchymal stem cells. Gene 2009;433(1–2):1–7.

[119] Lui PP, Lee YW, Wong YM, Zhang X, Dai K, Rolf CG. Expression of Wnt pathway mediators in metaplasic tissue in animal model and clinical samples of tendinopathy. Rheumatology (Oxford) 2013;52(9):1609–18.

[120] Lui PP, Cheuk YC, Lee YW, Chan KM. Ectopic chondro-ossification and erroneous extracellular matrix deposition in a tendon window injury model. J Orthop Res 2012;30(1):37–46.

[121] Lui PP, Kong SK, Lau PM, Wong YM, Lee YW, Tan C, et al. Allogeneic tendon-derived stem cells promote tendon healing and suppress immunoreactions in hosts: in vivo model. Tissue Eng Part A 2014;20(21–22):2998–3009.

[122] Dyment NA, Hagiwara Y, Matthews BG, Li Y, Kalajzic I, Rowe DW. Lineage tracing of resident tendon progenitor cells during growth and natural healing. PLoS ONE 2014;9(4):e96113.

[123] Utsunomiya H, Uchida S, Sekiya I, Sakai A, Moridera K, Nakamura T. Isolation and characterization of human mesenchymal stem cells derived from shoulder tissues involved in rotator cuff tears. Am J Sports Med 2013;41(3):657–68.

[124] Song N, Armstrong AD, Li F, Ouyang H, Niyibizi C. Multipotent mesenchymal stem cells from human subacromial bursa: potential for cell based tendon tissue engineering. Tissue Eng Part A 2014;20(1–2):239–49.

CHAPTER

# Bone mechanobiology in health and disease

# 6

Stefaan W. Verbruggen*,†, Laoise M. McNamara‡

*Department of Biomedical Engineering, Columbia University, New York, NY, United States*
†*Institute of Bioengineering, School of Engineering and Materials Science, Queen Mary University of London, London, United Kingdom*
‡*Biomedical Engineering, College of Engineering and Informatics, National University of Ireland Galway, Galway, Ireland*

## ABBREVIATIONS

| | |
|---|---|
| ALP | alkaline phosphatase |
| ATP | adenosine triphosphate |
| AVF | adipocyte volume fraction |
| CFD | computational fluid dynamics |
| COX-2 | cyclooxygenase-2 |
| ECM | extracellular matrix |
| ERK | extracellular signal-regulated kinase |
| FE | finite element |
| GPCR | G-protein-coupled receptor |
| HSC | hematopoietic stem cell |
| MSC | mesenchymal stem cell |
| MAPK | mitogen-activated protein kinase |
| MMP | matrix metalloproteinase |
| NO | nitric oxide |
| OA | osteoarthritis |
| OCN | osteocalcin |
| OPG | osteoprotegerin |
| OPN | osteopontin |
| $PGE_2$ | prostaglandin E2 |
| PKA | protein kinase A |
| PTH | parathyroid hormone |
| RUNX2 | runt-related transcription factor 2 |
| RANKL | receptor activator of nuclear factor kappa-B ligand |
| TEM | transmission electron microscopy |
| TGF-β | transforming growth factor-β |
| TRP | transient receptor potential |

Mechanobiology in Health and Disease. https://doi.org/10.1016/B978-0-12-812952-4.00006-4
© 2018 Elsevier Ltd. All rights reserved.

# 1 BONE

As the primary structural component of the skeleton, bone provides a support framework and protection for internal organs, allowing movement in concert with muscles, tendons, and ligaments. Bone is capable of efficiently performing these functions due to its high strength-to-weight ratio, arising from its unique composition including a stiff mineral phase and a softer organic phase [1]. These structural and material properties are all the more remarkable in that they are continually adapting and remodeling in response to mechanical loading, with the microarchitecture continually optimized at smaller levels to bear larger-scale macroscopic loads. Mechanobiology is crucial to this adaptive and regenerative nature. While it has long been known that bone can adapt its mass and structure to different loading conditions, only recently has significant progress been made in terms of understanding how external forces are transferred to skeletal tissues, how bone cells sense these loads, and how these signals are translated into a cascade of biochemical signals to produce cell expression or differentiation, ultimately resulting in macroscopic changes to bone structure. It is known that mechanical loading and disease alter cell mechanobiological function, tissue structure, and composition in various tissues [2–10]. However, how changes in mechanotransduction are associated with bone diseases, such as osteoporosis and osteoarthritis (OA), is not yet well understood but has recently gained attention. This chapter will systematically review our current understanding of bone mechanobiology and will outline key findings that have implicated changes in mechanobiology in the etiology of specific bone diseases.

## 1.1 BONE BIOMECHANICS

Bone comprises two distinct types of structure, with associated differences in microarchitecture. Cortical bone forms a dense outer shell of stiff, highly calcified tissue, representing over 80% of the mass of bones, dictating the external geometry and structure and providing support and protection to the human body [1,11,12]. Trabecular bone is found within this outer cortical structure and forms supportive struts, which act together with the cortical shell to distribute high stresses throughout the bone architecture [13–16].

Bone is therefore a complex material, with an inhomogeneous and anisotropic structure. Bone tissue composition, structural organization, and internal microarchitectures, such as blood vessels and cell pores, all contribute to this behavior and vary considerably across anatomical locations [17,18]. The mechanical properties of bone, obtained from testing of trabecular and cortical bone tissue, with elastic modulus measurements, range widely from 0.75 to 30 GPa [19–34]. Cortical and trabecular bone demonstrate distinct and separate mechanical behaviors, as result of differing tissue compositions and organizational microstructure [35–38]. Indeed, the degree of mineralization can be directly correlated with the stiffness and load-bearing strength of cortical bone [37], while collagen is intricately arranged so as to provide tensile strength, elasticity, and toughness [35]. Multiscale computational models have shown that variations in mineral volume fraction and mineral crystal size significantly affect tissue-level mechanical properties [39]. Separately, it has been shown that alternating orientations between lamellae give rise to different stiffness values [38], even with a similar degree of mineralization for all lamellae. Therefore, observed mechanical properties of bone are sensitive to both microstructural composition [35,37] and microstructural organization, that is, lamellar orientation [38].

## 1.2 HIERARCHICAL STRUCTURE OF BONE

The structure of bone is hierarchically organized into functional units over multiple scales, so that microarchitecture is optimized to bear high loads experienced in everyday situations. At the microscale, cortical bone is composed of functional units known as osteons (Fig. 1). These are cylindrical structures composed of concentric rings of osteocytes orientated about a vascular haversian canal. This compact network facilitates mechanosensation, communication, and nutrient supply to bone cells while providing the basic structural unit for cortical bone [41]. Indeed, it is thought that a major function of the osteon is to provide structural support to bone by slowing crack growth through their outer cement lines [42].

Similar structural subunits are found in trabecular bone, known as trabecular packets or hemiosteons [43]. While these packets are similar in size to osteons, they are highly porous and do not have a vascular supply [44], but instead receive nutrients and mechanical stimulus from the surrounding bone marrow [45,46]. These trabecular bone struts are thought to align themselves in order to optimize mechanical support for minimal bone mass [16].

## 1.3 BONE POROSITY AND FLUID FLOW

Bone is characterized by a number of different porosities across multiple scales (see Fig. 1): the trabecular porosity within the bone marrow cavity [47], the vascular porosity resulting from the network of blood vessels spread throughout bone [48], and the lacunar-canalicular porosity that houses more than 90% of bone cells [43].

The trabecular porosity consists of the space between trabecular struts, known as the medullary cavity, which contains bone marrow. This marrow displays viscosities up to two orders of magnitude greater than fluid in other bone porosities [49–51] and results in a unique mechanical environment that has been shown to be stimulatory to various bone cell types residing in the marrow [45,52]. The vascular porosity contains blood vessels at pressures largely similar to those found throughout the vasculature [43] and is highly permeable compared with the other porosities [48]. Finally, the lacunar-canalicular porosity contains interstitial fluid, which bathes the osteocytes within the system and has an ionic composition and viscosity similar to plasma or salt water [53]. Experimental tracer studies have shown that mechanical loading of the bone matrix surrounding this porosity generates a pressure differential that drives fluid through the lacunar-canalicular network [54]. Further tracer studies have

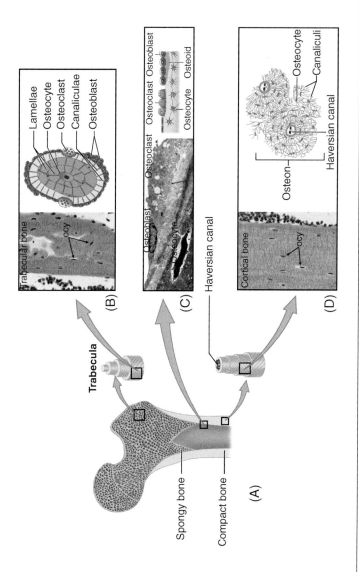

**FIG. 1**

The hierarchical organization of bone structure across multiple scales is integral to its mechanobiology. (A) The cross section of the femur highlighting (B) trabecular struts and the relatively disorganized arrangement of osteocytes within [40]; (C) the interplay between osteoblasts, osteoclasts, and osteocytes in close proximity at bone surfaces; and (D) the highly organized arrangement of osteocytes in compact cortical bone, aligned parallel to bone surfaces and concentrically around central vascular channels (haversian canals) [40].

observed interstitial fluid velocities of approximately 60 μm/s within individual canaliculi under applied mechanical loading [55].

These porosities contribute to the mechanical behavior of bone, by dictating the time-dependent response to loading, which is related to the time and energy required to drive fluid through these porosities [56]. While the more permeable vascular porosity can dissipate a loading-induced buildup of pressure rapidly through the vascular network, the lacunar-canalicular porosity requires a relaxation time that is three orders of magnitude greater [48]. The combination of the mechanical effects of these different porosities contributes to the ability of bone to not only absorb high loads [56] but also importantly generate fluid flow around osteocytes and other bone cells [54]. This fluid flow facilitates nutrient supply [28,57] and is also believed to play an important role in the mechanobiological behavior of bone [58].

## 2 BONE CELLS

Numerous cell types reside in bone tissues, which originate from either the osteogenic lineage or the monocyte-macrophage lineage and act in concert to maintain bone (Fig. 2). Bone marrow, encased within the cortical bone and surrounding the trabeculae, is the source of osteoprogenitor cells, which differentiate into osteoblasts and ultimately osteocytes [59]. Separately, the marrow and the vasculature within bone tissue itself act as a source of multinucleated cells and monocytes, which can develop into osteoclasts [60]. The development of these cells is illustrated in Fig. 2 and is described in this section.

### 2.1 MSCs AND OSTEOPROGENITORS

Mesenchymal stem cells (MSCs) are multipotent cells that have the potential to differentiate into numerous cell types. While they play a key role in early skeletal development, they can also be found in mature adult bone and are present in the periosteum, endosteum, and bone marrow. MSCs exhibit a basic cell morphology, defined by a small cell body containing a large, round nucleus, but they are capable of altering their morphology in response to their local environment [22]. MSCs can be influenced by growth factors, cytokines, and physical stimuli to differentiate to become osteoprogenitor cells, which have committed to the osteochondral lineage and are capable of differentiating along the chondrogenic (cartilage) or osteogenic (bone) pathways. Morphologically larger than MSCs, osteoprogenitors can be influenced by growth factors such as bone morphogenic proteins (BMPs) to differentiate further into osteoblasts and ultimately into osteocytes [59].

### 2.2 OSTEOBLASTS

Osteoblasts are bone-forming cells that are responsible for the production of bone tissue from birth and also the renewal of bone tissue through the processes of modeling and remodeling throughout life. Displaying a cuboidal geometry, the osteoblast

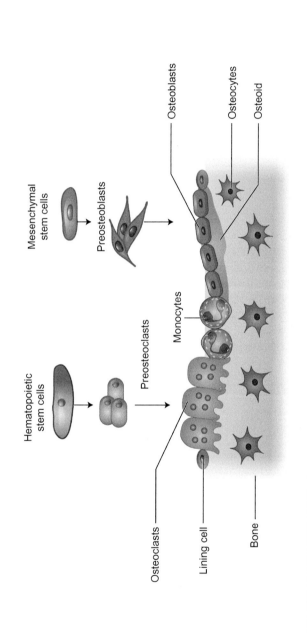

**FIG. 2**

Differentiation of bone cell types along the osteogenic lineage and the osteoclastic lineage.
*Adapted from Lian JB, Stein GS, van Wijnen AJ, Stein JL, Hassan MQ, Gaur T, et al. MicroRNA control of bone formation and homeostasis. Nat Rev Endocrinol 2012;8(4):212–27.*

possesses a large Golgi apparatus and rough endoplasmic reticulum when active [43] and organelles that play a key role in the cell's primary function of synthesis of osteoid [61,62]. The deposition and subsequent mineralization of this new bone matrix [63] form a key part of the bone remodeling cycle [64]. Osteoblasts actively control this process by deposition of fibronectin [65–67], which is responsible for mineralization, matrix formation, and osteoblast differentiation [65,66].

Quiescent osteoblasts on resting bone surfaces are known as bone-lining cells, displaying a flattened, elongated geometry and lying upon an organic collagen matrix approximately 1–2 μm thick [68]. These cells have been observed to form bone in response to both biochemical signals [69–71] and mechanical strain [72,73], suggesting involvement in maintaining bone homeostasis [71,74,75]. Furthermore, osteoblasts are thought to aid bone resorption by removing the organic collagen layer and contracting to allow osteoclasts to access the mineralized bone tissue [12,76].

## 2.3 OSTEOCYTES

Osteocytes are distributed abundantly and are encased within the bone matrix and represent the terminal differentiation of the osteogenic lineage [77]. Differentiation of osteoblasts into osteocytes occurs during the deposition of new bone matrix as part of the bone remodeling process [43]. Organized concentrically around blood vessels within osteons (see Figs. 1 and 3) and parallel to the bone surface, osteocytes are uniquely and ideally placed to sense mechanical loading.

The cells extend a network of dendritic processes outward, through a system of voids in the matrix known as canaliculi (see Fig. 3), which contact and communicate with other osteocytes via gap junctions [78,79]. These canaliculi are believed to provide a vital system for nutrient supply and waste disposal to the cells [43] and allow transduction of biochemical signals to other cells, both in the matrix and on the bone surface [80,81]. A mesh-like glycocalyx, or pericellular matrix (PCM), surrounds the osteocyte and tethers it to the extracellular matrix (ECM) [82], while punctate integrin attachments between the cell processes and the matrix are present in the canaliculi [83]. It has been proposed that both of these extracellular attachments may act to amplify strain signals to the osteocyte [82–86] through their connections with the cytoskeleton [87], and experimental studies have shown that osteocyte cell processes are significantly more mechanosensitive than the cell bodies [88–91].

Theoretical models were developed to investigate this canalicular environment, predicting the occurrence of pressure-driven interstitial fluid flow under applied global matrix strain (see Fig. 4A) [92] and resulting shear stresses and streaming potentials on the cell process membrane [94,95]. Further development led to the inclusion of an internal actin cytoskeleton in the cell process (see Fig. 4B), with tethering elements anchoring it to the surrounding ECM, greatly amplifying drag forces and strains to the cell process [82]. A refined theoretical model was then generated (see Fig. 4C), using contemporary experimental data on the flexural rigidity of the PCM tethering elements [86], and similarly found strain amplification occurring as a result of these tethering elements [93]. The most recent iteration of these models

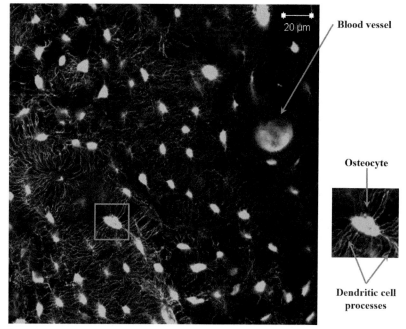

**FIG. 3**

Confocal scan of osteocyte network in cortical bone, with aspects of its mechanical environment highlighted.

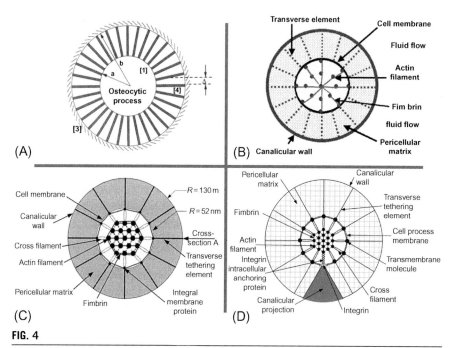

**FIG. 4**

Theoretical models of the canalicular environment by Cowin, Weinbaum, and coworkers [82,85,92,93], developing the strain amplification theory of osteocyte stimulation in vivo, to include (A) simple tethering elements, followed by (B and C) an actin cytoskeleton, and finally (D) canalicular projections.

incorporated the projections of the ECM (see Fig. 4D), which had recently been identified and shown to colocalize with proposed integrin attachments [85]. These attachments resulted in predicted strains an order of magnitude greater than those induced by the PCM tethering elements alone [85]. These models together represent the development of the strain amplification theory of osteocyte stimulation in vivo, later reinforced by computational models [96–99] and recently confirmed experimentally for the first time using confocal microscopy [100].

## 2.4 OSTEOCLASTS

Osteoclasts, derived from mononuclear precursor cells in the hematopoietic vascular channels in bone [60], are giant multinucleated cells that can range in diameter from 20 to 100 μm [101]. The main function of these cells is to break down and resorb bone matrix, and as such, they are usually found in temporary cavities on the bone surface known as Howship's lacuna during the bone remodeling process [102]. Osteoclast resorption is initiated by the activity of bone-lining cells, which retract from their location on the bone surface to provide access for the osteoclast to attach to the bone matrix [12]. The osteoclast then forms a sealing zone [103], which is a closed microenvironment [104] controlled by the cell and rendered highly acidic to digest bone tissue [105]. This is facilitated by a remarkable morphological feature known as the ruffled border (see Fig. 1), a complex folding of the cell's plasma membrane at the bone interface, providing an increased surface area through which acids and proteolytic enzymes are secreted to degrade the bone matrix [106].

## 3 BONE GROWTH AND ADAPTATION
## 3.1 BONE GROWTH

Bone formation occurs both during embryonic development and throughout life, adapting the geometry of bones, in response to changes in mechanical loading, or replacing aged or damaged bone. The bone formation process can occur via two specific pathways: endochondral ossification and intramembranous ossification. Both of these processes involve the initial deposition of an organic matrix, osteoid, or cartilage, which is then mineralized over time to produce the composite bone matrix. Mechanical loading can affect both of these processes, and the importance of loading during fetal bone development is evident from various animal models and human developmental conditions in which mechanical stimulation is absent or altered [107–110,110a], resulting in deficient bone development or malformed bone that has a propensity to fracture.

### 3.1.1 Endochondral ossification
Endochondral ossification is the process involved in embryonic bone formation and is initiated by MSCs creating a cartilage template upon which bone is formed [111]. This is performed by the differentiation of MSCs into chondroblasts, which then secrete a matrix composed of collagen and proteoglycans. Further differentiation

of these chondroblasts into chondrocytes results in the secretion of biochemicals and growth factors that initiate mineral deposition and promote vascularization of the template. Endothelial cells on the lining of these vessels produce essential growth factors that control the recruitment, proliferation, and differentiation of osteoblasts [112], emphasizing the importance of vascularization for bone formation [113,114]. This formation process through endochondral ossification is ongoing throughout childhood, particularly in the epiphyseal plate of long bones. New cartilage continues to be produced at this location during youth, which is replaced by bone, and thereby facilitates lengthening of bones.

### 3.1.2 Intramembranous ossification

Embryonic bone formation can also occur by means of intramembranous ossification, which regulates the formation of nonlong bones, such as the bones of the skull and clavicle. Critically, the intramembranous process does not rely on the formation of a cartilage template. Rather, embryonic stem cells within mesenchymal tissue of the embryo begin to proliferate and condense to form an aggregate of MSCs. This aggregate becomes surrounded by a membrane, and then MSCs within the membrane begin to differentiate directly to osteoprogenitor cells and then to osteoblasts. These osteoblasts line the aggregate and secrete the rudimentary collagenous extracellular matrix, which is known as the osteoid. Some of these osteoblasts become embedded within the newly formed matrix and become osteocytes. The cells on the outer surface form a periosteum, and mineralization begins to occur to form rudimentary bone tissue that is populated by osteocytes and lined by active osteoblasts [43]. This tissue eventually forms trabeculae, which then fuse to form woven bone that is then remodeled over time [43,115,116].

## 3.2 BONE MODELING AND REMODELING

The dynamic adaptive nature of bone is fundamental to its ability to protect and support the body throughout life, allowing it to optimize its structure and composition to provide maximum strength for minimal mass. Bone modeling is a process that facilitates growth and change in bone morphology and largely takes place in childhood and adolescence. Often described as uncoupled action of osteoclasts and osteoblasts, modeling occurs when bone resorption by osteoclasts occurs simultaneously to bone formation by osteoblasts but these process are active on different bone surfaces such that the dimensions of the bone are altered [14,117]. While the bulk of bone modeling occurs during early life, this process also continues in the adult skeleton, regulating both bone microstructure and overall bone geometry in response to changes in mechanical demands [118].

Bone remodeling is the process by which old bone is replaced with new tissue, enabling active repair of structural microdamage and adaptation of microarchitecture to local stress conditions [22]. The process of bone turnover begins at 2–3 years of age by replacing immature or primary bone present from infancy [43], and remodeling continues throughout life, resulting in renewal of bone every 20 years for cortical bone and 1–4 years for trabecular bone [44]. Bone remodeling, as the work of

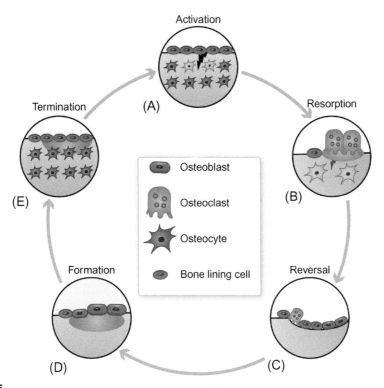

**FIG. 5**

Diagram showing the five nonresting phases of the bone remodeling cycle: (A) activation, (B) resorption, (C) reversal, (D) formation, and (E) termination.

*Adapted from Henriksen K, Neutzsky-Wulff AV, Bonewald LF, Karsdal MA. Local communication on and within bone controls bone remodeling. Bone 2009;44(6):1026–33.*

coupled osteoblasts and osteoclasts, is primarily a cell-controlled process [64]. Bone cells act in a coordinated fashion to remodel bone, in a cellular unit known as the bone multicellular unit (BMU) [119]. Within this unit, osteoclasts resorb unwanted/damaged bone to leave cavities, which are subsequently filled with new bone by osteoblasts [64]. This process, shown in Fig. 5, occurs in six successive phases: resting, activation, resorption, reversal, formation, and mineralization [43].

## 4 BONE MECHANOBIOLOGY

Crucial to the adaptive and regenerative nature of bone is bone cell mechanobiology, the effect of physical forces on cells and their ability to convert these forces into biochemical signals [120]. There is much experimental evidence of bone mass adapting to different loading conditions, with net bone resorption occurring at low strains and net bone formation occurring at high strains [121–126]. This process is governed by the cells, which work in concert as is outlined in Fig. 6. Each type of bone cell is subjected to mechanical cues due to skeletal loading [45,52,127] and can also elicit biochemical signaling to other cells in response to changes in their mechanical environment [85,128–138]. It is now known that many cells of the body have the ability to appraise their mechanical environment by means of specific molecule or protein complexes, known as mechanosensors. When these sensors are mechanically stimulated, bone cells can communicate the need for adaptation by producing certain biochemical signals (a process known as mechanotransduction), thereby initiating an adaptive response in other cells when the mechanical environment is not favorable. Multiple cellular mechanosensing mechanisms have been identified on osteoblasts and osteocytes including ion channels, gap junctions, integrins, primary cilia, and the actin cytoskeleton [135,139–142]. When these mechanosensors are mechanically

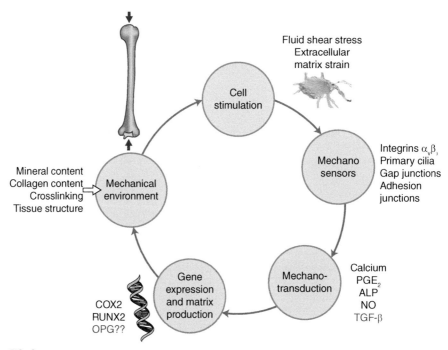

**FIG. 6**

The mechanical environment in bone is defined by the tissue composition (mineral, collagen, and cross-linking), the tissue microarchitecture or structure, and the extrinsic loads applied during daily activities. These loads are transmitted at the cell level as a combination of fluid shear stress and extracellular matrix strain. As introduced above, bone cells possess various mechanosensory organelles, which trigger intracellular signaling and activate mechanotransduction via extracellular signals and molecular-level changes in gene expression and matrix production. Such changes alter the tissue composition and structure, and the mechanobiological process continues until the material has adapted to the applied loading conditions and achieves a homeostatic mechanical environment.

stimulated, a number of important intracellular signaling cascades are activated [141,143] including proliferation, differentiation, matrix production and cell death, which govern cell behaviour. The resulting matrix production can alter the mechanical environment and thus the mechanical stimulation of the cell. Mechanosensation, mechanotransduction, and the mechanical environment of bone cells will be discussed in this section.

## 4.1 MECHANOSENSATION AND MECHANOSENSORS

Mechanosensors are specific protein or molecule complexes present on the surface of the cell that can detect a physical stimulus. They are broadly organized into three types [144]: (1) attachments between the individual cells, such as stretch-activated gap junctions and adhesion junctions; (2) structures on the cell membrane that can deform under fluid flow, such as the primary cilium; and (3) attachments between the membrane and the ECM, which are predominantly composed of integrins. In each of these cases, a mechanical stimulus is transmitted from the whole bone level down to the external cellular mechanosensor and thereby transmitted into the cytoskeleton or cytoplasm, with the potential to induce a biochemical cascade.

### *4.1.1 Integrins*

Integrins, which are transmembrane cell adhesion heterodimers comprising α and β subunits, bind directly to the ECM at anchoring points known as focal adhesions [117]. It is well-known that integrins are required for a load-induced bone remodeling response in osteocytes [145]. Of these integrins, the β1 subunit appears to be important for osteoblast cell signaling [146], due to the observation that knockdown of β1 results in decreased bone mass and increased cortical porosity. In osteocytes, expression of β1 has been reported in the cell body, while β3 is expressed primarily on the cell processes [83]. Upregulation of β1 has been reported in osteoblasts in response to fluid flow [130], and there is evidence of lower load-induced bone formation after osteocyte-specific β1 gene ablation in mice [140], which together indicate that β1 binding and signaling are a critical flow-sensitive pathway in bone cells. Separately, in osteocytes, it has been observed that punctate integrin attachments spaced along the canalicular wall exist and appear to colocalize with projections of the ECM into the pericellular space [83]. It has been proposed that osteocytes may sense matrix strain and fluid-induced deformation in bone canaliculi via integrin-based ($\alpha_V\beta_3$) focal attachments between their cell processes and these ECM projections (see Fig. 7) [83,85]. Interestingly, an in vitro study that blocked integrin $\alpha_V\beta_3$ in MLYO4 osteocytes reported that the osteocyte morphology was altered by means of a reduction in spread area and retraction of cell processes [147]. Moreover, integrin $\alpha_V\beta_3$ blocking also disrupted cyclooxygenase-2 (Cox-2) expression and prostaglandin E2 ($PGE_2$) release in response to fluid shear stress, which suggests that integrin $\alpha_V\beta_3$ is essential for the maintenance of osteocyte cell processes and also for mechanosensation and mechanotransduction by osteocytes [147].

Integrins bind to proteins known as ligands such as osteopontin (OPN), an important phosphorylated bone matrix protein located in the cement lines of osteons and

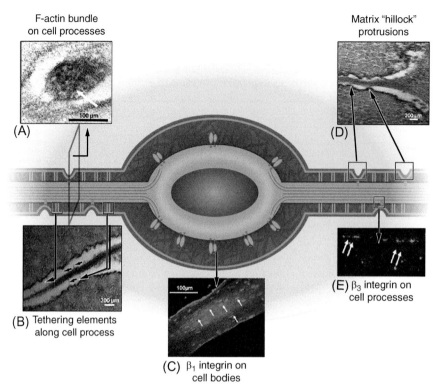

**FIG. 7**

Schematic representation of key mechanisms in the osteocyte environment implicated in mechanosensing: (A) TEM image of an osteocyte process displaying the actin cytoskeleton [86]; (B) TEM image of proteoglycan PCM tethering elements *(black arrows)* bridging an osteocyte cell process to the bony canalicular wall; (C) fluorescent immunohistochemical staining showing that $\beta_1$ integrins *(white arrows)* are located only on osteocyte cell bodies [83]; (D) TEM image demonstrating the discrete ECM projections from the canalicular wall that contact osteocyte processes; and (E) fluorescent immunohistochemical staining for $\beta_3$ integrins *(white arrows)* that are present in a punctate pattern along osteocyte processes, with similar periodicity and spacing pattern to ECM projections [83].

Adapted from Schaffler M, Cheung W-Y, Majeska R, Kennedy O. Osteocytes: master orchestrators of bone. Calcif Tissue Int 2013:1–20.

along the walls of canaliculi [23,148–151], which is regulated by calcium signaling [152]. Osteopontin is also believed to play a role in osteocyte mechanosensation by mediating their attachment to surfaces, and its absence has been shown to change the response of bone to loading [153].

### 4.1.2 Primary cilia

Primary cilia, nonmotile microtubule-based cellular organelles that project from the surface of many cell types, are known to be present in a number of cell types in bone tissue [154]. As these structures have been implicated as fluid flow sensors involved in mechanotransduction in other tissues [155], most prominently in the kidney [156], it has been proposed that primary cilia may allow bone cells to sense strain-derived fluid flow in vivo [157,158]. This hypothesis has been supported by in vitro studies identifying primary cilia on osteoblast- and osteocyte-like cell lines in culture [135,159], which have demonstrated osteogenic responses under in vitro fluid flow [135]. Additionally, the primary cilium has recently been shown to both mediate the recruitment of, and promote osteogenic differentiation by MSCs in bone [160,161]. Furthermore, knockdown of primary cilia in the bone marrow resulted in a decrease in the MSC population, indicating that their presence is necessary to maintain homeostasis [162]. However, in vivo primary cilia only exist in ~4% of osteocytes and bone-lining cells [163] and ~1% of bone marrow cells [163], and such a low presence of ciliated cells might limit their efficacy in a mechanosensory capacity.

### 4.1.3 Ion channels

Ion channels are pore-forming membrane proteins that allow ions to pass through the cell membrane, with various identified pressure-sensitive ion channels in osteoblastic cells [164,165]. The characteristics of ion channels themselves can be altered by mechanical loading, as has been demonstrated in vitro by means of experiments that reported increased numbers of open channels in response to cyclic stretch [166] and increased calcium signaling in response to fluid flow [167]. Furthermore, parathyroid hormone (PTH) is a significant activator of ion channels and a strong regulator of blood calcium levels [168]. PTH treatment has also been found to significantly enhance calcium signaling in response to fluid flow by activating both voltage-sensitive and mechanosensitive calcium channels [169]. While the manner in which mechanical stimuli mediate ion-channel behavior remains unknown, the strong cellular localization of the calcium-dependent binding protein annexin V when exposed to fluid flow suggests that exposure to flow changes the structure and activity of ion channels [170].

Members of the transient receptor potential (TRP) cation channel superfamily have been implicated due to the high calcium selectivity of TRPV4, 5, and 6 [171]. Interestingly, mice in which the TRPV4 gene was ablated displayed increased trabecular bone mass and a reduction in osteoclast activity [172], whereas TRPV5-null mice exhibited decreased cortical and trabecular bone thickness and hypercalciuria [173]. Moreover, TRPV4, which localizes to primary cilia [174], has been shown to be activated by mechanical stress and has been identified in human and murine osteoblast-like cells [175,176], suggesting it as a target for further study.

Finally, polycystin-1 (a transmembrane protein) and polycystin-2 (a TRP superfamily calcium channel) are expressed in both osteoblasts and osteocytes [159,177] and are known to localize on the primary ciliium [135]. Abnormal skeletal development after the ablation of genes encoding polycystin-1 [178,179] implicates polycystins as targets for further research in the understanding of ion channels for mechanosensation.

### 4.1.4 Cytoskeleton

As the cytoskeleton comprises a web of organized filaments (microtubules, actin microfilaments, and intermediate filaments), these fibers themselves act as crucial mechanoreceptors. The concept of tensegrity (tensional integrity), developed by Ingber [180], assumes that these elements are contractile stress fibers under tension, generating a prestressed network balanced by attachments to the ECM and thereby acting to directly connect focal adhesions to the cell nucleus. Osteoblasts exposed to steady flow have been observed to upregulate Cox-2 and c-fos while simultaneously rearranging their cytoskeleton [181]. Interestingly, under oscillatory flow, the cells did not rearrange their actin cytoskeleton [135], and the disruption of all three cytoskeletal components did not result in changes in $PGE_2$ release [182]. However, the disruption of microtubules abrogated flow-induced proliferation and matrix metalloproteinase (MMP) production [183,184], while disarrangement of the actin cytoskeleton attenuated flow-induced expression of Runx2 [185].

The pericellular matrix [82], which is directly connected to the cytoskeleton [86], has been proposed to transfer mechanical stimulation from fluid drag forces. Indeed, degradation of the PCM results in diminished responsiveness of osteocyte-like cells to fluid flow [186], whereas stress fiber formation and the number of cell processes were found to increase under fluid flow [187]. Cytoskeleton-mediated mechanosensing of fluid flow may vary between osteocytes and osteoblasts. Fluid flow increases calcium ($Ca^{2+}$) signaling more in osteoblasts than osteocytes [188] and induced recruitment of the actin-binding protein fimbrin in osteoblasts but not osteocytes [189]. Conversely, the calcium response of osteoblasts was affected by the disruption of focal adhesions whereas that of osteocytes was not [188]. Furthermore, $PGE_2$ production under fluid flow decreased with cytoskeletal disruption in osteocytic cells but increased in osteoblastic cells [190]. It is thought that some of these differences may be due to the greater presence of cytoskeleton-modulated stretch-activated ion channels in osteocytes, compared with osteoblasts [191]. Taken together, this evidence suggests that the cytoskeleton is critical for flow-induced osteogenic differentiation [117].

### 4.1.5 Gap junctions and hemichannels

Interestingly, recent in vitro experiments have observed the proliferation of both nitric oxide (NO) and intracellular calcium signaling through networks of bone cells [192–195], demonstrating their ability to communicate as a network. Moreover, osteocyte networks were found to be more mechanoresponsive in calcium signaling than osteoblast networks [193], illustrating the importance of the osteocyte network for mechanosensation and signal transduction in bone remodeling. Gap junctions are thought to provide the key mechanism for cell-cell communication, both between osteocytes and between osteocytes and osteoblasts [196,197], and appear to play an important role in signal transduction between cells on the bone surface to osteocytes deep within the matrix. While gap junctions are also present in osteoclasts, it is not known whether they provide a communication channel with other bone cells [198].

Each gap junction is an intercellular channel formed by two connexons; membrane proteins formed from six connexion subunits (primarily Cx43 in osteocytes) [199]. Gap junctions are known to directly connect adjacent osteoblasts and osteocytes, functionally coupling them both in vivo and in vitro [200]. It is also possible for them to form channels independent of contact with other cells, known as hemichannels, that provide communication between the internal cellular space and the surrounding extracellular environment [201]. Interestingly, studies have shown that mechanical loading in mice increases Cx43 expression in osteoblasts and more so in osteocytes [202]. These gap junctions and hemichannels appear closely related with flow-induced adenosine triphosphate (ATP) and $PGE_2$ release in osteocytes [203–206], as their release appears to be mediated by hemichannels [204] and hemichannels themselves also appear to be mediated by ATP release [207]. Furthermore, the flow-induced expression of the important osteoblast regulatory actors OPN and osteocalcin (OCN) is significantly reduced when gap junctions are blocked [208]. Therefore, the mechanical activation of gap junctions and hemichannels appears to constitute an important sensory mechanism and transduction pathway for both osteocytes and osteoblasts under fluid flow.

### 4.1.6 G-protein-coupled membrane receptors

Guanine nucleotide-binding G-protein-coupled receptors (GPCRs) are the largest transmembrane receptors [209,210], each crossing through the cell membrane seven times, detecting molecules outside the cell and activating internal signal transduction pathways in response. GPCRs activate both protein kinase A (PKA) and mitogen-activated protein kinase (MAPK) signaling, as well as Rho GTP-binding proteins [117]. In mice, ablation of GPCRs in osteoblasts resulted in decreased cortical and trabecular bone volume [211,212], while the reverse was true when they were overexpressed [213]. Given that GPCRs are linked to these osteogenic signaling pathways, and that they can be activated under fluid flow, they represent a potential target mechanoreceptor for further study [117].

## 4.2 MECHANOTRANSDUCTION AND SIGNALING PATHWAYS

All cells in the osteoblastic lineage can transduce mechanical strain signals into biochemical cues for osteogenesis [214]. It has also been observed that osteoblastic cells respond with biochemical signals to both direct matrix strain [215,216] and fluid flow in vitro [190,217,218], suggesting multiple mechanical mechanisms of bone cell stimulation in vivo. These mechanical stimuli are transduced into various biochemical cues, some early (e.g., intracellular calcium signaling) and some late (e.g., $PGE_2$ release and Cox-2 expression), which can orchestrate bone remodeling (Fig. 8).

Osteoblasts and osteocytes produce ATP, which activates receptors on neighboring cells to produce calcium, and a gap-junction influx of extracellular calcium occurs [218a]. This is followed by the production of various factors: alkaline phosphatase (ALP), $PGE_2$, NO, OPN, OCN, MMPs and extracellular signal-regulated

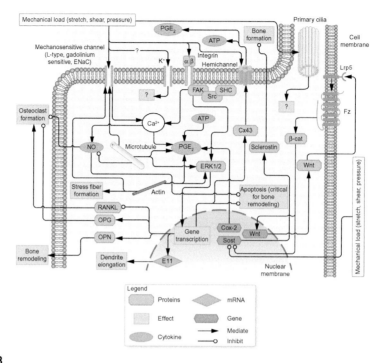

**FIG. 8**

Schematic representing much of the current understanding of the key mechanisms in osteocyte mechanobiology. While many of these mechanisms have been identified and defined, the complex interplay between mechanical stimulation, the subsequent signaling pathways, and the resulting bone remodeling are still unknown.

Adapted from Chen J-H, Liu C, You L, Simmons CA. Boning up on Wolff's Law: mechanical regulation of the cells that make and maintain bone. J Biomech 2010;43(1):108–18.

kinase (ERK1/ERK2), and MAPK [217,219–223]. These events are followed by the downregulation of sclerostin expression [224,225] and expression of Wnt proteins [226,227]. In vitro application of fluid shear stress to osteoblasts induces proliferation [130,228] and expression of osteogenic genes including c-fos [181,229], early growth response-1 (EGR-1) [229,230], and Cox-2 [228]. In vitro studies have demonstrated that $PGE_2$ release is facilitated by Cx43 hemichannels, under fluid flow [205]. In vivo mechanical loading stimulates the expression of growth factors (c-fos and IGF-1) by osteocytes, and NO and prostaglandin are crucial to the initial transduction of the mechanical stimulus into an osteogenic response [231]. Mechanical loading imposed on cell culture substrates at high strains (i.e., above those observed in vivo) leads to $Ca^{2+}$ ion flux [232] and production of collagen and noncollagenous proteins [233,234], as well as anabolic factors, such as $PGE_2$ [235], PTH [236], and insulin-like growth factor [216].

The most important pathways of this cascade and our current understanding of them are hereafter explained in more detail.

### 4.2.1 Calcium signaling

Intracellular calcium is an early signaling event in bone mechanotransduction and facilitates multiple important cell functions [237,238], including proliferation, differentiation, and cell motility [237]. Calcium is known to act as an early second messenger in bone cells, with both intracellular release and extracellular influxes occurring rapidly in response to strain, pressure, and fluid flow [216,222,239–241]. When calcium release is blocked in vitro, mechanically stimulated osteogenic gene expression is prevented [152,219,242]. Important signaling events occur downstream of calcium signaling in both osteoblasts and osteocytes, including PKA, MAPK, c-fos, and Cox-2 expression [243,244].

### 4.2.2 Wnt signaling

Canonical Wnt signaling is known to heavily influence the maintenance and development of bone by affecting cell differentiation, proliferation, migration, polarity, and gene expression [245]. Canonical Wnt signaling mediates Cox-2 and Runx2 gene expression and osteoblast mineralization [246–248]. While downregulation of canonical Wnt signaling encourages MSC renewal, upregulation has been shown to encourage osteogenic differentiation of MSCs [249–251]. β-Catenin is an important component of the canonical Wnt signaling pathway [252]. The canonical Wnt signaling pathway has multiple regulatory mechanisms, including adhesion junction sequestering of β-catenin, degradation of cytoplasmic β-catenin when canonical Wnt signaling is not active, and conformation changes in β-catenin and other receptors during active canonical Wnt signaling [253]. These steps all act in concert to regulate canonical Wnt signaling.

### 4.2.3 MAPK

A group of serin-/threonine-specific protein kinases known as mitogen-activated protein kinases (MAPKs) have been found to have a critical role in cell differentiation, proliferation, and apoptosis [254,255]. Mechanical strain is known to induce rapid initial phosphorylation of MAPKs [256–259]. In bone cells, MAPKs act through downstream activation of ERK1/ERK2, observed in response to fluid flow [260–263]. This phosphorylation of ERKs in response to fluid flow regulates receptor activator of nuclear factor kappa-B ligand (RANKL) expression [264], proliferation [265], NO production [266], and expression of MMP-13 [263] and appears to be modulated by both calcium and ATP [244].

### 4.2.4 G-Protein-related signaling

The Rho and Ras GTPases are subunits of G proteins [267] and under flow stimulus can activate PKA, resulting in the expression of Cox-2 [268]. Mechanical stimulation is known to activate G proteins [269,270], with this activation leading to $PGE_2$

release [271]. Furthermore, the inhibition of RhoA has been found to significantly reduce Runx2 expression and osteogenic differentiation in osteoblasts [185], implying a role in bone diseases.

### 4.2.5 Nitric oxide

NO is another second-messenger biochemical signal that plays a critical role in various biological processes throughout the body, maintaining vascular tension, neuronal activity, and function of the immune system [272]. NO is known to control mechanically stimulated bone formation in vivo [273,274], with inhibition of this signal suppressing bone formation [275]. Furthermore, in vivo mechanical loading elicits greater NO production in bone cells [276,277] and is linked to $PGE_2$ release and MAPK signaling [190,217,219,220,266,278–281].

## 4.3 MECHANICAL ENVIRONMENT OF BONE CELLS

The osteocyte has been implicated as the primary mechanosensory bone cell, given the fact that it has been shown to orchestrate osteogenic responses in other bone cells [128], and constitutes over 90% of bone cells in vivo [43]. The osteocyte's location deep within the bone tissue makes direct experimental observation challenging, without destructive interference with its environment. High-resolution imaging of the lacunar-canalicular network under mechanical loading was first limited to 2D imaging of lacunae on an exposed optical microscopy plane [282,283], and such studies demonstrated that applied strains at the whole bone level are amplified in the lacunar matrix [284]. Strain stimulation within ex vivo osteocytes was observed for on a confocal microscopy plane [100], and this study reported magnitudes of strain sufficient to elicit osteogenic signaling and observing localization of maximum strains along the cell processes. Furthermore, ex vivo imaging platforms have demonstrated mechanically stimulated intracellular calcium signaling in live osteoblasts and osteocytes [285,286]. As loading-induced fluid flow is thought to be highly stimulatory to osteocytes in vivo, fluorescent tracer studies have been performed to examine the fluid flow through the lacunar-canalicular network under mechanical loading, measuring the average fluid velocity and predicting resulting shear stresses [55,57,287].

Theoretical models and recent computational models of the environment have demonstrated increased strain and fluid shear stress stimulation of the cell projections of the ECM into the canaliculi [85,97,98], with a detailed review of computational studies found elsewhere [288] (Fig. 9). Another mechanism by which the osteocyte has been hypothesized to sense fluid flow is via flow-induced drag force on tethering elements of the pericellular glycocalyx, which attaches the cell to the lacunar-canalicular wall [82,86]. This theory is reinforced by the evidence that enzymatic removal of this glycocalyx in vitro reduces the ability of the osteocyte to respond via the $PGE_2$ pathway [186]. However, modeling of these tethering elements suggests that reduction in number does not significantly affect osteocyte mechanical stimulation [99]. Similarly, the importance of the primary cilium as an osteocyte

# 4 Bone mechanobiology

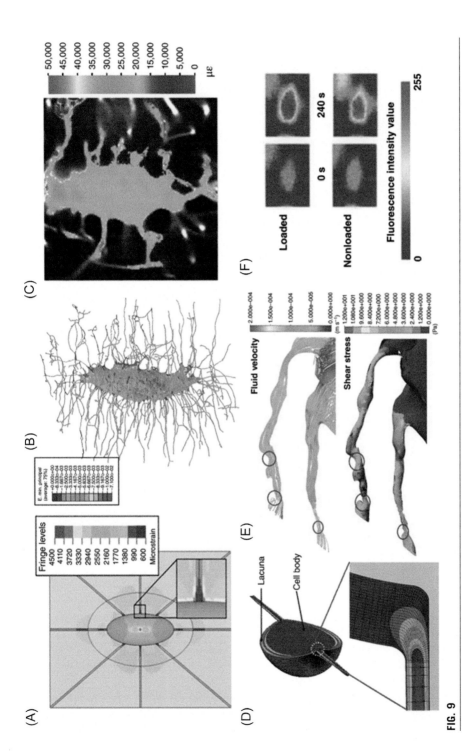

**FIG. 9**

Combinations of computational and experimental approaches, which have enabled the investigation of strain within the osteocyte and the effects of fluid flow around the cells in vivo: the evolution of FE models from (A) idealized lacunae [289] to (B) osteocyte geometries from X-ray nanotomography [290], predicting strain amplification that has been (C) validated experimentally [100]. Advances from (D) CFD models of osteocytes [96] with the development of (E) FSI techniques [98], predicting velocities and shear stresses that have been (F) validated by tracer studies [55].

*Adapted from Giorgi M, Verbruggen SW, Lacroix D. In silico bone mechanobiology: modeling a multifaceted biological system. Wiley Interdiscip Rev Syst Biol Med 2016;8(6):485–505.*

mechanosensor in vivo is uncertain as computational models showed that, while highly stimulated in vitro, primary cilia in vivo are less stimulated than attachments unless they are partially embedded in the ECM [291].

## 5 BONE MECHANOBIOLOGY AND DISEASE

Traumatic injuries, the process of aging, and various pathological diseases, such as osteoporosis, metastatic bone disease, osteogenesis imperfecta, and Paget's disease, can impair the normal function of bone, leading to bone fractures. Often, these fractures do not repair (nonunions) or cause immobility, severe pain, and deformity. This section will discuss what is known about bone mechanobiology in some of these conditions.

### 5.1 BONE MECHANOBIOLOGY AND OSTEOPOROSIS

Osteoporosis is a debilitating bone disease, in which an imbalance in normal bone cell remodeling results in severe bone loss and significantly reduced bone strength, leading to fractures of the hip, wrist, or vertebrae. The disease is most commonly manifested in women following the menopause when estrogen production not only is deficient (type I osteoporosis) but also is prevalent in aged men and women, arising from deficient bone turnover in senescence (type II osteoporosis) [292-296].

During osteoporosis, the levels of circulating estrogen in the blood are deficient, which is well understood to increase the number and resorption activity of osteoclasts, ultimately leading to bone loss [297,298]. The relationship between estrogen and osteoclast behavior (activation frequency, turnover rate, apoptosis, and resorption capacity) has been widely studied [297–301], and altered osteoclast behavior is known to increase propensity for fracture [302]. However, a study proposed that estrogen deficiency primarily alters osteoblasts and that osteoclast resorption may be a secondary effect mediated by the osteoblasts [303]. Moreover, various changes in osteoblast and osteocyte biology have been recently identified, which would appear to indicate that changes in bone mechanobiology also occur in osteoporosis, as is considered in-depth below.

#### *5.1.1 Mechanosensors*

Although the osteocyte was overlooked for many years with regard to the etiology of osteoporosis, it is now emerging that changes in osteocyte structure and biology do indeed arise at the onset of the disease. Estrogen deficiency induces osteocyte apoptosis [304,305], which might result in hypermineralization of the surrounding tissue [306–308]. In addition, the organization of the osteocyte network is altered [309], and osteocyte density is reduced [310]. In aged individuals, there is a diminished osteocyte network, characterized by fewer canaliculi per osteocyte lacuna, connections within the osteon and interosteon connections [311]. Furthermore, it has been shown that osteoclast resorption following estrogen loss does not occur without apoptosis of osteocytes, and therefore, osteocyte apoptosis is necessary to activate endocortical remodeling following estrogen loss [312]. Given the important role of the osteocyte network for mechanosensation, such changes may reduce the mechanoresponsiveness of the tissue [311,313].

A recent study sought to delineate whether the distribution of integrin-based mechanosensory complexes (i.e., $\beta_1$ or $\beta_3$-integrin receptors) was altered during osteoporosis [314]. Interestingly, while $\beta_1$ integrins were equally expressed in marrow and cortical bone of control and osteoporotic animals, the expression of $\beta_3$-integrin receptors decreased in the osteoporotic cortical bone tissue. In the bone marrow and at the endosteum, there was no difference in $\beta_3$-integrin expression, albeit that TRAP staining for osteoclasts showed a marked increased at the osteoporotic endosteum [314]. Such changes are likely to alter mechanotransduction, as integrins play a role in osteocyte mechanosensation by regulating Cox-2 expression and $PGE_2$ release [147]. In a separate study, the $\beta_1$ integrin was shown to play an important role in modulating signaling and gene expression in osteoblast-like cells in response to both estrogen treatment and fluid shear stress [315]. Interestingly, the estrogen induced $\beta_1$-integrin expression and the estrogen antagonist inhibited integrin expression.

Adhesion junctions are important for mechanosensation of fluid shear stress through the translocation of β-catenin into the nucleus of osteoblasts and MSCs following fluid flow stimulation [246,247]. P-cadherin deficiency results in limb defects [316,317]. N-cadherin adhesion junctions control osteoblast differentiation and bone mass in vitro and in vivo, via canonical Wnt signaling [318,319]. Adhesion junctions can sequester β-catenin, an important molecule for canonical Wnt signaling, in the adhesion junction complex, and genome-wide association studies have identified β-catenin as a gene likely associated with osteoporosis [318–320]. Additionally, N-cadherin plays a complex role in maintaining the hematopoietic stem cell (HSC) population; HSCs are maintained in a quiescent state in the MSC niche via attachment to N-cadherin-positive osteoblasts on the endosteal surface of the bone [321,322]. HSCs differentiate into an array of myeloid and lymphoid cells that together make up the cells of the blood system [323,324]. Overexpression of osteoblast N-cadherin inhibits the division of HSCs, while knockdown of HSC N-cadherin reduced the adhesion of HSCs onto the bone surfaces [325].

### *5.1.2 Mechanotransduction*

Mechanical stimulation of both osteoblasts and osteocytes activates specific signaling pathways [141], which stimulate the production of anabolic growth factors and synthesis of ECM proteins and mineral [214]. Studies have shown that estrogen plays an important role in the responsiveness of bone cells to in vitro mechanical stimuli [315,326–329]. Osteoblasts and osteocytes possess receptors for estrogen [330], and estrogen promotes $PGE_2$ and NO release in response to applied shear stress [326]. In vitro studies have demonstrated the importance of estrogen for modulating signal transduction (ERK and MAPK) and gene expression (c-fos and Cox-2) by osteoblast-like cells (MG63) and primary osteoblasts in response to flow-induced shear stress [315]. An antagonist of estrogen receptor (ICI 182,780) abrogated the

ERK, MAPK, c-fos, and Cox-2 responses in MG63 cells in response to fluid shear stress [315]. In vitro cell cultures of primary osteoblasts in an estrogen-depleted environment showed an impaired proliferative response to strain [328,329]. Human osteoblasts from osteoporotic donors show differences in proliferation and transforming growth factor-β (TGF-β) release in response to cyclic strain compared with those from nonosteoporotic donors [331]. It was proposed that osteoporotic osteoblasts require a different mechanical loading regime to become stimulated. Human osteoblasts from osteoporotic donors also demonstrated impaired $PGE_2$ release in response to flow-induced shear stress [332]. Recent studies have discovered alterations in the behavior of osteoblasts and osteocytes to estrogen deficiency and mechanical stimulation (apoptosis, calcium signaling, NO, $PGE_2$, osteogenic gene, and protein expression) [333–336]. In particular, murine MLO-Y4 osteocytes treated with estrogen and then exposed to osteoporotic conditions, either by fulvestrant treatment or by estrogen withdrawal, showed a significant change in the biological responses when exposed to mechanical loading [336] (see Fig. 10). Specifically, upon estrogen withdrawal or estrogen receptor inhibition in vitro, to simulate estrogen deficiency during postmenopausal osteoporosis, these osteocyte-like cells exhibited a reduction in NOS activity, NO and $PGE_2$ release, $[Ca^{2+}]_i$ oscillations, and actin stress fiber formation, compared with estrogen-treated controls. Moreover, the expression of bone-related genes in mechanically stimulated MLO-Y4 osteocytes was downregulated following estrogen withdrawal and inhibition. Thus, it was proposed that estrogen deficiency during postmenopausal osteoporosis may affect mechanotransduction in osteocytes via disturbed $[Ca^{2+}]_i$ oscillations (Figs. 2B and 3) and thereby alter osteocyte differentiation (Fig. 11).

### 5.1.3 Biomechanical changes and mechanical environment

In recent studies, it has been shown that the mechanical properties [23,337] and the tissue mineral content [23,337,338] are altered in bone tissue from animal models of osteoporosis. It is intriguing to speculate that such changes might be driven by perturbations in the normal mechanobiology of bone during estrogen deficiency. Firstly, changes in bone tissue mineral content might occur to compensate for bone loss, whereby osteocytes embedded in the remaining tissue experience altered mechanical stimuli, that is elevated strains or fluid flow, within the depleted trabecular architecture arising from bone loss. Indeed, computational studies have predicted higher strains within bone tissue of the osteoporotic femur during loading conditions that simulate a sideways fall [339]. Other computational studies have predicted a more heterogeneous strain state within individual trabeculae in osteoporotic cancellous bone [340]. Therefore, as a secondary response to this elevated mechanical stimulation, osteocytes might initiate an adaptive response by osteoblasts to bone tissue composition to compensate and return mechanical strains to normal levels.

Theoretical algorithms have been developed to predict bone remodeling in combination with finite element (FE) methods [337,341,342] and have been applied specifically to investigate whether mechanical stimuli arising during bone remodeling contribute to the rapid loss of bone trabeculae. It was found that a mechanobiological

5 Bone mechanobiology and disease 181

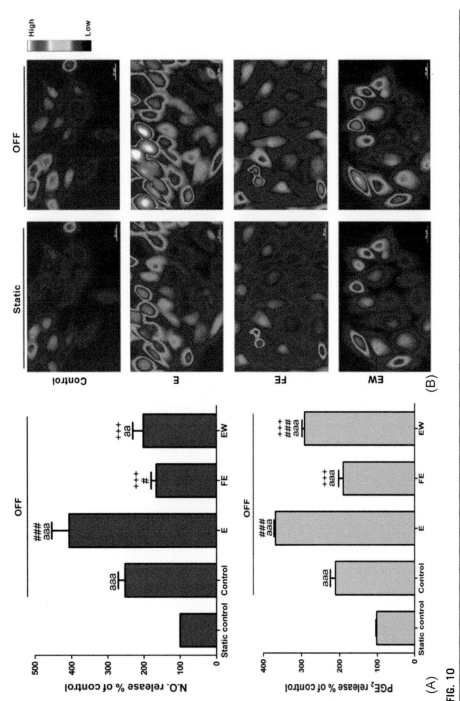

**FIG. 10**

Mechanobiological responses of MLO-Y4 osteocytes cultured with estrogen (E/E2, 10nM) or under simulated postmenopausal conditions (*LE*, low estrogen; *FE*, fulvestrant (100nM); *EW*, estrogen withdrawal) showing (A) secondary mediator release (NO and PGE$_2$) and (B) [Ca$^{2+}$]$_i$ oscillations after 1h of oscillatory fluid flow (OFF).

*From Deepak V, Kayastha P, McNamara LM. Estrogen deficiency attenuates fluid flow-induced [Ca2+]i oscillations and mechanoresponsiveness of MLO-Y4 osteocytes. FASEB J 2017.*

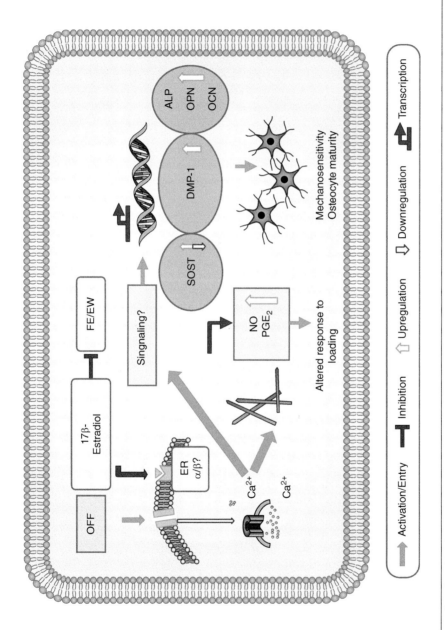

**FIG. 11**

Oscillatory fluid flow (OFF) stimulates the mechanoreceptor (e.g., integrin $\alpha_v\beta_3$, primary cilia, and adhesion junction) on osteocytes, which leads to intracellular calcium ($Ca^{2+}$) oscillations. Estrogen (17β-estradiol) acts in an additive manner by binding to $ER_{\alpha/\beta}$ receptors to increase the $Ca^{2+}$ oscillatory response. $Ca^{2+}$ oscillations lead to actin fiber formation, NO and $PGE_2$ release (secondary mediators of bone formation), and activation of signaling pathways and altered transcription of bone-specific genes (ALP, OPN, OCN, DMP-1, and SOST). Addition of estrogen receptor blocker fulvestrant in combination with estrogen or withdrawal of estrogen disrupts estrogen-mediated additive response on $Ca^{2+}$ oscillations, which in turn attenuates NO and $PGE_2$ release and expression of ALP, OPN, OCN, DMP-1, and SOST and thus diminishes osteocyte mechanosensitivity [336].

algorithm based on strain and microdamage can simulate the remodeling cycle in trabeculae. This model was applied to predict that there is a depth of resorption cavity deeper than which refilling of the resorption pits is impossible, and perforation inevitably occurs [341]. The simulation also predicted that perforations would occur more easily in trabeculae that are more highly mineralized (stiffer), such as those arising in osteoporosis [341,343]. Using a similar approach, it was also predicted that if cells become less mechanosensitive, the possibility of trabecular perforation and therefore the rapid loss of bone mass increase [343]. Others have shown that if bone tissue stiffness increases, bone structure adapts to achieve a lower mass and increased anisotropy similar to osteoporotic bone [344].

Multiscale-multiphysics computational models have been developed to account for the complex mechanical environment experienced in vivo and predicted the mechanical environment of various bone cells [39,52,98,291,345–347]. Such modeling approaches have been applied to understand the role of changes in mineral crystal dimensions for tissue-level changes in material properties of osteoporotic bone. A multiscale modeling approach, which discretely represents the cellular constituents of trabecular bone marrow in vivo, was applied to investigate how the increase in the fat content of bone marrow, arising in osteoporosis [348], affected mechanical stimulation in the cellular microenvironment of trabecular bone marrow [347]. It was predicted that mechanical stimulation levels in trabecular bone marrow cells during osteoporosis remained much higher than those predicted to occur under healthy conditions. This study also investigated whether increased trabecular stiffness and axial alignment of trabeculae would be effective in returning MSC stimulation in trabecular marrow to normal levels [347]. A multiscale computational scheme incorporating high-resolution, tissue-level, fluid-structure interaction simulations with discrete cell-level models was applied to characterize the potential effects of trabecular porosity and marrow composition on marrow mechanobiology in human femoral bone [349]. At the tissue level, induced shear stress in the marrow increased with bone volume fraction and strain rate. The maximum shear stress decreased when an adipocyte volume fraction (AVF) representative of osteoporotic patients (45% and 60%) was simulated, suggesting that increasing AVF similarly alters mechanobiological signaling in bone marrow [349].

Using a novel micromechanical loading and confocal imaging technique, direct experimental investigation of the local mechanical environment of osteocytes and osteoblasts in situ was conducted [99,100]. It was shown that the mechanical environment of osteoporotic bone cells is altered; osteocytes are exposed to higher strains than healthy bone cells after short durations of estrogen deficiency (5 weeks), whereas there is no significant difference in the mechanical stimulation of bone cells in healthy and osteoporotic bone in long-term estrogen deficiency (34 weeks) [100]. Thus, it was proposed that the mechanical environment of bone cells is altered during early-stage osteoporosis and that mechanobiological responses act to restore the mechanical environment of the bone tissue after it has been perturbed by ovariectomy [100], perhaps in an attempt to restore homeostasis [99,100]. It was previously proposed that such altered mechanical stimulation may elicit a secondary adaptive mechanobiological response from the osteocytes and osteoblasts to alter the local mineral composition of the bone tissue to compensate for this bone loss [350]. This would ultimately lead to more brittle bone, microdamage, and an associated greater risk of fracture, together with the depleted bone mass and architecture (see Fig. 12).

Taken together, recent studies have evolved scientific understanding of the mechanobiological mechanisms underlying osteoporosis and have revealed for the first time that there are complex changes in (a) bone mechanosensation mechanisms, (b) mechanical environment of bone cells, and (c) biochemical responses and matrix

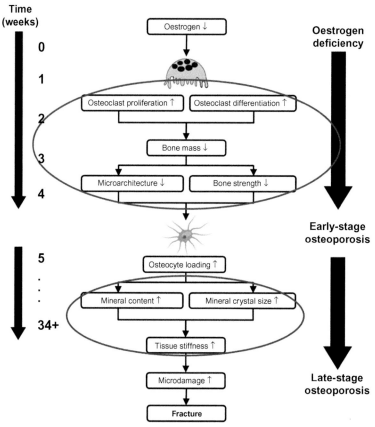

**FIG. 12**

Biomechanical changes and the proposed roles of the altered mechanical environment of osteocytes in the bone loss cascade during osteoporosis. Bone loss results in elevated mechanical stimuli in osteocytes, which may elicit a compensatory adaptive mechanobiological response to alter the local mineral of the bone tissue [350]. However, this ultimately leads to more brittle bone; microdamage; and, together with the depleted bone mass and architecture, a greater risk of fracture. Time scale indicates progression following ovariectomy in a rat model of osteoporosis.

production by bone cells during estrogen deficiency. These studies highlight the necessity for mechanical stimulation and hormone balance in unison to maintain bone mineralization and suggest that impaired mechanotransduction by osteoblasts during estrogen withdrawal may be a contributing factor in bone loss and increased fragility during osteoporosis.

## 5.2 BONE MECHANOBIOLOGY DURING OSTEOARTHRITIS

Osteoarthritis (OA) is a chronic, progressive, degenerative joint disease, which renders approximately 27 million sufferers in the United States disabled [351]. OA is primarily associated with degeneration of the articular cartilage, which also leads to a significant deterioration in the subchondral bone [352,353]. In particular, epiphyseal bone adaptation and osteophyte formation have been observed in an in vivo tibial loading model [354]. Most interesting sites of subchondral cortical bone thickening correspond to the regions of cartilage thinning, suggesting that each tissue responds to changes in the other tissue and the ensuing changes in the local mechanical environment [354]. Subchondral bone remodeling plays an important role in the pathogenesis of OA, and thus, inhibitors of bone resorption have been proposed a potential therapeutic for the treatment of OA [355,356]. OA is also common following alterations in the local mechanical loading environment in the joint, such as those arising from occupational heavy lifting or repetitive loading, obesity, and joint injury [357,358].

It has been shown that osteocyte density and morphology are altered in OA compared with control samples. There is an increase in the number of nonviable osteocytes in the femoral head of osteoarthritic patients, and it was observed that this necrosis was followed by new bone formation [359]. These changes are associated with dysregulated expression of osteocyte markers, apoptosis, and degradative enzymes. Interestingly, changes in the expression of osteocyte markers (decreased SOST and increased DMP1 expression) are correlated with subchondral bone changes (increased bone volume and altered bone mineral content) in OA samples [360]. Significant alteration of osteocytes identified in OA samples indicates a potential regulatory role of osteocytes in subchondral bone remodeling and mineral metabolism during OA pathogenesis [360]. Moreover, the connectivity of the osteocyte network is decreased in osteoarthritic bone, when compared with healthy bone [361]. It has also been reported that osteoarthritic human osteocytes are elongated and their lacunae smaller than those from osteopenic and osteopetrotic bone [362].

## 5.3 MECHANOBIOLOGY, OSTEOSARCOMA, AND METASTATIC BONE DISEASE

Osteosarcoma is a malignant tumor growth that occurs at areas of rapid growth (the knees and shoulders) of long bones of children and young adults. These tumors lead to severe pain and bone fracture and often metastasize to other tissues. How osteosarcoma tumors initiate, grow, and metastasize is not fully understood, and even with chemotherapy treatment, 5-year survival rates of patients are less than 70% [363], and those of patients suffering from metastatic osteosarcoma are less than 25%.

Bone metastases are initiated when tumor cells colonize the bone by expressing integrins, which facilitate tumor invasion, metastasis, and angiogenesis [364,365]. Bone is highly targeted by breast cancer, and it has been shown that osteoblasts express chemokines (CXCL12/SDF-1), which facilitate homing of tumor cells to bone [366]. The tumor cells produce factors (PtHRP, MMP-1, IL-11, and CTGF) that stimulate osteoblasts to secrete RANKL, which stimulates osteoclast resorption of bone matrix and provides a site for bone metastases [367]. This bone resorption releases growth factors stored in the mineralized matrix, such as TGF-β, which perpetuate osteolysis and tumor invasion [367]. Osteoblasts also produce specific growth factors (e.g., TGF-β and OPN) that stimulate tumor growth and activate osteoclasts, which contribute to the formation of a metastatic lesion. Thus, the interaction between prostate cancer cells, osteoblasts, and osteoclasts has been referred to as a "vicious cycle" [368,369].

Mechanobiological factors have been implicated in tumor invasion and metastasis of breast, brain, skin, kidney, and prostate tumors [7,364,370–375]. In particular, changes in ECM stiffness within the primary tumor contribute to growth and metastasis of breast cancer by promoting cell proliferation and transition of tumor cells to migrating cells [371–373,376–380]. It has been proposed that the mechanical properties of the bone tissue might provide a favorable microenvironment for tumor attachment and invasion [380–382]. The mechanical behavior of the tissue surrounding tumor cells has been shown to play a crucial role in the development, growth, and metastasis of tumors in many tissues of the body [7,370,376,378,379]. Interestingly, studies of bone specimens from metastatic tumors show that the tissue is *less* stiff compared with normal noncancer specimens [383,384], which is contrary to the changes in the breast and prostate.

### 5.3.1 Mechanosensors in metastatic bone disease

Altered cadherin expression has been linked to a variety of diseases, including metastatic cancer [385]. Alterations in cadherin expression have also been implicated in tumorigenesis [386,387]. For example, an increase in OB-cadherin expression promotes the metastasis of prostate cancer cells in bone [388]. Synthetic peptides of the His-Ala-Val (HAV) sequence bind to the first extracellular domains of classical cadherins and inhibit cell-cell adhesion [389,390]. Metastasis greatly decreases life expectancy [160], but treatment with cells with higher Young's modulus results in less metastasis [391,392].

### 5.3.2 Mechanical environment

As they evolve, both osteolytic and osteoblastic metastatic bone tumors become *less* stiff compared with normal noncancer specimens [383,384]. Given the well-established relationship between matrix properties and bone biology, such changes are likely to activate mechanobiological responses in both tumor cells and native bone cells, perpetuating the "vicious cycle" and ultimately leading to bone lysis and tumor growth, but this has never been established. Mathematical and computational modeling approaches have been applied to simulate and predict the role of biochemical processes and cell-cell signaling in tumor growth and to a lesser extent have considered biophysical processes [393–398].

## 5.4 MECHANOBIOLOGICAL THERAPY FOR BONE DISEASES

Clinical treatments of osteoporosis have traditionally used antiresorptive pharmaceutical agents to prevent progressive bone loss [399]. It is becoming increasingly clear that physical forces induced through exercise can prevent bone loss in osteoporosis [400–402]. Moreover, combination therapies that include aerobic exercise regimes can promote bone formation through mechanobiological means [403–405]. Significant improvements in bone mass can also occur as a result of low-impact exercise, strength training, and high-impact exercise of women with established osteoporosis [405–408]. Whole-body vibration can enhance bone formation in children with low bone density [409] and prevent a certain degree of bone loss in postmenopausal women [410]. These low-magnitude high-frequency (LMHF) signals are capable of enhancing bone quality and quantity [411] despite the fact that tissue-level strains induced are minimal [411].

The glycoprotein sclerostin has recently been identified as a potential target for osteoporosis therapy. Sclerostin is a negative regulator of bone growth, which inhibits Wnt signaling by interacting with the Wnt coreceptor LRP5/LRP6 to block the binding of Wnt factors. Thus, antisclerostin antibodies are being investigated as an osteoanabolic therapy for osteoporosis [412].

## 5.5 MECHANOBIOLOGY FOR BONE REGENERATION

Bone fractures cost the European economy €17.1 billion [413] excluding treatment, rehabilitation, and social welfare costs. Fracture nonunions occur in 10%–30% of patients [414], and the associated cost of further treatment is approximately €17,000 [415]. Clinical treatment for large bone fractures predominantly involves bone grafting, but the procedure is painful and demonstrates limited success, and major complications can occur including infections and death [416–419]. Tissue engineering and regenerative medicine are promising scientific fields that have significant potential to develop alternative strategies to regenerate deficient biological tissues. These approaches strive to regenerate tissues in the laboratory environment (in vitro) by exploiting the capacity for biological cells (progenitors) to grow and produce tissue constituents under conditions that emulate the body's biochemical and physical environment, and using such approaches, functional replacement skin and cartilage tissues have been developed for clinical use [420,421].

Abundant research has been dedicated to identifying specific factors and the desired properties of biocompatible scaffold materials to encourage bone regeneration in vitro [416–419,422–433], and these have provided evidence that such approaches have the potential to provide alternative treatments for bone defects arising due to disease or injury. However, MSC-seeded scaffolds show variable success with regard to the formation of a strong mineralized bone matrix [432], and once implanted, MSC-seeded scaffolds become encapsulated, and host vasculature is inhibited, which can lead to cell death (necrosis) and ultimately failure of the implant construct [431]. Recent studies have explored tissue regeneration strategies that mimic bone formation during embryogenesis (endochondral ossification) by in vitro chondrogenic priming, which has been shown to promote mineralization in vitro and in vivo [433–438], and striving to form of vasculature within biomaterial constructs to maintain cell viability and initiate mineralization [439,440]. However, mechanical stimulation is extremely important for bone formation. For example, during embryogenesis, mechanical loading imparted by movements and muscular activity of the mother and fetal muscle contractions is crucial for bone development [441–443]. A recent study showed the application of hydrostatic pressure induced chondrogenesis similar to that of chondrogenic priming. However, there was a reduction in hypertrophy (as assesed by collagen type X staining) under the application hydrostatic pressure when compared to chondrogenic priming alone. Chondrogenic priming and HP together accelerated the osteogenic potential of hMSCs [443a].

In vitro experiments have shown that mechanical stimulation enhances bone tissue regeneration to a certain extent [444]. Thus, recent studies have sought to develop mechanobiology-based tissue regeneration strategies. Multiscale and multiphysics (solid and fluid approach) computational models have been developed to determine the mechanical stimulation of cells in tissue engineering scaffolds under mechanical loading bioreactors. By implementing mechanoregulatory algorithms, in which osteogenic, chondrogenic, and fibrogenic differentiation of MSCs within the scaffolds was governed by mechanical stimuli, regeneration strategies have been compared to understand appropriate mechanical stimulation for bone regeneration [445–449]. Aged patients are most vulnerable to bone diseases and injury, and the ideal treatment approach would be to implant regenerated bone grown from the patients' own cells. However, the regeneration potential of human stem cells is diminished with age [450], and the increasing age of the donor may also diminish the effectiveness of donor MSC transplantation. Physical loading regimes have demonstrated anabolic effects in bone for both human and animal models [451]. Whether loading can enhance the regeneration potential of aged human stem cells is not fully understood.

## 6 CONCLUSION

The complex biological and mechanical behavior of bone is intriguing, and the pathogenesis of many bone diseases is not fully understood. Recent studies have evolved scientific understanding of the mechanobiological mechanisms underlying osteoporosis and have revealed for the first time that there are complex changes in (a) bone mechanosensation mechanisms, (b) mechanical environment of bone cells, and (c) biochemical responses and matrix production by bone cells during estrogen deficiency. There is also some evidence of the role of mechanobiological mechanisms in metastatic bone disease and OA. Further studies into the underlying mechanobiological changes are required to understand the pathogenesis of various bone diseases.

To date, the role of mechanobiology in bone cancer progression is not fully understood, and this may limit the potential for the development of therapeutic interventions for the treatment of osteosarcoma. In particular, it is not yet known whether the surrounding tissue mechanics are altered, whether bone cellular attachment or mechanosensation mechanisms are impaired, or whether the biochemical response to mechanical loading is altered. If these quantities were better understood, then it might be possible to develop novel treatments that could inhibit osteosarcoma growth and metastasis.

While implantation of cells/biomaterial constructs can regenerate bone tissue in nonload-bearing bones, there is a clinical need to produce tissue that can support loading and thereby treat fractures and large bone defects in load-bearing bones of aged patients. In the future, mechanobiology-based approaches for bone tissue regeneration might provide effective new strategies to enhance bone tissue regeneration in vitro.

## ACKNOWLEDGMENTS

Dr. Verbruggen's research is funded by the European Union's Horizon 2020 research and innovation program, as part of the Marie Sklodoska-Curie Actions (META-DORM). Work in Prof. McNamara's laboratory is funded by the European Research Council Grant ERC-2010-StG Grant 258992 BONEMECHBIO and the Science Foundation Ireland Investigators Grant co-funded under the European Regional Development fund (14/IA/2884).

## REFERENCES

[1] Currey JD. The mechanical adaptations of bones. Princeton: Princeton University Press; 1984.
[2] Arnoczky SP, Lavagnino M, Egerbacher M. The mechanobiological aetiopathogenesis of tendinopathy: is it the over-stimulation or the under-stimulation of tendon cells? Int J Exp Pathol 2007;88(4):217–26.
[3] Beaupre GS, Stevens SS, Carter DR. Mechanobiology in the development, maintenance, and degeneration of articular cartilage. J Rehabil Res Dev 2000;37(2):145–51.
[4] Chicurel ME, Chen CS, Ingber DE. Cellular control lies in the balance of forces. Curr Opin Cell Biol 1998;10(2):232–9.
[5] Eckes B, Krieg T. Regulation of connective tissue homeostasis in the skin by mechanical forces. Clin Exp Rheumatol 2004;22(3 Suppl. 33):S73–6.
[6] Grodzinsky AJ, Levenston ME, Jin M, Frank EH. Cartilage tissue remodeling in response to mechanical forces. Annu Rev Biomed Eng 2000;2:691–713.
[7] Ingber D. Mechanobiology and diseases of mechanotransduction. Ann Med 2003;35(8):564–77.
[8] Kolahi KS, Mofrad MR. Mechanotransduction: a major regulator of homeostasis and development. Wiley Interdiscip Rev Syst Biol Med 2010;2(6):625–39.
[9] Lammerding J, Kamm RD, Lee RT. Mechanotransduction in cardiac myocytes. Ann N Y Acad Sci 2004;1015:53–70.
[10] Thakar RG, Chown MG, Patel A, Peng L, Kumar S, Desai TA. Contractility-dependent modulation of cell proliferation and adhesion by microscale topographical cues. Small 2008;4(9):1416–24.

[11] Buckwalter JA, Glimcher MJ, Cooper RR, Recker R. Bone Biology. J Bone Joint Surg 1995;77(8):1276–89.
[12] Favus MJ, Christakos S. Primer on the metabolic bone diseases and disorders of mineral metabolism. Philadelphia, PA: Lippincott-Raven; 1996.
[13] Frost HM. The mechanostat: a proposed pathogenic mechanism of osteoporoses and the bone mass effects of mechanical and nonmechanical agents. Bone Miner 1987;2(2):73–85.
[14] Frost HM. Skeletal structural adaptations to mechanical usage (SATMU): 1. Redefining Wolff's Law: the bone modeling problem. Anat Rec 1990;226(4):403–13.
[15] Lemaire V, Tobin FL, Greller LD, Cho CR, Suva LJ. Modeling the interactions between osteoblast and osteoclast activities in bone remodeling. J Theor Biol 2004;229(3):293–309.
[16] Wolff J. The law of bone remodeling (translation of the German 1892 edition). New York, NY: Springer; 1986.
[17] Bromage TG, Goldman HM, McFarlin SC, Warshaw J, Boyde A, Riggs CM. Circularly polarized light standards for investigations of collagen fiber orientation in bone. Anat Rec 2003;274(1):157–68.
[18] Nazarian A, Muller J, Zurakowski D, Müller R, Snyder BD. Densitometric, morphometric and mechanical distributions in the human proximal femur. J Biomech 2007;40(11):2573–9.
[19] Choi K, Kuhn JL, Ciarelli MJ, Goldstein SA. The elastic moduli of human subchondral, trabecular, and cortical bone tissue and the size-dependency of cortical bone modulus. J Biomech 1990;23(11):1103–13.
[20] Ferguson VL, Bushby AJ, Boyde A. Nanomechanical properties and mineral concentration in articular calcified cartilage and subchondral bone. J Anat 2003;203(2):191–202.
[21] Hoffler C, Moore K, Kozloff K, Zysset P, Brown M, Goldstein S. Heterogeneity of bone lamellar-level elastic moduli. Bone 2000;26(6):603–9.
[22] McNamara LM. Bone as a material. In: Ducheyne P, editor. Comprehensive biomaterials. vol. 2. New York City, NY: Elsevier; 2011. p. 169–86.
[23] McNamara LM, Ederveen AG, Lyons CG, Price C, Schaffler MB, Weinans H, et al. Strength of cancellous bone trabecular tissue from normal, ovariectomized and drug-treated rats over the course of ageing. Bone 2006;39(2):392–400.
[24] Mente P, Lewis J. Experimental method for the measurement of the elastic modulus of trabecular bone tissue. J Orthop Res 1989;7(3):456–61.
[25] Ozcivici E, Ferreri S, Qin Y-X, Judex S. Determination of bone's mechanical matrix properties by nanoindentation. In: Osteoporosis: methods and protocols. Berlin, Heidelberg: Springer; 2008. p. 323–34.
[26] Rho JY, Ashman RB, Turner CH. Young's modulus of trabecular and cortical bone material: ultrasonic and microtensile measurements. J Biomech 1993;26(2):111–9.
[27] Rho JY, Roy ME, Tsui TY, Pharr GM. Elastic properties of microstructural components of human bone tissue as measured by nanoindentation. J Biomed Mater Res A 1999;45(1):48–54.
[28] Rho J-Y, Kuhn-Spearing L, Zioupos P. Mechanical properties and the hierarchical structure of bone. Med Eng Phys 1998;20(2):92–102.
[29] Rho J-Y, Tsui TY, Pharr GM. Elastic properties of human cortical and trabecular lamellar bone measured by nanoindentation. Biomaterials 1997;18(20):1325–30.

[30] Runkle J, Pugh J. The micro-mechanics of cancellous bone. II. Determination of the elastic modulus of individual trabeculae by a buckling analysis. Bull Hosp Joint Dis 1975;36(1):2.
[31] Ryan SD, Williams JL. Tensile testing of rodlike trabeculae excised from bovine femoral bone. J Biomech 1989;22(4):351–5.
[32] Samelin N, Köller W, Ascherl R, Gradinger R. A method for determining the biomechanical properties of trabecular and spongiosa bone tissue. Biomed Tech (Berl) 1996;41(7–8):203–8.
[33] Townsend PR, Rose RM, Radin EL. Buckling studies of single human trabeculae. J Biomech 1975;8(3–4):199IN5201–200IN6.
[34] Zysset PK, Guo XE, Hoffler CE, Moore KE, Goldstein SA. Elastic modulus and hardness of cortical and trabecular bone lamellae measured by nanoindentation in the human femur. J Biomech 1999;32(10):1005–12.
[35] Ascenzi A, Bonucci E, Benvenuti MA. Relationship between ultrastructure and "pin test" in osteons. Clin Orthop Relat Res 1976;(121):275–94.
[36] Brennan O, Kennedy OD, Lee TC, Rackard SM, O'Brien FJ, McNamara LM. The effects of estrogen deficiency and bisphosphonate treatment on tissue mineralisation and stiffness in an ovine model of osteoporosis. J Biomech 2011;44(3):386–90.
[37] Currey J. Effects of differences in mineralization on the mechanical properties of bone. Philos Trans R Soc Lond B: Biol Sci 1984;304(1121):509–18.
[38] Hofmann T, Heyroth F, Meinhard H, Fränzel W, Raum K. Assessment of composition and anisotropic elastic properties of secondary osteon lamellae. J Biomech 2006;39(12):2282–94.
[39] Vaughan TJ, McCarthy CT, McNamara LM. A three-scale finite element investigation into the effects of tissue mineralisation and lamellar organisation in human cortical and trabecular bone. J Mech Behav Biomed Mater 2012;12:50–62.
[40] Sasaki M, Hongo H, Hasegawa T, Suzuki R, Zhusheng L, de Freitas PHL, et al. Morphological aspects of the biological function of the osteocytic lacunar canalicular system and of osteocyte-derived factors. Oral Sci Int 2012;9(1):1–8.
[41] Eriksen EF, Axelrod DW, Melsen F. Bone histomorphometry. New York: Raven Press; 1994.
[42] Burr DB, Schaffler MB, Frederickson RG. Composition of the cement line and its possible mechanical role as a local interface in human compact bone. J Biomech 1988;21(11):939–45.
[43] Cowin S, editor. Bone mechanics handbook. 2nd ed. CRC Press LLC; 2001.
[44] Parfitt A. The physiologic and clinical significance of bone histomorphometric data. In: Bone histomorphometry: techniques and interpretation. Boca Raton, FL: CRC Press; 1983. p. 143–223.
[45] Coughlin TR, Niebur GL. Fluid shear stress in trabecular bone marrow due to low-magnitude high-frequency vibration. J Biomech 2012;45(13):2222–9.
[46] De Bruyn PPH, Breen PC, Thomas TB. The microcirculation of the bone marrow. Anat Rec 1970;168(1):55–68.
[47] Arramon YP, Cowin SC. Hydraulic stiffening of cancellous bone. Forma 1998;12(3):209–21.
[48] Zhang D, Weinbaum S, Cowin SC. Estimates of the peak pressures in bone pore water. J Biomech Eng 1998;120(6):697–703.

[49] Bryant JD. The effect of impact on the marrow pressure of long bones in vitro. J Biomech 1983;16(8):659–65.
[50] Bryant JD. On the mechanical function of marrow in long bones. Eng Med 1988;17(2):55–8.
[51] Hrubá A, Kiefman J, Sobotka Z. Rheological behaviour of bone marrow. In: Progress and trends in rheology II. Berlin, Heidelberg: Springer; 1988. p. 467–9.
[52] Birmingham E, Grogan JA, Niebur GL, McNamara LM, McHugh PE. Computational modelling of the mechanics of trabecular bone and marrow using fluid structure interaction techniques. Ann Biomed Eng 2013;41(4):814–26.
[53] Aukland K. Distribution of body fluids: local mechanisms guarding interstitial fluid volume. J Physiol 1984;79(6):395.
[54] Knothe Tate ML, Knothe U. An ex vivo model to study transport processes and fluid flow in loaded bone. J Biomech 2000;33(2):247–54.
[55] Price C, Zhou X, Li W, Wang L. Real-time measurement of solute transport within the lacunar-canalicular system of mechanically loaded bone: direct evidence for load-induced fluid flow. J Bone Miner Res 2011;26(2):277–85.
[56] Cowin SC. Bone poroelasticity. J Biomech 1999;32(3):217–38.
[57] Knothe Tate ML, Knothe U, Niederer P. Experimental elucidation of mechanical load-induced fluid flow and its potential role in bone metabolism and functional adaptation. Am J Med Sci 1998;316(3):189–95.
[58] Burger EH, Klein-Nulend J, Cowin SC. Mechanotransduction in bone. In: Advances in organ biology. vol. 5. New York City, NY: Elsevier; 1998. p. 123–36.
[59] Okazaki R, Inoue D, Shibata M, Saika M, Kido S, Ooka H, et al. Estrogen promotes early osteoblast differentiation and inhibits adipocyte differentiation in mouse bone marrow stromal cell lines that express estrogen receptor (ER) α or β. Endocrinology 2002;143(6):2349–56.
[60] Udagawa N, Takahashi N, Akatsu T, Tanaka H, Sasaki T, Nishihara T, et al. Origin of osteoclasts: mature monocytes and macrophages are capable of differentiating into osteoclasts under a suitable microenvironment prepared by bone marrow-derived stromal cells. Proc Natl Acad Sci U S A 1990;87(18):7260–4.
[61] Hammond C, Helenius A. Quality control in the secretory pathway. Curr Opin Cell Biol 1995;7(4):523–9.
[62] Rothman JE. Mechanisms of intracellular protein transport. Nature 1994;372:55–63.
[63] Ducy P, Schinke T, Karsenty G. The osteoblast: a sophisticated fibroblast under central surveillance. Science 2000;289(5484):1501–4.
[64] Parfitt AM. Osteonal and hemi-osteonal remodeling: the spatial and temporal framework for signal traffic in adult human bone. J Cell Biochem 1994;55(3):273–86.
[65] Gronowicz GA, Derome ME. Synthetic peptide containing Arg-Gly-Asp inhibits bone formation and resorption in a mineralizing organ culture system of fetal rat parietal bones. J Bone Miner Res 1994;9(2):193–201.
[66] Moursi AM, Damsky CH, Lull J, Zimmerman D, Doty SB, Aota S, et al. Fibronectin regulates calvarial osteoblast differentiation. J Cell Sci 1996;109(6):1369–80.
[67] Owen TA, Aronow M, Shalhoub V, Barone LM, Wilming L, Tassinari MS, et al. Progressive development of the rat osteoblast phenotype in vitro: reciprocal relationships in expression of genes associated with osteoblast proliferation and differentiation during formation of the bone extracellular matrix. J Cell Physiol 1990;143(3):420–30.

[68] Manolagas SC. Birth and death of bone cells: basic regulatory mechanisms and implications for the pathogenesis and treatment of osteoporosis. Endocr Rev 2000;21(2):115–37.
[69] Dobnig H, Turner RT. Evidence that intermittent treatment with parathyroid hormone increases bone formation in adult rats by activation of bone lining cells. Endocrinology 1995;136(8):3632–8.
[70] Merz WA, Schenk RK. A quantitative histological study on bone formation in human cancellous bone. Cells Tissues Organs 1970;76(1):1–15.
[71] Miller S, Jee WS. The bone lining cell: a distinct phenotype? Calcif Tissue Int 1987;41(1):1–5.
[72] Lanyon LE. Osteocytes, strain detection, bone modeling and remodeling. Calcif Tissue Int 1993;53(1):S102–7.
[73] Lanyon LE. Using functional loading to influence bone mass and architecture: objectives, mechanisms, and relationship with estrogen of the mechanically adaptive process in bone. Bone 1996;18(1 Suppl. 1):S37–43.
[74] Miller SC, Bowman BM, Smith JM, Jee WSS. Characterization of endosteal bone-lining cells from fatty marrow bone sites in adult beagles. Anat Rec 1980;198(2):163–73.
[75] Talmage RV. Calcium homeostasis-calcium transport-parathyroid action: the effects of parathyroid hormone on the movement of calcium between bone and fluid. Clin Orthop Relat Res 1969;67:210–24.
[76] Puzas E, Lewis GD. Biology of osteoclasts and osteoblasts. Boca Raton, FL: CRC Press; 1999.
[77] Frost HM. In vivo osteocyte death. J Bone Joint Surg 1960;42(1):138–43.
[78] Li M, Wang X, Meintzer MK, Laessig T, Birnbaum MJ, Heidenreich KA. Cyclic AMP promotes neuronal survival by phosphorylation of glycogen synthase kinase 3β. Mol Cell Biol 2000;20(24):9356–63.
[79] Xia X, Batra N, Shi Q, Bonewald LF, Sprague E, Jiang JX. Prostaglandin promotion of osteocyte gap junction function through transcriptional regulation of connexin 43 by glycogen synthase kinase 3/β-catenin signaling. Mol Cell Biol 2010;30(1):206–19.
[80] Lanyon LE. Amplification of the osteogenic stimulus of load-bearing as a logical therapy for the treatment and prevention of osteoporosis. In: Russell RG, Skerry T, Kollenkirchen U, editors. Novel approaches to treatment of osteoporosis. Ernst schering research foundation workshop. vol. 25. Springer Berlin Heidelberg; 1998. p. 199–209.
[81] Turner CH, Duncan RL, Pavalko FM. Mechanotransduction: an inevitable process for skeletal maintenance. In: Russell RG, Skerry T, Kollenkirchen U, editors. Novel approaches to treatment of osteoporosis. Ernst schering research foundation workshop. vol. 25. Springer Berlin Heidelberg; 1998. p. 157–77.
[82] You L, Cowin SC, Schaffler MB, Weinbaum S. A model for strain amplification in the actin cytoskeleton of osteocytes due to fluid drag on pericellular matrix. J Biomech 2001;34(11):1375–86.
[83] McNamara LM, Majeska RJ, Weinbaum S, Friedrich V, Schaffler MB. Attachment of osteocyte cell processes to the bone matrix. Anat Rec (Hoboken) 2009;292(3):355–63.
[84] Pavalko FM, Burridge K. Disruption of the actin cytoskeleton after microinjection of proteolytic fragments of alpha-actinin. J Cell Biol 1991;114(3):481–91.

[85] Wang Y, McNamara LM, Schaffler MB, Weinbaum S. A model for the role of integrins in flow induced mechanotransduction in osteocytes. Proc Natl Acad Sci U S A 2007;104 (40):15941–6.

[86] You L-D, Weinbaum S, Cowin SC, Schaffler MB. Ultrastructure of the osteocyte process and its pericellular matrix. Anat Rec A: Discov Mol Cell Evol Biol 2004;278A (2):505–13.

[87] Baik AD, Qiu J, Hillman EMC, Dong C, Guo XE. Simultaneous tracking of 3D actin and microtubule strains in individual MLO-Y4 osteocytes under oscillatory flow. Biochem Biophys Res Commun 2013;431(4):718–23.

[88] Adachi T, Aonuma Y, Tanaka M, Hojo M, Takano-Yamamoto T, Kamioka H. Calcium response in single osteocytes to locally applied mechanical stimulus: differences in cell process and cell body. J Biomech 2009;42(12):1989–95.

[89] Burra S, Nicolella DP, Francis WL, Freitas CJ, Mueschke NJ, Poole K, et al. Dendritic processes of osteocytes are mechanotransducers that induce the opening of hemichannels. Proc Natl Acad Sci U S A 2010;107(31):13648–53.

[90] Klein-Nulend J, Bakker AD, Bacabac RG, Vatsa A, Weinbaum S. Mechanosensation and transduction in osteocytes. Bone 2013;54(2):182–90.

[91] Wu D, Ganatos P, Spray DC, Weinbaum S. On the electrophysiological response of bone cells using a Stokesian fluid stimulus probe for delivery of quantifiable localized picoNewton level forces. J Biomech 2011;44(9):1702–8.

[92] Cowin S, Weinbaum S, Zeng Y. A case for bone canaliculi as the anatomical site of strain generated potentials. J Biomech 1995;28(11):1281–97.

[93] Han Y, Cowin SC, Schaffler MB, Weinbaum S. Mechanotransduction and strain amplification in osteocyte cell processes. Proc Natl Acad Sci U S A 2004;101(47):16689–94.

[94] Weinbaum S, Cowin S, Zeng Y. A model for the excitation of osteocytes by mechanical loading-induced bone fluid shear stresses. J Biomech 1994;27(3):339–60.

[95] Zeng Y, Cowin S, Weinbaum S. A fiber matrix model for fluid flow and streaming potentials in the canaliculi of an osteon. Ann Biomed Eng 1994;22(3):280–92.

[96] Anderson EJ, Knothe Tate ML. Idealization of pericellular fluid space geometry and dimension results in a profound underprediction of nano-microscale stresses imparted by fluid drag on osteocytes. J Biomech 2008;41(8):1736–46.

[97] Verbruggen SW, Vaughan TJ, McNamara LM. Strain amplification in bone mechanobiology: a computational investigation of the *in vivo* mechanics of osteocytes. J R Soc Interface 2012;9(75):2735–44.

[98] Verbruggen SW, Vaughan TJ, McNamara LM. Fluid flow in the osteocyte mechanical environment: a fluid-structure interaction approach. Biomech Model Mechanobiol 2014;13(1):85–97.

[99] Verbruggen SW, Vaughan TJ, McNamara LM. Mechanisms of osteocyte stimulation in osteoporosis. J Mech Behav Biomed Mater 2016;62:158–68.

[100] Verbruggen SW, Mc Garrigle MJ, Haugh MG, Voisin MC, McNamara LM. Altered mechanical environment of bone cells in an animal model of short- and long-term osteoporosis. Biophys J 2015;108(7):1587–98.

[101] Roodman GD. Advances in bone biology: the osteoclast. Endocr Rev 1996;17 (4):308–32.

[102] Watanabe H, Yanagisawa T, Sasaki J. Cytoskeletal architecture of rat calvarial osteoclasts: microfilaments, intermediate filaments, and nuclear matrix as demonstrated by detergent perfusion. Anat Rec 1995;243(2):165–74.

[103] Salo J, Lehenkari P, Mulari M, Metsikkö K, Väänänen HK. Removal of osteoclast bone resorption products by transcytosis. Science 1997;276(5310):270–3.

[104] Teitelbaum SL. Bone resorption by osteoclasts. Science 2000;289(5484):1504–8.
[105] Väänänen HK, Horton M. The osteoclast clear zone is a specialized cell-extracellular matrix adhesion structure. J Cell Sci 1995;108(8):2729–32.
[106] Palokangas H, Mulari M, Vaananen HK. Endocytic pathway from the basal plasma membrane to the ruffled border membrane in bone-resorbing osteoclasts. J Cell Sci 1997;110(15):1767–80.
[107] Nowlan N. Biomechanics of foetal movement. Eur Cell Mater 2015;29:1–21.
[108] Nowlan NC, Sharpe J, Roddy KA, Prendergast PJ, Murphy P. Mechanobiology of embryonic skeletal development: insights from animal models. Birth Defects Res C Embryo Today 2010;90(3):203–13.
[109] Verbruggen SW, Loo JHW, Hayat TTA, Hajnal JV, Rutherford MA, Phillips ATM, et al. Modeling the biomechanics of fetal movements. Biomech Model Mechanobiol 2016;15(4):995–1004.
[110] Verbruggen SW, Nowlan NC. Ontogeny of the human pelvis. Anat Rec 2017;300(4):643–52.
[110a] Verbruggen SW, Kainz B, Shelmerdine SC, Hajnal JV, Rutherford MA, Arthurs OJ, Phillips AT, Nowlan NC. Stresses and strains on the human fetal skeleton during development. J R Soc Interface 2018;15(138):20170593.
[111] Ortega N, Behonick DJ, Werb Z. Matrix remodeling during endochondral ossification. Trends Cell Biol 2004;14(2):86–93.
[112] Sumpio BE, Riley JT, Dardik A. Cells in focus: endothelial cell. Int J Biochem Cell Biol 2002;34(12):1508–12.
[113] Collin-Osdoby P. Role of vascular endothelial cells in bone biology. J Cell Biochem 1994;55(3):304–9.
[114] Gerber H-P, Ferrara N. Angiogenesis and bone growth. Trends Cardiovasc Med 2000;10(5):223–8.
[115] U-i C, Kawaguchi H, Takato T, Nakamura K. Distinct osteogenic mechanisms of bones of distinct origins. J Orthop Sci 2004;9(4):410–4.
[116] Kanczler J, Oreffo R. Osteogenesis and angiogenesis: the potential for engineering bone. Eur Cell Mater 2008;15(2):100–14.
[117] Castillo A, Jacobs CR. Skeletal mechanobiology. In: Mechanobiology handbook. Boca Raton, FL: CRC Press; 2011.
[118] Mosekilde L. Consequences of the remodelling process for vertebral trabecular bone structure: a scanning electron microscopy study (uncoupling of unloaded structures). Bone Miner 1990;10(1):13–35.
[119] Frost HM. Metabolism of bone. New Engl J Med 1973;289(16):864.
[120] Machwate M, Zerath E, Holy X, Hott M, Godet D, Lomri A, et al. Systemic administration of transforming growth factor-beta 2 prevents the impaired bone formation and osteopenia induced by unloading in rats. J Clin Invest 1995;96(3):1245.
[121] Carter D. Mechanical loading histories and cortical bone remodeling. Calcif Tissue Int 1984;36(1):S19–24.
[122] Forwood MR, Turner CH. Skeletal adaptations to mechanical usage: results from tibial loading studies in rats. Bone 1995;17(4 Suppl.):S197–205.
[123] Jee WSS, Li XJ, Schaffler MB. Adaptation of diaphyseal structure with aging and increased mechanical usage in the adult rat: a histomorphometrical and biomechanical study. Anat Rec 1991;230(3):332–8.
[124] Mosley JR, Lanyon LE. Strain rate as a controlling influence on adaptive modeling in response to dynamic loading of the ulna in growing male rats. Bone 1998;23(4):313–8.

[125] O'Connor JA, Lanyon LE, MacFie H. The influence of strain rate on adaptive bone remodelling. J Biomech 1982;15(10):767–81.
[126] Woo SL, Kuei SC, Amiel D, Gomez MA, Hayes WC, White FC, et al. The effect of prolonged physical training on the properties of long bone: a study of Wolff's Law. J Bone Joint Surg (Am Vol) 1981;63(5):780–7.
[127] Knothe Tate ML. "Whither flows the fluid in bone?" An osteocyte's perspective. J Biomech 2003;36(10):1409–24.
[128] Birmingham E, Niebur GL, McHugh PE, Shaw G, Barry FP, McNamara LM. Osteogenic differentiation of mesenchymal stem cells is regulated by osteocyte and osteoblast cells in a simplified bone niche. Eur Cell Mater 2012;23:13–27.
[129] Hoey DA, Chen JC, Jacobs CR. The primary cilium as a novel extracellular sensor in bone. Front Endocrinol (Lausanne) 2012;3:75.
[130] Kapur S, Baylink DJ, William Lau KH. Fluid flow shear stress stimulates human osteoblast proliferation and differentiation through multiple interacting and competing signal transduction pathways. Bone 2003;32(3):241–51.
[131] Klein-Nulend J, Semeins CM, Burger EH. Prostaglandin mediated modulation of transforming growth factor-beta metabolism in primary mouse osteoblastic cells in vitro. J Cell Physiol 1996;168(1):1–7.
[132] Lee KL, Hoey DA, Jacobs CR. Primary cilia-mediated mechanotransduction in bone. Clin Rev Bone Miner Metab 2010;8(4):201–12.
[133] Lee KL, Hoey DA, Spasic M, Tang T, Hammond HK, Jacobs CR. Adenylyl cyclase 6 mediates loading-induced bone adaptation in vivo. FASEB J 2014;28(3):1157–65.
[134] Li J, Rose E, Frances D, Sun Y, You L. Effect of oscillating fluid flow stimulation on osteocyte mRNA expression. J Biomech 2012;45:247–51.
[135] Malone AMD, Anderson CT, Tummala P, Kwon RY, Johnston TR, Stearns T, et al. Primary cilia mediate mechanosensing in bone cells by a calcium-independent mechanism. Proc Natl Acad Sci U S A 2007;104(33):13325–30.
[136] McAllister T. Fluid shear stress stimulates prostaglandin and nitric oxide release in bone marrow-derived preosteoclast-like cells. Biochem Biophys Res Commun 2000;270(2):643–8.
[137] Nauman EA, Satcher RL, Keaveny TM, Halloran BP, Bikle DD. Osteoblasts respond to pulsatile fluid flow with short- term increases in PGE2 but no change in mineralization. J Appl Physiol 2001;90:1849–54.
[138] Wang Y, McNamara LM, Schaffler MB, Weinbaum S. Strain amplification and integrin based signaling in osteocytes. J Musculoskelet Neuronal Interact 2008;8(4):332–4.
[139] Hoey DA, Tormey S, Ramcharan S, O'Brien FJ, Jacobs CR. Primary cilia-mediated mechanotransduction in human mesenchymal stem cells. Stem Cells 2012;30(11):2561–70.
[140] Litzenberger JB, Kim J-B, Tummala P, Jacobs CR. β1 integrins mediate mechanosensitive signaling pathways in osteocytes. Calcif Tissue Int 2010;86(4):325–32.
[141] Rubin J, Rubin C, Jacobs CR. Molecular pathways mediating mechanical signaling in bone. Gene 2006;367:1–16.
[142] Ziambaras K, Lecanda F, Steinberg TH, Civitelli R. Cyclic stretch enhances gap junctional communication between osteoblastic cells. J Bone Miner Res 1998;13(2):218–28.
[143] Jacobs CR, Temiyasathit S, Castillo AB. Osteocyte mechanobiology and pericellular mechanics. Annu Rev Biomed Eng 2010;12:369–400.

[144] Eyckmans J, Boudou T, Yu X, Chen CS. A hitchhiker's guide to mechanobiology. Dev Cell 2011;21(1):35–47.
[145] Chen J-H, Liu C, You L, Simmons CA. Boning up on Wolff's Law: mechanical regulation of the cells that make and maintain bone. J Biomech 2010;43(1):108–18.
[146] Zimmerman D, Jin F, Leboy P, Hardy S, Damsky C. Impaired bone formation in transgenic mice resulting from altered integrin function in osteoblasts. Dev Biol 2000;220(1):2–15.
[147] Haugh MG, Vaughan TJ, McNamara LM. The role of integrin alpha(V)beta(3) in osteocyte mechanotransduction. J Mech Behav Biomed Mater 2015;42:67–75.
[148] Devoll R, Pinero G, Appelbaum E, Dul E, Troncoso P, Butler W, et al. Improved immunohistochemical staining of osteopontin (OPN) in paraffin-embedded archival bone specimens following antigen retrieval: anti-human OPN antibody recognizes multiple molecular forms. Calcif Tissue Int 1997;60(4):380–6.
[149] McKee MD, Nanci A. Osteopontin: an interfacial extracellular matrix protein in mineralized tissues. Connect Tissue Res 1996;35(1–4):197–205.
[150] Nanci A. Content and distribution of noncollagenous matrix proteins in bone and cementum: relationship to speed of formation and collagen packing density. J Struct Biol 1999;126(3):256–69.
[151] Sodek J, McKee MD. Molecular and cellular biology of alveolar bone. Periodontology 2000;24(1):99–126.
[152] You J, Reilly GC, Zhen X, Yellowley CE, Chen Q, Donahue HJ, et al. Osteopontin gene regulation by oscillatory fluid flow via intracellular calcium mobilization and activation of mitogen-activated protein kinase in MC3T3–E1 osteoblasts. J Biol Chem 2001;276(16):13365–71.
[153] Yoshitake H, Rittling SR, Denhardt DT, Noda M. Osteopontin-deficient mice are resistant to ovariectomy-induced bone resorption. Proc Natl Acad Sci U S A 1999;96(14):8156–60.
[154] Tonna EA, Lampen NM. Electron microscopy of aging skeletal cells. I. Centrioles and solitary cilia. J Gerontol 1972;27(3):316–24.
[155] Eichholz KF, Hoey DA. The role of the primary cilium in cellular mechanotransduction. In: Mechanobiology. Hoboken, NJ: John Wiley & Sons, Inc.; 2017. p. 61–73.
[156] Schwartz EA, Leonard ML, Bizios R, Bowser SS. Analysis and modeling of the primary cilium bending response to fluid shear. Am J Physiol Ren Physiol 1997;272(1):F132–8.
[157] Whitfield JF. Primary cilium—is it an osteocyte's strain-sensing flowmeter? J Cell Biochem 2003;89(2):233–7.
[158] Whitfield JF. The solitary (primary) cilium—a mechanosensory toggle switch in bone and cartilage cells. Cell Signal 2008;20(6):1019–24.
[159] Xiao Z, Zhang S, Mahlios J, Zhou G, Magenheimer BS, Guo D, et al. Cilia-like structures and polycystin-1 in osteoblasts/osteocytes and associated abnormalities in skeletogenesis and Runx2 expression. J Biol Chem 2006;281(41):30884–95.
[160] Chen JC, Hoey DA, Chua M, Bellon R, Jacobs CR. Mechanical signals promote osteogenic fate through a primary cilia-mediated mechanism. FASEB J 2016;30(4):1504–11.
[161] Labour M-N, Riffault M, Christensen ST, Hoey DA. TGFβ1-induced recruitment of human bone mesenchymal stem cells is mediated by the primary cilium in a SMAD3-dependent manner. Sci Rep 2016;6:35542.

[162] Coughlin T, Schiavi J, Varsanik MA, Voisin M, Birmingham E, Haugh M, et al. Primary cilia expression in bone marrow in response to mechanical stimulation in explant bioreactor culture. Eur Cell Mater 2016;32:111–22.

[163] Coughlin TR, Voisin M, Schaffler MB, Niebur GL, McNamara LM. Primary cilia exist in a small fraction of cells in trabecular bone and marrow. Calcif Tissue Int 2015;96(1):65–72.

[164] Davidson RM. Membrane stretch activates a high-conductance K+ channel in G292 osteoblastic-like cells. J Membr Biol 1993;131(1):81–92.

[165] Davidson RM, Tatakis DW, Auerbach AL. Multiple forms of mechanosensitive ion channels in osteoblast-like cells. Pflügers Arch Eur J Physiol 1990;416(6):646–51.

[166] Duncan RL, Hruska KA. Chronic, intermittent loading alters mechanosensitive channel characteristics in osteoblast-like cells. Am J Physiol Ren Physiol 1994;267(6):F909–16.

[167] McDonald F, Somasundaram B, McCann T, Mason W, Meikle M. Calcium waves in fluid flow stimulated osteoblasts are G protein mediated. Arch Biochem Biophys 1996;326(1):31–8.

[168] Duncan RL, Hruska KA, Misler S. Parathyroid hormone activation of stretch-activated cation channels in osteosarcoma cells (UMR-106.01). FEBS Lett 1992;307(2):219–23.

[169] Ryder KD, Duncan RL. Parathyroid hormone enhances fluid shear-induced [Ca2+]i signaling in osteoblastic cells through activation of mechanosensitive and voltage-sensitive Ca2+ channels. J Bone Miner Res 2001;16(2):240–8.

[170] Donahue TH, Genetos D, Jacobs C, Donahue H, Yellowley C. Annexin V disruption impairs mechanically induced calcium signaling in osteoblastic cells. Bone 2004;35(3):656–63.

[171] Pedersen SF, Owsianik G, Nilius B. TRP channels: an overview. Cell Calcium 2005;38(3):233–52.

[172] Masuyama R, Vriens J, Voets T, Karashima Y, Owsianik G, Vennekens R, et al. TRPV4-mediated calcium influx regulates terminal differentiation of osteoclasts. Cell Metab 2008;8(3):257–65.

[173] Hoenderop JG, van Leeuwen JP, van der Eerden BC, Kersten FF, WCM van derKemp A, Mérillat A-M, et al. Renal Ca2+ wasting, hyperabsorption, and reduced bone thickness in mice lacking TRPV5. J Clin Invest 2003;112(12):1906.

[174] Gradilone SA, Masyuk AI, Splinter PL, Banales JM, Huang BQ, Tietz PS, et al. Cholangiocyte cilia express TRPV4 and detect changes in luminal tonicity inducing bicarbonate secretion. Proc Natl Acad Sci U S A 2007;104(48):19138–43.

[175] Abed E, Labelle D, Martineau C, Loghin A, Moreau R. Expression of transient receptor potential (TRP) channels in human and murine osteoblast-like cells. Mol Membr Biol 2009;26(3):146–58.

[176] Mochizuki T, Sokabe T, Araki I, Fujishita K, Shibasaki K, Uchida K, et al. The TRPV4 cation channel mediates stretch-evoked Ca2+ influx and ATP release in primary urothelial cell cultures. J Biol Chem 2009;284(32):21257–64.

[177] Delmas P. Polycystins: from mechanosensation to gene regulation. Cell 2004;118(2):145–8.

[178] Boulter C, Mulroy S, Webb S, Fleming S, Brindle K, Sandford R. Cardiovascular, skeletal, and renal defects in mice with a targeted disruption of the Pkd1 gene. Proc Natl Acad Sci U S A 2001;98(21):12174–9.

[179] Lu W, Shen X, Pavlova A, Lakkis M, Ward CJ, Pritchard L, et al. Comparison of Pkd1-targeted mutants reveals that loss of polycystin-1 causes cystogenesis and bone defects. Hum Mol Genet 2001;10(21):2385–96.

[180] Ingber DE. Tensegrity: the architectural basis of cellular mechanotransduction. Annu Rev Physiol 1997;59(1):575–99.
[181] Pavalko F, Chen N, Turner C, Burr D, Atkinson S, Hsieh Y, et al. Fluid shear-induced mechanical signaling in MC3T3-E1 osteoblasts requires cytoskeleton-integrin interactions. Am J Physiol 1998;275(6 Pt 1):C1591–601.
[182] Norvell SM, Ponik SM, Bowen DK, Gerard R, Pavalko FM. Fluid shear stress induction of COX-2 protein and prostaglandin release in cultured MC3T3-E1 osteoblasts does not require intact microfilaments or microtubules. J Appl Physiol 2004;96(3):957–66.
[183] Myers KA, Rattner JB, Shrive NG, Hart DA. Osteoblast-like cells and fluid flow: cytoskeleton-dependent shear sensitivity. Biochem Biophys Res Commun 2007;364(2):214–9.
[184] Rosenberg N. The role of the cytoskeleton in mechanotransduction in human osteoblast-like cells. Hum Exp Toxicol 2003;22(5):271–4.
[185] Arnsdorf EJ, Tummala P, Kwon RY, Jacobs CR. Mechanically induced osteogenic differentiation—the role of RhoA, ROCKII and cytoskeletal dynamics. J Cell Sci 2009;122(Pt 4):546–53.
[186] Reilly GC, Haut TR, Yellowley CE, Donahue HJ, Jacobs CR. Fluid flow induced PGE2 release by bone cells is reduced by glycocalyx degradation whereas calcium signals are not. Biorheology 2003;40(6):591–603.
[187] Ponik SM, Triplett JW, Pavalko FM. Osteoblasts and osteocytes respond differently to oscillatory and unidirectional fluid flow profiles. J Cell Biochem 2007;100(3):794–807.
[188] Kamioka H, Sugawara Y, Murshid SA, Ishihara Y, Honjo T, Takano-Yamamoto T. Fluid shear stress induces less calcium response in a single primary osteocyte than in a single osteoblast: implication of different focal adhesion formation. J Bone Miner Res 2006;21(7):1012–21.
[189] Kamioka H, Sugawara Y, Honjo T, Yamashiro T, Takano-Yamamoto T. Terminal differentiation of osteoblasts to osteocytes is accompanied by dramatic changes in the distribution of actin-binding proteins. J Bone Miner Res 2004;19(3):471–8.
[190] McGarry JG, Klein-Nulend J, Prendergast PJ. The effect of cytoskeletal disruption on pulsatile fluid flow-induced nitric oxide and prostaglandin E2 release in osteocytes and osteoblasts. Biochem Biophys Res Commun 2005;330(1):341–8.
[191] Rawlinson S, Pitsillides A, Lanyon L. Involvement of different ion channels in osteoblasts' and osteocytes' early responses to mechanical strain. Bone 1996;19(6):609–14.
[192] Jing D, Lu XL, Luo E, Sajda P, Leong PL, Guo XE. Spatiotemporal properties of intracellular calcium signaling in osteocytic and osteoblastic cell networks under fluid flow. Bone 2013;53(2):531–40.
[193] Lu XL, Huo B, Chiang V, Guo XE. Osteocytic network is more responsive in calcium signaling than osteoblastic network under fluid flow. J Bone Miner Res 2012;27(3):563–74.
[194] Lu XL, Huo B, Park M, Guo XE. Calcium response in osteocytic networks under steady and oscillatory fluid flow. Bone 2012;51(3):466–73.
[195] Vatsa A, Smit TH, Klein-Nulend J. Extracellular NO signalling from a mechanically stimulated osteocyte. J Biomech 2007;40(Suppl. 1):S89–95.
[196] Guo XE, Takai E, Jiang X, Xu Q, Whitesides GM, Yardley JT, et al. Intracellular calcium waves in bone cell networks under single cell nanoindentation. Mol Cell Biomech 2006;3(3):95.
[197] Yellowley CE, Li Z, Zhou Z, Jacobs CR, Donahue HJ. Functional gap junctions between osteocytic and osteoblastic cells. J Bone Miner Res 2000;15(2):209–17.

[198] Ilvesaro J, Väänänen K, Tuukkanen J. Bone-resorbing osteoclasts contain gap-junctional connexin-43. J Bone Miner Res 2000;15(5):919–26.
[199] Kato Y, Windle JJ, Koop BA, Mundy GR, Bonewald LF. Establishment of an osteocyte-like cell line, MLO-Y4. J Bone Miner Res 1997;12(12):2014–23.
[200] Doty SB. Morphological evidence of gap junctions between bone cells. Calcif Tissue Int 1981;33(1):509–12.
[201] Goodenough DA, Paul DL. Beyond the gap: functions of unpaired connexon channels. Nat Rev Mol Cell Biol 2003;4(4):285.
[202] Gluhak-Heinrich J, Gu S, Pavlin D, Jiang JX. Mechanical loading stimulates expression of connexin 43 in alveolar bone cells in the tooth movement model. Cell Commun Adhes 2006;13(1–2):115–25.
[203] Cherian PP, Siller-Jackson AJ, Gu S, Wang X, Bonewald LF, Sprague E, et al. Mechanical strain opens connexin 43 hemichannels in osteocytes: a novel mechanism for the release of prostaglandin. Mol Biol Cell 2005;16(7):3100–6.
[204] Genetos DC, Kephart CJ, Zhang Y, Yellowley CE, Donahue HJ. Oscillating fluid flow activation of gap junction hemichannels induces ATP release from MLO-Y4 osteocytes. J Cell Physiol 2007;212(1):207–14.
[205] Jiang JX, Cherian PP. Hemichannels formed by connexin 43 play an important role in the release of prostaglandin E(2) by osteocytes in response to mechanical strain. Cell Commun Adhes 2003;10(4–6):259–64.
[206] Siller-Jackson AJ, Burra S, Gu S, Xia X, Bonewald LF, Sprague E, et al. Adaptation of connexin 43-hemichannel prostaglandin release to mechanical loading. J Biol Chem 2008;283(39):26374–82.
[207] Li J, Liu D, Ke HZ, Duncan RL, Turner CH. The P2X7 nucleotide receptor mediates skeletal mechanotransduction. J Biol Chem 2005;280(52):42952–9.
[208] Jekir MG, Donahue HJ. Gap junctions and osteoblast-like cell gene expression in response to fluid flow. J Biomech Eng 2009;131(1):011005.
[209] Chachisvilis M, Zhang Y-L, Frangos JA. G protein-coupled receptors sense fluid shear stress in endothelial cells. Proc Natl Acad Sci U S A 2006;103(42):15463–8.
[210] Makino A, Prossnitz ER, Bünemann M, Wang JM, Yao W, Schmid-Schönbein GW. G protein-coupled receptors serve as mechanosensors for fluid shear stress in neutrophils. Am J Physiol Cell Physiol 2006;290(6):C1633–9.
[211] Peng J, Bencsik M, Louie A, Lu W, Millard S, Nguyen P, et al. Conditional expression of a Gi-coupled receptor in osteoblasts results in trabecular osteopenia. Endocrinology 2007;149(3):1329–37.
[212] Sakamoto A, Chen M, Nakamura T, Xie T, Karsenty G, Weinstein LS. Deficiency of the G-protein α-subunit Gsα in osteoblasts leads to differential effects on trabecular and cortical bone. J Biol Chem 2005;280(22):21369–75.
[213] Hsiao EC, Boudignon BM, Chang WC, Bencsik M, Peng J, Nguyen TD, et al. Osteoblast expression of an engineered Gs-coupled receptor dramatically increases bone mass. Proc Natl Acad Sci U S A 2008;105(4):1209–14.
[214] el Haj AJ, Minter SL, Rawlinson SCF, Suswillo R, Lanyon LE. Cellular responses to mechanical loading in vitro. J Bone Miner Res 1990;5(9):923–32.
[215] Owan I, Burr DB, Turner CH, Qiu J, Tu Y, Onyia JE, et al. Mechanotransduction in bone: osteoblasts are more responsive to fluid forces than mechanical strain. Am J Physiol Cell Physiol 1997;273(3):C810–5.
[216] You J, Yellowley C, Donahue H, Zhang Y, Chen Q, Jacobs C. Substrate deformation levels associated with routine physical activity are less stimulatory to bone cells relative to loading-induced oscillatory fluid flow. J Biomech Eng 2000;122(4):387–93.

## References

[217] Bakker AD, Soejima K, Klein-Nulend J, Burger EH. The production of nitric oxide and prostaglandin E(2) by primary bone cells is shear stress dependent. J Biomech 2001;34(5):671–7.

[218] McGarry JG, Klein-Nulend J, Mullender MG, Prendergast PJ. A comparison of strain and fluid shear stress in stimulating bone cell responses—a computational and experimental study. FASEB J 2005;19(3):482–4.

[218a] Orgensen NR, Henriksen Z, Brot C, Eriksen EF, Sorensen OH, Civitelli R, Steinberg TH. Human osteoblastic cells propagate intercellular calcium signals by two different mechanisms. J Bone Miner Res 2000;15:1024–32.

[219] Batra NN, Li YJ, Yellowley CE, You L, Malone AM, Kim CH, et al. Effects of short-term recovery periods on fluid-induced signaling in osteoblastic cells. J Biomech 2005;38(9):1909–17.

[220] Donahue TLH, Haut TR, Yellowley CE, Donahue HJ, Jacobs CR. Mechanosensitivity of bone cells to oscillating fluid flow induced shear stress may be modulated by chemotransport. J Biomech 2003;36(9):1363–71.

[221] Hillsley MV, Frangos JA. Alkaline phosphatase in osteoblasts is down-regulated by pulsatile fluid flow. Calcif Tissue Int 1997;60(1):48–53.

[222] Jacobs CR, Yellowley CE, Davis BR, Zhou Z, Cimbala JM, Donahue HJ. Differential effect of steady versus oscillatory flow on bone cells. J Biomech 1998;31:969–76.

[223] Jagodzinski M, Drescher M, Zeichen J, Hankemeier S, Krettek C, Bosch U, et al. Effects of cyclic longitudinal mechanical strain and dexamethasone on osteogenic differentiation of human bone marrow stromal cells. Eur Cell Mater 2004;7:35–41 [discussion].

[224] Robling AG, Castillo AB, Turner CH. Biomechanical and molecular regulation of bone remodeling. Annu Rev Biomed Eng 2006;8:455–98.

[225] Robling AG, Niziolek PJ, Baldridge LA, Condon KW, Allen MR, Alam I, et al. Mechanical stimulation of bone in vivo reduces osteocyte expression of Sost/sclerostin. J Biol Chem 2008;283(9):5866–75.

[226] Goldring SR, Goldring MB. Eating bone or adding it: the Wnt pathway decides. Nat Med 2007;13(2):133–4.

[227] Santos A, Bakker AD, Zandieh-Doulabi B, Semeins CM, Klein-Nulend J. Pulsating fluid flow modulates gene expression of proteins involved in Wnt signaling pathways in osteocytes. J Orthop Res 2009;.

[228] Lee DY, Li YS, Chang SF, Zhou J, Ho HM, Chiu JJ, et al. Oscillatory flow-induced proliferation of osteoblast-like cells is mediated by alphavbeta3 and beta1 integrins through synergistic interactions of focal adhesion kinase and Shc with phosphatidylinositol 3-kinase and the Akt/mTOR/p70S6K pathway. J Biol Chem 2010;285(1):30–42.

[229] Lee DY, Yeh CR, Chang SF, Lee PL, Chien S, Cheng CK, et al. Integrin-mediated expression of bone formation-related genes in osteoblast-like cells in response to fluid shear stress: roles of extracellular matrix, Shc, and mitogen-activated protein kinase. J Bone Miner Res 2008;23(7):1140–9.

[230] Ogata T. Fluid flow induces enhancement of the Egr-1 mRNA level in osteoblast-like cells: involvement of tyrosine kinase and serum. J Cell Physiol 1997;170(1):27–34.

[231] Chambers TJ, Fox S, Jagger CJ, Lean JM, Chow JW. The role of prostaglandins and nitric oxide in the response of bone to mechanical forces. Osteoarthritis Cartilage 1999;7(4):422–3.

[232] Vadiakas G, Banes A. Verapamil decreases cyclic load-induced calcium incorporation in ROS 17/2.8 osteosarcoma cell cultures. Matrix 1992;12(6):439–47.

[233] Harter L, Hruska K, Duncan R. Human osteoblast-like cells respond to mechanical strain with increased bone matrix protein production independent of hormonal regulation. Endocrinology 1995;136(2):528–35.

[234] Carvalho R, Scott J, Yen E. The effects of mechanical stimulation on the distribution of beta 1 integrin and expression of beta 1-integrin mRNA in TE-85 human osteosarcoma cells. Arch Oral Biol 1995;40(3):257–64.

[235] Rawlinson S, Mosley J, Suswillo R, Pitsillides A, Lanyon L. Calvarial and limb bone cells in organ and monolayer culture do not show the same early responses to dynamic mechanical strain. J Bone Miner Res 1995;10(8):1225–32.

[236] Carvalho R, Scott J, Suga D, Yen E. Stimulation of signal transduction pathways in osteoblasts by mechanical strain potentiated by parathyroid hormone. J Bone Miner Res 1994;9(7):999–1011.

[237] Berridge MJ, Lipp P, Bootman MD. The versatility and universality of calcium signalling. Nat Rev Mol Cell Biol 2000;1(1):11–21.

[238] Zayzafoon M. Calcium/calmodulin signaling controls osteoblast growth and differentiation. J Cell Biochem 2006;97(1):56–70.

[239] Donahue SW, Donahue HJ, Jacobs CR. Osteoblastic cells have refractory periods for fluid-flow-induced intracellular calcium oscillations for short bouts of flow and display multiple low-magnitude oscillations during long-term flow. J Biomech 2003;36 (1):35–43.

[240] Hung C, Allen F, Pollack S, Brighton C. Intracellular $Ca^{2+}$ stores and extracellular $Ca^{2+}$ are required in the real-time $Ca^{2+}$ response of bone cells experiencing fluid flow. J Biomech 1996;29(11):1411–7.

[241] Wiltink A, Nijweide PJ, Scheenen WJ, Ypey DL, Van Duijn B. Cell membrane stretch in osteoclasts triggers a self-reinforcing $Ca^{2+}$ entry pathway. Pflügers Arch 1995;429 (5):663–71.

[242] Chen NX, Ryder KD, Pavalko FM, Turner CH, Burr DB, Qiu J, et al. $Ca^{2+}$ regulates fluid shear-induced cytoskeletal reorganization and gene expression in osteoblasts. Am J Physiol Cell Physiol 2000;278(5):C989–97.

[243] Chen NX, Geist DJ, Genetos DC, Pavalko FM, Duncan RL. Fluid shear-induced NFκB translocation in osteoblasts is mediated by intracellular calcium release. Bone 2003;33 (3):399–410.

[244] Liu D, Genetos DC, Shao Y, Geist DJ, Li J, Ke HZ, et al. Activation of extracellular-signal regulated kinase (ERK1/2) by fluid shear is $Ca^{2+}$-and ATP-dependent in MC3T3-E1 osteoblasts. Bone 2008;42(4):644–52.

[245] Moon RT, Bowerman B, Boutros M, Perrimon N. The promise and perils of Wnt signaling through β-catenin. Science 2002;296(5573):1644–6.

[246] Arnsdorf EJ, Tummala P, Jacobs CR. Non-canonical Wnt signaling and N-cadherin related β-catenin signaling play a role in mechanically induced osteogenic cell fate. PLoS One 2009;4(4):e5388.

[247] Norvell SM, Alvarez M, Bidwell JP, Pavalko FM. Fluid shear stress induces ß-catenin signaling in osteoblasts. Calcif Tissue Int 2004;75(5):396–404.

[248] Rodda SJ, McMahon AP. Distinct roles for Hedgehog and canonical Wnt signaling in specification, differentiation and maintenance of osteoblast progenitors. Development 2006;133(16):3231–44.

[249] Bennett CN, Longo KA, Wright WS, Suva LJ, Lane TF, Hankenson KD, et al. Regulation of osteoblastogenesis and bone mass by Wnt10b. Proc Natl Acad Sci U S A 2005;102(9):3324–9.

[250] Gregory CA, Gunn WG, Reyes E, Smolarz AJ, Munoz J, Spees JL, et al. How Wnt signaling affects bone repair by mesenchymal stem cells from the bone marrow. Ann N Y Acad Sci 2005;1049(1):97–106.
[251] Jackson A, Vayssière B, Garcia T, Newell W, Baron R, Roman-Roman S, et al. Gene array analysis of Wnt-regulated genes in C3H10T1/2 cells. Bone 2005;36(4):585–98.
[252] Westendorf JJ, Kahler RA, Schroeder TM. Wnt signaling in osteoblasts and bone diseases. Gene 2004;341:19–39.
[253] Nusse R. Wnt signaling. Cold Spring Harb Perspect Biol 2012;4(5):a011163.
[254] Bonni A, Brunet A, West AE, Datta SR, Takasu MA, Greenberg ME. Cell survival promoted by the Ras-MAPK signaling pathway by transcription-dependent and-independent mechanisms. Science 1999;286(5443):1358–62.
[255] Pearson G, Robinson F, Beers Gibson T, Xu B-E, Karandikar M, Berman K, et al. Mitogen-activated protein (MAP) kinase pathways: regulation and physiological functions. Endocr Rev 2001;22(2):153–83.
[256] Fan D, Chen Z, Wang D, Guo Z, Qiang Q, Shang Y. Osterix is a key target for mechanical signals in human thoracic ligament flavum cells. J Cell Physiol 2007;211(3):577–84.
[257] Jansen J, Weyts F, Westbroek I, Jahr H, Chiba H, Pols H, et al. Stretch-induced phosphorylation of ERK1/2 depends on differentiation stage of osteoblasts. J Cell Biochem 2004;93(3):542–51.
[258] Simmons CA, Matlis S, Thornton AJ, Chen S, Wang C-Y, Mooney DJ. Cyclic strain enhances matrix mineralization by adult human mesenchymal stem cells via the extracellular signal-regulated kinase (ERK1/2) signaling pathway. J Biomech 2003;36(8):1087–96.
[259] Simmons CA, Nikolovski J, Thornton AJ, Matlis S, Mooney DJ. Mechanical stimulation and mitogen-activated protein kinase signaling independently regulate osteogenic differentiation and mineralization by calcifying vascular cells. J Biomech 2004;37(10):1531–41.
[260] Jessop H, Rawlinson S, Pitsillides A, Lanyon L. Mechanical strain and fluid movement both activate extracellular regulated kinase (ERK) in osteoblast-like cells but via different signaling pathways. Bone 2002;31(1):186–94.
[261] Kapur S, Chen S-T, Baylink DJ, Lau K-HW. Extracellular signal-regulated kinase-1 and-2 are both essential for the shear stress-induced human osteoblast proliferation. Bone 2004;35(2):525–34.
[262] Plotkin LI, Mathov I, Aguirre JI, Parfitt AM, Manolagas SC, Bellido T. Mechanical stimulation prevents osteocyte apoptosis: requirement of integrins, Src kinases, and ERKs. Am J Physiol Cell Physiol 2005;289(3):C633–43.
[263] Yang C-M, Chien C-S, Yao C-C, Hsiao L-D, Huang Y-C, Wu CB. Mechanical strain induces collagenase-3 (MMP-13) expression in MC3T3-E1 osteoblastic cells. J Biol Chem 2004;279(21):22158–65.
[264] Rubin J, Murphy TC, Fan X, Goldschmidt M, Taylor WR. Activation of extracellular signal-regulated kinase is involved in mechanical strain inhibition of RANKL expression in bone stromal cells. J Bone Miner Res 2002;17(8):1452–60.
[265] Boutahar N, Guignandon A, Vico L, Lafage-Proust M-H. Mechanical strain on osteoblasts activates autophosphorylation of focal adhesion kinase and proline-rich tyrosine kinase 2 tyrosine sites involved in ERK activation. J Biol Chem 2004;279(29):30588–99.

[266] Rubin J, Murphy TC, Zhu L, Roy E, Nanes MS, Fan X. Mechanical strain differentially regulates endothelial nitric-oxide synthase and receptor activator of nuclear κB ligand expression via ERK1/2 MAPK. J Biol Chem 2003;278(36):34018–25.

[267] Cotton M, Claing A. G protein-coupled receptors stimulation and the control of cell migration. Cell Signal 2009;21(7):1045–53.

[268] Ogasawara A, Arakawa T, Kaneda T, Takuma T, Sato T, Kaneko H, et al. Fluid shear stress-induced cyclooxygenase-2 expression is mediated by C/EBP β, cAMP-response element-binding protein, and AP-1 in osteoblastic MC3T3-E1 cells. J Biol Chem 2001;276(10):7048–54.

[269] Clark CB, McKnight NL, Frangos JA. Strain and strain rate activation of G proteins in human endothelial cells. Biochem Biophys Res Commun 2002;299(2):258–62.

[270] Gudi SR, Lee AA, Clark CB, Frangos JA. Equibiaxial strain and strain rate stimulate early activation of G proteins in cardiac fibroblasts. Am J Physiol Cell Physiol 1998;274(5):C1424–8.

[271] Reich KM, McAllister TN, Gudi S, Frangos JA. Activation of G proteins mediates flow-induced prostaglandin E2 production in osteoblasts. Endocrinology 1997;138(3):1014–8.

[272] Bryan NS, Bian K, Murad F. Discovery of the nitric oxide signaling pathway and targets for drug development. Front Biosci 2009;14(1):1–18.

[273] Chambers TJ, Chow JW, Fox SW, Jagger CJ, Lean JM. The role of prostaglandins and nitric oxide in the response of bone to mechanical stimulation. In: Recent advances in prostaglandin, thromboxane and leukotriene research. New York, NY: Plenum Press; 1998. p. 295–8.

[274] Fox S, Chambers T, Chow J. Nitric oxide is an early mediator of the increase in bone formation by mechanical stimulation. Am J Physiol Endocrinol Metab 1996;270(6):E955–60.

[275] Turner CH, Takano Y, Owan I, Murrell G. Nitric oxide inhibitor L-NAME suppresses mechanically induced bone formation in rats. Am J Physiol Endocrinol Metab 1996;270(4):E634–9.

[276] Basso N, Heersche JN. Effects of hind limb unloading and reloading on nitric oxide synthase expression and apoptosis of osteocytes and chondrocytes. Bone 2006;39(4):807–14.

[277] Zaman G, Pitsillides A, Rawlinson S, Suswillo R, Mosley J, Cheng M, et al. Mechanical strain stimulates nitric oxide production by rapid activation of endothelial nitric oxide synthase in osteocytes. J Bone Miner Res 1999;14(7):1123–31.

[278] Sikavitsas VI, Bancroft GN, Holtorf HL, Jansen JA, Mikos AG. Mineralized matrix deposition by marrow stromal osteoblasts in 3D perfusion culture increases with increasing fluid shear forces. Proc Natl Acad Sci U S A 2003;100(25):14683–8.

[279] Smalt R, Mitchell FT, Howard RL, Chambers TJ. Induction of NO and prostaglandin E2 in osteoblasts by wall-shear stress but not mechanical strain. Am J Physiol Endocrinol Metab 1997;273(4):E751–8.

[280] Tan S, Bakker A, Semeins C, Kuijpers-Jagtman A, Klein-Nulend J. Inhibition of osteocyte apoptosis by fluid flow is mediated by nitric oxide. Biochem Biophys Res Commun 2008;369(4):1150–4.

[281] van den Dolder J, Bancroft GN, Sikavitsas VI, Spauwen PHM, Jansen JA, Mikos AG. Flow perfusion culture of marrow stromal osteoblasts in titanium fiber mesh. J Biomed Mater Res A 2003;64A(2):235–41.

[282] Nicolella DP, Bonewald LF, Moravits DE, Lankford J. Measurement of microstructural strain in cortical bone. Eur J Morphol 2005;42(1–2):23–9.
[283] Nicolella DP, Nicholls AE, Lankford J, Davy DT. Machine vision photogrammetry: a technique for measurement of microstructural strain in cortical bone. J Biomech 2001;34(1):135–9.
[284] Nicolella DP, Moravits DE, Gale AM, Bonewald LF, Lankford J. Osteocyte lacunae tissue strain in cortical bone. J Biomech 2006;39(9):1735–43.
[285] Ishihara Y, Sugawara Y, Kamioka H, Kawanabe N, Hayano S, Balam TA, et al. Ex vivo real-time observation of Ca2+ signaling in living bone in response to shear stress applied on the bone surface. Bone 2013;53(1):204–15.
[286] Ishihara Y, Sugawara Y, Kamioka H, Kawanabe N, Kurosaka H, Naruse K, et al. In situ imaging of the autonomous intracellular Ca2+ oscillations of osteoblasts and osteocytes in bone. Bone 2012;50(4):842–52.
[287] Knothe Tate ML, Niederer P, Knothe U. In vivo tracer transport through the lacunocanalicular system of rat bone in an environment devoid of mechanical loading. Bone 1998;22(2):107–17.
[288] Giorgi M, Verbruggen SW, Lacroix D. In silico bone mechanobiology: modeling a multifaceted biological system. Wiley Interdiscip Rev Syst Biol Med 2016;8(6):485–505.
[289] Rath Bonivtch A, Bonewald LF, Nicolella DP. Tissue strain amplification at the osteocyte lacuna: a microstructural finite element analysis. J Biomech 2007;40(10):2199–206.
[290] Varga P, Hesse B, Langer M, Schrof S, Männicke N, Suhonen H, et al. Synchrotron X-ray phase nano-tomography-based analysis of the lacunar–canalicular network morphology and its relation to the strains experienced by osteocytes in situ as predicted by case-specific finite element analysis. Biomech Model Mechanobiol 2015;14(2):267–82.
[291] Vaughan TJ, Mullen CA, Verbruggen SW, McNamara LM. Bone cell mechanosensation of fluid flow stimulation: a fluid–structure interaction model characterising the role integrin attachments and primary cilia. Biomech Model Mechanobiol 2015;14(4):703–18.
[292] Melton 3rd LJ, Chrischilles EA, Cooper C, Lane AW, Riggs BL. How many women have osteoporosis? JBMR Anniversary Classic. JBMR, Volume 7, Number 9, 1992. J Bone Miner Res 2005;20(5):886–92.
[293] Kanis JA, McCloskey EV, Johansson H, Cooper C, Rizzoli R, Reginster JY, et al. European guidance for the diagnosis and management of osteoporosis in postmenopausal women. Osteoporos Int 2013;24(1):23–57.
[294] Randell KM, Honkanen RJ, Kroger H, Saarikoski S. Does hormone-replacement therapy prevent fractures in early postmenopausal women? J Bone Miner Res 2002;17(3):528–33.
[295] Compston JE, Rosen CJ. Osteoporosis. 3rd ed. Oxford, UK: Health Press Ltd.; 2002.
[296] Reginster JY, Burlet N. Osteoporosis: a still increasing prevalence. Bone 2006;38(2 Suppl. 1):S4–9.
[297] Rosen CJ. Pathogenesis of osteoporosis. Baillieres Best Pract Res Clin Endocrinol Metab 2000;14(2):181–93.
[298] Bell K, Loveridge N, Lunt M, Lindsay PC, Reeve J. Oestrogen suppression increases Haversian resorption depth as well as remodelling activity in women with endometriosis. Bone 1996;19(3(1)):131S.

[299] Brockstedt H, Kassem M, Eriksen EF, Mosekilde L, Melsen F. Age- and sex-related changes in iliac cortical bone mass and remodeling. Bone 1993;14(4):681–91.
[300] Eriksen EF, Langdahl B, Vesterby A, Rungby J, Kassem M. Hormone replacement therapy prevents osteoclastic hyperactivity: a histomorphometric study in early postmenopausal women. J Bone Miner Res 1999;14(7):1217–21.
[301] Hughes DE, Dai A, Tiffee JC, Li HH, Mundy GR, Boyce BF. Estrogen promotes apoptosis of murine osteoclasts mediated by TGF-beta. Nat Med 1996;2(10):1132–6.
[302] Dalle Carbonare L, Giannini S. Bone microarchitecture as an important determinant of bone strength. J Endocrinol Invest 2004;27(1):99–105.
[303] Michael H, Harkonen PL, Vaananen HK, Hentunen TA. Estrogen and testosterone use different cellular pathways to inhibit osteoclastogenesis and bone resorption. J Bone Miner Res 2005;20(12):2224–32.
[304] Kousteni S, Bellido T, Plotkin LI, O'Brien CA, Bodenner DL, Han L, et al. Nongenotropic, sex-nonspecific signaling through the estrogen or androgen receptors: dissociation from transcriptional activity. Cell 2001;104(5):719–30.
[305] Tomkinson A, Reeve J, Shaw RW, Noble BS. The death of osteocytes via apoptosis accompanies estrogen withdrawal in human bone. J Clin Endocrinol Metab 1997;82(9):3128–35.
[306] Frost HM. Micropetrosis. J Bone Joint Surg Am 1960;42-A:144–50.
[307] Boyde A. The real response of bone to exercise. J Anat 2003;203(2):173–89.
[308] Kingsmill VJ, Boyde A. Mineralisation density of human mandibular bone: quantitative backscattered electron image analysis. J Anat 1998;192(Pt 2):245–56.
[309] Knothe Tate ML, Adamson JR, Tami AE, Bauer TW. The osteocyte. Int J Biochem Cell Biol 2004;36(1):1–8.
[310] Mullender MG, Vandermeer DD, Huiskes R, Lips P. Osteocyte density changes in aging and osteoporosis. Bone 1996;18(2):109–13.
[311] Milovanovic P, Zimmermann EA, Hahn M, Djonic D, Puschel K, Djuric M, et al. Osteocytic canalicular networks: morphological implications for altered mechanosensitivity. ACS Nano 2013;7(9):7542–51.
[312] Emerton KB, Hu B, Woo AA, Sinofsky A, Hernandez C, Majeska RJ, et al. Osteocyte apoptosis and control of bone resorption following ovariectomy in mice. Bone 2010;46(3):577–83.
[313] Tatsumi S, Ishii K, Amizuka N, Li M, Kobayashi T, Kohno K, et al. Targeted ablation of osteocytes induces osteoporosis with defective mechanotransduction. Cell Metab 2007;5(6):464–75.
[314] Voisin M, McNamara LM. Differential $\beta_3$ and $\beta_1$ integrin expression in bone marrow and cortical bone of estrogen deficient rats. Anat Rec (Hoboken) 2015;298:1548–59.
[315] Yeh CR, Chiu JJ, Lee CI, Lee PL, Shih YT, Sun JS, et al. Estrogen augments shear stress-induced signaling and gene expression in osteoblast-like cells via estrogen receptor-mediated expression of beta1-integrin. J Bone Miner Res 2010;25(3):627–39.
[316] Shimomura Y, Wajid M, Shapiro L, Christiano AM. P-cadherin is a p63 target gene with a crucial role in the developing human limb bud and hair follicle. Development 2008;135(4):743–53.
[317] Kjær KW, Hansen L, Schwabe G, Marques-de-Faria A, Eiberg H, Mundlos S, et al. Distinct CDH3 mutations cause ectodermal dysplasia, ectrodactyly, macular dystrophy (EEM syndrome). J Med Genet 2005;42(4):292–8.

[318] Haÿ E, Laplantine E, Geoffroy V, Frain M, Kohler T, Müller R, et al. N-cadherin interacts with axin and LRP5 to negatively regulate Wnt/β-catenin signaling, osteoblast function, and bone formation. Mol Cell Biol 2009;29(4):953–64.
[319] Di Benedetto A, Watkins M, Grimston S, Salazar V, Donsante C, Mbalaviele G, et al. N-cadherin and cadherin 11 modulate postnatal bone growth and osteoblast differentiation by distinct mechanisms. J Cell Sci 2010;123(15):2640–8.
[320] Estrada K, Styrkarsdottir U, Evangelou E, Hsu Y-H, Duncan EL, Ntzani EE, et al. Genome-wide meta-analysis identifies 56 bone mineral density loci and reveals 14 loci associated with risk of fracture. Nat Genet 2012;44(5):491–501.
[321] Zhang J, Niu C, Ye L, Huang H, He X, Tong W-G, et al. Identification of the haematopoietic stem cell niche and control of the niche size. Nature 2003;425(6960):836–41.
[322] Xie Y, Yin T, Wiegraebe W, He XC, Miller D, Stark D, et al. Detection of functional haematopoietic stem cell niche using real-time imaging. Nature 2009;457(7225):97–101.
[323] Xu H, Zhang J, Wu J, Guan Y, Weng Y, Shang P. Oscillatory fluid flow elicits changes in morphology, cytoskeleton and integrin-associated molecules in MLO-Y4 cells, but not in MC3T3-E1 cells. Biol Res 2012;45(2):163–9.
[324] Barron MJ, Tsai C-J, Donahue SW. Mechanical stimulation mediates gene expression in MC3T3 osteoblastic cells differently in 2D and 3D environments. J Biomech Eng 2010;132(4):041005.
[325] Hosokawa K, Arai F, Yoshihara H, Iwasaki H, Nakamura Y, Gomei Y, et al. Knockdown of N-cadherin suppresses the long-term engraftment of hematopoietic stem cells. Blood 2010;116(4):554–63.
[326] Bakker AD, Klein-Nulend J, Tanck E, Albers GH, Lips P, Burger EH. Additive effects of estrogen and mechanical stress on nitric oxide and prostaglandin E2 production by bone cells from osteoporotic donors. Osteoporos Int 2005;16(8):983–9.
[327] Joldersma M, Klein-Nulend J, Oleksik AM, Heyligers IC, Burger EH. Estrogen enhances mechanical stress-induced prostaglandin production by bone cells from elderly women. Am J Physiol Endocrinol Metab 2001;280(3):E436–42.
[328] Damien E, Price JS, Lanyon LE. The estrogen receptor's involvement in osteoblasts' adaptive response to mechanical strain. J Bone Miner Res 1998;13(8):1275–82.
[329] Cheng MZ, Rawlinson SC, Pitsillides AA, Zaman G, Mohan S, Baylink DJ, et al. Human osteoblasts' proliferative responses to strain and 17beta-estradiol are mediated by the estrogen receptor and the receptor for insulin-like growth factor I. J Bone Miner Res 2002;17(4):593–602.
[330] Braidman IP, Hainey L, Batra G, Selby PL, Saunders PT, Hoyland JA. Localization of estrogen receptor beta protein expression in adult human bone. J Bone Miner Res 2001;16(2):214–20.
[331] Neidlinger-Wilke C, Stalla I, Claes L, Brand R, Hoellen I, Rubenacker S, et al. Human osteoblasts from younger normal and osteoporotic donors show differences in proliferation and TGF beta-release in response to cyclic strain. J Biomech 1995;28(12):1411–8.
[332] Sterck JG, Klein-Nulend J, Lips P, Burger EH. Response of normal and osteoporotic human bone cells to mechanical stress in vitro. Am J Physiol 1998;274(6 Pt 1):E1113–20.
[333] Brennan MA, Haugh MG, O'Brien FJ, McNamara LM. Estrogen withdrawal from osteoblasts and osteocytes causes increased mineralization and apoptosis. Horm Metab Res 2014;46(8):537–45.

[334] Brennan O, O'Brien F, McNamara L. Estrogen plus estrogen receptor antagonists alter mineral production by osteoblasts in vitro. Horm Metab Res 2012;44(1):47–53.

[335] Brennan MA, McDermott A, McNamara LM, editors. The impact of shear stress during estrogen withdrawal on cell viability and mineral production of osteoblasts. San Francisco, USA: Orthopaedic Research Society; 2012.

[336] Deepak V, Kayastha P, McNamara LM. Estrogen deficiency attenuates fluid flow-induced [Ca2+]i oscillations and mechanoresponsiveness of MLO-Y4 osteocytes. FASEB J 2017;31(7):3027–39.

[337] McNamara LM, Prendergast PJ. Perforation of cancellous bone trabeculae by damage-stimulated remodelling at resorption pits: a computational analysis. Eur J Morphol 2005;42(1–2):99–109.

[338] Brennan MA, Gleeson JP, Browne M, O'Brien FJ, Thurner PJ, McNamara LM. Site specific increase in heterogeneity of trabecular bone tissue mineral during oestrogen deficiency. Eur Cell Mater 2011;21:396–406.

[339] Verhulp E, van Rietbergen B, Huiskes R. Load distribution in the healthy and osteoporotic human proximal femur during a fall to the side. Bone 2008;42(1):30–5.

[340] Gefen A, Portnoy S, Diamant I. Inhomogeneity of tissue-level strain distributions in individual trabeculae: mathematical model studies of normal and osteoporosis cases. Med Eng Phys 2008;30(5):624–30.

[341] Mulvihill BM, McNamara LM, Prendergast PJ. Loss of trabeculae by mechanobiological means may explain rapid bone loss in osteoporosis. J R Soc Interface 2008;5(27):1243–53.

[342] McNamara LM, Prendergast PJ. Bone remodelling algorithms incorporating both strain and microdamage stimuli. J Biomech 2007;40(6):1381–91.

[343] Mulvihill BM, Prendergast PJ. Mechanobiological regulation of the remodelling cycle in trabecular bone and possible biomechanical pathways for osteoporosis. Clin Biomech (Bristol, Avon) 2010;25(5):491–8.

[344] van der Linden JC, Day JS, Verhaar JA, Weinans H. Altered tissue properties induce changes in cancellous bone architecture in aging and diseases. J Biomech 2004;37(3):367–74.

[345] Metzger TA, Kreipke TC, Vaughan TJ, McNamara LM, Niebur GL. The in situ mechanics of trabecular bone marrow: the potential for mechanobiological response. J Biomech Eng 2015;137(1).

[346] Vaughan TJ, Verbruggen SW, McNamara LM. Are all osteocytes equal? Multiscale modelling of cortical bone to characterise the mechanical stimulation of osteocytes. Int J Numer Methods Biomed Eng 2013;29(12):1361–72.

[347] Vaughan TJ, Voisin M, Niebur GL, McNamara LM. Multiscale modeling of trabecular bone marrow: understanding the micromechanical environment of mesenchymal stem cells during osteoporosis. J Biomech Eng 2015;137(1).

[348] Gurkan UA, Akkus O. The mechanical environment of bone marrow: a review. Ann Biomed Eng 2008;36(12):1978–91.

[349] Metzger TA, Vaughan TJ, McNamara LM, Niebur GL. Altered architecture and cell populations affect bone marrow mechanobiology in the osteoporotic human femur. Biomech Model Mechanobiol 2017;16(3):841–50.

[350] McNamara LM. Perspective on post-menopausal osteoporosis: establishing an interdisciplinary understanding of the sequence of events from the molecular level to whole bone fractures. J R Soc Interface 2010;7(44):353–72.

[351] Lawrence RC, Felson DT, Helmick CG, Arnold LM, Choi H, Deyo RA, et al. Estimates of the prevalence of arthritis and other rheumatic conditions in the United States. Part II. Arthritis Rheum 2008;58(1):26–35.
[352] Goldring MB, Goldring SR. Osteoarthritis. J Cell Physiol 2007;213(3):626–34.
[353] Goldring MB, Goldring SR. Articular cartilage and subchondral bone in the pathogenesis of osteoarthritis. Ann N Y Acad Sci 2010;1192:230–7.
[354] Ko FC, Dragomir C, Plumb DA, Goldring SR, Wright TM, Goldring MB, et al. In vivo cyclic compression causes cartilage degeneration and subchondral bone changes in mouse tibiae. Arthritis Rheum 2013;65(6):1569–78.
[355] Hayami T, Pickarski M, Wesolowski GA, McLane J, Bone A, Destefano J, et al. The role of subchondral bone remodeling in osteoarthritis: reduction of cartilage degeneration and prevention of osteophyte formation by alendronate in the rat anterior cruciate ligament transection model. Arthritis Rheum 2004;50(4):1193–206.
[356] Hayami T, Pickarski M, Zhuo Y, Wesolowski GA, Rodan GA, Duong LT. Characterization of articular cartilage and subchondral bone changes in the rat anterior cruciate ligament transection and meniscectomized models of osteoarthritis. Bone 2006;38(2):234–43.
[357] Felson DT, Lawrence RC, Dieppe PA, Hirsch R, Helmick CG, Jordan JM, et al. Osteoarthritis: new insights. Part 1: the disease and its risk factors. Ann Intern Med 2000;133(8):635–46.
[358] Kaila-Kangas L, Arokoski J, Impivaara O, Viikari-Juntura E, Leino-Arjas P, Luukkonen R, et al. Associations of hip osteoarthritis with history of recurrent exposure to manual handling of loads over 20 kg and work participation: a population-based study of men and women. Occup Environ Med 2011;68(10):734–8.
[359] Wong SY, Evans RA, Needs C, Dunstan CR, Hills E, Garvan J. The pathogenesis of osteoarthritis of the hip. Evidence for primary osteocyte death. Clin Orthop Relat Res 1987;214:305–12.
[360] Jaiprakash A, Prasadam I, Feng JQ, Liu Y, Crawford R, Xiao Y. Phenotypic characterization of osteoarthritic osteocytes from the sclerotic zones: a possible pathological role in subchondral bone sclerosis. Int J Biol Sci 2012;8(3):406–17.
[361] Tate MK, Tami A, Bauer T, Knothe U. Micropathoanatomy of osteoporosis: indications for a cellular basis of bone disease. Adv Osteopor Fract Manage 2002;2(1):9–14.
[362] van Hove RP, Nolte PA, Vatsa A, Semeins CM, Salmon PL, Smit TH, et al. Osteocyte morphology in human tibiae of different bone pathologies with different bone mineral density—is there a role for mechanosensing? Bone 2009;45(2):321–9.
[363] Meyers P, Schwartz C, Krailo M, Kleinerman E, Betcher D, Bernstein M, et al. Osteosarcoma: a randomized, prospective trial of the addition of ifosfamide and/or muramyl tripeptide to cisplatin, doxorubicin, and high-dose methotrexate. J Clin Oncol 2005;23(9):2004–11.
[364] Guo W, Giancotti F. Integrin signalling during tumour progression. Nat Rev Mol Cell Biol 2004;5(10):816–26.
[365] Ramsay A, Marshall J, Hart I. Integrin trafficking and its role in cancer metastasis. Cancer Metastasis Rev 2007;26(3–4):567–78.
[366] Jung Y, Wang J, Schneider A, Sun YX, Koh-Paige AJ, Osman NI, et al. Regulation of SDF-1 (CXCL12) production by osteoblasts; a possible mechanism for stem cell homing. Bone 2006;38(4):497–508.

[367] Coughlin T, Moreno RR, Mason D, Nystrom L, Boerckel J, Niebur G, et al. Bone: a fertile soil for cancer metastasis. Curr Drug Targets 2016;18(11):1281–95.
[368] Guise TA. Molecular mechanisms of osteolytic bone metastases. Cancer 2000;88 (12 Suppl.):2892–8.
[369] Waning DL, Guise TA. Molecular mechanisms of bone metastasis and associated muscle weakness. Clin Cancer Res 2014;20(12):3071–7.
[370] Paszek M, Weaver V. The tension mounts: mechanics meets morphogenesis and malignancy. J Mammary Gland Biol Neoplasia 2004;9(4):325–42.
[371] Makale M. Cellular mechanobiology and cancer metastasis. Birth Defects Res C Embryo Today 2007;81(4):329–43.
[372] Suresh S. Biomechanics and biophysics of cancer cells. Acta Biomater 2007;3 (4):413–38.
[373] Zaman M, Trapani L, Sieminski A, Siemeski A, Mackellar D, Gong H, et al. Migration of tumor cells in 3D matrices is governed by matrix stiffness along with cell-matrix adhesion and proteolysis. Proc Natl Acad Sci U S A 2006;103(29):10889–94.
[374] Pignatelli M, Cardillo M, Hanby A, Stamp G. Integrins and their accessory adhesion molecules in mammary carcinomas: loss of polarization in poorly differentiated tumors. Hum Pathol 1992;23(10):1159–66.
[375] Albelda S, Mette S, Elder D, Stewart R, Damjanovich L, Herlyn M, et al. Integrin distribution in malignant melanoma: association of the beta 3 subunit with tumor progression. Cancer Res 1990;50(20):6757–64.
[376] Paszek M, Zahir N, Johnson K, Lakins J, Rozenberg G, Gefen A, et al. Tensional homeostasis and the malignant phenotype. Cancer Cell 2005;8(3):241–54.
[377] Ulrich T, de Juan Pardo E, Kumar S. The mechanical rigidity of the extracellular matrix regulates the structure, motility, and proliferation of glioma cells. Cancer Res 2009;69 (10):4167–74.
[378] Krouskop T, Wheeler T, Kallel F, Garra B, Hall T. Elastic moduli of breast and prostate tissues under compression. Ultrason Imaging 1998;20(4):260–74.
[379] Plewes D, Bishop J, Samani A, Sciarretta J. Visualization and quantification of breast cancer biomechanical properties with magnetic resonance elastography. Phys Med Biol 2000;45(6):1591–610.
[380] Fenner J, Stacer AC, Winterroth F, Johnson TD, Luker KE, Luker GD. Macroscopic stiffness of breast tumors predicts metastasis. Sci Rep 2014;4:5512.
[381] Steeg PS. Tumor metastasis: mechanistic insights and clinical challenges. Nat Med 2006;12(8):895–904.
[382] Fidler IJ. The pathogenesis of cancer metastasis: the 'seed and soil' hypothesis revisited. Nat Rev Cancer 2003;3(6):453–8.
[383] Nazarian A, von Stechow D, Zurakowski D, Müller R, Snyder B. Bone volume fraction explains the variation in strength and stiffness of cancellous bone affected by metastatic cancer and osteoporosis. Calcif Tissue Int 2008;83(6):368–79.
[384] Kaneko T, Bell J, Pejcic M, Tehranzadeh J, Keyak J. Mechanical properties, density and quantitative CT scan data of trabecular bone with and without metastases. J Biomech 2004;37(4):523–30.
[385] Van Roy F. Beyond E-cadherin: roles of other cadherin superfamily members in cancer. Nat Rev Cancer 2014;14(2):121–34.
[386] Wheelock MJ, Johnson KR. Cadherins as modulators of cellular phenotype. Annu Rev Cell Dev Biol 2003;19(1):207–35.

[387] Nakajima S, Doi R, Toyoda E, Tsuji S, Wada M, Koizumi M, et al. N-cadherin expression and epithelial-mesenchymal transition in pancreatic carcinoma. Clin Cancer Res 2004;10(12):4125–33.

[388] Chu K, Cheng C-J, Ye X, Lee Y-C, Zurita AJ, Chen D-T, et al. Cadherin-11 promotes the metastasis of prostate cancer cells to bone. Mol Cancer Res 2008;6(8):1259–67.

[389] Blaschuk OW, Sullivan R, David S, Pouliot Y. Identification of a cadherin cell adhesion recognition sequence. Dev Biol 1990;139(1):227–9.

[390] Williams E, Williams G, Gour BJ, Blaschuk OW, Doherty P. A novel family of cyclic peptide antagonists suggests that N-cadherin specificity is determined by amino acids that flank the HAV motif. J Biol Chem 2000;275(6):4007–12.

[391] Watanabe T, Kuramochi H, Takahashi A, Imai K, Katsuta N, Nakayama T, et al. Higher cell stiffness indicating lower metastatic potential in B16 melanoma cell variants and in (−)-epigallocatechin gallate-treated cells. J Cancer Res Clin Oncol 2012;138(5):859–66.

[392] Swaminathan V, Mythreye K, O'Brien ET, Berchuck A, Blobe GC, Superfine R. Mechanical stiffness grades metastatic potential in patient tumor cells and in cancer cell lines. Cancer Res 2011;71(15):5075–80.

[393] Kim Y, Othmer HG. Hybrid models of cell and tissue dynamics in tumor growth. Math Biosci Eng 2015;12(6):1141–56.

[394] Rejniak KA, Anderson AR. Hybrid models of tumor growth. Wiley Interdiscip Rev Syst Biol Med 2011;3(1):115–25.

[395] Zhou X, Liu J. A computational model to predict bone metastasis in breast cancer by integrating the dysregulated pathways. BMC Cancer 2014;14:618.

[396] Cook LM, Araujo A, Pow-Sang JM, Budzevich MM, Basanta D, Lynch CC. Predictive computational modeling to define effective treatment strategies for bone metastatic prostate cancer. Sci Rep 2016;6:29384.

[397] Araujo A, Cook LM, Lynch CC, Basanta D. An integrated computational model of the bone microenvironment in bone-metastatic prostate cancer. Cancer Res 2014;74(9):2391–401.

[398] Tracqui P. Biophysical models of tumour growth. Rep Prog Phys 2009;72(5):056701.

[399] Khajuria DK, Razdan R, Mahapatra DR. Drugs for the management of osteoporosis: a review. Rev Bras Reumatol 2011;51(4):365–71 [79–82].

[400] Todd JA, Robinson RJ. Osteoporosis and exercise. Postgrad Med J 2003;79(932):320–3.

[401] Marcus R. Role of exercise in preventing and treating osteoporosis. Rheum Dis Clin North Am 2001;27(1):131–41 [vi].

[402] Kemmler W, Lauber D, Weineck J, Hensen J, Kalender W, Engelke K. Benefits of 2 years of intense exercise on bone density, physical fitness, and blood lipids in early postmenopausal osteopenic women: results of the Erlangen Fitness Osteoporosis Prevention Study (EFOPS). Arch Intern Med 2004;164(10):1084–91.

[403] Bonaiuti D, Shea B, Iovine R, Negrini S, Robinson V, Kemper HC, et al. Exercise for preventing and treating osteoporosis in postmenopausal women. Cochrane Database Syst Rev 2002;3:CD000333.

[404] Howe TE, Shea B, Dawson LJ, Downie F, Murray A, Ross C, et al. Exercise for preventing and treating osteoporosis in postmenopausal women. Cochrane Database Syst Rev 2011;7:CD000333.

[405] Chien MY, Wu YT, Hsu AT, Yang RS, Lai JS. Efficacy of a 24-week aerobic exercise program for osteopenic postmenopausal women. Calcif Tissue Int 2000;67(6):443–8.
[406] Iwamoto J, Takeda T, Ichimura S. Effect of exercise training and detraining on bone mineral density in postmenopausal women with osteoporosis. J Orthop Sci 2001;6(2):128–32.
[407] Hartard M, Haber P, Ilieva D, Preisinger E, Seidl G, Huber J. Systematic strength training as a model of therapeutic intervention. A controlled trial in postmenopausal women with osteopenia. Am J Phys Med Rehabil 1996;75(1):21–8.
[408] Chow R, Harrison J, Dornan J. Prevention and rehabilitation of osteoporosis program: exercise and osteoporosis. Int J Rehabil Res 1989;12(1):49–56.
[409] Ward K, Alsop C, Caulton J, Rubin C, Adams J, Mughal Z. Low magnitude mechanical loading is osteogenic in children with disabling conditions. J Bone Miner Res 2004;19(3):360–9.
[410] Rubin C, Recker R, Cullen D, Ryaby J, McCabe J, McLeod K. Prevention of postmenopausal bone loss by a low-magnitude, high-frequency mechanical stimuli: a clinical trial assessing compliance, efficacy, and safety. J Bone Miner Res 2004;19(3):343–51.
[411] Rubin C, Judex S, Qin YX. Low-level mechanical signals and their potential as a nonpharmacological intervention for osteoporosis. Age Ageing 2006;35(Suppl. 2):ii32–6.
[412] Boschert V, Frisch C, Back JW, van Pee K, Weidauer SE, Muth EM, et al. The sclerostin-neutralizing antibody AbD09097 recognizes an epitope adjacent to sclerostin's binding site for the Wnt co-receptor LRP6. Open Biol 2016;6(8).
[413] Osteoporosis in the workplace: the social, economic and human costs of osteoporosis on employees, employers and governments. Liege, Belgium: International Osteoporosis Foundation Study Group; 2002.
[414] Tzioupis C, Giannoudis PV. Prevalence of long-bone non-unions. Injury 2007;38(Suppl. 2):S3–9.
[415] Kanakaris NK, Giannoudis PV. The health economics of the treatment of long-bone non-unions. Injury 2007;38(Suppl. 2):S77–84.
[416] Mesenchymal stem cells as a potent cell source for bone regeneration. Stem Cells Int 2012;2012:1–9.
[417] Dawson JI, Oreffo RO. Bridging the regeneration gap: stem cells, biomaterials and clinical translation in bone tissue engineering. Arch Biochem Biophys 2008;473(2):124–31.
[418] Rose FR, Oreffo RO. Bone tissue engineering: hope vs hype. Biochem Biophys Res Commun 2002;292(1):1–7.
[419] Cancedda R, Giannoni P, Mastrogiacomo M. A tissue engineering approach to bone repair in large animal models and in clinical practice. Biomaterials 2007;28(29):4240–50.
[420] Horch RE, Kopp J, Kneser U, Beier J, Bach AD. Tissue engineering of cultured skin substitutes. J Cell Mol Med 2005;9(3):592–608.
[421] Kreuz PC, Muller S, Ossendorf C, Kaps C, Erggelet C. Treatment of focal degenerative cartilage defects with polymer-based autologous chondrocyte grafts: four-year clinical results. Arthritis Res Ther 2009;11(2):R33.
[422] Ohgushi H, Goldberg VM, Caplan AI. Repair of bone defects with marrow cells and porous ceramic. Experiments in rats. Acta Orthop Scand 1989;60(3):334–9.
[423] Bruder SP, Kraus KH, Goldberg VM, Kadiyala S. The effect of implants loaded with autologous mesenchymal stem cells on the healing of canine segmental bone defects. J Bone Joint Surg Am 1998;80(7):985–96.

[424] Williams JM, Adewunmi A, Schek RM, Flanagan CL, Krebsbach PH, Feinberg SE, et al. Bone tissue engineering using polycaprolactone scaffolds fabricated via selective laser sintering. Biomaterials 2005;26(23):4817–27.
[425] Mathieu LM, Mueller TL, Bourban PE, Pioletti DP, Muller R, Manson JA. Architecture and properties of anisotropic polymer composite scaffolds for bone tissue engineering. Biomaterials 2006;27(6):905–16.
[426] Marra KG, Szem JW, Kumta PN, DiMilla PA, Weiss LE. In vitro analysis of biodegradable polymer blend/hydroxyapatite composites for bone tissue engineering. J Biomed Mater Res 1999;47(3):324–35.
[427] Uemura T, Dong J, Wang Y, Kojima H, Saito T, Iejima D, et al. Transplantation of cultured bone cells using combinations of scaffolds and culture techniques. Biomaterials 2003;24(13):2277–86.
[428] Yang XB, Roach HI, Clarke NM, Howdle SM, Quirk R, Shakesheff KM, et al. Human osteoprogenitor growth and differentiation on synthetic biodegradable structures after surface modification. Bone 2001;29(6):523–31.
[429] Yang XB, Bhatnagar RS, Li S, Oreffo RO. Biomimetic collagen scaffolds for human bone cell growth and differentiation. Tissue Eng 2004;10(7-8):1148–59.
[430] MacArthur BD, Oreffo RO. Bridging the gap. Nature 2005;433(7021):19.
[431] Lyons FG, Al-Munajjed AA, Kieran SM, Toner ME, Murphy CM, Duffy GP, et al. The healing of bony defects by cell-free collagen-based scaffolds compared to stem cell-seeded tissue engineered constructs. Biomaterials 2010;31(35):9232–43.
[432] Meijer GJ, de Bruijn JD, Koole R, van Blitterswijk CA. Cell based bone tissue engineering in jaw defects. Biomaterials 2008;29(21):3053–61.
[433] Farrell E, Both SK, Odorfer KI, Koevoet W, Kops N, O'Brien FJ, et al. In-vivo generation of bone via endochondral ossification by in-vitro chondrogenic priming of adult human and rat mesenchymal stem cells. BMC Musculoskelet Disord 2011;12:31.
[434] Miot S, Brehm W, Dickinson S, Sims T, Wixmerten A, Longinotti C, et al. Influence of in vitro maturation of engineered cartilage on the outcome of osteochondral repair in a goat model. Eur Cell Mater 2012;23:222–36.
[435] Jukes JM, Both SK, Leusink A, Sterk LMT, van Blitterswijk CA, de Boer J. Endochondral bone tissue engineering using embryonic stem cells. Proc Natl Acad Sci U S A 2008;105(19):6840–5.
[436] Farrell E, van der Jagt OP, Koevoet W, Kops N, van Manen CJ, Hellingman CA, et al. Chondrogenic priming of human bone marrow stromal cells: a better route to bone repair? Tissue Eng Part C Methods 2009;15(2):285–95.
[437] Scotti C, Tonnarelli B, Papadimitropoulos A, Scherberich A, Schaeren S, Schauerte A, et al. Recapitulation of endochondral bone formation using human adult mesenchymal stem cells as a paradigm for developmental engineering. Proc Natl Acad Sci U S A 2010;107(16):7251–6.
[438] Freeman FE, Haugh MG, McNamara LM. Investigation of the optimal timing for chondrogenic priming of MSCs to enhance osteogenic differentiation in vitro as a bone tissue engineering strategy. J Tissue Eng Regen Med 2013;.
[439] Freeman FE, Haugh MG, McNamara L. An in vitro bone tissue regeneration strategy combining chondrogenic and vascular priming enhances the mineralisation potential of MSCs in vitro whilst also allowing for vessel formation. Tissue Eng Part A 2015;21(7–8):1320–32.

[440] Freeman FE, Stevens H, Owens P, Guldberg R, McNamara L. Osteogenic differentiation of MSCs by mimicking the cellular niche of the endochondral template. Tissue Eng Part A 2016;22(19-20).

[441] Nowlan NC, Murphy P, Prendergast PJ. A dynamic pattern of mechanical stimulation promotes ossification in avian embryonic long bones. J Biomech 2008;41(2):249–58.

[442] Carter DR, Orr TE, Fyhrie DP, Schurman DJ. Influences of mechanical stress on prenatal and postnatal skeletal development. Clin Orthop Relat Res 1987;(219):237–50.

[443] Carter DR. Mechanical loading history and skeletal biology. J Biomech 1987;20(11–12):1095–109.

[443a] Freeman FE, Schiavi J, Brennan MA, Owens P, Layrolle P, McNamara LM[1]. Mimicking the biochemical and mechanical extracellular environment of the endochondral ossification process to enhance the in vitro mineralization potential of human mesenchymal stem cells. Tissue Eng Part A 2017;23(23–24):1466–78.

[444] Sittichockechaiwut A, Scutt AM, Ryan AJ, Bonewald LF, Reilly GC. Use of rapidly mineralising osteoblasts and short periods of mechanical loading to accelerate matrix maturation in 3D scaffolds. Bone 2009;44(5):822–9.

[445] Zhao F, Vaughan TJ, McNamara LM. Multiscale fluid structure interaction modelling to determine the mechanical stimulation of bone cells in a tissue engineered scaffold. Biomech Model Mechanobiol 2014; [in press].

[446] Stops AJ, Heraty KB, Browne M, O'Brien FJ, McHugh PE. A prediction of cell differentiation and proliferation within a collagen-glycosaminoglycan scaffold subjected to mechanical strain and perfusive fluid flow. J Biomech 2010;43(4):618–26.

[447] Sandino C, Lacroix D. A dynamical study of the mechanical stimuli and tissue differentiation within a CaP scaffold based on micro-CT finite element models. Biomech Model Mechanobiol 2011;10(4):565–76.

[448] Guyot Y, Luyten FP, Schrooten J, Papantoniou I, Geris L. A three-dimensional computational fluid dynamics model of shear stress distribution during neotissue growth in a perfusion bioreactor. Biotechnol Bioeng 2015;112(12):2591–600.

[449] Guyot Y, Papantoniou I, Chai YC, Van Bael S, Schrooten J, Geris L. A computational model for cell/ECM growth on 3D surfaces using the level set method: a bone tissue engineering case study. Biomech Model Mechanobiol 2014;13(6):1361–71.

[450] Mareschi K, Ferrero I, Rustichelli D, Aschero S, Gammaitoni L, Aglietta M, et al. Expansion of mesenchymal stem cells isolated from pediatric and adult donor bone marrow. J Cell Biochem 2006;97(4):744–54.

[451] Rubin CT, Lanyon LE. Regulation of bone formation by applied dynamic loads. J Bone Joint Surg 1984;66-A:397–402.

# FURTHER READING

[452] Lian JB, Stein GS, van Wijnen AJ, Stein JL, Hassan MQ, Gaur T, et al. MicroRNA control of bone formation and homeostasis. Nat Rev Endocrinol 2012;8(4):212–27.

[453] Henriksen K, Neutzsky-Wulff AV, Bonewald LF, Karsdal MA. Local communication on and within bone controls bone remodeling. Bone 2009;44(6):1026–33.

[454] Schaffler M, Cheung W-Y, Majeska R, Kennedy O. Osteocytes: master orchestrators of bone. Calcif Tissue Int 2013;94(1):1–20.

# CHAPTER 7

# Vascular mechanobiology, immunobiology, and arterial growth and remodeling

**Alexander W. Caulk[*], George Tellides[†,‡], Jay D. Humphrey[*,†]**

*Department of Biomedical Engineering, Yale University, New Haven, CT, United States[*]*
*Department of Surgery, Yale University, New Haven, CT, United States[†] Vascular Biology and Therapeutics Program, Yale University, New Haven, CT, United States[‡]*

## ABBREVIATIONS

| | |
|---|---|
| **ADAM** | a disintegrin and metalloproteinase |
| **ADAMTS** | ADAM with thrombospondin motifs |
| **AngII** | angiotensin II |
| **APC** | antigen-presenting cell |
| **CCL** | C-C chemokine ligand |
| **CCR** | C-C chemokine receptor |
| **CSF** | colony-stimulating factor |
| **CXCL** | CXC chemokine ligand |
| **CXCR** | CXC chemokine receptor |
| **DC** | dendritic cell |
| **EC** | endothelial cell |
| **ECM** | extracellular matrix |
| **G&R** | growth and remodeling |
| **GF** | growth factor |
| **GM-CSF** | granulocyte-macrophage colony-stimulating factor |
| **IFN** | interferon |
| **IL** | interleukin |
| **LPS** | lipopolysaccharide |
| **MHC** | major histocompatibility complex |
| **MMP** | matrix metalloproteinase |
| **NF-κB** | nuclear factor κB |
| **NKT** | natural killer T cells |
| **RGD** | Arginine-Glycine-Aspartic Acid |
| **ROS** | reactive oxygen species |
| **TCR** | T-cell receptor |
| **TGFβ** | transforming growth factor β |
| **TIMP** | tissue inhibitor of metalloproteinases |

| | |
|---|---|
| **TNF** | tumor necrosis factor |
| **Treg** | T regulatory |
| **TSLP** | thymic stromal lymphopoietin |
| **VCAM-1** | vascular cell adhesion molecule-1 |

## 1 INTRODUCTION

Cardiovascular disease remains the leading cause of disability and death in the developed world and is becoming increasingly problematic in many other regions as well. Although advances in medical technology, pharmacotherapies, and clinical practice have dramatically improved our understanding and treatment of these conditions, a full understanding of the complex interplay among the many contributing factors remains wanting. Importantly, growth and remodeling (G&R) in health and disease represent general, complementary processes by which vascular microstructure, material properties, and function change during normal adaptations and disease progression. By growth, we mean a change in mass; by remodeling, we mean a change in microstructure. Both processes can result in changes of vessel and/or vessel compartment size. G&R thus contribute to many clinically observed conditions, including arterial stiffening, dilatation, and stenosis. Identifying common underlying pathways in these processes thus offers potential for scientific and therapeutic benefit. At the intersection of many of these processes are vascular cells sensing and responding to changing mechanical stimuli and inflammatory cells infiltrating the vascular wall and exerting their effects via autocrine and paracrine mechanisms. In particular, turnover (i.e., synthesis and degradation) of extracellular matrix (ECM) resulting from mechano- and immuno-mediated processes is fundamental to outcomes in many vascular adaptations and maladaptations. Although these two aspects of the (patho)biology are usually discussed separately, the goal of this chapter is to consider them together. That mechano- and immuno-mediated processes are inextricably linked can be appreciated by a simple case. Inflammation can drive either fibrosis or atrophy, depending on the relative cell-mediated synthesis and degradation of matrix. Both of these effects change the environment of the intramural cells that mechanosense and mechanoregulate the matrix, thus coupling the mechanobiology and immunobiology within the context of vascular mechanics.

## 2 VASCULAR STRUCTURE AND FUNCTION

To understand better how cardiovascular disease develops, it is helpful first to understand the physical structure and physiological role of the vascular system under normal (homeostatic) conditions. Arteries within the systemic circulation serve as conduits that facilitate the transport of oxygenated blood from the heart to the peripheral tissues, beginning with the largest artery in the body (the aorta) and terminating with arterioles that regulate the exchange of gases and nutrients within local organ systems. Disparate local physiological needs necessarily require a regionally specific structure and function of arteries. Central arteries have a microstructure that enables elastic energy to be stored during systole and used during diastole to augment blood flow; peripheral arteries have a more muscular structure that contributes to a finely tuned regulation of vessel caliber and thus blood perfusion of tissues locally. Contributing to these functions are three distinct layers, the tunica intima, tunica media, and tunica adventitia, each consisting of specific types of highly organized ECM and cells. Together these layers endow the artery with highly coordinated mechanical properties (e.g., stiffness and strength), as, for example, the middle layer serving as the parenchymal layer and the outer layer serving as a protective sheath.

## 2.1 INTIMA

The innermost layer, or tunica intima, consists of a monolayer of endothelial cells (ECs) adhered to a basement membrane consisting primarily of laminin and type IV collagen. Although the intima is typically not structurally significant, the ECs yet serve multiple critical roles that affect the mechanics and thus the state of health or disease. The endothelium serves as a nonthrombogenic surface between the flowing blood and contents of the wall, it is a selective, permeable barrier that modulates the transport of cells and nutrients into the wall, it helps to control blood pressure and blood flow by signaling the vasoregulatory smooth muscle cells of the media, and it even facilitates the formation of new blood vessels via angiogenesis [1]. Of particular importance herein, while transcellular endothelial transport regulates the movement of plasma proteins from the bloodstream to the perivascular space, paracellular transport regulates the infiltration and extravasation of leukocytes across the endothelium via the alteration of adherens and tight junctions. These leukocytes, in turn, play key roles in vascular inflammation, particularly in the development of atherosclerosis. The initial capture of leukocytes is often due to the expression of endothelial surface proteins (e.g., vascular cell adhesion molecule-1 (VCAM-1)), which are regulated in part by mechanotransduction mechanisms. In general, low wall shear stress or high oscillatory shear stress increase the expression of adhesion molecules by the ECs and thereby promote atherosclerosis and other inflammatory diseases. Finally, it is important to note that ECs can undergo a mesenchymal transition, so-called Endo-MT, whereby these cells can cross the basement membrane and contribute to intimal remodeling such as neointimal hyperplasia, which includes additional ECM deposition [2].

The endothelium is also critical in maintaining blood pressure (the primary mechanical load) and tissue perfusion (the primary functional role) because the ECs are able to sense the mechanical consequences of flowing blood (wall shear stress) and synthesize and release diverse vasoactive molecules. In particular, these cells release nitric oxide, a potent vasodilator, in response to high wall shear stress; conversely, they release endothelin-1, a potent vasoconstrictor, in response to low wall shear stress. These vasoactive molecules diffuse into the medial layer and cause vascular smooth muscle cell relaxation and contraction, respectively, which thereby adjusts the caliber of the artery to achieve desired values of blood flow and, in peripheral vessels, tissue perfusion. Importantly, it appears that this vasocontrol attempts to maintain a constant local value of wall shear stress, which in large arteries in humans is typically $\sim 1.5$ Pa. The precise mechanisms by which the ECs sense changes in mechanical stimuli remain unknown, but VE-cadherin and platelet and endothelial cell adhesion molecule (PECAM-1) clearly play a role [3]. Loss of proper mechanosensing may contribute, for example, to a condition referred to as endothelial dysfunction, often defined by a decrease in nitric oxide bioavailability and an associated inability of arteries to dilate in response to increased flow. Because nitric oxide is antiinflammatory, not just vasodilatory, endothelial dysfunction can contribute to inflammation of the intima and subintima, including conditions such as neointimal hyperplasia and atherosclerosis. Note, too, that endothelin-1 is proinflammatory, not just vasoactive. It is not surprising, therefore, that endothelial (dys)function is an independent predictor of arterial remodeling and future cardiovascular events.

## 2.2 MEDIA

The middle layer of the arterial wall, or tunica media, contains elastic fibers, smooth muscle cells, collagens, and glycosaminoglycans organized differently depending on location within the vascular tree. For example, large (elastic) arteries have a highly organized lamellar structure, with single layers of smooth muscle cells organized within the nearly concentric layers of elastic fibers, or laminae. The collagens, not only type III but also types I and V, and the glycosaminoglycans (often versican) colocalize with the smooth muscle between the elastic laminae. The number of lamellar layers scales linearly with the diameter of the vessel, with the thoracic aorta having more than 60 layers in the human and about 6 layers in mice [4]. In contrast, most medium-sized (muscular) arteries have only two prominent elastic laminae, the internal and external, which together delimit multiple layers of smooth muscle cells and admixed matrix constituents. As the parenchymal layer of the wall, the media in large arteries endows these vessels with an ability to store elastic energy during systole that can be used during diastole to work on the blood and augment blood flow, a function particularly important in the thoracic aorta [5]. This energy is stored mainly within the elastic fibers, which are unique among arterial proteins in that they are produced primarily during the perinatal period and have an extremely long half-life, on the order of decades. Whereas smooth muscle cells produce the elastic fibers during development in large arteries they serve mainly in maturity to maintain the collagen and glycosaminoglycans. That is, they exhibit a "maintenance phenotype," mainly synthetic but slightly contractile. Conversely, in medium-sized muscular arteries, the smooth muscle cells exhibit mainly a contractile phenotype, to vasoregulate the wall, but a maintenance phenotype as well.

In particular, the smooth muscle cells within muscular arteries regulate local blood flow primarily by controlling luminal diameter through actomyosin-mediated contraction and relaxation, an action that is controlled in large part by the aforementioned vasoactive molecules released by ECs in response to changes in flow, that is, fluid shear stress. Note further that small arteries, or arterioles, are particularly important in controlling both local blood flow and systemic blood pressure. Indeed, these arterioles are referred to as resistance vessels for they dominate the total peripheral resistance (because of their small size and large number) that controls mean blood pressure. Interestingly, the caliber of arterioles is controlled by both endothelial derived vasoactive molecules and a distinct myogenic response wherein the smooth muscle cells contract in response to increased blood pressure (i.e., intramural stress). It appears that this myogenic response is protective in that large pulse pressures are attenuated, which can prevent damage of the capillary bed in otherwise low-resistance, high-flow organs such as the kidney and brain. Pressure-induced microvascular damage is increasingly viewed as an important cause of organ damage [6]. There is, therefore, a gradient in smooth muscle phenotype from large (maintenance) to medium (contractile) to small (myogenic) vessels, which likely contributes to differential mechano-mediated adaptations in conditions such as hypertension. All three types of arteries increase wall thickness in hypertension, but elastic arteries do so while luminal diameter increases, muscular arteries do so while luminal diameter remains nearly the same, and arterioles do so while luminal diameter decreases. That is, muscular arteries alone appear to be able to maintain both intramural and wall shear stress near normal values over a range of conditions.

Regardless of location, contraction of smooth muscle relies on the presence of calcium, the mechanisms of which have been extensively reviewed [7]. Although many forms of cardiovascular disease are driven in part by inflammatory processes, the tunica media appears to be affected by inflammation only in select diseases, particularly those wherein the innermost or outermost layers of elastic fibers are compromised. Rather, inflammation plays important roles in the intima, in atherosclerosis, and in the adventitia in forms of arteriosclerosis, while the media remains largely immunoprivileged [8]. In contrast, the loss of elastic fiber integrity associated with multiple forms of aneurysms results in an inflamed media, at least to the extent that the media still exists.

## 2.3 ADVENTITIA

The outermost layer, or tunica adventitia, consists primarily of collagen and fibroblasts, with some admixed elastin. It thereby provides mechanical support to the vascular wall; that is, it serves largely as a protective sheath that prevents acute overdistension of the more vulnerable elastic fibers and smooth muscle cells of the media. Likewise, the high collagen content of this layer contributes significantly to the overall structural strength of the wall, preventing rupture under normal conditions. Fibroblasts are responsible for maintaining the normal adventitia, via regulated deposition (synthesis and cell-mediated organization) and removal (proteolytic degradation, often via matrix metalloproteinases (MMPs)) of matrix, mainly collagen. It is important to note, therefore, that in contrast to elastin, which has a half-life of multiple decades, the half-life of vascular collagen is on the order of only a few months. Hence, there is considerable normal turnover of adventitial collagen. In response to increased mechanical loading, the fibroblasts increase their deposition of collagen so as to increase load bearing and protect the wall mechanically. Although the adventitia was long considered to be predominantly a protective mechanical sheath, it is now also recognized to be very important biologically. Not only does the adventitia of large arteries contain the vasa vasorum (small vessels of the vessel, which provide oxygen and nutrients to the outer part of a thick wall), but also there are nerves and lymphatic vessels. Indeed, in addition to the fibroblasts, the adventitia also contains resident progenitor cells and immune cells [9]. As such, the adventitia plays a critical role in the regulation of inflammatory cascades and in interactions with the perivascular tissue. Many functions of fibroblasts depend on the stress within or the stiffness of the matrix to which they adhere. Thus, a feedback loop is established in which fibroblasts sense and regulate their mechanical environment and the mechanical environment regulates the function of the fibroblasts and, via paracrine mechanisms, other cells with which the fibroblasts communicate. In this way, mechano- and immuno-mediated mechanisms of wall G&R are highly coupled in the adventitia, which is thus a critical biologically active regulator of vascular structure, material properties, and function.

## 3 QUANTIFICATION OF ARTERIAL MECHANICS

The mechanics of the arterial wall is important for two fundamental reasons. First, the primary function of an artery is to serve as a mechanical conduit, facilitating transport of blood and the many substances it carries. With respect to this role, arteries are subject to mechanical failures (e.g., dissection or rupture) when wall stress exceeds wall strength. There is, therefore, a pressing need to understand local mechanics and associated mechanisms of failure. Second, since vascular cells sense and respond to their mechanical environment [10], there is a pressing need to understand both the nature of the extracellular mechanical environment and the mechanisms by which the cells probe this environment. The cellular machinery responsible for sensing and responding to mechanical stimuli varies depending on cell type. For example, the monolayer of ECs is subjected directly to external forces, namely, flow-induced shear stresses, whereas intramural smooth muscle cells and fibroblasts experience the effects of external forces primarily through adhesions to the ECM. Dispersed cells within the wall, such as infiltrating inflammatory cells, may also be affected by the changes in external loads, though load transmission and sensing may be different. That is, cell-matrix and cell-cell interactions can be varied, though generally important nonetheless. Finally, there is a need to define the primary type of loading, whether it be tensile or compressive, normal or shear, and hydrostatic or not. Only via mechanics—micromechanics and continuum mechanics—can we relate cellular responses to mechanical stimuli. Fundamental to describing the mechanics is, of course, quantification of the material properties. The requisite data can be obtained via diverse methods, which are briefly enumerated here.

## 3.1 IN VIVO METHODOLOGIES

Quantification of vascular mechanical properties from in vivo data is particularly important; only in this way can we contribute to clinical diagnosis, prognosis, and treatment. Most current in vivo methods seek to characterize the structural stiffness of arteries, which includes both geometric (wall thickness) and intrinsic (material stiffness) information. It is the structural stiffness, of course, that affects directly the hemodynamics. The current gold standard for measuring systemic arterial stiffness is the pulse wave velocity (PWV), that is, the velocity at which the pressure pulse travels along the arterial tree [11]. PWV can be measured using applanation tonometry, ultrasound, or invasive pressure measurement; it measures the time required for a systolic pulse to travel between two locations of known separation distance, often from the carotid artery to the femoral artery. This cf-PWV has been shown to predict future cardiovascular events and is thus widely employed to quantify central arterial stiffness. Albeit not theoretically applicable, the Moens-Korteweg equation is often used to understand interrelations between PWV and the structural stiffness, namely,

$$c_0 = \sqrt{\frac{Eh}{2\rho a}} \qquad (1)$$

where $c_0$ is the measured PWV, $E$ is the incremental elastic modulus, $h$ is the wall thickness, $a$ is the inner radius of the vessel, and $\rho$ is the mass density of blood; the product $Eh$ represents the structural stiffness. By measuring PWV and luminal radius, one can estimate the structural stiffness. Knowledge of wall thickness further allows one to estimate the incremental elastic modulus. Based on this result, we see that an increase in the intrinsic material stiffness, an increase in wall thickness, or both, can increase the PWV. An increase in PWV, in turn, is thought to result in an earlier return of the reflected pressure wave within the proximal aorta, which augments the pulse pressure and thus load on the heart. Notwithstanding general insight gleaned from the Moens-Korteweg equation, its limitations must be understood: The material of the wall is assumed to be linear, elastic, and isotropic, the geometry is assumed to remain constant along the length of the vascular segment considered, and the blood is assumed to be inviscid. The resulting estimates of stiffness are thus averaged over the segment considered and approximate. Importantly, calculations of arterial stiffness based on carotid-femoral artery PWV exclude assessment of the ascending aorta, whose function is greatly impacted by the compliance of its elastic wall. Direct clinical measurements of PWV, independent of any need for a theoretical correlate and despite being averaged over space, remain useful in practice.

Other commonly used clinical metrics of local, rather than spatially averaged, stiffness include the distensibility and Peterson's elastic modulus. The latter is defined by

$$E_p = \frac{A \Delta P}{\Delta A} \qquad (2)$$

where $A$ is the luminal cross-sectional area at systolic pressure, $\Delta P$ is the pulse pressure defined as the difference between the systolic and diastolic pressures, and $\Delta A$ is the change in cross-sectional area from systole to diastole. Many also attempt to infer the mean circumferential stress and strain using similar data. The calculation of these metrics similarly embodies inherent limitations that should be considered. Ultrasound, which is commonly the imaging modality of choice, can only detect the thickness of the intimal and medial layers, not the adventitia. Significant thickening of the adventitia has been observed in multiple vascular pathologies, making the interpretation of a local stress and strain tenuous at present.

A general problem affecting all in vivo methods is the inability to measure the unloaded dimensions of the vasculature, that is, in the absence of luminal pressure and axial extension. This situation has led to significant research in search for nonclassical methods of data analysis and interpretation, but these are beyond the current scope. Rather, let us consider ex vivo methods that provide greater control over the applied loads and access to diverse configurations, including the unloaded and nearly stress-free.

## 3.2 EX VIVO METHODOLOGIES

Ex vivo testing methods seek to maintain the blood vessel or a part thereof biologically viable during testing, hence allowing one to assess both passive and active properties. Such methods have been used widely to study the mechanics of vessels from many animal models and, to a lesser extent, samples from humans. These methods also allow assessments of properties more locally, focusing on a small region of a particular segment (e.g., infrarenal aorta vs suprarenal aorta). Truly local or focal properties are yet difficult to measure, whether it be local to a particular layer of the wall, a particular region around the circumference, or along the length of a lesion (e.g., plaque or aneurysm). Again, the need for increased information has stimulated the development of advanced experimental (e.g., digital image correlation) and computational (e.g., inverse methods) methods, but their description is beyond the current scope (see, e.g., Refs. [12,13]). Rather, for purposes of illustration, the focus here is characterization of vascular mechanical properties when it is reasonable to consider a segment as a uniform cylinder, either layered or transmurally homogenized. Below is a summary of a general framework typically employed when characterizing vascular tissue ex vivo.

### 3.2.1 Basic assumptions

Like most biological soft tissues, arteries exhibit nearly incompressible and highly nonlinear, anisotropic, pseudoelastic behaviors over finite strains. It is thought, for example, that this nonlinear response results from the combination of a nearly linear and compliant response at low pressures due to the elastic fibers plus the gradual recruitment of initially undulated but then stiff collagen fibers at high pressures; these collagen fibers are distributed in different directions in different layers within different vessels, thus resulting in tissue anisotropy. The pseudoelastic response under cyclic loading likely results from collagen and elastic fibers moving within a viscous ground substance and dissipating some energy (evidenced by hysteresis in the pressure-diameter or stress-stretch curves), with glycosaminoglycans contributing to such inelasticity. The dominant elasticity is due, in large part, to the elastic fibers, and the near incompressibility is due to the high water content. An appropriate general theoretical framework (e.g., incompressible nonlinear elasticity) for analysis is widely available [14], though there are numerous forms of possible constitutive relations depending on the question of interest.

### 3.2.2 Experimental approaches

Testing conditions for vascular tissue may be classified by the number of loading directions (e.g., uniaxial, biaxial, or triaxial), the configuration in which the tissue is tested (e.g., ring, planar, or cylindrical), and the parameters rendered variable (e.g., fixed load with changing geometry, termed isotonic, or fixed geometry with changing load, termed isometric). Because arteries experience multiaxial loads and deformations in vivo, uniaxial tests are less useful even though much easier to perform and interpret. Planar biaxial tests enable tensile forces to be applied to a square piece of tissue in directions corresponding to circumferential or axial and thus can be very useful. The primary limitation of these tests, however, is that excised segments often retain a curved geometry in the unloaded configuration and the need for applied moments to flatten the specimen for in-plane biaxial loading complicates the experiment and its interpretation. Hence, the most natural and useful experiment is combined pressurization and axial loading of a cylindrical segment. This test is not without complications, however, for the governing equilibrium equations involve integrals over the wall thickness. Triaxial tests add torsion to the inflation and extension, which can provide important information on shearing behavior. Biaxial tests are the most common, however. See Ref. [14] for more details, including formal methods of nonlinear mechanical analyses.

### 3.2.3 Stress/strain determination

Stress and strain are mathematical abstractions that are defined pointwise. Since we cannot measure the requisite quantities at a point, we seek experimental situations in which values averaged over small regions represent well the values at all points within the region of interest. In biaxial tests on cylindrical segments, it is common to calculate the mean circumferential and axial stretch (components of a stretch tensor, $\lambda_\theta$ and $\lambda_z$, respectively) based on ratios of finite lengths, namely,

$$\lambda_\theta = \frac{r_m}{R_m} \tag{3}$$

$$\lambda_z = \frac{l}{L} \tag{4}$$

where $R_m$ and $r_m$ are the unloaded and loaded midwall radius, respectively, and $L$ and $l$ are the unloaded and loaded length, respectively. Assuming isochoric deformations, the radial stretch can be calculated as the inverse of the product of circumferential and axial stretch. Green strain is simply one-half the square of the stretch minus 1 as long as torsion is not applied.

The mean values of Cauchy (or true) biaxial stress can be obtained via simple force balances, which yield the following:

$$\sigma_\theta = \frac{Pa}{h} \tag{5}$$

$$\sigma_z = \frac{f}{\pi h(2a+h)} \tag{6}$$

where $\sigma_\theta$ and $\sigma_z$ are circumferential and axial stress, respectively, $P$ is the luminal pressure, $f$ is the axial force applied to the vessel, $a$ is the inner radius, and $h$ is the thickness. These relations are universal, that is, exact representations of the mean value for any material in equilibrium. The key question, however, is whether the mean value describes well the actual (transmural) values. In general, the mean value is an accurate representation of the actual stress distribution only when the wall is thin. Because of the existence of residual stresses in arteries (i.e., stresses that exist in the traction-free configuration) [15], however, the mean value can yet represent reasonably well the transmural distribution. The primary exception, however, is when there is a desire to delineate stresses within the different layers of the wall, particularly the media and the adventitia. In this case, more general analyses are needed (cf. Ref. [16]). Nevertheless, the mean values are useful when one seeks a radially homogenized descriptor, particularly when seeking to determine properties for analyses of fluid-solid interactions. Whether based on radially homogenized values or pointwise values of stress and strain, the ultimate goal is to identify appropriate (nonlinear) stress-strain relations (e.g., $\sigma_\theta = \sigma_\theta(\lambda_\theta, \lambda_z)$) and similarly for the axial component. In practice, however, it is typically preferred to identify a stored energy function $W$ for the material that describes the pseudoelastic passive behavior. An appropriate derivative of this function with respect to an appropriate strain yields a conjugate stress; an appropriate second derivative with respect to strain yields the associated material stiffness. Toward this end, it is generally most convenient to use the second Piola-Kirchhoff stress $S$ and Green strain $E$, namely,

$$S_{AB} = \frac{\partial W}{\partial E_{AB}}$$

where $W$ is the stored energy function. This relation can be modified by introducing the kinematic constraint for incompressibility, which is common in arterial mechanics. In general, however, the material stiffness is

$$\mathbb{E}_{ABCD} = \frac{\partial^2 W}{\partial E_{AB} \partial E_{CD}}$$

Given these quantities (stress, strain, and stiffness) computed relative to a reference configuration, one can use the deformation gradient to compute similar quantities in the current configuration, that is, in terms of the Cauchy stress used above (cf. Ref. [17]) (see, e.g., Fig. 1). Regardless of formulation, identifying the stored energy function is fundamental to quantifying arterial mechanics.

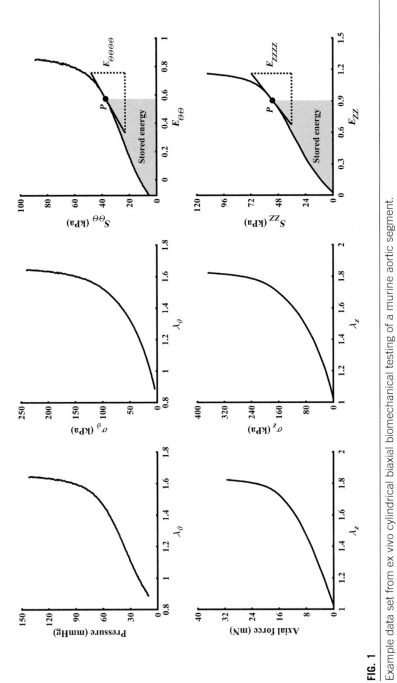

**FIG. 1**

Example data set from ex vivo cylindrical biaxial biomechanical testing of a murine aortic segment.

### 3.2.4 Mechanotargets

It has long been thought that ECs attempt to regulate the vascular lumen so as to maintain nearly constant the wall shear stress, and that intramural cells attempt to regulate the wall thickness so as to maintain the circumferential wall stress nearly constant, both at presumably homeostatic values [10]. Interestingly, it also appears that the intramural cells fashion the composition of the wall to achieve a target circumferential stiffness [18]. Such a target value could represent a favorable mechanical environment in which the cells prefer to reside. There is, therefore, strong motivation to quantify both stress and intrinsic material stiffness. Because material stiffness is a measure of how stress changes with strain, many have quantified the slope of a local tangent at various points along the Cauchy stress-stretch curve. There are, therefore, many reports of stiffness at a "low pressure" and a "high pressure" [19]. As noted above, however, it is best to compute stiffness directly given a stored energy function. Because material stiffness is particularly useful in the analyses of solid-fluid interactions, wherein the stiffness varies from diastolic to systolic pressures, the concept of "small deformations superimposed on large" has also been used to compute a mean value of stiffness that has in vivo relevance [20]. This value has similarly proved useful in comparing values across multiple mouse models [21,22].

## 3.3 APPLICATION

The exquisite control of biaxial mechanical loading and deformations renders ex vivo analyses of vascular mechanics of great utility in understanding mechanisms that drive normal mechanobiological responses and those that associate with disease and disease progression. Vessels from humans and many different animal models have been so tested, including bovine, swine, canine, and rodent. Over the past 15 years in particular, murine models have proved useful given the ease of longitudinal studies in surgical, pharmacological, and especially genetic models. As an example, a recent study comparing the mechanical properties of common carotid arteries across seven different murine models confirmed that circumferential material stiffness tends to be maintained near a target value in induced hypertension and in the cases of diverse mutations, including those to dystrophin (related to muscular dystrophy), fibrillin-1 (related to Marfan syndrome), and fibulin-5 (related to an early, moderate aging phenotype)—see Ref. [22]. Such data are difficult to obtain from humans. Associated quantification of the wall microstructure further allows detailed histomechanical correlations, which are fundamental to understanding mechanobiological responses and those in disease progression. That is, quantifying how the microstructural and mechanical environment changes is fundamental to correlating with changes in the cell mechanobiology.

## 4 CELLULAR SENSING OF ITS MECHANICAL ENVIRONMENT

It has long been known that vascular cells sense and respond to their local mechanical environment [23–25]. Indeed, such mechanobiological activity is essential in both mechanical homeostasis [10] and responses to perturbations in hemodynamics or disease progression [26]. Regarding the former, it is important to note that there is a slow but continual turnover of most ECM components under normal conditions. Consequently, the intramural cells must replace that which is removed with equivalently stressed matrix in order to maintain geometry and properties—this process requires that the cells both mechanosense (i.e., assess) and mechanoregulate (i.e., organize) the matrix [5]. Fundamental to such mechanobiological responses are heterodimeric transmembrane proteins, called integrins, that connect directly to the ECM and indirectly to the intracellular cytoskeleton via focal adhesion-associated proteins. Integrins (or clusters of integrins known as focal adhesions) thereby enable both outside-in (sensing of loads borne by matrix) and inside-out (regulating matrix by imposing loads) actions. The heterodimeric protein structure is denoted by $\alpha$ and $\beta$ subunits, which selectively bind different ECM proteins. For example, $\alpha_5\beta_1$ binds fibronectin, whereas $\alpha_2\beta_1$ binds collagen; there are over 20 different $\alpha_i\beta_j$ pairs with specific and overlapping ligand targets. Another important subunit is $\alpha_4$, which allows T lymphocytes to bind selectins on ECs, which initiates their rolling, adhesion, and then migration into or through the wall. Other proteins have also been implicated in sensing and regulating the mechanical environment. They include cadherins and "stretch-activated" ion channels. For a general review of possible mechanosensors in the vasculature, see Ref. [27].

### 4.1 MECHANOSENSING IN ENDOTHELIAL CELLS

ECs experience diverse mechanical stimuli, including shearing forces due to blood flow (likely sensed via proteins within the glycocalyx, ion channels, cadherins, and other processes), circumferential stretch due to pressure-induced distensions (likely sensed via cadherins but also integrins), and in some regions axial extensions (again sensed via cadherins and integrins) due to gross motions imposed at the organ level, as in the heart and lungs. In each case, of course, the adherence of the ECs to the underlying matrix enables these cells to resist the applied forces. Under normal conditions, the basement layer upon which the ECs rest consists primarily of collagen IV, laminin, and proteoglycans; due to injury or disease, the composition of this layer can change to include a proportion of fibronectin and fibrinogen, effectively shifting the signaling burden from $\alpha_4\beta_1$ (laminin) and $\alpha_2\beta_1$ (collagen) integrins to $\alpha_5\beta_1$ (fibronectin) and $\alpha_v\beta_3$ (fibrinogen) integrins. There is a need to understand better how the associated mechanosignaling depends on both changes in loading and changes in substrate.

## 4.2 MECHANOSENSING IN VASCULAR SMOOTH MUSCLE CELLS

Vascular smooth muscle cells do not experience the effects of the flowing blood directly, but there is yet a continuity of the associated shear stresses throughout the wall [28] and another fluid-induced shear stress associated with a slow transmural flow of interstitial fluid [29]. Gross, torsion-induced shear stresses are also found in select arteries, including the ascending aorta and coronary arteries. In addition, of course, the smooth muscle cells also experience considerable intramural normal stresses due to pressure-induced circumferential and axial stresses as well as axial prestresses. The latter are induced largely during somatic growth, though axial stress can also be changed by body motion, as, for example, effects of movement of the head and neck on the carotid arteries or bending of the knee on the popliteal arteries.

Although it is not known precisely how intramural cells sense their complex time-varying mechanical environment, integrins must play particularly important roles. In this regard, we note further that it is not known precisely how the mechanical loads partition within the vascular wall, as, for example, how much of the load is borne by elastic fibers, collagen I fibers, collagen III fibers, and so forth. There are, of course, on the order of 100 different proteins, glycoproteins, and glycosaminoglycans constituting the arterial wall, hence many different structurally significant constituents could convey mechanical stimuli. Nevertheless, 11 particular integrins have attracted most attention in vascular smooth muscle cells, with $\alpha_1\beta_1$, $\alpha_5\beta_1$, and $\alpha_v\beta_3$ tending to have the highest expressions, thus suggesting dominant roles of collagen and fibronectin in many cases [30]. In vitro studies have shown further that smooth muscle phenotype depends on matrix content and stiffness, with cells demonstrating a mitogenic response to cyclic stretch on matrices of fibronectin, vitronectin, and collagen, but not laminin or elastin [31]. Proliferation has similarly been observed to increase with increasing matrix stiffness [32].

Importantly, integrins also play additional roles in regulation of vascular smooth muscle behavior depending on the vascular bed in which the cells reside. For example, smooth muscle cells in peripheral arterioles can exhibit a myogenic response in which the cells contract over a period of seconds to minutes in response to an acute increase in blood pressure, thus maintaining appropriate rates of blood perfusion in peripheral tissues in the face of elevated pressures or pulse pressures. Experiments utilizing Arg-Gly-Asp (RGD) containing peptides, a sequence found in multiple matrix proteins and recognized by four predominant integrins in vascular smooth muscle, demonstrate that integrin subunits help to regulate myogenic tone since cremaster and mesenteric arterioles exposed to such peptides transiently relax. Conversely, renal arterioles constrict when exposed to RGD-containing peptides.

In addition to regulating contractile function, integrins have also been widely implicated in maintaining or modulating smooth muscle cell phenotype. Integrin $\alpha_5\beta_1$ is increased in smooth muscle cells in response to inflammatory stimuli and mediates migration and apoptosis, behaviors that may also be regulated by the cadherin-catenin complex. As with contractility, phenotypic modulation varies by region and correlates with differential integrin expression. For example, integrin

$\alpha_5\beta_1$ is expressed more highly in coronary smooth muscle cells than other types; it also mediates stronger binding to fibronectin and higher rates of proliferation in coronary compared with carotid smooth muscle cells when cultured on fibronectin [33]. Much of the cellular signaling downstream of integrins observed in cell culture depends on matrix stiffness, thus implicating integrin-mediated mechanosensing in diverse conditions.

Adding both to the complexity and uncertainty of the local mechanical environment, most blood vessels are surrounded by perivascular tissue, yet it is not known precisely how much of the applied loads are borne or imposed by these tissues. For example, subepicardial coronary arteries are bounded below by myocardium and above by epicardium, both of which experience significant loads due to passive filling of the heart with blood and active ejection of the blood. Finally, we note that smooth muscle cells within different parts of the vascular tree experience different degrees of cyclic loading, with proximal vessels tending to experience the highest pulsatile loads and the most peripheral vessels the least. In this regard, it is important to note that the frequency of loading can vary markedly across species, with heart rates being $\sim 1\,\text{Hz}$ in humans and $\sim 10\,\text{Hz}$ in mice, as examples.

### 4.2.1 Mechanosensing in fibroblasts

The contribution of adventitial fibroblasts, indeed the adventitia as a whole, has long been thought to serve primarily as a mechanical support for the media. Fibroblasts produce collagen fibers that have significant stiffness and strength and thus can protect the wall from overdistension due to acute increases in luminal pressure. Again, these cells appear to sense and respond to changes in mechanical stimulus through the action of multiple integrins. As with smooth muscle cells, these fibroblasts are responsible for controlling both matrix production (synthesis and incorporation within extant matrix) and removal (via a host of MMPs) and thereby are significant contributors to vascular homeostasis, not just responses to perturbations in hemodynamic loading or disease progression. It is also likely that there is transcompartmental communication, via cytokines or other biomolecules, wherein the activity of smooth muscle cells and fibroblasts can influence that of the other cell type (cf. Ref. [34]). Importantly, it also appears that chemokine/cytokine production by adventitial fibroblasts can contribute to inflammatory activation in the vascular wall by stimulating tissue-resident immune cells, such as macrophages and dendritic cells (DCs), which can then further recruit other immune cells to the site of inflammation [35].

It is perhaps easy to see the profound implication of mechano-mediated fibroblast function, as altered adventitial stress, stiffness, or strength may initiate a cascade of matrix G&R that ultimately affects the entire wall, particularly via increased stress shielding of the media. Although studies of mechanotransduction in adventitial fibroblasts are fewer than those in smooth muscle cells and ECs, there is a significant body of literature on the mechanobiology of fibroblasts from diverse tissues, including the heart, tendons, and skin [23,36,37]. Such studies reveal that fibroblasts are highly mechanosensitive, sharing many signaling pathways with smooth muscle cells and many characteristic responses. For example, proliferation, matrix synthesis, and inflammatory activation all depend on matrix stiffness [38–40]. Again, many of the stiffness-mediated responses are integrin-mediated processes. Finally, it should be noted that many processes alter integrin expression in fibroblasts, possibly changing the mechanosensing and mechanoregulating machinery to alter sensitivity to biomechanical or biochemical changes in matrix composition.

## 5 REGULATION OF THE INFLAMMATORY RESPONSE—CELLULAR FUNCTION AND PHENOTYPE

Inflammation is typically a highly coordinated process with many players involved. Note, therefore, that the immune system is divided into (i) the innate system, comprising granulocytes (i.e., neutrophils, eosinophils, and basophils), monocytes, macrophages, mast cells, and DCs, and (ii) the adaptive system, comprising T lymphocytes and B lymphocytes [41]. Cells of the innate immune system are usually activated via pattern recognition receptors (PRRs) by identifying pathogen-associated molecular patterns (PAMPs) that exist on many types of foreign bodies. In contrast, cells of the adaptive immune system are typically activated upon contact with antigens from antigen-presenting cells (APCs). Cells of both systems communicate with each other using a variety of signaling proteins known as cytokines, which are typically divided into interleukins (ILs), interferons (IFNs), tumor necrosis factors (TNFs), growth factors (GFs), colony-stimulating factors (CSFs), and chemotactic cytokines, or chemokines. The local biochemical milieu can shift between proinflammatory and profibrotic to antiinflammatory depending on the combinations of cytokines that are produced by the immune and tissue-resident cells; such shifts can induce a wide variety of effector responses from each cell type. Thus, it is helpful to define an inflammatory response in multiple steps.

First, an inflammatory stimulus may be considered simply to be a perturbation in a regulated variable from its homeostatic value [42]. Such stimuli can be difficult to identify in cases of chronic inflammation, but relevant examples include altered stiffness of the adventitia, low or oscillatory shear stresses on the endothelium, or marked increases in endogenous substances such as angiotensin II (AngII). Inflammatory stimuli proceed to activate specific subsets of immune cells in an attempt to resolve the perturbation. Typically, resident cells secrete chemokines and cytokines that induce a wave of immune cell infiltration that leads to further secretion of chemokines and cytokines to promote a feed-forward inflammatory response that persists until the perturbation is resolved. Such resolution can be driven, in part, by the subsequent secretion and activity of diverse bioactive enzymes (e.g., granzymes by granulocytes and MMPs by macrophages) that facilitate the breakdown and removal of foreign bodies. An example of transient inflammation leading to adaptive changes and restoring homeostatic properties of the vessel wall is the inward remodeling of the murine carotid artery in response to decreased blood flow [43]. Chronic inflammation, in contrast, results when the stimulus cannot be resolved; in this case, the system tends to adapt to the stimulus rather than remove it, often resulting in a persistent inflammatory response. Thus, characterizing an inflammatory response requires the identification of (i) the stimulus, (ii) the effector cells (and possibly the subsequently recruited effector cells), (iii) the effector cytokines, and (iv) the effector bioactive enzymes. We here present members of components (ii)–(iv) to provide a foundation of general effector responses before addressing specific stimuli and subsequent inflammatory cascades. To simplify, we focus primarily on cell types that have been implicated in chronic inflammation and vascular remodeling: monocytes/macrophages, T cells, and DCs (Table 1), but it is noteworthy to mention that all immune cell types have been observed in the context of cardiovascular disease.

## 5.1 MONOCYTES/MACROPHAGES

Monocytes derive from the bone marrow and egress into the bloodstream where they circulate until migrating into peripheral tissues. Monocytes in mice that express C-C chemokine receptor 2 (CCR2), $CX_3C$ chemokine receptor 1 ($CX_3CR1$), and the granulocytic marker GR1 (also known as Ly6C) are characterized as inflammatory monocytes; those expressing CD14 but not CD16 in humans are characterized similarly. Alternatively, monocytes negative for these markers are qualified as resident monocytes and are delineated from inflammatory monocytes by the greater time that they spend circulating in the bloodstream prior to entering peripheral tissue. Monocytes adhere to the vascular endothelium by binding surface proteins on ECs such as VCAM-1, which can be mechanoregulated, then subsequently migrate into or through the vascular wall and differentiate into macrophages (noting a usual distinction between infiltrating macrophages and constitutively present tissue-resident macrophages, which self-proliferate and are derived from the yolk sac) or DCs. Such infiltration is, for example, a critical first step in the development of atherosclerosis [44]. Differentiation of monocytes into macrophages or DCs also confers specialized function depending on the location of the extravasated cells (e.g., the skin vs the lung), though this may be explained partially by the phenotypic heterogeneity of circulating monocytes [45].

### 5.1.1 Classically activated (M1) macrophages

"Classical activation" of macrophages, resulting in the so-called M1 phenotype, defines a pronounced proinflammatory phenotype in which the cells secrete inflammatory cytokines including interleukins IL-1β and IL-6 and/or tumor necrosis factor-α (TNFα) and interferon γ (IFNγ). The M1 activation state can be difficult to describe uniformly as the biomarker profile differs depending on the tissue location where the macrophage was activated [46]. Yet, common characteristics describe most M1 macrophages, including the expression of F4/80 and CD68. Induction of the classical activation state occurs primarily through exposure to lipopolysaccharide (LPS), granulocyte-macrophage colony-stimulating factor (GM-CSF), IFNγ, TNFα, or IL-1β, which may be produced by other immune cells such as Th1 cells, neutrophils,

**Table 1** List of Select Immune Cells and Associated Phenotypes, Extracellular Markers, Polarizing Cytokines, and Effector Molecules and Enzymes

|  | Dendritic Cells | | | Macrophages | | | | T Cells | | | |
|---|---|---|---|---|---|---|---|---|---|---|---|
|  | Type I (Conventional) | Type II (Conventional) | Regulatory | M1 | M2a | M2b | M2c | Th1 | Th2 | Th17 | Treg |
| Phenotype | Induction of Th1 cells, pro-inflammatory | Induction of Th2 cells, pro-inflammatory | Induction of regulatory T cells, anti-inflammatory | Classically activated proinflammatory | Alternatively activated, wound healing | Alternatively activated | Alternatively activated, regulatory | Pro-inflammatory | Profibrotic | Pro-inflammatory | Anti-inflammatory |
| Selected polarizing cytokines | IFNγ | TSLP | IL-10 | IFNγ | IL-4 | LPS | IL-10 | IL-12 | IL-2 | TGFβ plus IL-6 | IL-10 |
|  |  | CCR2L Histamine | TGFβ | TNFα IL-1β GM-CSF | IL-13 | IL-1β | TGFβ | IL-18 IFNγ TNFα | IL-4 | IL-1β | TGFβ |
| Selected extracellular markers (mouse) | CD11c | CD11c | CD11c | CD68 | Arg I | CD86 | CD163 | CD4 | CD4 | CD4 | CD4 |
|  | CD18 | CD18 | CD18 | F4/80 | CD163 |  |  | CXCR3 | CCR4 | CCR6 | CD25 plus Foxp3[a] |
| Selected effector cytokines and chemokines | IFNγ | IL-4 | IL-10 | TNFα | IL-4 | IL-10 | IL-10 | IFNγ | IL-4 | IL-17A (IL-17) | TGFβ |
|  | IL-18 | IL-5 | TGFβ | IL-1β | CCL17 | IL-12 | CCL1 | TNFα | IL-5 | IL-17F | IL-10 |
|  |  | IL-13 |  | IL-6 | CCL18 | CCL1 | CCL13 |  | IL-6 | IL-6 |  |
|  |  |  |  | IFNγ | CCL22 |  | CCL16 |  | IL-13 | IL-22 |  |
|  |  |  |  |  |  |  | CCL18 |  |  | TNFα |  |

[a] Foxp3 is a transcription factor, not an extracellular marker, utilized to identify Treg cells in conjunction with CD25 expression.

basophils, natural killer cells, and other monocytes/macrophages and by vascular cells such as endothelial cells, smooth muscle cells, and fibroblasts. Production of inflammatory cytokines by M1 macrophages induces polarization of other cell types; for instance, production of IL-1β and IL-6 by M1 macrophages induces the differentiation of naive T cells into Th17 cells. Many of the inflammatory cytokines produced by M1 macrophages play a critical role in cell-mediated turnover of ECM, making these macrophages an important contributor to preserved or altered vascular structure and function.

### *5.1.2 Alternatively activated (M2) macrophages*

Since the identification of two distinct subsets of macrophages (M1 and M2), ongoing research has revealed that alternatively activated (M2) macrophages comprise a heterogeneous population of macrophages that are often divided into four subclasses, M2a-d [47]. Others have suggested a division based on function that include "wound-healing" types, likely associated with M2a macrophages, and "regulatory" types, likely associated with M2c macrophages. Macrophages are directed to differentiate into the M2a phenotype via exposure to IL-4 and IL-13, of which basophils and mast cells are key early contributors while Th2 cells may provide a more sustained exposure. M2a activation is characterized by the expression of proteins such as CD163 and arginase I (in mice) and secretion of chitinase and similar molecules as well as chemokines such as CCL17, CCL18, and CCL22. IL-4 stimulates the production of ECM, which has led to the hypothesis that these macrophages are primarily responsible for wound healing.

Polarization to the M2b phenotype occurs upon the presentation of IL-1β or LPS and is characterized by the expression of CD86 and secretion of IL-10, low levels of IL-12, and CCL1. M2c polarization occurs in response to IL-10 and transforming growth factor β (TGFβ) and is characterized by the expression of CD163 and secretion of IL-10, CCL1, CCL13, CCL16, and CCL18. IL-10 is a potent antiinflammatory cytokine and thus endows the M2c phenotype with the "regulatory" classification. M2d macrophages are commonly called tumor-associated macrophages and are not discussed here.

The distribution of macrophage subsets within the vasculature has been studied extensively in atherosclerotic plaques. Generally, M2 macrophages appear to be more concentrated near the fibrous cap, possibly endowing the plaque with mechanical stability given the profibrotic and antiinflammatory phenotype (excepting M2b macrophages). Conversely, M1 macrophages, which have a potently proinflammatory phenotype, are more abundant near the shoulder of the plaque, which is considered the most mechanically unstable location [48]. Though atherosclerosis is one of the most extensively studied cardiovascular diseases, macrophage activity can also contribute to hypertensive G&R, aneurysm, dissection and rupture, and even neovessel development in tissue-engineered vascular grafts [49–52].

## 5.2 T CELLS

T cells originate in the bone marrow as hematopoietic stem cells and migrate to the thymus where they become immature ($CD4^-CD8^-$) thymocytes. Over the course of their development, thymocytes become double-positive cells ($CD4^+CD8^+$) before becoming single-positive cells ($CD4^-CD8^+$ or $CD4^+CD8^-$), which are defined as immunocompetent T lymphocytes. These cells then emigrate from the thymus to lymph nodes throughout the body. Activation of T cells leads to a variety of phenotypes based on surface protein expression and cytokine production that may be classified as helper, cytotoxic, memory, regulatory, natural killer, and gamma delta T cells.

### 5.2.1 T helper cells

Thymocytes that mature to $CD4^+CD8^-$ single-positive cells are termed $CD4^+$ T helper cells; they serve myriad functions in the adaptive immune system. Within the T helper cell domain, a further subdivision includes Th1, Th2, Th17, and T regulatory (Treg) cells. Activation of $CD4^+$ T cells occurs when the cells are presented with an antigen by an APC such as a DC, B cell, or macrophage at which point the T-cell receptor (TCR) complex, combined with a costimulatory molecule such as CD28, binds to major histocompatibility complex (MHC) class II associated with antigen and an additional costimulatory protein on the APC. Following activation, the milieu of local cytokines determines the polarization fate of the activated T cell.

Th1 cells are polarized by IL-12 and IL-18 as well as IFNγ and are marked by surface expression of CXCR3. These cells primarily produce IFNγ and TNFα to assist in classical activation of macrophages. Conversely, Th2 cells are polarized by IL-2 and IL-4 and are marked by the expression of CCR4. These cells produce IL-4, thereby promoting autocrine polarization, and IL-5, IL-6, and IL-13, which promote a fibrotic response from intramural cells such as fibroblasts and alternatively activated macrophages. Studies have suggested that a proper balance between the Th1 and Th2 response from the adaptive immune system is critical for preventing tissue damage and autoimmune disease. Indeed, chronic inflammation is generally characterized by a more pronounced activation of the adaptive immune response, but IFNγ-producing Th1 cell-dominated inflammation produces a potent inflammatory response with very little fibrosis, whereas Th2-dominated inflammation induces significant matrix remodeling [53].

More recently characterized are Th17 cells that are polarized by TGFβ and IL-6 and express CCR6 [54]. Polarization is tightly regulated and requires both TGFβ and IL-6, whereas TGFβ alone induces polarization toward the Treg phenotype. This process may be promoted further by the presence of IL-1β. Th17 cells are so named because of their prolific production of IL-17, which is a potent proinflammatory cytokine that promotes secretion of other proinflammatory cytokines by both immune cells and intramural cells. They also produce IL-6, IL-22, and TNFα, thereby contributing overall to a highly proinflammatory milieu. Conversely, without the presence of IL-6, TGFβ promotes the Treg phenotype that facilitates an antiinflammatory environment by producing IL-10 and TGFβ [55]. These cells express

CD25 and are further subdivided by their expression of the transcription factor FoxP3 (or lack thereof). Treg cells can directly suppress the action of T cells by their production of IL-10, which suppresses inflammation by downregulating Th1 cytokine production and nuclear factor κB (NF-κB) activity, or they can inhibit T cells indirectly by suppressing costimulatory molecules on APCs, namely, CD80 and CD86, on DCs and macrophages. Treg cells play an important role in the resolution of inflammation, and their dysfunction may contribute to the pathogenesis of autoimmune diseases.

### 5.2.2 Cytotoxic T cells

Thymocytes that mature to $CD4^-CD8^+$ single-positive cells are termed $CD8^+$ T cells or cytotoxic T cells. These cells are activated when the TCR complex together with CD28 binds to MHC class I associated with antigen and costimulatory proteins such as CD80 or CD86 on APCs. Activated $CD8^+$ T cells produce IFNγ and TNFα that facilitate signaling of other immune and intramural cells. Additionally, they release cytotoxins such as perforin and granzymes to induce apoptosis of target cells. Cytotoxic T cells have a less defined role in vascular G&R induced by chronic inflammation, but they have been implicated in acute coronary syndrome, allograft arteriosclerosis, and other types of chronic diseases predisposing to cardiovascular disease such as obesity [56].

### 5.2.3 Other T cell subsets

Memory T cells are T cells that have previously encountered their antigen and may be either $CD4^+$ or $CD8^+$ cells. They can be further subdivided into stem memory, central memory, effector memory, or tissue-resident memory T cells and are characterized by their expression of CCR7 and L-selectin and whether they continue to circulate in the bloodstream (as with stem, central, and effector cells) or reside in peripheral tissues until activated (as with tissue-resident cells). These cells are defined by their ability to mobilize quickly into activated T cells to initiate a sufficient immune reaction in response to a previously encountered immune challenge.

Natural killer T (NKT) cells (separate and distinct from natural killer cells, which are part of the innate immune system) may be either $CD4^+$ or $CD8^+$ or they may be double negative, expressing neither CD4 nor CD8. They recognize and respond to glycolipid antigens such as CD1d rather than peptide antigens presented by MHC class I. Activation of NKT cells leads to the production of IFNγ, which polarizes Th1 cells, and IL-4, which polarizes Th2 cells. The rapid production of T helper cells enhances the response of the immune system to challenge by mobilizing DCs, cytotoxic T cells, B cells, and macrophages.

## 5.3 DENDRITIC CELLS

DCs are produced in the bone marrow and reside primarily in the skin and mucosal tissues to serve as a first line of defense against invading pathogens. Activation of T cells is critically dependent on the function of DCs as they are responsible for antigen presentation to effector cells [57]. DCs can be subdivided into different types based on the expression of various extracellular proteins. Conventional DCs express CD11c/CD18 and may be subdivided into three phenotypes based on their ability to polarize T cells to the Th1, Th2, or Treg phenotype. Each of the subtypes is polarized by differing factors. Type I DCs are polarized by IFNγ, which is also produced by Th1 cells, suggesting a possible feed-forward response. Type II DCs are polarized primarily by thymic stromal lymphopoietin (TSLP), CCR2L, and histamine and produce IL-4, IL-5, and IL-13 that direct Th2 polarization. Regulatory DCs are polarized by IL-10 and TGFβ, both of which are also produced by Treg cells, suggesting a similar feed-forward response to that of Type I DCs and Th1 cells.

Plasmacytoid DCs do not express high levels of CD11c but rather are defined by their expression of blood DC antigen 2 (in humans) or sialic acid binding immunoglobulin-like lectin H (in mice). Plasmacytoid DCs express CXCR3 and produce proinflammatory cytokines such as IL-12 and IFNγ. These cell types provide an effector response more suited to resolution of viral infection and thus do not appear to play a significant role in vascular inflammation.

Although DCs do not appear to have a significant direct role in vascular G&R, their critical contribution to the polarization of other cell types and thus the general inflammatory milieu makes them key regulators of both vascular homeostasis and disease. Indeed, DCs are found in pulmonary hypertension, they contribute to atherosclerotic plaque progression, and they localize in aneurysmal tissue.

## 6 ECM AND INFLAMMATION

Potent effects of inflammatory cell recruitment and activation include the synthesis and deposition of ECM proteins and production or activation of MMPs and reactive oxygen species (ROS), each of which contribute to the modification of the local environment. G&R can occur either as an intentional response by immune cells or as a side effect of the primary function of removing the source of the inflammatory response (e.g., pathogen or necrotic tissue). Both immune and intramural cells produce a wide variety of bioactive enzymes and biomolecules with matrix-altering capabilities; thus, homeostatic maintenance of ECM content and structure depends critically on the coordinated effect of resident and invading cells. This section will focus on cell secretion of matrix proteins and proteases. Although ROS plays a significant role in matrix remodeling, their role is largely as a mediator of the production of matrix proteins and proteases; thus, ROS will not be a focus here. Rather, the interested reader is referred to a detailed review summarizing the role of ROS in mechanically mediated vascular remodeling [58].

### 6.1 MATRIX METALLOPROTEINASES

The aforementioned MMPs are a family of catalytic enzymes comprising a prodomain approximately 80 amino acids long and a catalytic domain typically containing a zinc ion in the active site. MMPs are first secreted as a neutral proenzyme and are subsequently activated by factors including other MMPs and ROS. To date, 24 forms of MMPs have been identified in mammals, each having an enzymatic effect on a specific extracellular protein. Several MMPs play a significant role in vascular G&R [59]. MMP1, MMP8, and MMP13 are collagenases, MMP2 and MMP9 are gelatinases, and MMP12 is an elastase. MMPs may be secreted by immune cells (generally of the innate immune system) or intramural cells of the vasculature, especially adventitial fibroblasts and medial smooth muscle cells. MMPs also regulate the function of other immune cells by modulating second messengers associated with immune cell recruitment and activation [60].

### 6.1.1 Collagenases

MMP1, also known as collagenase-1, is secreted by fibroblasts, among other cells, and binds most types of fibrillar collagen. MMP1 is downregulated by the presence of IFNγ and is upregulated by the presence of IL-1β, which is predominantly produced by classically activated M1 macrophages. MMP1 also promotes the production of CCL2 that recruits inflammatory macrophages. MMP1 indirectly affects other matrix molecules by binding to and regulating the activation of MMP2 and MMP9. MMP8, also known as collagenase-2, is primarily secreted by neutrophils and binds many of the same types of fibrillar collagen. Interestingly, cleavage of collagen I by MMP8 results in collagen fragments that function as chemokines by acting on G-protein-coupled receptors to promote further accumulation of neutrophils. In mice, MMP8 binds CXCL5, a protein expressed by eosinophils and promotes neutrophil accumulation, suggesting a regulatory function of the enzyme for neutrophil activation. MMP13, or collagenase-3, degrades fibrillar collagens I and II. Given their roles in enzymatic digestion of collagen, MMPs 1, 8, and 13 have each been implicated in the progression of abdominal aortic aneurysm [61–63] as they contribute to the loss of structural strength of the tissue. They also are implicated in atherogenesis and hypertensive remodeling.

### 6.1.2 Gelatinases and metalloelastases

Two gelatinases exist in the MMP family—MMP2 (gelatinase A) and MMP9 (gelatinase B). These enzymes primarily digest denatured collagen, but they cleave other ECM proteins as well. MMP2 cleaves many types of extracellular proteins, including fibrillar collagen, elastin, fibronectin, and laminin, as well as other MMPs, including MMP9 and MMP13. MMP2 also binds and inactivates chemokines such as CCL7 and CXCL12. MMP2 expression is downregulated by IFNγ, suggesting a regulatory role of Th1 cells in the function of MMP2. MMP9 is similarly downregulated by IFNγ but is upregulated by IL-1β. MMP9 cleaves many of the same proteins as MMP2 and also inactivates CXCL12; however, it also activates other chemokines including CXCL5, which promotes neutrophil accumulation. MMP2 and MMP9 have both been extensively implicated in a variety of forms of disease including atherosclerosis, aneurysm and dissection, and hypertensive remodeling.

MMP12 is a metalloelastase that degrades primarily elastin and binds to latent TNFα. It is secreted by both classically and alternatively activated macrophages. MMP12, like the gelatinases, has also been extensively implicated in aneurysm, atherosclerosis, and hypertensive remodeling.

## 6.2 OTHER PROTEASES

Other types of metalloproteinases exist, namely, a disintegrin and metalloproteinase (ADAMs, also known as adamlysins) and ADAMs with a thrombospondin-binding motif (ADAMTS). ADAMs are membrane-bound MMPs and serve similar functions to those of the previously enumerated MMPs in that they bind and degrade various ECM proteins. They maintain similar proteolytic function as the MMPs, but they are also unique in that their disintegrin domain binds integrins and may thus interrupt cell-matrix interactions and possibly mechanosensing (though this has not been proved). Like other MMPs, there are a large number of ADAM isoforms, but the number directly contributing to cardiovascular disease is lower, and their function is less characterized, though their upregulation has been demonstrated in some types of disease.

Although MMPs are the most recognized mediators of vascular G&R and contributors to cardiovascular disease, intramural and immune cells also secrete other types of proteases, including serine proteases, cysteine proteases, and aspartate proteases. Cathepsins are a family of enzymatic proteins that can fall into the serine (cathepsins A and G) or cysteine (cathepsins B, C, F, H, K, L, O, S, V, W, and Z) protease families. Cathepsins are secreted as preproenzymes and subsequently cleaved by activating molecules to induce enzymatic activity that degrades matrix proteins. Activation of cathepsins can occur in the presence of ROS, cytokines, and other proteases. Various cathepsins have been implicated in the progression of multiple forms of disease, including atherosclerosis and aneurysm development [64]. Secretion of cathepsins in the context of disease has been attributed largely to macrophages, smooth muscle cells, and ECs, although other cell types have been implicated to a lesser extent.

## 6.3 TISSUE INHIBITORS OF METALLOPROTEINASES

Tissue inhibitors of metalloproteinases (TIMPs) are a family of proteins that bind and inactivate MMPs. There are four TIMPs, namely TIMP1, TIMP2, TIMP3, and TIMP4. TIMP1, TIMP2, and TIMP4 have inhibitory activity over nearly all MMPs, while TIMP3 functions primarily over ADAMs. TIMP1 interacts with both collagenases and elastases to reduce the formation of atherosclerotic plaques, and TIMP2 has similar function in preventing atherosclerosis. Because of their inhibitory action, TIMPs have been investigated as potential therapeutic mechanisms for treating cardiovascular disease, but progress in this area remains underdeveloped [59].

## 7 MATRIX G&R IN CARDIOVASCULAR DISEASE

There is a clear interaction among mechanosensing of the ECM by intramural cells, inflammation, and mechanoregulating the ECM that may be illustrated as a feedback loop in which intramural cells sense their mechanical environment and produce a

# 7 Matrix G&R in cardiovascular disease

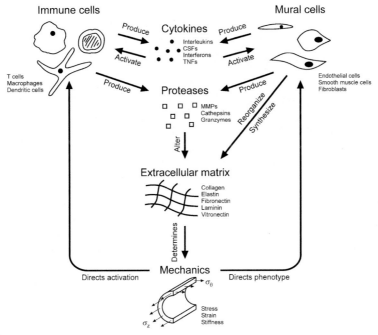

**FIG. 2**

Schematic diagram illustrating cellular regulation of and response to the local mechanical environment.

response that adjusts the mechanical environment, often to preferred, or homeostatic, conditions (Fig. 2). The response may consist of a biomechanical change (e.g., cells exert forces to physically remodel the ECM in a way that alters the "perceived" stiffness of the matrix), or it may consist of a biochemical response (e.g., matrix-altering enzymes are produced either directly by intramural cells or by immune cells that are recruited via the secretion of inflammatory cytokines by the intramural cells). Additional factors such as exposure to neuroendocrine factors have also been shown to modulate matrix G&R in cardiovascular disease [65], but this is beyond the scope of this chapter. It should also be noted that effector responses that alter the ECM are not limited to mechanical stimuli; indeed, other modes of activation exist, but the focus here will be on mechanically mediated responses.

Two possible types of mechanically mediated disease progression exist: First, cells may correctly sense and evaluate their mechanical environment but respond maladaptively in response to altered loading conditions, perhaps because of genetic defects, or second, cells may suffer from dysfunctional mechanosensing machinery and thus incorrectly sense a normal or abnormal mechanical environment and thereby respond inappropriately. While evidence is accumulating for dysfunctional mechanosensing in the context of different cardiovascular diseases, including thoracic aneurysms [5], we focus on the former, altered mechanical conditions that are correctly sensed by intramural cells and yet lead to pathological responses.

Because of the wide variety of mechanical forces experienced by the different cell types of the vascular wall, myriad cardiovascular diseases can develop due to altered mechanical conditions. Among the best characterized conditions is the development of atherosclerosis in response to altered fluid shear stress on the endothelium. Since the early 1980s, many have demonstrated that atherosclerosis develops preferentially in locations of disturbed blood flow [66]; namely, locations in the vasculature with complex geometry, including bifurcations (such as that of the common carotid artery) or marked curvature (such as that of the aortic arch), can lead to regions of low values of time-averaged wall shear stress or high values of oscillatory shear stress. These abnormal shear stress fields, in turn, induce an inflammatory response in part via NF-κB and $\alpha_5\beta_1$ integrin signaling. For a detailed review, see Ref. [27]. Another disease at the intersection of mechanosensing, inflammation, and matrix G&R is the development of abdominal aortic aneurysms. Again, there is an association with altered flow conditions and similarly associated mechanisms (e.g., altered NF-κB signaling), but changes in intramural stress are implicated to a greater extent. For a detailed review, see Ref. [67]. The remainder of this section will focus, however, on two related conditions that serve as risk factors for many cardiovascular diseases, hypertension, and aging.

## 7.1 MECHANICS AND INFLAMMATION IN HYPERTENSION

Systemic blood pressure is typically measured in the clinic using a sphygmomanometer that provides a value of systolic pressure (i.e., the pressure at the time of contraction of the heart) and diastolic pressure (i.e., the pressure at the time of relaxation). Typical values of blood pressure in healthy adults are approximately 120 (systolic)/80 mmHg (diastolic). Elevated pressures, typically in excess of 140/90 mmHg, indicate hypertension, which is thought to arise due to constriction of peripheral arterioles, thus increasing the resistance to blood flow and forcing the heart to raise the blood pressure to maintain necessary perfusion rates to peripheral tissues. The increased pressure has multiple effects on the wall mechanics: it initially increases the cyclic wall stress and thus stiffness and increases the cyclic wall strain, thus significantly altering the mechanical environment of the intramural cells. Such a dramatic change in the mechanical environment induces a wide variety of effector responses from vascular cells.

At a macroscopic level, hypertension leads to thickening of the vascular wall, especially in the central vasculature (i.e., the aorta), apparently in an attempt to restore circumferential wall stress to a preferred homeostatic value (cf. Eq. 5). Clinically, thickening of the arterial wall (e.g., as measured in the carotid artery) is a strong predictor of future cardiovascular events and thus a focus of significant attention [68]. Wall thickening occurs via multiple means, including medial thickening (by hypertrophy of smooth muscle cells and matrix deposition) and adventitial thickening (by increased collagen deposition). Both processes have been demonstrated in animal models, with the earliest studies linking mechanics and hypertensive vascular remodeling in spontaneously hypertensive rats. Medial thickening was demonstrated under hypertensive conditions many years ago [4], whereas effects on the adventitia were quantified more recently [69]. Importantly, the latter was due to both mechano- and immuno-mediated mechanisms.

Polarization of T cells has been widely implicated in the development of hypertension in models leading to either medial or adventitial thickening. Mice lacking recombinant activating gene-1 ($Rag1^{-/-}$) do not develop mature T or B cells and also do not develop hypertension in response to chronic infusion with the vasoconstrictor AngII [70]. Interestingly, AngII-induced hypertension is restored upon adoptive transfer of mature T cells, but not B cells, suggesting a role for T-cell activation in the development of AngII-induced hypertension. The requirement of T-cell activation in AngII-induced hypertension development has been associated with production of IL-17 (implicating Th17 cells), IL-6 production (implicating Th17 cells or macrophages), and TNFα (implicating any number of immune cells). Interestingly, an alternative subset of T cells protects against AngII-induced hypertensive remodeling (though not the development of hypertension). Tregs produce IL-10 that promotes the resolution of inflammation, and deficiency in this cytokine elevates AngII-induced superoxide production in mice [71].

While T cells have been widely suggested as a mediator of developing hypertension, the resulting vascular G&R appears to involve a much more complex inflammatory environment. The majority of structural changes in AngII-induced hypertension occur largely in the adventitia and may be mediated by macrophages (likely alternatively activated) and IL-6 production [52] in addition to a wide variety of T-cell subsets, including Th2 and Th17 cells, a function that may be sex-dependent [72]. The intersection of this type of G&R with mechanobiology has been investigated in various capacities. Integrin-dependent changes in small arteries have been demonstrated in spontaneous hypertension, with integrin subunit $\alpha_v$ necessary for inward remodeling, while functional blocking results in compensatory hypertrophy of the vessel [73]. Similarly, $\alpha_5\beta_1$ integrin and fibronectin expression are increased in large arteries of spontaneously hypertensive rats (SHR) [74]. Recent evidence suggests that circumferential material stiffness is a target of homeostatic regulation in the vasculature. Indeed, chronic AngII infusion in mice results in a transient increase in material stiffness in a subset of regions in the aorta that subsequently returns to normal values in an apparently mechano-mediated fashion. Interestingly, an inability to regulate circumferential material stiffness appears to predispose to aneurysm development in the model of AngII infusion and others [21], an apparent example of the aforementioned dysfunctional mechanosensing. A mouse model of integrin deficiency ($\alpha_1$ subunit knockout) shows a blunted increase in blood pressure and no adventitial remodeling of the carotid artery in response to AngII infusion [75]. Although not investigated in that study, the results may suggest that proper mechanosensing lies upstream of inflammatory cascades that lead to vascular remodeling in hypertension. Such a hypothesis should be treated with caution, however, as the link between mechanosensing of arterial stiffness and subsequent inflammatory activation has not been proven.

The linking of hypertension, inflammation, and vascular G&R is becoming well accepted, but linking hypertension, mechanosensing, and vascular G&R is in its nascent stage. We know, however, that immune cells harbor mechanosensitive proteins that facilitate rolling and adhesion to the endothelium as well as migration through ECM to sites of infection or damage. What is less understood (in the context of hypertension) is whether intramural cells first "sense" a change in their mechanical environment due to the initial increase in blood pressure and then produce inflammatory cytokines in response, or if hypertension development is accompanied by inflammatory activation that recruits immune cells to the vascular wall independent of mechanosensing machinery. Our most recent study suggests that inflammatory activation lies downstream of mechanosensing as material stiffness (when evaluated at systolic blood pressure) increases as blood pressure increases, peaking at 14 days of infusion, whereas inflammatory cell infiltration into the adventitia of various segments of the aorta does not begin until ~14 days [69]. Little evidence exists to address this question; thus, much remains to be accomplished in better understanding the relationship between hypertensive vascular remodeling and mechanobiology.

## 7.2 AGING, INFLAMMATION, AND VASCULAR MECHANICS

Nearly all forms of cardiovascular disease become more prevalent with increased age. Indeed, Sir William Osler suggested early in the 20th century that we (humans) are "as old as our arteries." Aging remains one of the leading risk factors, for example, for hypertension, atherosclerosis, aneurysm and dissection, and associated cardiovascular events including myocardial infarction and stroke [76]. Aging affects all organ systems. With respect to the vasculature, however, aging tends to lead to an increased intimal-medial thickness and, in large arteries, luminal dilatation associated with increased structural stiffness as measured by PWV. The structural stiffening due to the intimal thickening may be due in part to a decrease in nitric oxide bioavailability, a condition attributed to endothelial dysfunction, which is aging-related and an independent risk factor for future cardiovascular events. Loss of endothelial nitric oxide production could increase vascular tone and decrease diameter (i.e., affects the mechanics and hemodynamics) as well as increase intimal inflammation since nitric oxide is antiinflammatory. Aging in the vasculature has also been associated with the production of ROS, which may be generated in part by infiltrating immune cells under chronic inflammatory conditions. Thus, there is again a potential feed-forward cycle in which inflammation leads to arterial stiffening, which induces inflammatory activation of intramural cells via mechanosensing of the altered matrix stiffness, which then facilitates additional recruitment of immune cells.

Aged patients have increased systemic expression of inflammatory cytokines such as IL-1β, IL-6, and TNFα. Many studies have suggested an intersection of such inflammation with the NF-κB pathway, which has been implicated in various cardiovascular diseases including atherosclerosis, aneurysm and dissection, and hypertension. Interestingly, the concept of "immunosenescence" has existed since the mid-20th century; it suggests that the immune system can have a decreased or delayed responsiveness to specific antigens. Indeed, various deficiencies are observed in immune cells with aging, as, for example, a decreased chemotaxis in neutrophils and decreased cytokine production by macrophages [77]. This theory has been used to explain the observation of increased susceptibility to infection in elderly populations. Nevertheless, vascular aging seems to occur primarily due to chronic, low-grade inflammation. Although the aged immune system appears compromised in its ability to respond to external challenges, vascular cells express inflammatory cytokines that promote immune cell infiltration in aging. For example, vascular smooth muscle cells increase production of IL-6 and CCR2 in aging [78].

Accompanying chronic inflammation in aging is an alteration in the production and activation of matrix-altering proteins, including MMPs and other proteases [79]. These proteins likely contribute to the increase in central arterial stiffness observed in aging by increasing collagen deposition despite some degradation. Studies utilizing atomic force microscopy show that increased stiffness of the subendothelial matrix in aging is associated with increased endothelial permeability and subsequent transmigration of neutrophils [80]. Increased vascular stiffness seems uniformly observed across models (i.e., human and mouse models) regardless of the methods employed to quantify stiffness (e.g., PWV or ex vivo biomechanical testing).

## 8 CONCLUSIONS AND FUTURE DIRECTIONS

The wealth of data associating vascular G&R with inflammation is staggering, and it is well accepted that both contribute to the occurrence of future cardiovascular events. Understanding the complex mechanical environment of vascular cells in health and disease as well as their effector responses to changes in the mechanical environment may offer useful tools for identifying and targeting key contributors to the progression of some of the most common contributors to cardiovascular events, including hypertension and aging. Critical remaining needs include accurately quantifying the mechanical environment of vascular cells in vivo, experimentally measuring inflammatory responses as a function of mechanical changes, and characterizing vascular structural and functional changes in response to mechanically mediated inflammatory cascades. Successful realization of these tasks will lend insight into important links between known contributors to debilitating and life-threatening cardiovascular events, thus paving the way for novel therapeutic treatment.

## ACKNOWLEDGMENT

This work was supported, in part, by current grants from the NIH (R01 HL086418, R01 HL105297, R01 HL128602, and R01 HL134712).

## REFERENCES

[1] Chiu J-J, Chien S. Effects of disturbed flow on vascular endothelium: pathophysiological basis and clinical perspectives. Physiol Rev 2011;91:327–87. https://doi.org/10.1152/physrev.00047.2009.

[2] Chen P-Y, Qin L, Baeyens N, Li G, Afolabi T, Budatha M, et al. Endothelial-to-mesenchymal transition drives atherosclerosis progression. J Clin Invest 2015;125: 4514–28. https://doi.org/10.1172/JCI82719.

[3] Conway DE, Schwartz MA. Mechanotransduction of shear stress occurs through changes in VE-cadherin and PECAM-1 tension: implications for cell migration. Cell Adh Migr 2015;9:335–9. https://doi.org/10.4161/19336918.2014.968498.

[4] Wolinsky H, Glagov S. A lamellar unit of aortic medial structure and function in mammals. Circ Res 1967;20:99–111. https://doi.org/10.1161/01.RES.20.1.99.

[5] Humphrey JD, Schwartz MA, Tellides G, Milewicz DM. Role of mechanotransduction in vascular biology focus on thoracic aortic aneurysms and dissections. Circ Res 2015;116: 1448–61. https://doi.org/10.1161/CIRCRESAHA.114.304936.

[6] Laurent S, Boutouyrie P. The structural factor of hypertension. Circ Res 2015;116: 1007–21. https://doi.org/10.1161/CIRCRESAHA.116.303596.

[7] Brozovich FV, Nicholson CJ, Degen CV, Gao YZ, Aggarwal M, Morgan KG. Mechanisms of vascular smooth muscle contraction and the basis for pharmacologic treatment of smooth muscle disorders. Pharmacol Rev 2016;68:476–532. https://doi.org/10.1124/pr.115.010652.

[8] Tellides G, Pober JS. Inflammatory and immune responses in the arterial media. Circ Res 2015;116:312–22. https://doi.org/10.1161/CIRCRESAHA.116.301312.

[9] Stenmark KR, Yeager ME, Kasmi KCE, Nozik-Grayck E, Gerasimovskaya EV, Li M, et al. The adventitia: essential regulator of vascular wall structure and function. Annu Rev Physiol 2013;75:23–47. https://doi.org/10.1146/annurev-physiol-030212-183802.

[10] Humphrey JD. Mechanisms of arterial remodeling in hypertension coupled roles of wall shear and intramural stress. Hypertension 2008;52:195–200. https://doi.org/10.1161/HYPERTENSIONAHA.107.103440.

[11] Laurent S, Cockcroft J, Van Bortel L, Boutouyrie P, Giannattasio C, Hayoz D, et al. Expert consensus document on arterial stiffness: methodological issues and clinical applications. Eur Heart J 2006;27:2588–605. https://doi.org/10.1093/eurheartj/ehl254.

[12] Genovese K, Lee Y-U, Lee AY, Humphrey JD. An improved panoramic digital image correlation method for vascular strain analysis and material characterization. J Mech Behav Biomed Mater 2013;27:132–42. https://doi.org/10.1016/j.jmbbm.2012.11.015.

[13] Bersi M, Bellini C, Di Achille P, Humphrey JD, Genovese K, Avril S. Novel methodology for characterizing regional variations in material properties of murine aortas. J Biomech Eng 2016. https://doi.org/10.1115/1.4033674.

[14] Humphrey JD. Experimental methods. In: Cardiovasc. solid mech. New York: Springer; 2002. p. 158–210. https://doi.org/10.1007/978-0-387-21576-1_5.

[15] Chuong CJ, Fung YC. Residual stress in arteries. In: Schmid-Schönbein GW, Woo SL-Y, Zweifach BW, editors. Front. biomech. New York: Springer; 1986. p. 117–29. https://doi.org/10.1007/978-1-4612-4866-8_9.

[16] Bellini C, Ferruzzi J, Roccabianca S, Martino ESD, Humphrey JD. A microstructurally motivated model of arterial wall mechanics with mechanobiological implications. Ann Biomed Eng 2014;42:488–502. https://doi.org/10.1007/s10439-013-0928-x.

[17] Holzapfel GA. Nonlinear solid mechanics: continuum mechanics approach. Chichester: John Wiley & Sons; 2000.
[18] Shadwick RE. Mechanical design in arteries. J Exp Biol 1999;202:3305–13.
[19] Tabima DM, Chesler NC. The effects of vasoactivity and hypoxic pulmonary hypertension on extralobar pulmonary artery biomechanics. J Biomech 2010;43:1864–9. https://doi.org/10.1016/j.jbiomech.2010.03.033.
[20] Baek S, Gleason RL, Rajagopal KR, Humphrey JD. Theory of small on large: potential utility in computations of fluid–solid interactions in arteries. Comput Methods Appl Mech Eng 2007;196:3070–8. https://doi.org/10.1016/j.cma.2006.06.018.
[21] Bellini C, Bersi MR, Caulk AW, Ferruzzi J, Milewicz DM, Ramirez F, et al. Comparison of 10 murine models reveals a distinct biomechanical phenotype in thoracic aortic aneurysms. J R Soc Interface 2017;14:20161036. https://doi.org/10.1098/rsif.2016.1036.
[22] Bersi MR, Ferruzzi J, Eberth JF, Jr RLG, Humphrey JD. Consistent biomechanical phenotyping of common carotid arteries from seven genetic, pharmacological, and surgical mouse models. Ann Biomed Eng 2014;42:1207–23. https://doi.org/10.1007/s10439-014-0988-6.
[23] Chiquet M, Gelman L, Lutz R, Maier S. From mechanotransduction to extracellular matrix gene expression in fibroblasts. Biochim Biophys Acta Mol Cell Res 2009;1793: 911–20. https://doi.org/10.1016/j.bbamcr.2009.01.012.
[24] Davies PF. Flow-mediated endothelial mechanotransduction. Physiol Rev 1995;75: 519–60.
[25] Haga JH, Li Y-SJ, Chien S. Molecular basis of the effects of mechanical stretch on vascular smooth muscle cells. J Biomech 2007;40:947–60. https://doi.org/10.1016/j.jbiomech.2006.04.011.
[26] Dajnowiec D, Langille BL. Arterial adaptations to chronic changes in haemodynamic function: coupling vasomotor tone to structural remodelling. Clin Sci 2007;113: 15–23. https://doi.org/10.1042/CS20060337.
[27] Hahn C, Schwartz MA. Mechanotransduction in vascular physiology and atherogenesis. Nat Rev Mol Cell Biol 2009;10:53–62. https://doi.org/10.1038/nrm2596.
[28] Humphrey JD, Na S. Elastodynamics and arterial wall stress. Ann Biomed Eng 2002;30:509–23. https://doi.org/10.1114/1.1467676.
[29] Shi Z-D, Tarbell JM. Fluid flow mechanotransduction in vascular smooth muscle cells and fibroblasts. Ann Biomed Eng 2011;39:1608–19. https://doi.org/10.1007/s10439-011-0309-2.
[30] Davis MJ, Wu X, Nurkiewicz TR, Kawasaki J, Davis GE, Hill MA, et al. Integrins and mechanotransduction of the vascular myogenic response. Am J Physiol Heart Circ Physiol 2001;280:H1427–33.
[31] Wilson E, Sudhir K, Ives HE. Mechanical strain of rat vascular smooth muscle cells is sensed by specific extracellular matrix/integrin interactions. J Clin Invest 1995;96: 2364–72.
[32] Brown XQ, Bartolak-Suki E, Williams C, Walker ML, Weaver VM, Wong JY. Effect of substrate stiffness and PDGF on the behavior of vascular smooth muscle cells: implications for atherosclerosis. J Cell Physiol 2010;225:115–22. https://doi.org/10.1002/jcp.22202.
[33] Davenpeck KL, Marcinkiewicz C, Wang D, Niculescu R, Shi Y, Martin JL, et al. Regional differences in integrin expression role of $\alpha 5\beta 1$ in regulating smooth muscle cell functions. Circ Res 2001;88:352–8. https://doi.org/10.1161/01.RES.88.3.352.

[34] Li W, Li Q, Jiao Y, Qin L, Ali R, Zhou J, et al. Tgfbr2 disruption in postnatal smooth muscle impairs aortic wall homeostasis. J Clin Invest 2014;124:755–67. https://doi.org/10.1172/JCI69942.

[35] Tieu BC, Ju X, Lee C, Sun H, Lejeune W, Adrian Recinos III, et al. Aortic adventitial fibroblasts participate in angiotensin-induced vascular wall inflammation and remodeling. J Vasc Res 2011;48:261–72. https://doi.org/10.1159/000320358.

[36] Hinz B. The myofibroblast: paradigm for a mechanically active cell. J Biomech 2010;43:146–55. https://doi.org/10.1016/j.jbiomech.2009.09.020.

[37] Tomasek JJ, Gabbiani G, Hinz B, Chaponnier C, Brown RA. Myofibroblasts and mechano-regulation of connective tissue remodelling. Nat Rev Mol Cell Biol 2002;3: 349–63. https://doi.org/10.1038/nrm809.

[38] Branco da Cunha C, Klumpers DD, Li WA, Koshy ST, Weaver JC, Chaudhuri O, et al. Influence of the stiffness of three-dimensional alginate/collagen-I interpenetrating networks on fibroblast biology. Biomaterials 2014;35:8927–36. https://doi.org/10.1016/j.biomaterials.2014.06.047.

[39] Hadjipanayi E, Mudera V, Brown RA. Close dependence of fibroblast proliferation on collagen scaffold matrix stiffness. J Tissue Eng Regen Med 2009;3:77–84. https://doi.org/10.1002/term.136.

[40] Xie J, Zhang Q, Zhu T, Zhang Y, Liu B, Xu J, et al. Substrate stiffness-regulated matrix metalloproteinase output in myocardial cells and cardiac fibroblasts: implications for myocardial fibrosis. Acta Biomater 2014;10:2463–72. https://doi.org/10.1016/j.actbio.2014.01.031.

[41] Murphy KM. Janeway's immunobiology. New York and London: Taylor & Francis Group; 2011.

[42] Chovatiya R, Medzhitov R. Stress, inflammation, and defense of homeostasis. Mol Cell 2014;54:281–8. https://doi.org/10.1016/j.molcel.2014.03.030.

[43] Tang PCY, Qin L, Zielonka J, Zhou J, Matte-Martone C, Bergaya S, et al. MyD88-dependent, superoxide-initiated inflammation is necessary for flow-mediated inward remodeling of conduit arteries. J Exp Med 2008;205:3159–71. https://doi.org/10.1084/jem.20081298.

[44] Ross R. Atherosclerosis—an inflammatory disease. N Engl J Med 1999;340:115–26. https://doi.org/10.1056/NEJM199901143400207.

[45] Gordon S, Taylor PR. Monocyte and macrophage heterogeneity. Nat Rev Immunol 2005;5:953–64. https://doi.org/10.1038/nri1733.

[46] Mosser DM, Edwards JP. Exploring the full spectrum of macrophage activation. Nat Rev Immunol 2008;8:958–69. https://doi.org/10.1038/nri2448.

[47] Mantovani A, Sica A, Sozzani S, Allavena P, Vecchi A, Locati M. The chemokine system in diverse forms of macrophage activation and polarization. Trends Immunol 2004;25:677–86. https://doi.org/10.1016/j.it.2004.09.015.

[48] Chinetti-Gbaguidi G, Colin S, Staels B. Macrophage subsets in atherosclerosis. Nat Rev Cardiol 2015;12:10–7. https://doi.org/10.1038/nrcardio.2014.173.

[49] Bush E, Maeda N, Kuziel WA, Dawson TC, Wilcox JN, DeLeon H, et al. CC chemokine receptor 2 is required for macrophage infiltration and vascular hypertrophy in angiotensin II-induced hypertension. Hypertension 2000;36:360–3. https://doi.org/10.1161/01.HYP.36.3.360.

[50] Choke E, Cockerill G, Wilson WRW, Sayed S, Dawson J, Loftus I, et al. A review of biological factors implicated in abdominal aortic aneurysm rupture. Eur J Vasc Endovasc Surg 2005;30:227–44. https://doi.org/10.1016/j.ejvs.2005.03.009.

[51] Hibino N, Yi T, Duncan DR, Rathore A, Dean E, Naito Y, et al. A critical role for macrophages in neovessel formation and the development of stenosis in tissue-engineered vascular grafts. FASEB J 2011;25:4253–63. https://doi.org/10.1096/fj.11-186585.

[52] Tieu BC, Lee C, Sun H, LeJeune W, Recinos A, Ju X, et al. An adventitial IL-6/MCP1 amplification loop accelerates macrophage-mediated vascular inflammation leading to aortic dissection in mice. J Clin Invest 2009;119:3637–51. https://doi.org/10.1172/JCI38308.

[53] Wynn TA. Fibrotic disease and the TH1/TH2 paradigm. Nat Rev Immunol 2004;4:583–94. https://doi.org/10.1038/nri1412.

[54] Korn T, Bettelli E, Oukka M, Kuchroo VK. IL-17 and Th17 cells. Annu Rev Immunol 2009;27:485–517. https://doi.org/10.1146/annurev.immunol.021908.132710.

[55] Meng X, Yang J, Dong M, Zhang K, Tu E, Gao Q, et al. Regulatory T cells in cardiovascular diseases. Nat Rev Cardiol 2016;13:167–79. https://doi.org/10.1038/nrcardio.2015.169.

[56] Nishimura S, Manabe I, Nagasaki M, Eto K, Yamashita H, Ohsugi M, et al. CD8+ effector T cells contribute to macrophage recruitment and adipose tissue inflammation in obesity. Nat Med 2009;15:914–20. https://doi.org/10.1038/nm.1964.

[57] Kapsenberg ML. Dendritic-cell control of pathogen-driven T-cell polarization. Nat Rev Immunol 2003;3:984–93. https://doi.org/10.1038/nri1246.

[58] Birukov KG. Cyclic stretch, reactive oxygen species, and vascular remodeling. Antioxid Redox Signal 2009;11:1651–67. https://doi.org/10.1089/ars.2008.2390.

[59] Raffetto JD, Khalil RA. Matrix metalloproteinases and their inhibitors in vascular remodeling and vascular disease. Biochem Pharmacol 2008;75:346–59. https://doi.org/10.1016/j.bcp.2007.07.004.

[60] Parks WC, Wilson CL, López-Boado YS. Matrix metalloproteinases as modulators of inflammation and innate immunity. Nat Rev Immunol 2004;4:617–29. https://doi.org/10.1038/nri1418.

[61] Mao D, Lee JK, VanVickle SJ, Thompson RW. Expression of collagenase-3 (MMP-13) in human abdominal aortic aneurysms and vascular smooth muscle cells in culture. Biochem Biophys Res Commun 1999;261:904–10. https://doi.org/10.1006/bbrc.1999.1142.

[62] Tamarina NA, McMillan WD, Shively VP, Pearce WH. Expression of matrix metalloproteinases and their inhibitors in aneurysms and normal aorta. Surgery 1997;122:264–72. https://doi.org/10.1016/S0039-6060(97)90017-9.

[63] Wilson WRW, Schwalbe EC, Jones JL, Bell PRF, Thompson MM. Matrix metalloproteinase 8 (neutrophil collagenase) in the pathogenesis of abdominal aortic aneurysm. Br J Surg 2005;92:828–33. https://doi.org/10.1002/bjs.4993.

[64] Lutgens SPM, Cleutjens KBJM, Daemen MJAP, Heeneman S. Cathepsin cysteine proteases in cardiovascular disease. FASEB J 2007;21:3029–41. https://doi.org/10.1096/fj.06-7924com.

[65] Harrison DG. The immune system in hypertension. Trans Am Clin Climatol Assoc 2014;125:130–40.

[66] Ku DN, Giddens DP, Zarins CK, Glagov S. Pulsatile flow and atherosclerosis in the human carotid bifurcation. Positive correlation between plaque location and low oscillating shear stress. Arterioscler Thromb Vasc Biol 1985;5:293–302. https://doi.org/10.1161/01.ATV.5.3.293.

[67] Humphrey JD, Holzapfel GA. Mechanics, mechanobiology, and modeling of human abdominal aorta and aneurysms. J Biomech 2012;45:805–14. https://doi.org/10.1016/j.jbiomech.2011.11.021.

[68] Lorenz MW, Markus HS, Bots ML, Rosvall M, Sitzer M. Prediction of clinical cardiovascular events with carotid intima-media thickness. Circulation 2007;115:459–67. https://doi.org/10.1161/CIRCULATIONAHA.106.628875.

[69] Bersi MR, Khosravi R, Wujciak AJ, Harrison DG, Humphrey JD. Differential cell-matrix mechanoadaptations and inflammation drive regional propensities to aortic fibrosis, aneurysm or dissection in hypertension. J R Soc Interface 2017. https://doi.org/10.1098/rsif.2017.0327.

[70] Wu J, Thabet SR, Kirabo A, Trott DW, Saleh MA, Xiao L, et al. Inflammation and mechanical stretch promote aortic stiffening in hypertension through activation of p38 mitogen-activated protein kinase. Circ Res 2014;114:616–25. https://doi.org/10.1161/CIRCRESAHA.114.302157.

[71] Barhoumi T, Kasal DA, Li MW, Shbat L, Laurant P, Neves MF, et al. T regulatory lymphocytes prevent angiotensin II-induced hypertension and vascular injury. Hypertension 2011;57:469–76. https://doi.org/10.1161/HYPERTENSIONAHA.110.162941.

[72] Ji H, Zheng W, Li X, Liu J, Wu X, Zhang MA, et al. Sex-specific T-cell regulation of angiotensin II-dependent hypertension. Hypertension 2014;64:573–82. https://doi.org/10.1161/HYPERTENSIONAHA.114.03663.

[73] Heerkens EHJ, Shaw L, Ryding A, Brooker G, Mullins JJ, Austin C, et al. αV integrins are necessary for eutrophic inward remodeling of small arteries in hypertension. Hypertension 2006;47:281–7. https://doi.org/10.1161/01.HYP.0000198428.45132.02.

[74] Bézie Y, Lamazière J-MD, Laurent S, Challande P, Cunha RS, Bonnet J, et al. Fibronectin expression and aortic wall elastic Modulus in spontaneously hypertensive rats. Arterioscler Thromb Vasc Biol 1998;18:1027–34. https://doi.org/10.1161/01.ATV.18.7.1027.

[75] Louis H, Kakou A, Regnault V, Labat C, Bressenot A, Gao-Li J, et al. Role of α1β1-integrin in arterial stiffness and angiotensin-induced arterial wall hypertrophy in mice. Am J Physiol Heart Circ Physiol 2007;293:H2597–604. https://doi.org/10.1152/ajpheart.00299.2007.

[76] Lakatta EG, Wang M, Najjar SS. Arterial aging and subclinical arterial disease are fundamentally intertwined at macroscopic and molecular levels. Med Clin North Am 2009;93:583–604. https://doi.org/10.1016/j.mcna.2009.02.008.

[77] Shaw AC, Goldstein DR, Montgomery RR. Age-dependent dysregulation of innate immunity. Nat Rev Immunol 2013;13:875–87. https://doi.org/10.1038/nri3547.

[78] Song Y, Shen H, Schenten D, Shan P, Lee PJ, Goldstein DR. Aging enhances the basal production of IL-6 and CCL2 in vascular smooth muscle cells. Arterioscler Thromb Vasc Biol 2012;32:103–9. https://doi.org/10.1161/ATVBAHA.111.236349.

[79] Jacob MP. Extracellular matrix remodeling and matrix metalloproteinases in the vascular wall during aging and in pathological conditions. Biomed Pharmacother 2003;57:195–202. https://doi.org/10.1016/S0753-3322(03)00065-9.

[80] Huynh J, Nishimura N, Rana K, Peloquin JM, Califano JP, Montague CR, et al. Age-related intimal stiffening enhances endothelial permeability and leukocyte transmigration. Sci Transl Med 2011;3:112ra122. https://doi.org/10.1126/scitranslmed.3002761.

# CHAPTER 8

# Mechanobiology of the heart valve interstitial cell: Simulation, experiment, and discovery

Alex Khang*, Rachel M. Buchanan*, Salma Ayoub*, Bruno V. Rego*, Chung-Hao Lee[†], Giovanni Ferrari[‡], Kristi S. Anseth[§¶], Michael S. Sacks*

*Willerson Center for Cardiovascular Modeling and Simulation, Institute for Computational Engineering and Sciences, Department of Biomedical Engineering, The University of Texas at Austin, Austin, TX, United States* School of Aerospace and Mechanical Engineering, The University of Oklahoma, Norman, OK, United States[†] Department of Surgery, Columbia University, New York, NY, United States[‡] Department of Chemical and Biological Engineering, BioFrontiers Institute, University of Colorado, Boulder, CO, United States[§] Howard Hughes Medical Institute, University of Colorado, Boulder, CO, United States[¶]

## ABBREVIATIONS

| | |
|---|---|
| AC | against curvature |
| AFM | atomic force microscopy |
| AV | aortic valve |
| AVIC | aortic valve interstitial cell |
| CAVD | calcific aortic valve disease |
| CFA | collagen fiber architecture |
| CRGDS | adhesive peptide sequence (Cys-Arg-Gly-Asp-Ser) |
| ECM | extracellular matrix |
| $E_{eff}$ | effective stiffness |
| FE | finite element |
| FIB-SEM | focused ion beam scanning electron microscope |
| HSP47 | heat shock protein 47 |
| MA | micropipette aspiration |
| MMP | matrix metalloproteinase |
| MV | mitral valve |
| MVAL | mitral valve anterior leaflet |
| MVIC | mitral valve interstitial cell |
| NAR | nuclear aspect ratio |
| NOI | normalized orientation index |
| PEG | poly(ethylene glycol) |

| | |
|---|---|
| **PV** | pulmonary valve |
| **PVIC** | pulmonary valve interstitial cell |
| **RVE** | representative volume element |
| **α-SMA** | alpha-smooth muscle actin |
| **TGF-β1** | transforming growth factor beta 1 |
| **TV** | tricuspid valve |
| **TVIC** | tricuspid valve interstitial cell |
| **TVP** | transvalvular pressure |
| **VEC** | valve endothelial cell |
| **VIC** | valve interstitial cell |
| **WC** | with curvature |

# 1 INTRODUCTION

Proper blood flow through the heart is made possible through healthy and functioning heart valves. The heart has four valves: the mitral (MV), aortic (AV), pulmonary (PV), and tricuspid valve (TV). The MV and TV are atrioventricular valves and control blood flow from the atria to the ventricles, whereas the PV and AV are semilunar valves and regulate blood flow through the pulmonary artery and aorta, respectively. The MV and AV are on the left side of the heart and help in transporting oxygenated blood from the lungs to the rest of the body. On the right side of the heart, the TV and PV guide the flow of deoxygenated blood from the body to the lungs. Heart valves perform within a mechanically demanding environment and undergo tension, flexure, and shear stress throughout the cardiac cycle [1]. At the left side of the heart, the AV and MV experience greater transvalvular pressures (TVPs) (80 and 120 mmHg, respectively) than the PV and TV do at the right side of the heart (10 and 25 mmHg, respectively) [2]. Previous work has shown that heart valves experience mean peak strain rates of 300–400% $s^{-1}$ in the radial direction and 100–130% $s^{-1}$ in the circumferential direction [3–6]. The complex mechanical demands of the heart system highlight the need for repair and remodeling mechanisms of valvular tissues.

Heart valves are known to remodel and repair throughout life [7]. All heart valves contain leaflets composed of multilayered, high-functioning tissues designed to withstand the above driving stresses. For example, the AV has three distinct layers: the collagen-rich fibrosa, the spongiosa that contains a high concentration of proteoglycans, and the ventricularis that features dense collagen and elastin fiber networks [7,8]. In contrast, the MV has four distinct layers in the central leaflet region: the spongiosa, the fibrosa, the ventricularis, and a fourth layer called the atrialis that faces toward the atria. Recent evidence has shown that the spongiosa is essentially a specialized extension of the fibrosa and ventricularis layers, in terms of both mechanical functionality and extracellular matrix (ECM) components [9,10]. The unique organization of valve leaflet layers and underlying ECM components allow for the simultaneous and proper function of semilunar and atrioventricular heart valves, accounting for the differences in location within the heart and orientation to the direction of blood flow. Previous studies have quantified in detail the transmural structural heterogeneity in heart valve leaflet tissues and have elucidated the role that this unique and finely regulated ECM architecture plays in maintaining a homeostatic distribution of stress in vivo [11,12]. These regulatory functions occur at the hands of the underlying valve interstitial cells (VICs) embedded throughout all layers. VICs are mechanocytes and are sensitive to their mechanical environment. They display multiple, reversible phenotypes and remain quiescent under normal conditions and behave similarly to fibroblasts. In response to pathological alterations in force and hemodynamics, VICs display a myofibroblast phenotype. Valvular damage causes VICs to upregulate ECM biosynthesis and become contractile through the expression of alpha-smooth muscle actin (α-SMA). Moreover, further investigations have shown that the geometry, deformation, and biosynthetic behavior of VICs depend highly on local mechanical properties of the surrounding ECM [12–15].

Recent work has reinforced the hypothesis that VIC deformation is a major driver in regulating heart valve repair and remodeling mechanisms [16–18]. VIC response to deformation may be motivated by biological pressure to return to a homeostatic state through remodeling of the leaflet ECM [12,18]. However, little is known about the underlying mechanisms that make this possible. For example, it is unclear whether the intrinsic stiffness of VICs or the stiffness of the environment plays a role in this process. In addition, more work is necessary to link cellular deformations to biosynthetic activity and relate these observations to in vivo function. Multiscale approaches that utilize both experimental and computational techniques are thus required to study the complex biomechanical states of VICs [19].

## 2 MAJOR QUESTIONS AND CHALLENGES

It has long been hypothesized that deformation plays a critical role in regulating the biosynthetic activity of mechanocytes and, therefore, governs catabolic mechanisms. In the case of heart valves, cyclic loading causes the deformation of VICs, which in turn provides the mechanoregulation necessary for maintaining homeostasis. Alterations in force and hemodynamics can cause VICs to display an activated, myofibroblast phenotype and increase matrix remodeling activity. If activation persists, drastic changes in valve leaflet structure and function can occur, furthering the onset of disease.

In certain disease states, such as myxomatous MV degeneration and calcific aortic valve disease (CAVD), the pathological tissue structure plays a large role in the mechanoregulation of VIC biosynthetic activity. Myxomatous MV degeneration can occur due to alterations in mechanical stress and genetic abnormalities [20]. Some hallmarks of this disease include an increase in glycosaminoglycans within the spongiosa layer, degradation of collagen, and fragmentation of elastin networks. The net effect of the ECM remodeling causes the leaflet tissue to thicken and weaken. This in turn induces VICs to become activated and increase expression of proteolytic enzymes. In addition, VICs begin to proliferate within the spongiosa layer. These symptoms can result in MV regurgitation due to "floppy" MV leaflets. On the other hand, CAVD is caused by calcium accumulation onto AV leaflets. As the disease progresses, the valve leaflets thicken, and calcium nodules begin to form. This occurs due to VICs undergoing phenotypic transitions to an osteoblastic phenotype leading to the secretion of bone-like ECM [21]. Late stages of CAVD can cause the AV to be less efficient, resulting in aortic stenosis.

Myxomatous MV degeneration and CAVD highlight the role of VIC-ECM interactions in valvular disease. VIC deformation and thus biosynthetic regulation are dependent upon these interactions. Due to these reasons, a need exists to study VIC-ECM coupling to better understand valve disease in the hope of engineering effective therapeutics. A need also exists to elucidate the intricate relationship between VICs and valve endothelial cells (VECs). VIC phenotype is highly dependent upon chemokines and cytokines excreted by VECs. Factors like endothelin-1 and transforming growth factor beta 1 (TGF-β1) can directly affect the contractility and activation levels of VICs, respectively. Future efforts should also be directed toward investigating the synergistic effects of mechanical stimulations, VIC-ECM coupling, and VIC-VEC interactions on repair and remodeling mechanisms.

Studying the behavior of VICs under physiologically relevant conditions is of the utmost importance for developing accurate and trustworthy models. Popular methods among researchers to study VIC biology and mechanics include two-dimensional cultures and assays. Although these studies are insightful, they fall short of mimicking the complexity of the native VIC microenvironment and fail to elicit accurate behavior of VICs. To circumvent this issue, some researchers have resorted to using valve leaflet explants for mechanobiological studies and utilizing soft hydrogel matrices to seed and study VICs in 3D. Such environments allow for the examination of cellular deformation, specifically nuclear deformation, in response to mechanical loading. In recent years, nuclear aspect ratio (NAR), the dimensionless ratio between the nuclear major and minor axes, has been used as a cell-scale metric for correlating mechanical stimulation with biological function [13,18,22,23]. Analysis of NAR has been tremendously valuable in studying the underlying mechanobiology responsible for valve tissue remodeling and repair. However, rigorous elucidation of the role that cellular deformation plays in homeostatic regulation of normal and diseased valves remain a challenge. In addition, more work is needed in studying the role of pathological tissue structure in local VIC mechanoregulation. In this chapter, contemporary approaches for the study of VIC mechanobiology and the associated cellular phenomena are summarized.

# 3 ADVANCES IN INVESTIGATING VIC MECHANOBIOLOGY
## 3.1 ISOLATED CELL STUDIES
### 3.1.1 Overview
The mechanosensitive nature of VICs allow for the maintenance and repair of valvular tissue. Ex situ observations of valve leaflets from the four heart valves reveal that leaflets belonging to the left side of the heart (AV and MV) are far stiffer than

those from the right side of the heart (PV and TV) [24–26]. This is most likely due to the stark difference in the TVP between the two sides. It is hypothesized that since the left side of the heart experiences higher TVP, larger stresses are imposed on the underlying VICs, thus inducing them to form more developed cytoskeletal networks and increase collagen production. To test this hypothesis, techniques such as micropipette aspiration (MA) and atomic force microscopy (AFM) are used to measure the stiffness of VICs isolated from the leaflets of all four valves [24,25,27]. Protein assays are performed to complement the single-cell mechanical tests and provide insight into how VIC biosynthetic behavior varies between valves. These studies are a crucial first step in studying the biophysical state of VICs and how it is altered with respect to the mechanical demands of the cellular milieu.

### 3.1.2 MA studies of VIC biomechanical behaviors

VICs isolated from all four heart valves were cultured and used for MA stiffness measurements and protein quantification assays. For MA measurements, VICs were suspended in culture media and tested with micropipettes with inner diameters ranging from 6 to 9 μm (Fig. 1A). The pressure through the micropipette was adjustable and used to achieve consistent testing of VICs using the following protocol: (1) An initial tare pressure (~50 Pa for 60 s) was applied to capture cells and form a seal in between the cell and micropipette, (2) the pressure was increased to ~250 Pa and held constant for 120 s, and (3) the pressure was then increased to ~500 Pa and held for the last 120 s of the test. Extended experimental details can be found here [27]. The actual pressure applied at each phase was recorded, and images throughout the test were captured with a camera coupled with a bright-field microscope. From the images, the aspiration length of the cell was measured. VIC effective stiffness ($E_{eff}$) was determined through the use of a half-space model (punch model) that models the cell as an isotropic, elastic, incompressible half-space material [28] and was determined from the following equation:

$$E_{\text{eff}} = \Phi(\eta)[(3r)/(2\pi)](\Delta P/L) \tag{1}$$

where $\Phi(\eta)$ is a dimensionless parameter determined from the ratio of the micropipette inner radius to the wall thickness with a set value of 2.1, $r$ is the radius of the micropipette inner radius, and $\Delta P/L$ is the change in applied pressure divided by the aspiration length. Here, $E_{eff}$ is used as a convenient parameter to compare the intrinsic stiffness of the different VIC types.

It was observed experimentally that pulmonary and tricuspid valve interstitial cells (PVIC and TVIC, respectively) displayed a larger aspiration length in response to the applied pressure compared with mitral valve interstitial cells (MVICs) and aortic valve interstitial cells (AVICs). This translates to AVICs and MVICs being significantly stiffer than PVICs and TVICs (Fig. 1B). However, VICs from the same side of the heart (i.e., AVICs and MVICs) did not vary significantly in $E_{eff}$. It was found that the $E_{eff}$ of various VIC types was linearly related to the maximum TVP felt by each valve suggesting that VIC stiffness is, at the least, correlated to the stress present within the microenvironment (Fig. 1C).

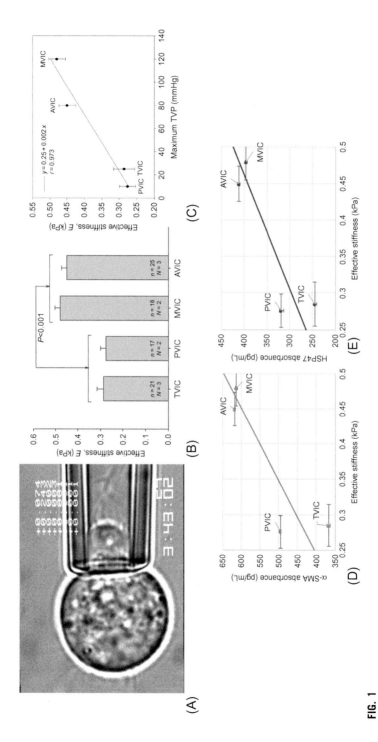

**FIG. 1**

(A) Mechanical evaluation of a VIC via micropipette aspiration. (B) The effective stiffness of the various VIC types evaluated from micropipette aspiration. (C) VIC effective stiffness with respect to the maximum transvalvular pressure (TVP) of the different heart valves. (D) α-SMA and (E) HSP47 absorbance determined via ELISA plotted against the effective stiffness of the various VIC types.

*Modified from Merryman WD, Youn I, Lukoff HD, Krueger PM, Guilak F, Hopkins RA, et al. Correlation between heart valve interstitial cell stiffness and transvalvular pressure: implications for collagen biosynthesis. Am J Physiol Heart Circ Physiol 2006;290(1):H224–31.*

ELISA protein assays were performed on both in vitro and in situ VICs specifically to assess the levels of α-SMA and heat shock protein 47 (HSP47) among the different valve types. α-SMA is present within VICs displaying an activated phenotype and is hypothesized to alter the cellular stiffness and play a role in collagen biosynthesis, thus making it a protein of interest. Production of type I collagen has been shown to be dependent upon HSP47, and therefore, HSP47 serves as a surrogate for the level of collagen biosynthesis [29]. Analysis of the in vitro ELISA results show higher levels of SMA and HSP47 among valve cells from the left side of the heart (Fig. 1D and E, respectively). In situ results display a similar trend as well. When taken together, these results suggest that VICs residing within the MV and AV adapt to the higher levels of stress through the regulation of cytoskeletal organization and higher collagen production rates.

### *3.1.3 Microindentation studies*

AFM is another method used to measure the stiffness of VICs through deforming the cells with a cantilever [30]. In short, the test samples were prepared by seeding VICs in a monolayer onto collagen-coated glass coverslips.

AFM measurements were performed using the "tapping mode" in which the cantilever makes ~70 indentations in a rectilinear grid sampling pattern for every VIC (Fig. 2A). Stiffness ($E$) is calculated from this method using the following equation:

$$E = \frac{F(1-v^2)}{\pi \cdot \Phi(\delta)} \quad (2)$$

where $v$ is Poisson's ratio (0.5 due to incompressibility assumption), $F$ is the force applied on the VIC by the AFM tip, and $\Phi(\delta)$ is a function of the conical tip probe geometry defined as [31]:

$$\Phi(\delta) = \delta^2 \left[ \frac{2 \cdot \tan(\alpha)}{\pi^2} \right] \quad (3)$$

where $\alpha$ represents the probe opening angle (35 degrees) and $\delta$ denotes the indentation depth of the probe. Incorporation of Eq. (3) simplifies Eq. (2) to the following:

$$E = \frac{F}{0.594 \cdot \delta^2} \quad (4)$$

$E$ was then determined from fitting the experimental data to Eq. (4) (Fig. 2B). Similar to the previously mentioned MA studies, $E$ serves as a parameter for comparison between the various VIC types.

The $E$ values obtained from ~70 AFM measurements (Fig. 2C) were averaged for each VIC, and it was shown that AVICs were approximately twice as stiff as PVICs (Fig. 2D), consistent with the observations made with MA. However, VIC stiffness reported from the AFM measurements was approximately 100 times larger than those obtained from MA and this observation is consistent with studies performed on other cell types [32–35]. The drastic differences may be owing to the assumptions of the models used to describe VIC mechanical behavior and the difference in

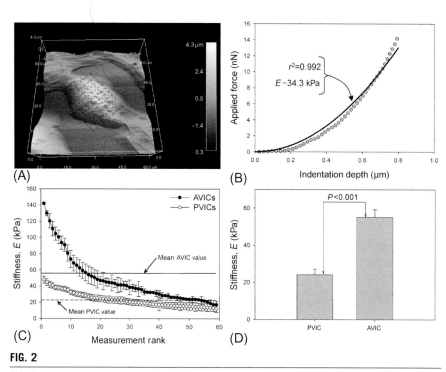

**FIG. 2**

(A) Height map developed from atomic force microscopy measurements of a VIC. (B) Representative applied force versus indentation depth generated from atomic force microscopy measurements. The experimental data is fit to Eq. (4), and E is backed out from the resulting fit. (C) AVIC and PVIC stiffness computed from atomic force microscopy measurements ranked in order of magnitude. (D) Average AVIC and PVIC stiffness.

*From Merryman WD, Liao J, Parekh A, Candiello JE, Lin H, Sacks MS. Differences in tissue-remodeling potential of aortic and pulmonary heart valve interstitial cells. Tissue Eng 2007;13(9):2281–9.*

deformation methods. AFM utilizes a localized force to make measurements, whereas MA tests a larger region. Intuitively, a localized measurement would be largely affected by the stiffness of nearby cellular components like local stress fibers and the nucleus, whereas measurements made in larger regions would consider cell membrane properties and the local cytoskeleton network. Although deviations are present in the absolute stiffness values from both techniques, the general trend observed among different types of VICs is consistent.

### 3.1.4 Isolated VIC mechanical models

The disparity between stiffness measurements from MA and AFM may arise from not only the difference in deformation modes but also the activation state of the VICs and the corresponding levels of α-SMA expression during the experiment. During MA, VICs are presumably in an inactivated state due to the lack of adhesion, and during AFM experiments, they are assumed to be highly activated resulting from adhesion on a 2D substrate. The disparity between stiffness measurements arising from different experimental techniques and VIC activation states suggest that analysis of the effective stiffness of VICs alone is insufficient and a need exists for more rigorous methods to better understand VIC mechanobiology. To accomplish this, isolated VIC mechanical models were developed, which incorporated the orientation of the cytoskeletal network, the relative expression of α-SMA stress fibers, and the mechanical properties of the nucleus [24,25].

The model was designed to estimate the effective mechanical behaviors of the major VIC subcellular components by integrating the MA and AFM experimental data from both AVICs and PVICs (Fig. 3). Analysis of the MA data required consideration of the cellular components assumed to most likely contribute during the experiments, namely, the passive stress fibers and the cytoplasm. Modeling the respective contributions of these subcellular components using a shear modulus allows for a single parameter to represent their mechanical state, simplifying determination of differences between VIC types. The first step in developing this model is

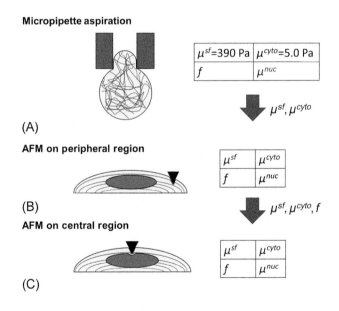

**FIG. 3**

(A) The data from micropipette aspiration experiments is used to calibrate the shear modulus of the stress fibers ($\mu^{sf}$) and the cytoplasm ($\mu^{cyto}$). (B) Using the previously determined parameters and atomic force microscopy data from peripheral regions of VICs, the contraction force ($f$) is computed. (C) The shear modulus of the nucleus ($\mu^{nuc}$) is calculated last, incorporating all the previously determined parameters and atomic force microscopy measurements made over the nucleus.

*From Sakamoto Y, Buchanan RM, Sacks MS. On intrinsic stress fiber contractile forces in semilunar heart valve interstitial cells using a continuum mixture model. J Mech Behav Biomed Mater 2016;54:244–58.*

taken from the MA data analysis [27], where the total Cauchy stress $T$ was considered the sum of the passive cytoskeletal and passive (i.e., inactive) stress fibers:

$$T = T_{cyto} + T_{sf} \tag{5}$$

Next, from the AFM experiments, contributions from the cytoplasm, passive stress fibers, stress fiber active stress ($f$), and the nucleus were considered in the total Cauchy stress using

$$T = T_{cyto} + T_{sf} + T_{f} + T_{nuc} \tag{6}$$

For parameter optimization, the data from the MA measurements was first used to back out the shear modulus of the stress fibers ($\mu^{sf}$) and the cytoplasm ($\mu^{cyto}$) (Fig. 3A). Briefly, the cytoskeleton was modeled as an isotropic hyperelastic material using a nearly incompressible neo-Hookean model. The passive component of individual α-SMA fibers was modeled as a transversely isotropic material, and both these components were incorporated into the expanded total Cauchy stress equation used for MA analysis:

$$T = 2\frac{1}{J}F\frac{\partial}{\partial C}\left[\frac{\mu^{cyto}}{2}(\bar{I}_1 - 3) + \frac{1}{2}K(\ln J)^2\right]F^T$$
$$+ \frac{1}{4\pi}\int_0^{2\pi}\int_0^{\pi}\left[H(I_4 - 1)2\frac{I_4}{J}\frac{\partial}{\partial I_4}\left(\frac{\mu^{sf}\overline{\varphi}^{sf}}{2}(I_4 - 1)^2\right)m\otimes m\right]\sin\theta d\theta d\phi. \tag{7}$$

The parameters $\mu^{cyto}$ and $\mu^{sf}$ were calibrated through minimization of the least square error between the simulated and experimental aspiration lengths while taking into consideration the applied pressure and normalized α-SMA expression levels ($\overline{\varphi}^{sf}$).

After a proper calibration of these parameters, AFM data obtained on peripheral regions of VICs were used to back out the contractile force ($f$) of the stress fibers (Fig. 3B) using the following expanded total Cauchy stress equation for AFM analysis:

$$T = 2\frac{1}{J}F\frac{\partial}{\partial C}\left[\frac{\mu^{cyto}}{2}(\bar{I}_1 - 3) + \frac{1}{2}K(\ln J)^2\right]F^T$$
$$+ \int_0^{2\pi}\int_0^{\pi}\left[\Gamma_t(m)H(I_4 - 1)2\frac{I_4}{J}\frac{\partial}{\partial I_4}\left(\frac{\mu^{sf}\overline{\varphi}^{sf}}{2}(I_4 - 1)^2\right)m\otimes m\right]\sin\theta d\theta d\phi \tag{8}$$
$$+ \int_0^{2\pi}\int_0^{\pi}\left[\Gamma_t(m)f\cdot\overline{\varphi}^{sf}\frac{I_4}{J}m\otimes m\right]\sin\theta d\theta d\phi$$

After finding the parameter $f$, AFM data on the central region (above the nucleus) is used to calibrate the respective shear modulus of the nuclei ($\mu^{nuc}$) (Fig. 3C) by replacing the first term of Eq. (8) with

$$T = 2\frac{1}{J}F\frac{\partial}{\partial C}\left[\frac{\mu^{nuc}}{2}(\bar{I}_1 - 3) + \frac{1}{2}K(\ln J)^2\right]F^T \tag{9}$$

and minimizing the least square error between simulated and experimental indentation depth versus force data. Detailed information on the formulation of the model can be found in Ref. [24].

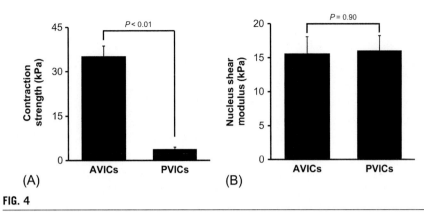

**FIG. 4**

(A) The contraction strength of AVICs and PVICs backed out from the isolated VIC mechanical models. The model predicts that AVICs display a larger contraction strength than PVICs. (B) There was no significant difference between the nucleus shear modulus of AVICs and PVICs as determined from the model.

*From Ayoub S, Ferrari G, Gorman RC, Gorman JH, Schoen FJ, Sacks MS. Heart valve biomechanics and underlying mechanobiology. Compr Physiol 2016;6(4):1743–80.*

Simulations demonstrated that AVICs displayed a ~10 times stronger contractile force than PVICs, highlighting the model's ability to account for the higher levels of α-SMA expression within AVICs (Fig. 4A). Similarities in nuclear stiffness between AVICs and PVICs were also captured by the model (Fig. 4B). This suggests that any major difference in stiffness between the VIC types is most likely due to the stiffness of α-SMA and not the nucleus. Modifications to the model were made to consider F-actin and α-SMA expression levels along with stress fiber orientation, strain rate effects, and the relation between stress fiber length and tension [25]. The model outputted AVIC contraction strength under various activation scenarios, and it was found that contraction strength was greatest when the cells were treated with activation inducing TGF-β1 and then with 90mM of potassium chloride (KCl). The adjusted model was also able to showcase that inhibiting stress fiber formation through the use of cytochalasin D was effective in reducing cellular contraction strength.

These studies demonstrate that isolated cell models are beneficial tools for discovery and offer an explanation for the difference in stiffness levels observed between VIC types and also between testing methods. These models mark a crucial first step in capturing the mechanical behavior of VICs. In addition, they have the potential to be incorporated into multiscale models and be adapted to represent the mechanics of VIC populations. Future work will focus on coupling VIC mechanical models with agent-based models to thoroughly capture the biomechanical and biosynthetic cellular events occurring within valve leaflet tissues.

## 3.2 IN SITU TISSUE LEVEL EVALUATION OF VIC CONTRACTION BEHAVIORS

### 3.2.1 Overview

To elicit accurate biological behavior, physiologically relevant forms of deformation are desirable in mechanobiological studies. Planar membrane tension and flexural deformation testing are the two most suitable methods that aim to mimic the natural strain experienced by heart valve leaflets. While biaxial tensile testing has proved highly effective for characterizing the mechanical properties of the native leaflet and individual tissue layers [10,11], flexure-based approaches are advantageous in two regards: (1) They are extremely sensitive at low stresses and strains, and (2) they can reflect differences in ECM components and architecture throughout the different layers of the valve leaflets. In addition, beam-bending models, such as the Euler-Bernoulli relation [36], can be utilized to determine an instantaneous effective modulus $E_{eff}$ of the valve leaflet tissue to gauge VIC contraction behavior and biophysical state from tissue-level studies.

### 3.2.2 VIC-ECM coupling and bidirectional valve leaflet bending response

Flexural deformation tests have been used to show significant increases in $E_{eff}$ arising from cellular contraction (90 mM KCl). This form of testing has also been used to capture a decrease in $E_{eff}$ with the loss of the cellular basal tonus (Fig. 5) [37].

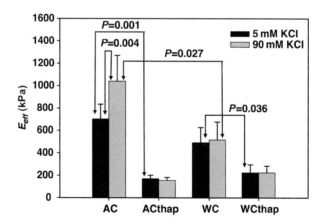

**FIG. 5**

Tissue effective stiffness of normal and contraction inhibited (thap) aortic valve leaflet test specimens bent with (WC) and against (AC) curvature after treatment with 5 mM KCl (normal) and 90 mM KCl (hypertensive). The tissue effective stiffness was computed from the derivative of the moment versus curvature plots. The effective stiffness increased significantly in the against curvature direction after treatment with 90 mM KCl.

*From Merryman WD, Huang HY, Schoen FJ, Sacks MS. The effects of cellular contraction on aortic valve leaflet flexural stiffness. J Biomech 2006;39(1):88–96.*

In short, this is done through tracking dots placed along the transmural cross section of the test specimen excised from AV leaflets (Fig. 6A) to keep track of its change in curvature ($\Delta \kappa$) from an initial reference configuration. The moment ($M$) for each $\Delta \kappa$ is calculated throughout the test, and $M$ versus $\Delta \kappa$ plots are generated. From these plots, the tissue effective stiffness $E_{eff}$ is determined through the use of the Euler-Bernoulli relation:

$$\frac{M}{I} = E_{eff} \Delta \kappa \tag{10}$$

where $I$ is the second moment of inertia and a function of the sample dimensions $\left(I = \frac{t^3 w}{12}\right)$. For more experimental details, the reader is referred to the following references [37–39]. The moment-curvature plots also revealed the direction-dependent bending response of AV leaflets (Fig. 6B and D). Valve leaflet tissues were bent "with" and "against" the natural leaflet curvature (WC and AC, respectively, Fig. 6C), and it was observed that the tissue modulus was greatest when bent in the AC direction. This is so due to the highly organized and collagen-rich fibrosa layer being under tension, causing a rise in $E_{eff}$. When bent in the WC direction, the ventricularis is under tension, and as a result, the tissue effective modulus was much lower owing to the collagen and compliant elastin fibers residing within that layer. The bending response of AV leaflets was found to be linear in both bending directions (Fig. 6B and D), which is unlike that of the nonlinear planar tensile tissue response [40,41].

### 3.2.3 Interlayer micromechanics of the AV leaflet

Subsequent flexure studies have focused on examining the intricate interlayer micromechanics of the AV leaflet [9]. The behavior of the layers with respect to one another was of special interest to delineate the way valve leaflets deform, either as a bonded unit or separately. To better understand this phenomenon, analysis of the transmural strain of AV leaflets was conducted. Microscopic India ink tracking dots were airbrushed onto the transmural wall of AV leaflet testing specimens. Opposite of this side were macrolevel tracking dots used for the assessment of the bending test and development of moment-curvature plots. Both sets of tracking dots were tracked throughout the flexural deformation test at $\Delta \kappa$ ranges of 0 (reference image), 0.1, 0.2, and 0.3 mm$^{-1}$. The tissue-level deformation gradient **F** was computed from the reference and deformed coordinates of the microscopic markers and decomposed into stretch and rotation tensors to remove rigid body deformations, leaving only axial stretch information. The total axial stretch ($\Lambda_1$) was assessed across the normalized transmural cross section of the test specimens with $\Lambda_1 > 1$ representing tension, $\Lambda_1 < 1$ representing compression, and $\Lambda_1 = 0$ representing the neutral axis in which no deformations occur.

By plotting $\Lambda_1$ against the normalized leaflet thickness, a linear trend was observed (Fig. 7A and B) for both WC and AC directions. This observation confirms that the spongiosa does not allow for shearing to occur between the fibrosa and ventricularis under physiologically relevant deformations, a phenomenon that some

**262 CHAPTER 8** Mechanobiology of the heart valve interstitial cell

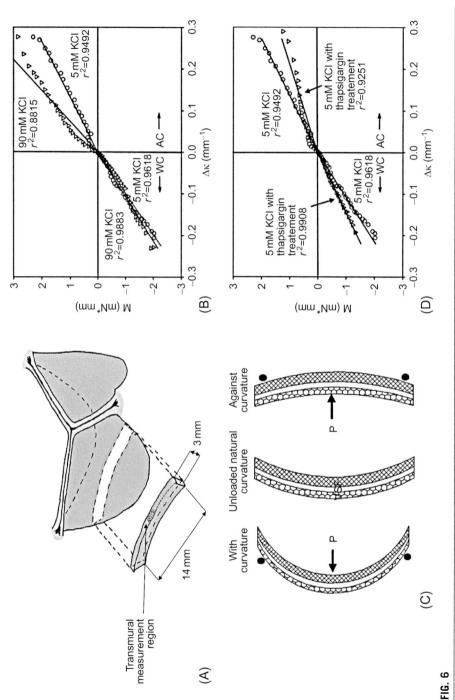

**FIG. 6**

(A) Testing specimens are excised from aortic valve leaflets. (B) Representative moment versus curvature plot of an aortic valve leaflet specimen under normal (5mM KCl) and hypertensive (90mM KCl) conditions. (C) Broad overview of mechanical testing experimental protocol subjecting leaflet specimens to bending deformations with and against curvature. The aortic valve layers (V, ventricularis; S, spongiosa; and F, fibrosa) were noted to ensure proper orientation of the valve specimen while testing. (D) Representative moment versus curvature plot of aortic valve leaflet specimen treated with a stress fiber inhibiting solution (thapsigargin) under normal (5mM KCl) and hypertensive (90mM KCl) conditions.

*From Merryman WD, Huang HY, Schoen FJ, Sacks MS. The effects of cellular contraction on aortic valve leaflet flexural stiffness. J Biomech 2006;39(1):88–96.*

# 3 Advances in investigating VIC mechanobiology 263

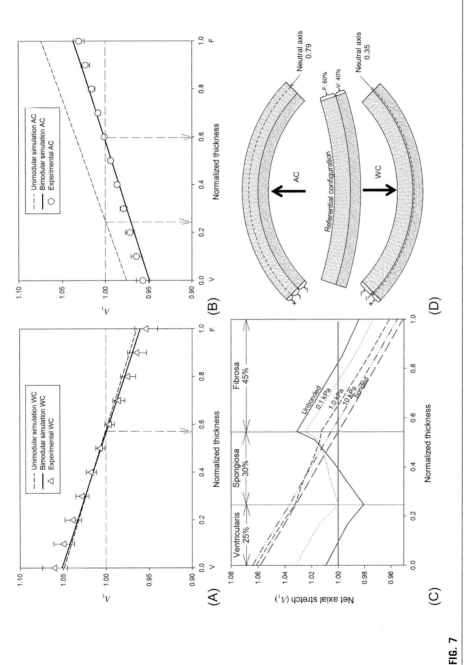

**FIG. 7**

Axial stretch plotted against the normalized leaflet thickness in the (A) with curvature and (B) against curvature directions. Note that a unimodular material model is insufficient in capturing the bending response in both directions. Thus, a bimodular material model is needed. (C) Results from the simulated interlayer bonding parametric study. For measurable differences to occur between the transmural strain of the heart valve layers, notably the ventricularis and the fibrosa, the spongiosa must have a shear modulus of <1 kPa. (D) The location of the neutral axis in both the against curvature and with curvature bending directions.

*Modified from Buchanan RM, Sacks MS. Interlayer micromechanics of the aortic heart valve leaflet. Biomech Model Mechanobiol 2013.*

believed to occur before [42–44]. If the spongiosa did allow for shearing between the layers, each layer would exhibit its own neutral axis when undergoing flexure. This was discovered through performing interlayer bonding simulations of the AV leaflet. In short, an AV leaflet finite element (FE) model that considered the relative thickness of each layer (fibrosa, 45%; spongiosa, 30%; and ventricularis, 25%) as determined from histological measurements was developed in COMSOL Multiphysics v4.3 (COMSOL, Burlington, MA). A parametric study was performed on the shear modulus of the spongiosa layer ($\mu_s$). $\mu_s$ was set at 1 Pa, 0.1, 1, 10, and 45 kPa, and it was determined that for the spongiosa to have a significant difference in axial stretch from the fibrosa and ventricularis, $\mu_s$ would have to be <1 kPa (Fig. 7C). No discernable difference was found experimentally, and instead, evidence suggests that AV leaflets function as a single bonded unit. Furthermore, analysis of histological sections revealed gradual transmural changes in ECM components throughout the AV leaflet layers that further support the conclusion that AV leaflets behave as a functionally graded bonded unit containing contiguous features [10].

It was also discovered that a unimodular material model was insufficient in capturing the bidirectional bending behavior of AV leaflets. Instead, a bimodular material model is needed to accomplish this, which further highlights the complex, heterogeneous ECM components and architecture across valve tissue layers. In addition, it suggests that the spongiosa can be viewed as a contiguous extension of both the fibrosa and ventricularis in that it does not play a major mechanical role in bending deformations. The final bimodular incompressible neo-Hookean material model used is as follows:

$$W^\pm = \frac{\mu_L^\pm}{2}(I_1 - 3) - p(I_3 - 1) \tag{11}$$

where $\mu_L^\pm$ represents four total moduli. $\mu_F^\pm$ and $\mu_V^\pm$ are the fibrosa and ventricularis shear moduli, respectively, in tension (+) and compression (−). Eq. (11) successfully captured the bidirectional bending behavior of the AV leaflet and was able to estimate layer- and deformation-specific shear moduli. This bimodular model would later be adjusted and incorporated into a downscale model of the AV leaflet, which we will discuss in the next section.

Computation of the axial stretch during AV bending tests revealed that the neutral axis (NA) is located at different locations depending on whether the test is performed WC or AC (Fig. 7D). In the WC direction, the NA shifts toward the ventricularis, ~35% of the normalized thickness. This implies that the entire fibrosa, located above the NA, is in compression. In the AC direction, the NA shifts toward the fibrosa, ~79% of the normalized thickness. This suggests that the fibrosa is far stiffer while under tension than in compression. The following results are due to the distinct ECM components within each layer that allow the leaflets to be less stiff when the valve opens (WC) and stiffer when the valve closes (AC) to ensure proper blood flow through the heart.

### 3.2.4 Downscale model of the VIC within ECM

Elucidating the biomechanical state of the VIC in its natural environment is desirable to shed light on how it effects the valve leaflet tissue. However, it is extremely difficult to observe how VICs are responding mechanically in situ. To circumvent this issue, a FE model of the AV leaflet under bending deformations was developed in Abaqus version 6.14 (Dassault Systemes, Johnston, RI, United States) to estimate layer-specific VIC contractility, connectivity, and stiffness [26]. The FE model was designed to consider microstructural details like AVIC size, shape, distribution, and orientation as determined from histological data.

The model features a macro and micro (downscale) component (Fig. 8). Representative volume elements (RVEs) were optimized statistically for each layer to represent the native 3D structure. A brief overview of the formulation of the FE model follows.

The macro component is formulated to estimate layer- and contractile state-dependent mechanics of the AV leaflet under bending deformations. The AV leaflet specimen is simulated as a bilayered and bimodular neo-Hookean isotropic nearly incompressible material

$$W^{\pm} = \frac{{}_{state}\mu_{Macro}^{layer\pm}}{2}(I_1 - 3) - p(I_3 - 1) \tag{12}$$

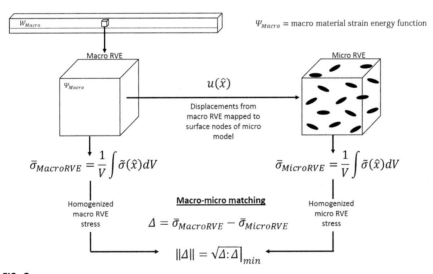

**FIG. 8**

Overview of macro-micro finite element methodology. Macrolevel simulations are performed, and the resulting displacements ($u(\hat{x})$) from the macro RVE are mapped to the micro RVE as boundary conditions. The average stress across the macro ($\tilde{\sigma}_{MacroRVE}$) and micro ($\tilde{\sigma}_{MicroRVE}$) RVEs are then compared, and finite element simulations are performed until the difference between both values are minimized.

From Buchanan RM. An integrated computational-experimental approach for the in situ estimation of valve interstitial cell biomechanical state. Austin, USA: The University of Texas at Austin; 2016.

where $_{state}\mu_{Macro}^{layer\pm}$ is the shear modulus of a specific AV leaflet layer under compression (−) and tension (+), $I_1$ and $I_3$ are the first and third invariants of the left Cauchy-Green deformation tensor $\mathbf{C} = \mathbf{F}^T\mathbf{F}$, respectively, and $p$ is the Lagrange multiplier used to enforce incompressibility. Here, the subscript "state" denotes the contractile state of the AVICs (normal, hypertensive, and inactivated) during testing and is used for the comparison of shear moduli between the contractile states. From the macro model, $_{state}\mu_{Macro}^{layer\pm}$ is determined from matching experimental moment-curvature data obtained from flexure tests to Eq. (12). The simulated displacements of the macro RVE from the center of the specimen (the point of greatest moment and curvature) are then mapped to the surface nodes of the micro model (Fig. 8).

Within the framework of the micro model, AVIC stiffness, contractility, and connectivity are built in as variable parameters (Fig. 9). AVICs were modeled using neo-Hookean, ellipsoidal inclusions within the micro RVE. The contractile capability of AVICs was incorporated through the use of isometric thermal expansion to mimic cellular contraction by inducing isothermal deformations of the ellipsoidal inclusions [45]. Consistent with the macro model, a bilayered and bimodular neo-Hookean

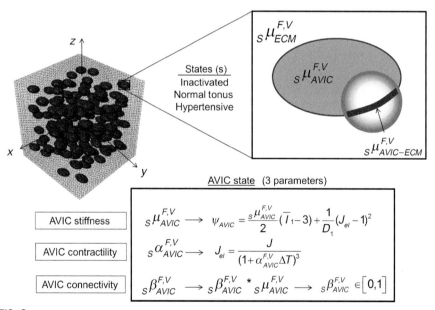

**FIG. 9**

The micro RVE with ellipsoidal VIC inclusions used to back out AVIC stiffness, contractility, and connectivity.

From Buchanan RM. An integrated computational-experimental approach for the in situ estimation of valve interstitial cell biomechanical state. Austin, USA: The University of Texas at Austin; 2016.

isotropic nearly incompressible material model was used in the micro component as well

$$W^{\pm} = \frac{\mu_{AVIC}^{layer}}{2}(I_1 - 3) + \frac{1}{D_1}(J^{el} - 1)^2 \quad (13)$$

where $\mu_{AVIC}^{layer}$ is the layer-specific shear modulus of AVIC inclusions, $D_1$ is the constant defined by $\frac{2}{\kappa}$ where $\kappa$ is the bulk modulus of the material, and $J^{el}$ is the elastic volume ratio computed through

$$J^{el} = \frac{J}{(1+\varepsilon^{th})^3} \quad (14)$$

where $\varepsilon^{th}$ is the thermal expansion strain as a function of temperature ($T$) and the thermal expansion coefficient ($\alpha$)

$$\varepsilon^{th} = \alpha \Delta T \quad (15)$$

In the simulation, $\alpha$ is correlated with AVIC contraction and is parametrically altered, while $\Delta T$ is kept constant.

AVIC connectivity to the ECM is represented with a thin interface boundary around each AVIC inclusion (Fig. 10). The modulus of this interface is equivalent to $\mu_{AVIC}^{layer}$ and is parametrically altered by multiplying it with the variable $\beta_{AVIC}^{F,V\pm}$ ($\beta_{AVIC}^{F,V\pm} \in [0,1]$) with $\beta = 1$ denoting complete binding of the AVIC inclusions to the ECM and $\beta = 0$ denoting no binding. The average stress across the macro and micro RVE are calculated and a FE simulation is performed by adjusting AVIC stiffness ($\mu_{AVIC}^{layer}$), contractility ($\alpha$), and connectivity ($\beta_{AVIC}^{F,V\pm}$) until the average stress across both RVEs match. Once an acceptable match is achieved, the estimated parameters are reported. For more detailed information regarding the formulation of the model, readers are directed to the following source [26].

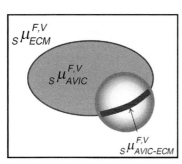

**FIG. 10**

The interfacial layer between the VIC inclusions and the simulated ECM. The modulus of the interfacial layer is multiplied by the term $\beta$ ($\beta \in [0,1]$) to simulate various levels of VIC connectivity throughout the different layers of the AV leaflet.

From Buchanan RM. An integrated computational-experimental approach for the in situ estimation of valve interstitial cell biomechanical state. Austin, USA: The University of Texas at Austin; 2016.

These FE models were highly beneficial tools to estimate mechanical parameters that are impractical or even impossible to measure experimentally. Such tools are especially useful in the studies of native tissues with poor optical properties that disallow the direct assessment of VIC behavior. In addition, modeling also provides techniques to assess the layer-specific response of VICs.

### 3.2.5 Use of 3D hydrogels for mechanobiological studies

Native valve leaflet tissue studies are highly desirable because they elicit the most accurate behavior of VICs. Although extremely insightful, native tissue environments are limited mainly due to (1) their inability to be tuned to answer specific mechanobiological questions and (2) the difficulty involved with detailed imaging of cellular and subcellular components. Thus, a need exists for a 3D platform that mimics the native micromechanical environment while featuring favorable optical properties for visualization. One avenue that has shown promise in this regard are peptide-modified poly(ethylene glycol) (PEG) hydrogels [46]. These matrices consist of norbornene functionalized PEG molecules that bind covalently with peptides that contain cysteine residues via thiol:ene reactions. Incorporation of the adhesive peptide sequence Cys-Arg-Gly-Asp-Ser (CRGDS) and matrix metalloproteinase (MMP) degradable peptide cross-linkers allows for cellular binding, growth, and microenvironmental remodeling. Through the use of the following equation

$$G = \frac{\rho_{xL} RT}{Q^{1/3}} \tag{16}$$

where $\rho_{xL}$ is the cross-linking density, $R$ is the ideal gas constant, $T$ is the temperature, and $Q$ is the volumetric swelling ratio, the shear modulus ($G$) can be estimated and fine-tuned by adjusting the amount of MMP degradable cross-linkers and essentially the thiol:ene ratio of the hydrogel chemistry [47]. From $G$, the elastic modulus ($E$) can be computed from the equation:

$$E = 2G(1 + v) \tag{17}$$

where $v$ is Poisson's ratio of the hydrogel material, which is assumed to be 0.5 (incompressible). In addition to tunable mechanics, the PEG hydrogel matrices also offer modulation of adhesion levels through adjustment of the concentration of CRGDS peptide sequences. Coupled with tunable mechanics, the hydrogel system can be utilized to mimic healthy and diseased environments.

Preliminary work has been done to quantify the contractility of AVICs within 10 kPa hydrogels through the use of end-loading, flexural deformation testing. The experimental testing protocol was adapted from Ref. [37]. The VIC-embedded hydrogels were treated with a normal (5 mM KCl) solution and subjected to flexural deformations. The testing solution was then switched for a hypertensive solution (90 mM KCl) to elicit VIC active contraction, and the specimens were tested again. Finally, contractility was halted by essentially killing the cells with 70% methanol before testing the gels a final time.

The moment-curvature plots (Fig. 11A) produced from these tests display higher moment values resulting from the hypercontractile state and lower moment values in the inactive state, compared with the normal condition. Initial observations revealed that the hydrogel material displayed a nonlinear moment-curvature response. Thus, the moment-curvature data were modeled using a second-order polynomial of the form

$$\frac{M}{I} = a\Delta\kappa^2 + b\Delta\kappa + c \qquad (18)$$

This analysis technique was borrowed from Ref. [38]. Through the use of the Euler-Bernoulli relation (Eq. 10) and the derivative of the second-order polynomial (Eq. 18), a linear relation between the hydrogel $E_{eff}$ and $\Delta\kappa$ was established. $E_{eff}$ at small strains ($\Delta\kappa = 0$ mm$^{-1}$) was used as a parameter for comparison between the contractile states, and it was observed that the hydrogel was stiffer when the embedded VICs were in a hypertensive state and less stiff when they were inactivated, compared with the control condition (Fig. 11B). These preliminary results reflect the appreciable cell-material interactions and show great promise in using tunable hydrogel environments for future mechanobiological studies. In addition, hydrogel environments are an intriguing tool for the development of downscale FEM models similar to the one mentioned in the prior section to estimate the biomechanical state of VICs. Another avenue of interest involves taking advantage of the optical clarity of the hydrogels to observe cellular and subcellular events and capturing them within agent-based models.

## 3.3 UNIAXIAL PLANAR STRETCH BIOREACTORS

### 3.3.1 Overview

Planar tensile bioreactors allow for highly controlled, mechanical conditioning of valve leaflet tissues [18,48–51]. Such systems allow for the emulation of hypophysiological, normal, and hyperphysiological strain levels to study how tissue-level deformations drive the biosynthetic response of VICs. To maximize relevance of these studies, a macro-micro FE model is employed to interpret in vitro findings and relate them to in vivo functional states. In the following section, we summarize a novel experimental-computational approach to link VIC biosynthetic response to cellular deformation and explain the implications this may have on surgically repaired MVs.

### 3.3.2 Bioreactor design

MV anterior leaflets (MVALs) were excised from porcine hearts, and rectangular specimens were dissected from the clear zone measuring 11.5 mm in the circumferential direction and 7.5 mm in the radial direction (Fig. 12A). A tissue strip bioreactor was engineered based off previous designs [48–51]. The system featured an environmental specimen chamber that housed the tissue specimens (Fig. 12B). Metallic springs were used to pierce the MV leaflet specimens along the radial width on each

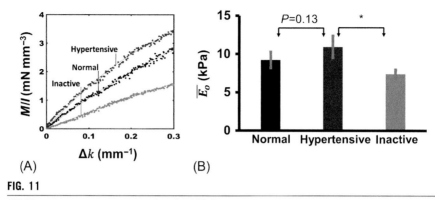

**FIG. 11**

(A) Moment versus curvature plots for VIC hydrogels under hypertensive, normal, and inactive conditions. (B) The average initial effective stiffness for VIC hydrogels under hypertensive, normal, and inactive conditions obtained from the initial slope of nonlinear moment versus curvature plots.

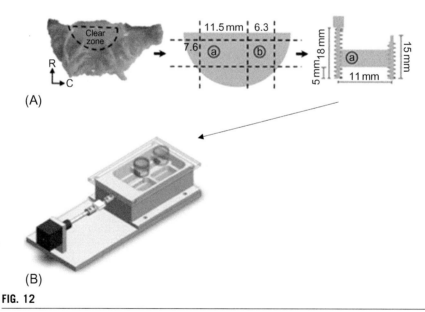

**FIG. 12**

(A) A mitral valve anterior leaflet excised into two distinct sections used for (a) mechanical conditioning (b) protein quantification. (B) A uniaxial tissue strip tensile bioreactor used to impose controlled deformations of test specimens.

*From Ayoub S, Lee C-H, Driesbaugh KH, Anselmo W, Hughes CT, Ferrari G, et al. Regulation of valve interstitial cell homeostasis by mechanical deformation: implications for heart valve disease and surgical repair. J R Soc Interface 2017;14 (135):20170580.*

side to establish five points of contact with proximity to one another. The spring was used to attach one side of the specimen to a stationary post within the environmental chamber and the other side to a metal rod connected to a linear actuator used to impose deformations. Cyclic stretch tests were performed for 48 h before downstream assessment of the test specimens.

### 3.3.3 Collagen fiber architecture alters nuclear and cytoplasmic geometry

Small-angle light scattering was used to assess the collagen fiber architecture (CFA) of the test samples and return a normalized orientation index (NOI) with an NOI value of 100% representing uniformly aligned collagen networks and 0% denoting a completely random fiber orientation. Extended experimental details can be found here [52].

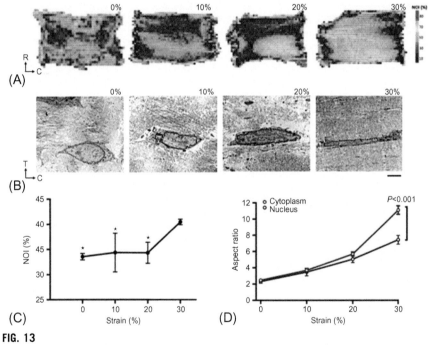

**FIG. 13**

(A) Normalized orientation index of the collagen fiber architecture with respect to increasing strain levels. (B) Transmission electron micrographs of MVICs displaying nuclear and cytoplasmic deformation due to increasing strains. (C) Quantification of the normalized orientation index shows that a drastic increase in collagen alignment occurs at 30% strain. (D) The nuclear and cytoplasmic aspect ratio of MVICs with respect to strain level. Note how the cytoplasmic aspect ratio decouples from the nuclear aspect ratio at 30% strain.

*From Ayoub S, Lee C-H, Driesbaugh KH, Anselmo W, Hughes CT, Ferrari G, et al. Regulation of valve interstitial cell homeostasis by mechanical deformation: implications for heart valve disease and surgical repair. J R Soc Interface 2017;14(135):20170580.*

CFA orientation appeared to increase drastically at 30% strain (Fig. 13A and C). Transmission electron microscopy revealed that NAR (ratio between major and minor axis) and cytoplasmic aspect ratio increased in a similar fashion at 30% strain (Fig. 13B and D). It was also observed that the cytoplasmic aspect ratio closely resembled the NAR at lower strains (0%–20%) but decouples from the NAR at hyperphysiological strain levels (30%, Fig. 13D). This phenomenon provides evidence that MVICs may behave differently under nonphysiological strain levels.

### 3.3.4 Large strains cause MVIC activation

RT-qPCR revealed a linear trend between α-SMA and type I collagen gene expression in relation to increasing strain levels suggesting MVICs become activated in response to large strains (Fig. 14A and B). Consistent with these observations, colorimetric assays displayed a significant increase in collagen production within the 30% strain group after 48 h compared with measurements made postmortem (Fig. 14D). An increase in sulfated GAG content was also observed in this group (Fig. 14C). Interestingly, elastin content decreased at higher strains possibly due to an increase in proteolytic activity of MMPs, TIMPs, and cathepsins (Fig. 14E) [49]. When taken together, these results provide evidence to support the phenotypic switch of MVICs from a normal, quiescent state to an activated state exhibiting increased ECM remodeling activity. These results provide insight to how MVICs may respond to disease-induced stress alterations in vivo.

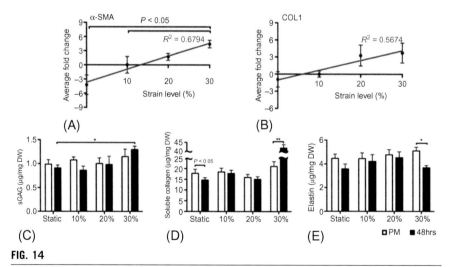

**FIG. 14**

The average fold change of (A) α-SMA and (B) type I collagen normalized to the 10% strain as determined from RT-qPCR. Colorimetric assay results at different strain levels for (C) sulfated GAGs, (D) soluble collagen, and (E) elastin content of both postmortem samples and samples subjected to cyclic strain for 48 h.

*From Ayoub S, Lee C-H, Driesbaugh KH, Anselmo W, Hughes CT, Ferrari G, et al. Regulation of valve interstitial cell homeostasis by mechanical deformation: implications for heart valve disease and surgical repair. J R Soc Interface 2017;14(135):20170580.*

### 3.3.5 Clinical relevance: Relating in vitro observations to in vivo function

The deformations of normal and surgically repaired MVAL have been measured previously through the use of sonomicrometry array techniques (Fig. 15A and B) [53]. This information was incorporated into a previously developed MVIC microenvironment model [13]. In short, an RVE was used to model the fibrosa layer of the MVAL including MVICs as ellipsoidal inclusions (Fig. 15C). Cell dimensions, density, and orientations were determined previously via histology and were reflected within the model [13,54]. ECM was modeled using a simplified structural constitutive model, and the cellular inclusions were modeled using a modified Saint Venant-Kirchhoff material model [13]. FE simulations were performed inside ABAQUS 6.13 framework (Simulia, Dassault Systemes, Providence, RI, United States) by assigning time-dependent tissue deformations from sonomicrometry arrays as boundary conditions on the RVE. The predicted deformation field was used to compute the in vivo deformation of the MVIC inclusions through the analysis of the NAR. It was found that the NAR of MVICs at peak end-systole is 4.92 in normal MVs (Fig. 15D). The model predicted a decrease in NAR to 3.28 in surgically repaired valves due to the decrease in circumferential deformation (Fig. 15D) [53]. The results suggest that nuclear deformation is closely linked to the circumferential deformation of MVALs. More recently, the same simulation framework was applied in a novel study of MV leaflet remodeling during pregnancy [12]. Remarkably, changes in NAR were found to correlate closely with geometric changes to the valve apparatus that occur in early pregnancy. This initial perturbation of VIC geometry was then found to trigger a complex layer-specific cascade of fiber remodeling mechanisms in the ECM, which work in concert to restore VIC homeostasis by the end of gestation.

Through combining the experimental data and the simulated NARs, a link between in vivo cell deformations and biosynthetic response is made (Fig. 16). The NAR is used as a dimensionless parameter to loosely bracket ranges of biological behavior within hypophysiological ($<3.30$), normal ($3.30-5.0$), and hyperphysiological ($>5.0$) states. From the analysis of the bracketed ranges, it is speculated that deformation scenarios leading to hypo- and hyperphysiological NAR values may cause a change in VIC phenotype and essentially alter biosynthetic response in attempt to return to homeostasis.

## 3.4 MODEL-DRIVEN EXPERIMENTAL DESIGN

### 3.4.1 FE models relating tissue-level and MVIC nuclear deformations

It has been shown experimentally that increased cellular deformation correlates with abnormal biosynthetic activity. To delineate the role that tissue-level deformations have on cellular biomechanical state and geometry, experimental and computational techniques were utilized [13].

MVALs were loaded under equibiaxial tension, and the resulting layer-specific deformations of MVICs were measured using multiphoton microscopy (Fig. 17A). A FE MVIC microenvironment model was developed to describe the relation between planar tension and layer-specific MVIC nuclear deformation (Fig. 17B),

**274 CHAPTER 8** Mechanobiology of the heart valve interstitial cell

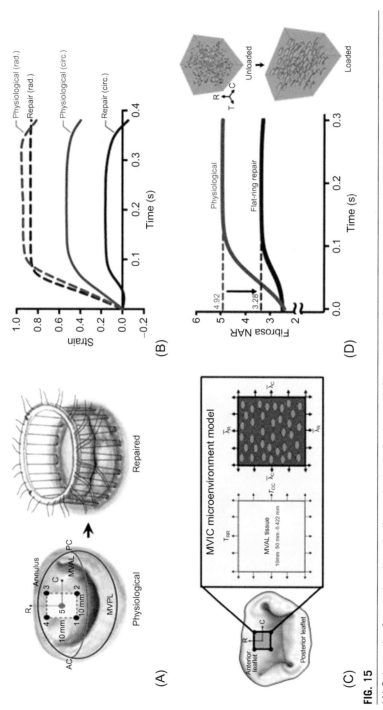

**FIG. 15**

(A) Schematic of normal and surgically repaired mitral valves with five sonocrystals placed on the anterior leaflets. (B) Circumferential (*solid line*) and radial (*dashed line*) strain for both physiological and repaired mitral valves. Surgical repair leads to a decrease in circumferential strain. (C) Schematic of the RVE used to model the anterior leaflet. The tissue-level strains from (B) are applied as boundary conditions. Finite element simulations are performed to predict deformation fields and essentially to estimate the nuclear aspect ratio of the ellipsoidal MVIC inclusions. (D) The estimated NAR of MVICs within the fibrosa layer from the finite element simulations. The model estimates that a decrease in NAR occurs due to surgical repair.

*From Ayoub S, Lee C-H, Driesbaugh KH, Anselmo W, Hughes CT, Ferrari G, et al. Regulation of valve interstitial cell homeostasis by mechanical deformation: implications for heart valve disease and surgical repair. J R Soc Interface, 2017;14(135):20170580.*

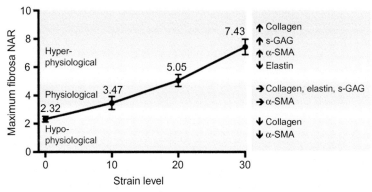

**FIG. 16**

MVIC biosynthetic behavior is bracketed into hyperphysiological, physiological, and hypophysiological ranges based off the simulated NARs for the normal and surgically repaired case. NAR below 3.28 is deemed as hypophysiological. An NAR between 3.28 and 4.92 is considered normal, and above 4.92 is hyperphysiological.

*From Ayoub S, Lee C-H, Driesbaugh KH, Anselmo W, Hughes CT, Ferrari G, et al. Regulation of valve interstitial cell homeostasis by mechanical deformation: implications for heart valve disease and surgical repair. J R Soc Interface 2017;14(135):20170580.*

and it was observed that the stiffness of MVICs varied little between the different MVAL layers (Fig. 17C–F). This suggests that differences in MVIC NAR across MVAL layers are mainly due to the deformation of the surrounding ECM and not the intrinsic stiffness of the cells. Documenting the nuclear deformation of MVICs in response to in vitro equibiaxial loading was a crucial first step toward the analysis of cellular function within surgically repaired MVs.

Additional MV models focus on estimating the in vivo stress and strain through sonomicrometry arrays [19].

Through tracking of sonocrystals placed on Dorset sheep MVALs, the deformation between several kinematic states of normal and surgically repaired MVs was measured making it possible to map in vitro deformations back to the in vivo state (Fig. 18A–C). In addition, in vivo stresses were characterized through the use of MVAL kinematic data and inverse modeling techniques. The in vivo stresses were estimated to be ~360 kPa in the circumferential direction and ~450 kPa in the radial direction when the effects of prestrain were not accounted for. With prestrain, the stresses were estimated to increase to ~510 kPa in the circumferential direction and ~740 kPa in the radial direction.

Through coupling both previously discussed MV models by applying the in vivo tissue deformations as boundary conditions within the MVIC microenvironment model, in vivo layer-specific NAR was estimated for a representative cardiac cycle (Fig. 19A). Using kinematic data gathered on surgically repaired MVs from Ref. [53], NAR estimates were also made for surgically repaired MVs (Fig. 19C). The layer-specific NAR rate of change is also reported (Fig. 19B and D), and it was observed that the fibrosa displayed the highest rate of change within normal

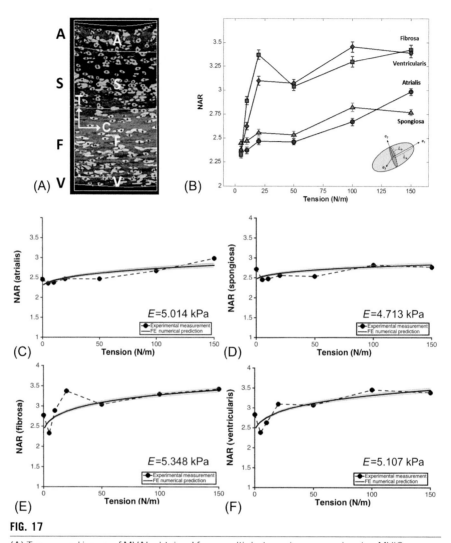

**FIG. 17**

(A) Transmural image of MVAL obtained from multiphoton microscopy showing MVICs across all layers. (B) Layer-specific nuclear aspect ratio measurements made at different equibiaxial membrane tensions. (C–F) Experimentally determined nuclear aspect ratio is fit to the FE model, and layer-specific stiffness of MVICs is determined.

*Modified from Lee CH, Carruthers CA, Ayoub S, Gorman RC, Gorman 3rd JH, Sacks MS. Quantification and simulation of layer-specific mitral valve interstitial cells deformation under physiological loading. J Theor Biol 2015;373:26–39.*

valves. Interestingly, among repaired valves, the atrialis displayed the highest NAR values and rate of change. This is hypothesized to be a result of leaflet contraction in the circumferential direction and expansion in the radial direction after flat-ring repair. This notable deviation from normal valve geometry most likely causes altered distribution of pressure throughout the anterior leaflet, thus resulting in altered,

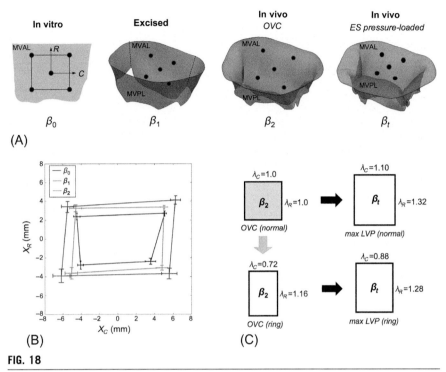

**FIG. 18**

(A) Schematic of the different MVAL kinematic states. Sonocrystals are tracked and their displacements are recorded across the in vitro configuration ($\beta_o$), the excised configuration ($\beta_1$), the in vivo configuration just before the onset of ventricular contraction (OVC) where the MV leaflets are coapted ($\beta_2$), and the in vivo configuration at the end-systolic (ES) time point where the MV is under transvalvular pressure ($\beta_t$). (B) The population averaged location of the sonocrystals with respect to the radial ($X_R$) and circumferential ($X_C$) directions for key kinematic states. (C) The circumferential ($\lambda_C$) and radial ($\lambda_R$) deformation of the center region of MVALs due to surgical ring repair.

*From Lee CH, Zhang W, Feaver K, Gorman RC, Gorman 3rd JH, Sacks MS. On the in vivo function of the mitral heart valve leaflet: insights into tissue-interstitial cell biomechanical coupling. Biomech Model Mechanobiol 2017.*

layer-specific MVIC deformations. The ability to estimate in vivo stresses and deformations for normal and surgically repaired MVALs highlights the benefits of taking an experimental-computational approach to studying VIC mechanobiology.

## 4 FUTURE DIRECTIONS

A need exists for a full multiscale approach to study and capture the complexity of the heart system. Combining experimental and computational methods is key to the refinement of our current knowledge base on VIC mechanobiology. Linking the cell-, tissue-, and organ-level response of the heart system not only widens our

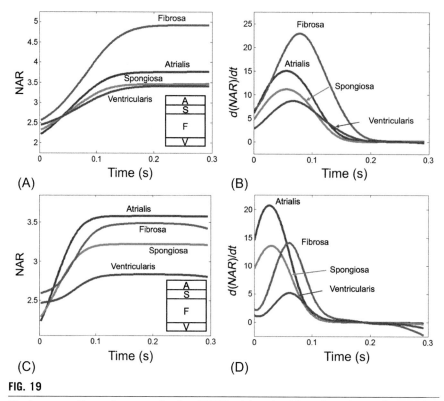

**FIG. 19**

The layer-specific nuclear aspect ratio of MVICs determined from the microenvironment model for the (A) normal and (C) surgical repair scenarios. The NAR rate of change for (B) normal and (D) surgical repair highlights the differences in layer-specific cell deformations as a result of altered tissue-level deformations.

*From Lee CH, Zhang W, Feaver K, Gorman RC, Gorman 3rd JH, Sacks MS. On the in vivo function of the mitral heart valve leaflet: insights into tissue-interstitial cell biomechanical coupling. Biomech Model Mechanobiol 2017.*

understanding of the heart's physiology but also offers a unique opportunity to simulate diseased states. Recent computational modeling studies have elucidated the importance of including microstructurally informed tissue properties in predictive simulation frameworks, as both CFA and macroscale geometric properties vary considerably over local tissue regions in heart valves [55–59]. In light of recent findings quantifying the link between local ECM structural properties and VIC mechanical and biosynthetic behavior [18], it has become increasingly clear that a full understanding of valvular function, maintenance, and adaptation must be rooted in a detailed knowledge of the interrelationships between organ-level physiology and tissue/cell coupling.

Currently, a promising avenue for investigating complex valvular tissue systems is through the development of agent-based models that can combine the VIC

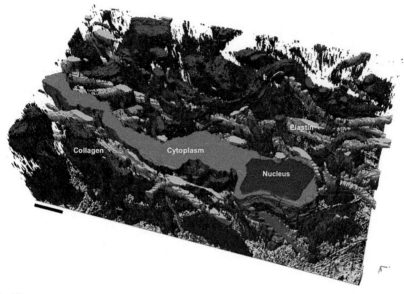

**FIG. 20**

Three-dimensional reconstruction of the VIC microenvironment from FIB-SEM images. Note how the VIC is embedded within networks of collagen and elastin fibers.

mechanical environment, biomechanical states, and their resulting biosynthetic activity within a cohesive modeling pipeline. For this, high detailed cellular-level observations are crucial toward the formulation of VIC models.

State-of-the-art imaging technologies like the focused ion beam scanning electron microscope (FIB-SEM) have produced high-quality images of the native valve leaflet structure and help shed light on the intricate VIC-ECM interactions (Fig. 20). This preliminary work reflects the complex VIC microenvironment and highlights considerations to be made when formulating agent-based models.

Linking tissue-level deformations to changes in MVIC NAR in the in vivo state marks a huge success in gauging the biophysical state of these cells. To take these studies further, quantification of MVIC biosynthetic response within the model-prescribed physiological stresses is currently underway. With the use of biaxial bioreactor tissue strip systems, MVALs will undergo mechanical conditioning, and the ECM components will be quantified for the normal and valve repair cases. From these experiments, we hope to validate the NAR values estimated from the model and document the biosynthetic behavior of MVICs at the predicted stress levels. This model-driven study serves as an example of how experimental-computational techniques can broaden our understanding of the underlying mechanobiology within living systems.

To increase predictive capabilities of VIC models, parameters correlating physical cellular characteristics and downstream expression of genes are needed.

One avenue that researchers have used to do this is through analysis of VIC nuclear geometry, specifically NAR [16–18]. Preliminary correlations have been made to associate NAR with specific stress environments and biosynthesis rates. However, nuclear deformations may play a larger role in controlling intracellular pathways and regulating the behavior of VICs. Future efforts should focus on closely examining the cellular and subcellular events of VIC biological function and relating it back to measurable parameters (e.g., NAR). Advances in hydrogel chemistry with favorable optical properties and tunable characteristics show great promise in this endeavor. Hydrogel environments that offer tunable mechanics and peptide modifications present a unique opportunity to emulate healthy and diseased states. In addition, the ability to gather mechanical and optical data from these matrices makes them extremely attractive as a tool for future 3D mechanobiological studies.

## ACKNOWLEDGMENTS

This work was supported by the National Institutes of Health (Grant No. R01-HL108330 and R01-HL119297 to MSS; Grant No. F31-HL137328 to SA) and the National Science Foundation (Grant No. DGE-1610403 to AK and BVR). RMB and SA were each supported in part by American Heart Association (AHA) Predoctoral Fellowships. CHL was supported in part by start-up funds from the School of Aerospace and Mechanical Engineering at the University of Oklahoma, as well as an AHA Scientist Development Grant (16S-DG27760143).

## REFERENCES

[1] Merryman WD, Lukoff HD, Long RA, Engelmayr Jr GC, Hopkins RA, Sacks MS. Synergistic effects of cyclic tension and transforming growth factor-beta1 on the aortic valve myofibroblast. Cardiovasc Pathol 2007;16(5):268–76.

[2] Guyton AC. Textbook of medical physiology. 5th ed. Philadelphia: W.B. Saunders Company; 1976.

[3] He Z, Ritchie J, Grashow JS, Sacks MS, Yoganathan AP. In vitro dynamic strain behavior of the mitral valve posterior leaflet. J Biomech Eng 2005;127(3):504–11.

[4] He Z, Sacks MS, Baijens L, Wanant S, Shah P, Yoganathan AP. Effects of papillary muscle position on in-vitro dynamic strain on the porcine mitral valve. J Heart Valve Dis 2003;12(4):488–94.

[5] Sacks MS, Enomoto Y, Graybill JR, Merryman WD, Zeeshan A, Yoganathan AP, et al. In-vivo dynamic deformation of the mitral valve anterior leaflet. Ann Thorac Surg 2006;82(4):1369–77.

[6] Sacks MS, He Z, Baijens L, Wanant S, Shah P, Sugimoto H, et al. Surface strains in the anterior leaflet of the functioning mitral valve. Ann Biomed Eng 2002;30(10):1281–90.

[7] Ayoub S, Ferrari G, Gorman RC, Gorman JH, Schoen FJ, Sacks MS. Heart valve biomechanics and underlying mechanobiology. Compr Physiol 2016;6(4):1743–80.

[8] Sacks MS, Yoganathan AP. Heart valve function: a biomechanical perspective. Philos Trans R Soc Lond Ser B Biol Sci 2007;362(1484):1369–91.

[9] Buchanan RM, Sacks MS. Interlayer micromechanics of the aortic heart valve leaflet. Biomech Model Mechanobiol 2013;13:813–26.
[10] Rego BV, Sacks MS. A functionally graded material model for the transmural stress distribution of the aortic valve leaflet. J Biomech 2017;54:88–95.
[11] Stella JA, Sacks MS. On the biaxial mechanical properties of the layers of the aortic valve leaflet. J Biomech Eng 2007;129(5):757–66.
[12] Rego BV, Wells SM, Lee CH, Sacks MS. Mitral valve leaflet remodelling during pregnancy: insights into cell-mediated recovery of tissue homeostasis. J R Soc Interface 2016;13(125):20160709.
[13] Lee CH, Carruthers CA, Ayoub S, Gorman RC, Gorman 3rd JH, Sacks MS. Quantification and simulation of layer-specific mitral valve interstitial cells deformation under physiological loading. J Theor Biol 2015;373:26–39.
[14] Pierlot CM, Lee JM, Amini R, Sacks MS, Wells SM. Pregnancy-induced remodeling of collagen architecture and content in the mitral valve. Ann Biomed Eng 2014;42(10):2058–71.
[15] Pierlot CM, Moeller AD, Lee JM, Wells SM. Pregnancy-induced remodeling of heart valves. Am J Physiol Heart Circ Physiol 2015;309(9):H1565–78.
[16] Lam NT, Muldoon TJ, Quinn KP, Rajaram N, Balachandran K. Valve interstitial cell contractile strength and metabolic state are dependent on its shape. Integr Biol 2016;8(10):1079–89.
[17] Tandon I, Razavi A, Ravishankar P, Walker A, Sturdivant NM, Lam NT, et al. Valve interstitial cell shape modulates cell contractility independent of cell phenotype. J Biomech 2016;49(14):3289–97.
[18] Ayoub S, Lee C-H, Driesbaugh KH, Anselmo W, Hughes CT, Ferrari G, et al. Regulation of valve interstitial cell homeostasis by mechanical deformation: implications for heart valve disease and surgical repair. J R Soc Interface 2017;14(135):20170580.
[19] Lee CH, Zhang W, Feaver K, Gorman RC, Gorman 3rd JH, Sacks MS. On the in vivo function of the mitral heart valve leaflet: insights into tissue-interstitial cell biomechanical coupling. Biomech Model Mechanobiol 2017;16:1613–32.
[20] Rabkin E, Aikawa M, Stone JR, Fukumoto Y, Libby P, Schoen FJ. Activated interstitial myofibroblasts express catabolic enzymes and mediate matrix remodeling in myxomatous heart valves. Circulation 2001;104(21):2525–32.
[21] Rajamannan NM, Evans FJ, Aikawa E, Grande-Allen KJ, Demer LL, Heistad DD, et al. Calcific aortic valve disease: not simply a degenerative process: a review and agenda for research from the national heart and lung and blood institute aortic stenosis working group. Executive summary: calcific aortic valve disease-2011 update. Circulation 2011;124(16):1783–91.
[22] Sacks MS, Merryman WD, Schmidt DE. On the biomechanics of heart valve function. J Biomech 2009;42(12):1804–24.
[23] Carruthers CA, Good B, D'Amore A, Liao J, Amini R, Watkins SC, et al., Alterations in the microstructure of the anterior mitral valve leaflet under physiological stress. ASME 2012 summer bioengineering conferenceAmerican Society of Mechanical Engineers; 2012.
[24] Sakamoto Y, Buchanan RM, Sacks MS. On intrinsic stress fiber contractile forces in semilunar heart valve interstitial cells using a continuum mixture model. J Mech Behav Biomed Mater 2016;54:244–58.

[25] Sakamoto Y, Buchanan RM, Sanchez-Adams J, Guilak F, Sacks MS. On the functional role of valve interstitial cell stress fibers: a continuum modeling approach. J Biomech Eng 2017;139(2):021007.

[26] Buchanan RM. An integrated computational-experimental approach for the in situ estimation of valve interstitial cell biomechanical state. Austin, USA: The University of Texas at Austin; 2016.

[27] Merryman WD, Youn I, Lukoff HD, Krueger PM, Guilak F, Hopkins RA, et al. Correlation between heart valve interstitial cell stiffness and transvalvular pressure: implications for collagen biosynthesis. Am J Physiol Heart Circ Physiol 2006;290(1):H224–31.

[28] Theret DP, Levesque MJ, Sato M, Nerem RM, Wheeler LT. The application of a homogeneous half-space model in the analysis of endothelial cell micropipette measurements. J Biomech Eng 1988;110(3):190–9.

[29] Rocnik EF, van der Veer E, Cao H, Hegele RA, Pickering JG. Functional linkage between the endoplasmic reticulum protein Hsp47 and procollagen expression in human vascular smooth muscle cells. J Biol Chem 2002;277(41):38571–8.

[30] Merryman WD, Liao J, Parekh A, Candiello JE, Lin H, Sacks MS. Differences in tissue-remodeling potential of aortic and pulmonary heart valve interstitial cells. Tissue Eng 2007;13(9):2281–9.

[31] Costa KD, Yin FC. Analysis of indentation: implications for measuring mechanical properties with atomic force microscopy. J Biomech Eng 1999;121(5):462–71.

[32] Mathur AB, Collinsworth AM, Reichert WM, Kraus WE, Truskey GA. Endothelial, cardiac muscle and skeletal muscle exhibit different viscous and elastic properties as determined by atomic force microscopy. J Biomech 2001;34(12):1545–53.

[33] Sato M, Theret DP, Wheeler LT, Ohshima N, Nerem RM. Application of the micropipette technique to the measurement of cultured porcine aortic endothelial cell viscoelastic properties. J Biomech Eng 1990;112(3):263–8.

[34] Guilak F, Ting-Beall HP, Baer AE, Trickey WR, Erickson GR, Setton LA. Viscoelastic properties of intervertebral disc cells. Identification of two biomechanically distinct cell populations. Spine 1999;24(23):2475–83.

[35] Na S, Sun Z, Meininger GA, Humphrey JD. On atomic force microscopy and the constitutive behavior of living cells. Biomech Model Mechanobiol 2004;3(2):75–84.

[36] Frisch-Fay R. Flexible bars. Washington, DC: Butterworths; 1962220.

[37] Merryman WD, Huang HY, Schoen FJ, Sacks MS. The effects of cellular contraction on aortic valve leaflet flexural stiffness. J Biomech 2006;39(1):88–96.

[38] Mirnajafi A, Raymer J, Scott MJ, Sacks MS. The effects of collagen fiber orientation on the flexural properties of pericardial heterograft biomaterials. Biomaterials 2005;26(7):795–804.

[39] Mirnajafi A, Raymer JM, McClure LR, Sacks MS. The flexural rigidity of the aortic valve leaflet in the commissural region. J Biomech 2006;39(16):2966–73.

[40] Billiar KL, Sacks MS. Biaxial mechanical properties of the native and glutaraldehyde-treated aortic valve cusp: part II—a structural constitutive model. J Biomech Eng 2000;122(4):327–35.

[41] Billiar KL, Sacks MS. Biaxial mechanical properties of the natural and glutaraldehyde treated aortic valve cusp: part I—experimental results. J Biomech Eng 2000;122(1):23–30.

[42] Mohri H, Reichenback D, Merendino K. Biology of homologous and heterologous aortic valves. In: Ionescu M, Ross D, Wooler G, editors. Biological tissue in heart valve replacement. London: Butterworths; 1972. p. 137.

[43] Vesely I, Boughner D. Analysis of the bending behaviour of porcine xenograft leaflets and of natural aortic valve material: bending stiffness, neutral axis and shear measurements. J Biomech 1989;22(6/7):655–71.
[44] Song T, Vesely I, Boughner D. Effects of dynamic fixation on shear behavior of porcine xenograft valves. Biomaterials 1990;11:191–6.
[45] Lu SCH, Pister KS. Decomposition of deformation and representation of the free energy function for isotropic thermoelastic solids. Int J Solids Struct 1975;11(7–8):927–34.
[46] Benton JA, Fairbanks BD, Anseth KS. Characterization of valvular interstitial cell function in three dimensional matrix metalloproteinase degradable PEG hydrogels. Biomaterials 2009;30(34):6593–603.
[47] Byrant SJ, Anseth KS. Photopolymerization of hydrogel scaffolds. In: Ma PX, Elisseeff J, editors. Scaffolding in tissue engineering. Boca Raton, FL: CRC Press, Taylor & Francis Group; 2005. p. 71–90.
[48] Balachandran K, Alford PW, Wylie-Sears J, Goss JA, Grosberg A, Bischoff J, et al. Cyclic strain induces dual-mode endothelial-mesenchymal transformation of the cardiac valve. Proc Natl Acad Sci U S A 2011;108(50):19943–8.
[49] Balachandran K, Hussain S, Yap CH, Padala M, Chester AH, Yoganathan AP. Elevated cyclic stretch and serotonin result in altered aortic valve remodeling via a mechanosensitive 5-HT(2A) receptor-dependent pathway. Cardiovasc Pathol 2012;21(3):206–13.
[50] Balachandran K, Konduri S, Sucosky P, Jo H, Yoganathan A. An ex vivo study of the biological properties of porcine aortic valves in response to circumferential cyclic stretch. Ann Biomed Eng 2006;34(11):1655–65.
[51] Balachandran K, Sucosky P, Jo H, Yoganathan AP. Elevated cyclic stretch alters matrix remodeling in aortic valve cusps: implications for degenerative aortic valve disease. Am J Physiol Heart Circ Physiol 2009;296(3):H756–64.
[52] Sacks MS, Smith DB, Hiester ED. A small angle light scattering device for planar connective tissue microstructural analysis. Ann Biomed Eng 1997;25(4):678–89.
[53] Amini R, Eckert CE, Koomalsingh K, McGarvey J, Minakawa M, Gorman JH, et al. On the in vivo deformation of the mitral valve anterior leaflet: effects of annular geometry and referential configuration. Ann Biomed Eng 2012;40(7):1455–67.
[54] Carruthers CA, Alfieri CM, Joyce EM, Watkins SC, Yutzey KE, Sacks MS. Gene expression and collagen fiber micromechanical interactions of the semilunar heart valve interstitial cell. Cell Mol Bioeng 2012;5(3):254–65.
[55] Lee CH, Rabbah JP, Yoganathan AP, Gorman RC, Gorman 3rd JH, Sacks MS. On the effects of leaflet microstructure and constitutive model on the closing behavior of the mitral valve. Biomech Model Mechanobiol 2015;14(6):1281–302.
[56] Khalighi AH, Drach A, CHt B, Pierce EL, Yoganathan AP, Gorman RC, et al. Mitral valve chordae tendineae: topological and geometrical characterization. Ann Biomed Eng 2017;45(2):378–93.
[57] Khalighi AH, Drach A, Gorman RC, Gorman 3rd JH, Sacks MS. Multi-resolution geometric modeling of the mitral heart valve leaflets. Biomech Model Mechanobiol 2017;17(2):351–6.
[58] Drach A, Khalighi AH, Sacks MS. A comprehensive pipeline for multi-resolution modeling of the mitral valve: validation, computational efficiency, and predictive capability. Int J Numer Methods Biomed Eng 2017;34(2):e2921.
[59] Sacks MS, Khalighi A, Rego B, Ayoub S, Drach A. On the need for multi-scale geometric modelling of the mitral heart valve. Healthc Technol Lett 2017;4(5):150.

CHAPTER

# Platelet receptor-mediated mechanosensing and thrombosis

9

Lining A. Ju*,†,a, Yunfeng Chen†,a, Zhenhai Li§,a, Cheng Zhu¶

*Heart Research Institute, The University of Sydney, Camperdown, NSW, Australia* *Charles Perkins Centre, The University of Sydney, Camperdown, NSW, Australia*† *Department of Molecular Medicine, MERU-Roon Research Center on Vascular Biology, The Scripps Research Institute, La Jolla, CA, United States*‡ *Molecular Modeling and Simulation Group, National Institutes for Quantum and Radiological Science and Technology, Kyoto, Japan*§ *Coulter Department of Biomedical Engineering, Georgia Institute of Technology, Atlanta, GA, United States*¶

## ABBREVIATIONS

| | |
|---|---|
| **ADAMTS13** | a disintegrin and metalloproteinase with a thrombospondin type 1 motif, member 13 |
| **AFM** | atomic force microscopy |
| **fBFP** | fluorescence BFP |
| **GPIb** | glycoprotein Ib |
| **LRRD** | leucine-rich repeat domain |
| **MD** | molecular dynamics |
| **MP** | macroglycopeptide |
| **MSD** | mechanosensitive domain |
| **MT** | magnetic tweezers |
| **PI3K** | PI3-kinase |
| **VWD** | von Willebrand disease |
| **VWF** | von Willebrand factor |
| **WT** | wild-type |

## 1 INTRODUCTION

Platelets, cells that circulate in the blood, have been known for decades for their central role in forming a hemostatic plug at the site of vessel breach (hemostasis) or occlusive arterial thrombi at the site of atherosclerosis plaque (thrombosis) [1].

---

aThese primary authors contributed equally to this work

After vascular injury, a number of subendothelial matrix proteins become exposed to blood or are deposited from blood, including von Willebrand factor (VWF), collagens, fibronectin, and fibrinogen, which support platelet adhesion through the engagement of specific receptors [1,2]. In the current view, under conditions of rapid blood flow, as occurs in arterioles or atherosclerotic arteries with restricted lumen, which represents the majority of acute cases of myocardial infarction and stroke, platelet adhesion and aggregation is critically dependent on blood-borne VWF, a shear-sensitive glycoprotein composed of 240 kDa subunits that forms large disulfide-linked multimers of up to 50,000 kDa [3,4]. When VWF is first released from the storage granules of endothelial cells, they are rich in the ultralarge forms (ULVWF, over 10,000 kDa) [5]. Its multimeric size and adhesive activity are regulated through proteolysis cleavage of the VWF-A2 domain induced by a metalloprotease *a* *d*isintegrin and *m*etalloprotease with *t*hrombo*s*pondin motifs-*13* (ADAMTS13). The VWF-A1 domain contains the binding site for platelet glycoprotein Ibα (GPIbα) that is dependent upon conformational activation by elevated shear stress [6,7] or association with the extracellular matrix (ECM), mainly collagen [8,9] (Fig. 1). The GPIbα contains an N-terminal leucine-rich repeat domain (LRRD) that

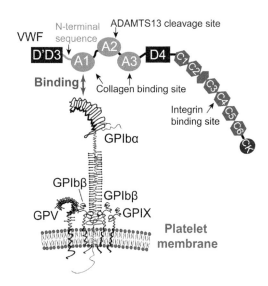

**FIG. 1**

VWF-GPIbα interaction axis. *Top*: The VWF domain organization [4]. Each mature VWF monomer contains a triplicate repeat sequence of A domains in the central portion of the 2050-residue mature subunit (D'-D3-A1-A2-A3-D4-C1-6-CK). The A1 domain (the disulfide loop C1272-C1458) contains contact sites for the platelet GPIbα and collagen. Its homologous A3 domain binds only collagen. The central A2 domain contains a proteolytic site for the metalloprotease ADAMTS13. *Bottom*: The latest model of the GPIb-IX-V complex. It consists of four subunits, GPIbα, GPIbβ × 2, GPIX, and GPV. The N-terminal autoinhibitory sequence of VWF-A1 is indicated.

interacts with the VWF-A1, a highly glycosylated long stalk region (macroglycopeptide region, MP) [10], a juxtamembrane mechanosensitive domain (MSD) [11], a single-span transmembrane segment, and a short cytoplasmic tail (Fig. 1). Moreover, GPIbα is covalently linked to GPIbβ through disulfides, and together, they associate tightly with GPIX to form the GPIb-IX complex [12,13] and interact with cytoplasmic adaptor proteins such as 14-3-3. Upon the VWF-GPIbα interaction under shear, platelet activation is triggered (mechanosensing), which arrests the translocating platelets on the disrupted vessel wall and form platelet plugs through integrin-ligand interactions [14]. Notably, all these processes may be altered by mutations in VWF and/or GPIbα, which cause bleeding disorders as found in patients with von Willebrand disease (VWD) [15].

Although it seems clear how shear regulates platelet adhesion at the level of cellular function, the detailed molecular events as to how VWF interacts with the platelet (especially under physiological shear forces) and how a mechanical stimulus is translated into biochemical signals have remained elusive for decades, until single-molecule biomechanical approaches were recently successfully applied to live cells [16]. In this chapter, we will review the latest advancement of single-molecule methods and their application to understanding the VWF-GPIbα-dependent platelet mechanosensing. Kinetics of receptor-ligand interaction and mechanics of the molecular players are discussed. In conclusion, a four-step model is summarized to describe the general receptor-mediated cell mechanosensing principle, which may inspire a potential mechanomedicine approach for antithrombotic therapeutics.

## 2 ULTRASENSITIVE FORCE TECHNIQUES

Electron microscopy, crystallography, and antibody mapping are the most commonly used approaches to visualize conformations and characterize behaviors of purified proteins [17]. However, these approaches only take snapshots of proteins' stable states and lack real-time details of transitional processes. With the exception of antibody mapping, these methods cannot be used to investigate the coupling of protein conformational changes with subsequent signaling events on live cells. Over the past two decades, ultrasensitive force techniques, that is, dynamic force spectroscopy (DFS) (Fig. 2A) such as atomic force microscopy (AFM, Fig. 2B), optical tweezer (OT, Fig. 2C), magnetic tweezer (MT, Fig. 2D), and biomembrane force probe (BFP, Fig. 2E), have provided various biomechanical approaches for manipulation, characterization, and visualization of single receptor-ligand interactions and conformational change with tunable force [16,18,19]. With nanometer spatial, submillisecond temporal, and piconewton force resolutions, these techniques can be used to induce, follow, and analyze single-molecule mechanical events in real time, thus revealing individual molecular details inaccessible by conventional methods based on ensemble averaging. In a typical experiment, automated precise movement brings together ligands and receptors on the respective force probe (i.e., an AFM cantilever in Fig. 2B or a mechanically-controlled bead in Fig. 2C–E) and target (i.e., a bead

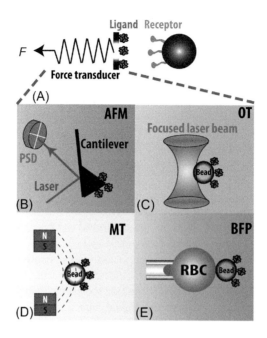

**FIG. 2**

Ultrasensitive force techniques. (A) A generic force probe that applies forces, $F$, to the receptor-ligand bond spanned between a surface and a force transducer. (B) Atomic force microscopy (AFM). Force is applied to individual molecules tethered between a functionalized cantilever and a surface. (C) Optical tweezer (OT). A protein-coated bead is held by a laser beam. (D) Magnetic tweezer (MT). Permanent/electric magnets are used to manipulate a protein-coated magnetic bead. (E) Biomembrane force probe (BFP). The protein-coated bead is attached to the apex of a micropipette-aspirated red blood cell (RBC). (B–E) Force is determined, respectively, by cantilever deflection (B), bead displacement (C), gradient of the magnetic field (D), and RBC deformation (E).

Reprinted and modified from Chen Y, Ju L, Rushdi MN, Ge C, Zhu C. Receptor-mediated cell mechanosensing. Mol Biol Cell 2017;28(23):3134–3155. https://doi.org/10.1091/mbc.E17-04-0228.

in Fig. 2A *right* or a cell). The contact time, area, and force are preset to allow for bond formation under controlled conditions. The following separation of the two surfaces applies a piconewton-level force to the receptor-ligand bond to induce mechanical changes in the molecules and modulate bond dissociation. These techniques are usually used to perform either force-ramp or force-clamp DFS. Force-ramp DFS exerts a linearly increasing force on the receptor-ligand bond with a range of force application rates (Fig. 4C), whereas force-clamp DFS exerts a range of constant forces on the bond (Fig. 4E) [19a]. However, other modes of operation can be used to exert different force waveforms on the molecular bond, for example, jump-and-ramp DFS [20] and cyclic force DFS [21,22].

These DFS techniques are commonly used to analyze force-induced receptor-ligand unbinding and protein domain unfolding. Both are usually treated as state transition processes governed by off rates or unfolding rates as modulated by force. However, the off rate versus force relationships determined by force-ramp and force-clamp DFS sometimes differ substantially, indicating that off-rates are not always a single-valued function of the instantaneous level of force at the present time, but may also depend on the past history of force application [21,23,24]. Furthermore, it was noticed that only ramped, but not clamped, forces are able to unfold LRRD of GPIbα, whereas both ramped and clamped forces are able to unfold MSD of GPIbα [25]. Moreover, clamped forces are much more effective than ramped forces in receptor-mediated calcium triggering in T cells and platelets [25,26]. Thus, the way how force is applied to the receptor affects how frequent a mechanical event may occur in the molecule and how rapidly it dissociates from the ligand. These in turn affect the time course of force that can be exerted on the bond, thereby impacting cell signaling.

To directly observe single receptor-mediated mechanosensing events on a live cell requires running DFS analysis with concurrent imaging of intracellular signaling (e.g., calcium flux), as was done by us using a fluorescence BFP (fBFP) [25–27,27a] and others using fluorescence OT [28,29]. This ability to apply a controlled force waveform and concurrently analyze the resulting receptor-ligand unbinding, receptor mechanical events, and intracellular signaling enables one to ask important questions in mechanosensing: (1) How is receptor dissociation modulated by force? (2) What molecular events are induced by a receptor and how are these events modulated mechanically? (3) Whether and, if so, how changes in the presentation of force and/or ligand structure affect the mechanopresentation? (4) What features of the force waveform could be sensed during the mechanoreception? (5) What proximal events are responsible for transducing force into a biochemical signal?

As demonstrated in recent studies on platelets, only durable forces applied via force clamp, but not transient forces applied by force ramp, on the VWF-GPIbα bonds are able to effectively trigger calcium signaling in platelets [25]. To induce calcium signaling in platelets requires only a single VWF-GPIbα bond with an above-threshold lifetime under 22–25 pN tensile force. Thus, platelets can distinguish different force waveforms and respond by eliciting different calcium signals.

## 3 FORCE-INDUCED VWF ACTIVATION

Several mechanisms contribute to the force-induced activation of VWF. In the blood plasma, multimeric VWF adopts an intertwined conformation, which tends to bury its functional sites and reduces its chance to contact its interacting partners, for example, collagen on the subendothelial surface. An emerging concept is that shear stress gradient may have an elongational effect—highly efficient to unfold and extend the VWF [30]. Shear stress gradient mostly occurs at the upstream of stenosed blood vessels, where the flow rate increases along the direction of flow as the luminal area narrows [31,32]. This special hydrodynamic environment quickly transforms the plasma VWF from a globular to a more linear shape, thereby exposing its functional epitopes [33]. Alternatively, pathologically fast flow with shear stress constant serves as another way to extend plasma VWF independent of shear stress gradient, which facilitates its attachment to collagen surface and fiber formation [34].

Force can unfold and unmask a specific protein domain to expose cryptic binding or enzymatic cleaving sites. The plasma VWF-A1 domain has a limited binding potential for GPIbα in circulation due to autoinhibitory mechanisms that mask the GPIbα binding site on VWF-A1 by the adjacent D'D3 domain [35], the N-terminal flanking region of the A1 domain (residues 1238–1260) [36–38], and the A2 domain [39,40]. Once immobilized onto subendothelial collagen and subjected to flow, however, hemodynamic drag forces on VWF stretch the macromolecule to change the local structure surrounding the A1 domain, which removes the autoinhibition and thereby facilitates A1 engagement by GPIbα [41–43].

In addition to being stretched to undergo deformation or unfolding, VWF was observed to undergo self-association, which also requires shear force [44]. Under shear rate $> 2000\,s^{-1}$, VWF molecules in solution are able to attach to the ones that are already immobilized on a surface, for example, a collagen matrix or platelets, and form hemolytic interactions [45]. This allows the VWF to form bundles that enhance the adhesion function and facilitate platelet adhesion [44] and possibly platelet activation as well [46]. The underlying molecular mechanism of VWF self-association and the domains responsible for the interactions are still unclear.

## 4 VWF-GPIbα BINDING KINETICS

A central feature of hemostatic responses is that platelet adhesion to the vascular surface occurs in a mechanically stressful environment of the circulation. This requires rapid VWF-GPIbα bond association to capture flowing platelets and rapid bond dissociation to allow platelet translocation (rolling) in a stop-and-go fashion on the subendothelial surface [37,47–50]. In blood flow, physical transport—convection and diffusion—drives platelets to collide with vascular surface, bringing interacting receptors and ligands into close proximity. The three distinct steps of transport mechanism have been demonstrated to regulate the VWF-GPIbα association: tethering of platelet to the vascular surface, Brownian motion of the platelet, and rotational diffusion of the interacting molecules [51].

Moreover, shear flow results in force on VWF-GPIbα bonds to modulate their dissociation kinetics, eliciting the counterintuitive catch bond to prolong bond lifetime at forces $<22\,pN$ [9,37,47]. At higher forces, the VWF-GPIbα interaction displays the ordinary slip bond behavior, where force shortens bond lifetime. Such force-dependent kinetics governs the counterintuitive flow enhancement of platelet translocation on VWF where increasing shear first slows down, reaches a minimum, and then speeds up platelet translocation [52,53]. Interestingly, VWF-A1 mutations exhibiting the phenotype of type 2B VWD form slip-only bonds with GPIbα and eliminate flow-enhanced platelet translocation [37,47]. Furthermore, immobilized at sites of vascular injury, force-dependent VWF activation is also affected by its interaction with collagen. DFS and 2D affinity measurements suggested relative contributions of distinct VWF domains such that the initial VWF capture is mediated by collagen interaction with the A3 domain while the subsequent VWF activation is mediated by the interaction with the A1 domain [9]. Given that VWF represents an excellent model for mechanopresenter (Fig. 1), the relative contributions of its distinct domains and their synergy to its mechanopresentation are interesting questions for future studies.

The above studies have established the causal relationship between VWF-GPIbα catch bond and flow-enhanced platelet translocation, which is crucial to hemostasis in the arterioles where considerable hemodynamic forces constantly apply. Taken together, the flow-/force-enhanced VWF-GPIbα-dependent platelet adhesion involves at least three steps that directly affect the binding kinetics: (1) relief of VWF autoinhibitory mechanism involving interdomain associations within the A1A2A3 tridomain and between A1 and D'D3 to expose GPIbα-binding site, (2) enhancement of VWF-GPIbα association by transport, and (3) upregulation of A1 binding affinity to GPIbα by catch bond. As discussed below, more recent studies have further elucidated the implications of these force-dependent kinetic properties to platelet mechanosensing via GPIb, showing that now different components of VWF-GPIbα axis serve their respective roles.

## 5 FORCE-INDUCED VWF CLEAVAGE BY ADAMTS13

Activated endothelial cells secrete hyperactive ULVWF, which are quickly cleaved by the plasma metalloprotease ADAMTS13 to yield smaller and functionally less active VWF multimers [54]. The mutations in the ADAMTS13 gene or autoantibodies against the metalloprotease lead to the catastrophic accumulation of hyperactive ULVWF multimers on endothelial cells and in plasma [55]. As a result, these poorly cleaved ULVWF agglutinate/aggregate platelets to cause systemic microvascular thrombosis that is enriched in VWF and platelets and lead to a rare blood disorder, thrombotic thrombocytopenic purpura (TTP). Thus, ADAMTS13 is essential to maintain the plasma VWF in the normal multimeric size range.

ADAMTS13 consists of an N-terminal metalloprotease domain followed by a disintegrin-like domain; a thrombospondin type 1 motif (TSP); a cysteine-rich domain; a spacer domain; seven additional TSP domains; and two unique components C1r/C1s, urchin epidermal growth factor, and bone morphogenic protein-1 (CUB) domains at the C-terminus (Fig. 3A) [56]. ADAMTS13 specifically cleaves the Tyr1605-Met1606 peptide bond [57], which is buried in the VWF A2 domain [58]. As the cryptic site is exposed with the force-induced VWF unfolding and extension, the proteolysis reaction is regulated by hemodynamic forces [3,59].

Moreover, the interaction between VWF and ADAMTS13 is also regulated by hemodynamic forces. Multiple exosites or binding sites on ADAMTS13 were reported to interact with VWF: spacer domain Arg660, Tyr661, and Tyr665 interact with a region flanking the residues Glu1660-Arg1668 in VWF-A2 domain [60,61]; cysteine-rich domain may interact with another exosite in VWF-A2 adjacent to Glu1660-Arg1668 [61,62]; an exosite involving Arg349 in disintegrin-like domain may interact with an exosite involving Asp1614 in VWF-A2 domain [61,63];

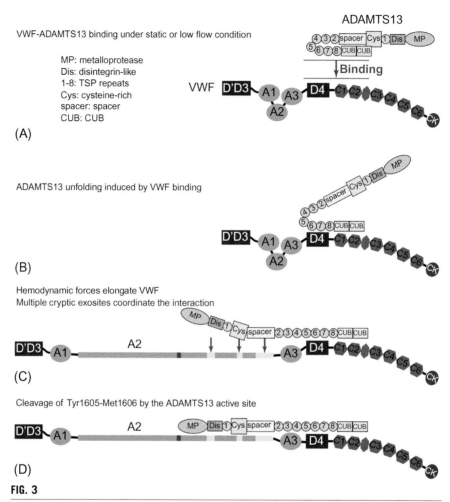

FIG. 3

VWF-ADATMS13 interaction and VWF cleavage. (A) ADAMTS13 interacts with globular VWF under static or low flow condition through the binding between the C-terminal domains, T5-CUB in ADAMTS13, and D4-C2 domains in VWF. The domain labeling of ADAMTS13 is as follows: MP (metalloprotease), Dis (disintegrin-like), 1–8 (TSP repeats), Cys (cysteine-rich), spacer (spacer), and CUB. (B) The initial binding induces the ADAMTS13 unfolding. (C) VWF is elongated by hemodynamic force and exposes the cryptic exosites and the cleavage site. These exosites bind to the complementary exosites on ADAMTS13 step by step. (D) These interactions precisely guide the active site on ADAMTS13 to the cleavage site on VWF, resulting in the eventual cleavage.

the TSP5-8 and CUB domains may interact with VWF-D4 domain [64,65]. Except for the last interaction that occurs between ADAMTS13 and globular VWF, all the other three are buried in globular VWF and require VWF-A2 domain unfolding by hemodynamic forces (Fig. 3A). All these studies lead to a model, in which ADAMTS13 recognizes VWF under static or low flow condition through the interaction between TSP5-CUB in ADAMTS13 and D4 in VWF, colocalizing ADAMTS13 and VWF (Fig. 3B); when high hemodynamic forces elongate VWF, the cryptic exosites are exposed to complementary sites on ADAMTS13 in a step-by-step manner (Fig. 3C); eventually, these interactions guide the Tyr1605-Met1606 scissile bond to the ADAMTS13 active site, allowing the cleavage to take place (Fig. 3D) [66]. On the other hand, ADAMTS13 adopts a folded conformation, in which CUB domains interact with spacer domains [67,68]. This interaction impedes VWF cleavage by ADAMTS13, indicating that VWF-ADAMTS13 may also be regulated by the ADAMTS13 conformational change. Binding to VWF or monoclonal antibody to ADAMTS13 C-terminal domains releases the interaction between spacer and CUB domains (Fig. 3B) [68]. However, existing data cannot rule out the possibility of force-induced ADAMTS13 conformational change.

## 6 FORCE-INDUCED GPIbα DOMAIN UNFOLDING

Two types of mechanical events have been observed when the GPIbα molecule is pulled by force, namely, unfolding of the LRRD and MSD [11,69] (Fig. 4A and B). The former was identified using combined molecular dynamics (MD) simulations and BFP experiments [25,69]. The latter was identified by OT experiments and mutagenesis studies [11]. Pulling GPIbα on the platelet surface, both types of unfolding events are possible, but the two can be distinguished by their different unfolding lengths and by their differential susceptibilities to varying force waveforms. Unfolding of LRRD and MSD generates a length around 36 and 20 nm, respectively. Ramped force can unfold both LRRD and MSD (Fig. 4C and D), whereas clamped force can only unfold MSD (Fig. 4E and F).

## 7 COUPLING BETWEEN UNBINDING KINETICS AND THE UNFOLDING MECHANICS

Interestingly, the occurrence frequency of MSD unfolding was found to depend on the level of clamped force in the same fashion as does the VWF-GPIbα bond lifetime. This is true for both cases of the wild-type (WT) VWF that forms a catch-slip bond with GPIbα and the type 2B VWD mutant VWF that forms a slip-only bond [25], despite that MSD unfolding rate is accelerated by force, that is, behaving as a slip bond regardless of whether GPIbα is pulled by the WT or the mutant VWF. These data suggest a coupling between the unbinding kinetics and the unfolding mechanics. Indeed, BFP experiments showed that LRRD unfolding prolonged VWF-GPIbα

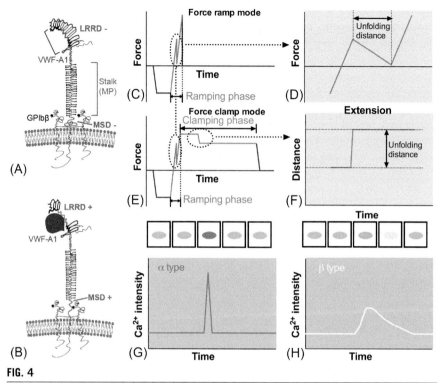

**FIG. 4**

Identification and characterization of GPIbα mechanosensing mechanism. (A and B) Schematics of GPIbα on the platelet membrane, highlighting the folded (−) and unfolded (+) LRRD and MSD. Ligand-binding domain for VWF-A1 and other regions is indicated. (C–F) Illustrative BFP force traces showing zoom-in views of unfolding signatures in both ramping and clamping phases. (G and H) Illustrative analysis of GPIb-mediated single-platelet $Ca^{2+}$ flux. *Top*: Pseudocolored images of intracellular $Ca^{2+}$ in platelets in time sequences. *Bottom*: Time courses of normalized intracellular $Ca^{2+}$ intensity of the α (G) and β (H) types.

*Modified from Chen Y, Ju L, Rushdi MN, Ge C, Zhu C. Receptor-mediated cell mechanosensing. Mol Biol Cell 2017;28(23):3134–3155. https://doi.org/10.1091/mbc.E17-04-0228.*

bond lifetime [25,69]. A simple interpretation for this is a force-induced fit mechanism where LRRD unfolding exposes cryptic binding site(s) for VWF-A1, as suggested by MD simulations [69].

In contrast to the membrane-distal LRRD, the juxtamembrane MSD is separated from the VWF binding site by the >30 nm MP region. The highly glycosylated MP region is thought to be poorly structured and not known to propagate conformational changes allosterically [25]. Without doubt, the MP region can transmit tensile force from the LRRD to MSD. The simplest explanation for the coupling between VWF unbinding kinetics and MSD unfolding kinetics is that to unfold MSD requires sustained force and it cannot occur after force removal by VWF unbinding. Indeed, a mathematical model based on this idea predicts the MSD unfolding frequencies very well [25]. The model has zero freely adjustable fitting parameters and calculates the MSD unfolding frequency entirely using actual measurements of force-dependent VWF-GPIbα unbinding rates and MSD unfolding rates. The model predictions agree with experimental data across the entire tested force range for both WT and type 2B VWD mutant VWF to pull GPIbα and unfold MSD, providing strong support for this hypothetical mechanism.

Importantly, the coupling between ligand-binding kinetics and domain unfolding mechanics generates cooperativity between LRRD and MSD unfolding, such that LRRD unfolding greatly increases the probability of MSD unfolding. This cooperativity occurs only when GPIbα is pulled by WT VWF but not by type 2B VWD mutant VWF and the force reaches maximum at 25 pN where the WT VWF-GPIbα bond lifetime is the longest [25].

## 8 MSD UNFOLDING AND GPIb MECHANOTRANSDUCTION

Recent work has demonstrated that unfolding of MSD results in platelet intracellular signaling [70,71]. The Li and Zhang groups have combined protein engineering and structural analysis with force-ramp DFS via OT to identify the MSD [11] and its trigger sequence [71]. Binding of VWF under physiological shear induced MSD unfolding and intracellular signaling in the platelet. Furthermore, mutations that unfolded the MSD and the juxtamembrane trigger sequence therein induced calcium fluxes, filopodia formation, and P-selectin expression in the absence of ligand binding [71].

At the level of single-molecular interactions, the role of MSD unfolding in GPIb-mediated mechanosensing has been demonstrated using a fBFP, which enables real-time single-molecule, single-cell analysis of ligand-binding kinetics, receptor domain unfolding, and intracellular calcium imaging concurrently [25]. Single-platelet calcium imaging has revealed two types of intracellular $Ca^{2+}$ fluxes induced in platelets interrogated by a VWF-coated probe that forms infrequent, sequential, and intermittent single bonds with GPIbα (Fig. 4G and H): (i) α-type, featured by an initial latent phase followed by a high spike with a quick decay (Fig. 4G), and (ii) β-type, featured by fluctuating signals around the baseline or gradually increasing signals to an intermediate level followed by a gradual decay to baseline (Fig. 4H) [25]. In-depth time-lapse correlation analysis has revealed that MSD unfolding is required to trigger α-type $Ca^{2+}$. Furthermore, the GPIb system meets the criterion for mechanosensing via a cell surface receptor, in which differential information embedded in the force waveform received by the receptor is transduced into distinct biochemical signals. Indeed, the sustainability of force at an optimal level, not the magnitude of the transient force, generates maximum calcium. Moreover, an interfering peptide that disrupts the attachment of the cytoplasmic adaptor protein 14-3-3ζ to GPIbα [72] abolishes the α-type $Ca^{2+}$, implicating the role of 14-3-3ζ as a signal-transducing protein for GPIb-mediated mechanosensing [25].

It is worth noting that the detailed mechanotransduction mechanism is still not fully understood. Based on their spatial proximity [12], potential interaction of GPIbβ with the trigger sequence of GPIbα upon MSD unfolding has been suggested to relay the mechanical information from GPIbα to GPIbβ [71]. However, pulling GPIbβ directly by an antibody did not induce intraplatelet calcium [25]. Although 14-3-3ζ has been shown to bind the cytoplasmic tails of both GPIbα and β subunits [73], how the interfering peptide works is not clear nor are the roles of other GPIb-associated cytoplasmic proteins (i.e., filamin, PI3K, and Src) [74–77]. In a mouse model, mutated GPIbα with normal VWF binding (but lacking anchorage to filamin) causes membrane disruption when platelets translocate on VWF [78], signifying that force is transmitted to the cytoskeleton through the GPIbα bond under normal conditions. Moreover, traction force microscopy demonstrated that filamin is involved in GPIb mechanotransduction and that contractile forces generated from cytoskeleton transmit through GPIbα to strengthen platelet adhesion via an inside-out catch-bond mechanism [79]. Furthermore, the intracellular signaling enzyme PI3-kinase (PI3K) is well known to mediate numerous signaling pathways of platelet function, including integrin upregulation; protease-activated receptor $P_2Y$ and GPVI-mediated platelet activation; and apoptosis [80–83]. It has also been shown to associate with the GPIbα cytoplasmic tail [76] and required for GPIb-IX transfected CHO cells to spread on a VWF surface [84], suggesting that it also plays a role in GPIb-mediated platelet mechanosensing.

## 9 FOUR-STEP MODEL FOR RECEPTOR-MEDIATED MECHANOSENSING

Upon ligand engagement, cell surface receptors can transduce signals intracellularly. When force is exerted on such receptor-ligand bonds, this can initiate cell mechanosensing. Being anuclear cells with simplified signaling machinery and rapid responses to highly variable mechanical environments, platelets represent a natural model for studying receptor-mediated mechanosensing. The single-molecule studies on VWF-GPIbα axis summarize a generic four-step model of the receptor-mediated cell mechanosensing:

**Step 1.** *Mechanopresentation.* In this step, mechanical cues are presented by a ligand for a cell to sense. If sensing requires applying force to a receptor, mechanopresentation requires a ligand that is anchored on a surface to support the force upon its exertion (Fig. 5A step 1). For example, when VWF is immobilized onto the subendothelium following vascular injury, it becomes a ligand for mechanopresentation, or mechanopresenter, because it presents the binding epitope to embarking platelets under shear flow (Fig. 5B step 1). A soluble ligand that does not sustain force cannot present mechanical cues; for example, no force is exerted on the platelet receptors when soluble gain-of-function mutant VWF binds platelets, so the VWF does not serve a mechanopresenter role in this case.

**Step 2.** *Mechanoreception.* In this step, the ligand is engaged with a cell surface receptor onto which force is exerted (Fig. 5A step 2). As a mechanoreceptor, GPIbα receives the mechanical signal and induces conformational changes

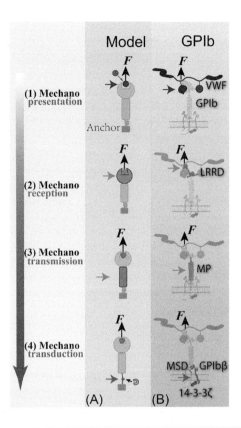

**FIG. 5**

Four-step model of GPIb mechanosensing: (1) mechanopresentation, (2) mechanoreception, (3) mechanotransmission, and (4) mechanotransduction. Arrows indicate the location of each step that is carried out by a distinct molecular component or molecular assembly. *Black arrow* indicates external force, *F*. Steps of mechanosensing of a generic model (A) and the VWF-GPIbα axis system (B) were depicted in correspondence to the proposed model.

*Modified from Chen Y, Ju L, Rushdi MN, Ge C, Zhu C. Receptor-mediated cell mechanosensing. Mol Biol Cell 2017;28(23):3134–3155. https://doi.org/10.1091/mbc.E17-04-0228.*

in the LRRD to alter the bond properties. Thus, the GPIbα LRRD plays the mechanoreception role (Fig. 5B step 2).

**Step 3.** *Mechanotransmission.* In this step, the mechanical signal propagates from the ligand-binding site to the site of mechanotransduction (Fig. 5A step 3). The long stalk MP region of GPIbα plays the mechanotransmission role because the force is transmitted through this linker to the juxtamembrane MSD to induce its unfolding (Fig. 5B step 3).

**Step 4.** *Mechanotransduction.* In this step, certain molecular domain(s) undergo(es) conformational change(s) in response to the force, enabling a chemical event to occur, usually in a location away from where the ligand binds. In this way, the information embedded in the force waveform is carried over biochemically, initiating the downstream biochemical signal transduction cascade. The mechanically sensitive domains, plus other participants of the chemical event, play the mechanotransduction role (Fig. 5A step 4). As an example, force that unfolds GPIbα MSD thereafter exposes the "trigger" sequence, which somehow induces association of 14-3-3ζ to GPIbα cytoplasmic tail and initiates calcium flux (Fig. 5B step 4). Thus, the GPIbα MSD plays the mechanotransduction role in converting mechanical cues to biochemical signals.

It is worth noting that many details of the mechanotransduction step still remain unclear. This is largely due to the complexity of transmembrane and intracellular environment, where even screening for the key players in a biological pathway becomes extremely difficult. Therefore, more efforts will have to be shifted from the cell exterior to interior on discovering the "magical finger snap" of mechanotransduction that converts the signal from one form to another across the membrane.

It should be noted that a single molecule may play multiple roles of mechanoreception, transmission, and transduction for a cell. Conversely, each of these steps may also require multiple molecules to carry out.

## 10 CONCLUSIONS

Researchers in the past decades have focused on developing biochemical and molecular biological approaches, which have allowed scientists to identify and characterize molecular players of mechanosensing pathways. Notably, ultrasensitive force techniques have prompted the mechanobiology field into a new era. It has become possible to address certain fundamental questions: how mechanical cues are presented, received, transmitted, and transduced by extracellular and intracellular proteins. More importantly, the combined kinetic analysis of molecular interactions and mechanical analysis of the molecular conformational changes reveal the inner workings of mechanosensing molecules and their MSDs. Now is the time to elucidate the design and working principles of the mechanosensing apparatus. In the future, full characterization of the mechanosensing machineries will provide synthetic biology design principles for customization of a generic mechanosensing machine.

Unlike the conventional view that platelet thrombus development is driven by agonist diffusion, under rapid and disturbed blood flow conditions, which are typically caused by a partial luminal obstruction (a developing thrombus, an atherosclerotic plaque, or an intravascular device), a change in vessel geometry (extrinsic constriction of blood vessels, vascular bifurcation, or aneurysm), or sudden flow changes (vessel hypoperfusion due to shunting or upstream obstruction), the platelet thrombus has been shown to be mechanically initiated by the shear stress gradients [31]. However, other than the requirement for VWF-GPIbα binding [31,32,85], the molecular insights on the prothrombotic effects of shear stress gradients still remain elusive. Interestingly, the LRRD unfolding may play a role in this scenario because it requires a ramped force (resembles shear stress gradient) [25] and strengthens VWF-GPIbα bonds [69]. The validation of such hypotheses warrants future investigation that combines experiments at both single-molecular and cellular scales.

The prothrombotic effects of shear stress gradients are fundamentally linked to thrombus growth because the three-dimensional geometry of a developing thrombus progressively increases its surrounding shear stress gradients into a feedforward loop of further thrombus formation, which has been increasingly recognized as a key feature that distinguishes thrombosis from normal hemostasis [2]. Clinically, the occurrence of bleeding complications remains a major limitation of current antithrombotic therapy and prevents the use of higher, more effective drug doses [2]. As a result, the efficacy of antithrombotics in preventing secondary acute coronary syndrome events is proved but remains disappointingly low and highly variable in the individual patient [86]. From a clinical perspective, understanding the VWF-GPIbα mechanosensing pathway and its association with prothrombotic effect of shear stress gradients promises the discovery of a potential mechanomedicine "magic bullet" that selectively targets pathological thrombus but does not interfere with normal hemostasis and therefore would lead to less bleeding side effects and would thus be safer and more efficacious than existing antithrombotic therapy.

## NOTES AND ACKNOWLEDGMENTS

This chapter is an extension from our recent review [87]. We thank Shaun P. Jackson for the helpful discussion and providing environment to finish this work. *Funding information*—Dr. Ju was supported by a postdoctoral fellowship with Paul Korner Innovation Award (#101285) from the National Heart Foundation of Australia. This work was supported by the National Institutes of Health/National Heart, Lung, and Blood Institute, the United States (HL132019, C. Z.), and the Royal College of Pathologists of Australasia (Kanematsu research award, L. J.). *Author contributions*—L. J., Y. C., and Z. L. designed the study, prepared figures, and cowrote the chapter; L. J. and C. Z. cosupervised the study. *Conflicting interests*—The authors have no conflicting interests to declare.

## REFERENCES

[1] Ruggeri ZM. Platelets in atherothrombosis. Nat Med 2002;8(11):1227–34.

[2] Jackson SP. Arterial thrombosis—insidious, unpredictable and deadly. Nat Med 2011;17(11):1423–36.

[3] Zhang X, Halvorsen K, Zhang CZ, Wong WP, Springer TA. Mechanoenzymatic cleavage of the ultralarge vascular protein von Willebrand factor. Science 2009;324(5932):1330–4.

[4] Sadler JE. Biochemistry and genetics of von Willebrand factor. Annu Rev Biochem 1998;67:395–424.

[5] Dong JF. Cleavage of ultra-large von Willebrand factor by ADAMTS-13 under flow conditions. J Thromb Haemost 2005;3(8):1710–6.

[6] Mazzucato M, Santomaso A, Canu P, Ruggeri ZM, De Marco L. Flow dynamics and haemostasis. Ann Ist Super Sanita 2007;43(2):130–8.
[7] Ju L, Zhu C. Biomechanical role of von Willebrand factor and glycoprotein Ibα interaction in platelet thrombus formation. Int Rev Thromb 2014;9(3):30–8.
[8] Ruggeri ZM. Platelet and von Willebrand factor interactions at the vessel wall. Hamostaseologie 2004;24(1):1–11.
[9] Ju L, Chen Y, Zhou F, Lu H, Cruz MA, Zhu C. Von Willebrand factor-A1 domain binds platelet glycoprotein Ibalpha in multiple states with distinctive force-dependent dissociation kinetics. Thromb Res 2015;136(3):606–12.
[10] Fox JE, Aggerbeck LP, Berndt MC. Structure of the glycoprotein Ib.IX complex from platelet membranes. J Biol Chem 1988;263(10):4882–90.
[11] Zhang W, Deng W, Zhou L, Xu Y, Yang W, Liang X, et al. Identification of a juxtamembrane mechanosensitive domain in the platelet mechanosensor glycoprotein Ib-IX complex. Blood 2015;125(3):562–9.
[12] McEwan PA, Yang W, Carr KH, Mo X, Zheng X, Li R, et al. Quaternary organization of GPIb-IX complex and insights into Bernard-Soulier syndrome revealed by the structures of GPIbβ and a GPIbβ/GPIX chimera. Blood 2011;118(19):5292–301.
[13] Luo S-Z, Mo X, Afshar-Kharghan V, Srinivasan S, López JA, Li R. Glycoprotein Ibalpha forms disulfide bonds with 2 glycoprotein Ibbeta subunits in the resting platelet. Blood 2007;109(2):603–9.
[14] Ruggeri ZM, Mendolicchio GL. Adhesion mechanisms in platelet function. Circ Res 2007;100(12):1673–85.
[15] Sadler J. New concepts in von Willebrand disease. Annu Rev Med 2005;56(1):173.
[16] Liu B, Chen W, Zhu C. Molecular force spectroscopy on cells. Annu Rev Phys Chem 2015;66(1):427–51.
[17] Springer TA, Dustin ML. Integrin inside-out signaling and the immunological synapse. Curr Opin Cell Biol 2012;24(1):107–15.
[18] Neuman KC, Nagy A. Single-molecule force spectroscopy: optical tweezers, magnetic tweezers and atomic force microscopy. Nat Methods 2008;5(6):491–505.
[19] Dulin D, Lipfert J, Moolman MC, Dekker NH. Studying genomic processes at the single-molecule level: introducing the tools and applications. Nat Rev Genet 2013;14(1):9–22.
[19a] Ju L, Chen Y, Xue L, Du X, Zhu C. Cooperative unfolding of distinctive mechanoreceptor domains transduces force into signals. Elife 2016;5.
[20] Evans E, Leung A, Heinrich V, Zhu C. Mechanical switching and coupling between two dissociation pathways in a P-selectin adhesion bond. Proc Natl Acad Sci U S A 2004;101(31):11281–6.
[21] Kong F, Li Z, Parks WM, Dumbauld DW, García AJ, Mould AP, et al. Cyclic mechanical reinforcement of integrin-ligand interactions. Mol Cell 2013;49:1060–8. https://doi.org/10.1016/j.molcel.2013.01.015.
[22] Li Z, Kong F, Zhu C. A model for cyclic mechanical reinforcement. Sci Rep 2016;6:35954.
[23] Marshall BT, Sarangapani KK, Lou J, McEver RP, Zhu C. Force history dependence of receptor-ligand dissociation. Biophys J 2005;88(2):1458–66.
[24] Sarangapani KK, Qian J, Chen W, Zarnitsyna VI, Mehta P, Yago T, et al. Regulation of catch bonds by rate of force application. J Biol Chem 2011;286(37):32749–61.
[25] Ju L, Chen Y, Xue L, Du X, Zhu C. Cooperative unfolding of distinctive mechanoreceptor domains transduces force into signals. elife 2016;5:e15447. https://doi.org/10.7554/eLife.15447.

[26] Liu B, Chen W, Evavold BD, Zhu C. Accumulation of dynamic catch bonds between TCR and agonist peptide-MHC triggers T cell signaling. Cell 2014;157(2):357–68.
[27] Chen Y, Liu B, Ju L, Hong J, Ji Q, Chen W, et al. Fluorescence biomembrane force probe: concurrent quantitation of receptor-ligand kinetics and binding-induced intracellular signaling on a single cell. J Vis Exp 2015;102:e52975. https://doi.org/10.3791/52975.
[27a] Ju L, Chen Y, Li K, Yuan Z, Liu B, Jackson SP, et al. Dual Biomembrane Force Probe enables single-cell mechanical analysis of signal crosstalk between multiple molecular species. Sci Rep 2017;7(1):14185.
[28] Kim ST, Takeuchi K, Sun ZY, Touma M, Castro CE, Fahmy A, et al. The alphabeta T cell receptor is an anisotropic mechanosensor. J Biol Chem 2009;284(45):31028–37.
[29] Feng Y, Brazin KN, Kobayashi E, Mallis RJ, Reinherz EL, Lang MJ. Mechanosensing drives acuity of alphabeta T-cell recognition. Proc Natl Acad Sci U S A 2017;114(39): E8204-E13.
[30] Sing CE, Alexander-Katz A. Elongational flow induces the unfolding of von Willebrand factor at physiological flow rates. Biophys J 2010;98(9):L35–7.
[31] Nesbitt W, Westein E, Tovar-Lopez F, Tolouei E, Mitchell A, Fu J, et al. A shear gradient-dependent platelet aggregation mechanism drives thrombus formation. Nat Med 2009;15(6):665–73.
[32] Westein E, van der Meer AD, Kuijpers MJE, Frimat J-P, van den Berg A, Heemskerk JWM. Atherosclerotic geometries exacerbate pathological thrombus formation poststenosis in a von Willebrand factor-dependent manner. Proc Natl Acad Sci U S A 2013;110(4):1357–62.
[33] Springer TA. von Willebrand factor, Jedi knight of the bloodstream. Blood 2014; 124(9):1412–25.
[34] Colace TV, Diamond SL. Direct observation of von Willebrand factor elongation and fiber formation on collagen during acute whole blood exposure to pathological flow. Arterioscler Thromb Vasc Biol 2013;33(1):105–13.
[35] Ulrichts H, Udvardy M, Lenting PJ, Pareyn I, Vandeputte N, Vanhoorelbeke K, et al. Shielding of the A1 domain by the D'D3 domains of von Willebrand factor modulates its interaction with platelet glycoprotein Ib-IX-V. J Biol Chem 2006;281(8):4699–707.
[36] Auton M, Sowa KE, Behymer M, Cruz MA. N-terminal flanking region of A1 domain in von Willebrand factor stabilizes structure of A1A2A3 complex and modulates platelet activation under shear stress. J Biol Chem 2012;287(18):14579–85.
[37] Ju L, Dong J-F, Cruz MA, Zhu C. The N-terminal flanking region of the A1 domain regulates the force-dependent binding of von Willebrand factor to platelet glycoprotein Ibα. J Biol Chem 2013;288(45):32289–301.
[38] Deng W, Wang Y, Druzak SA, Healey JF, Syed AK, Lollar P, et al. A discontinuous autoinhibitory module masks the A1 domain of von Willebrand factor. J Thromb Haemost 2017;15(9):1867–77.
[39] Martin C, Morales LD, Cruz MA. Purified A2 domain of von Willebrand factor binds to the active conformation of von Willebrand factor and blocks the interaction with platelet glycoprotein Ibalpha. J Thromb Haemost 2007;5(7):1363–70.
[40] Aponte-Santamaria C, Huck V, Posch S, Bronowska AK, Grassle S, Brehm MA, et al. Force-sensitive autoinhibition of the von Willebrand factor is mediated by interdomain interactions. Biophys J 2015;108(9):2312–21.
[40a] Butera D, Passam F, Ju L, Cook KM, Woon H, Aponte-Santamaria C, et al. Autoregulation of von Willebrand factor function by a disulfide bond switch. Sci Adv 2018;4(2): eaaq1477.

[41] Barg A, Ossig R, Goerge T, Schneider MF, Schillers H, Oberleithner H, et al. Soluble plasma-derived von Willebrand factor assembles to a haemostatically active filamentous network. Thromb Haemost 2007;97(4):514–26.

[42] Schneider SW, Nuschele S, Wixforth A, Gorzelanny C, Alexander-Katz A, Netz RR, et al. Shear-induced unfolding triggers adhesion of von Willebrand factor fibers. Proc Natl Acad Sci U S A 2007;104(19):7899–903.

[43] Fu H, Jiang Y, Yang D, Scheiflinger F, Wong WP, Springer TA. Flow-induced elongation of von Willebrand factor precedes tension-dependent activation. Nat Commun 2017;8(1):324.

[44] Savage B, Sixma JJ, Ruggeri ZM. Functional self-association of von Willebrand factor during platelet adhesion under flow. Proc Natl Acad Sci U S A 2001;99(1):425–30.

[45] Shankaran H, Alexandridis P, Neelamegham S. Aspects of hydrodynamic shear regulating shear-induced platelet activation and self-association of von Willebrand factor in suspension. Blood 2003;101(7):2637–45.

[46] Dayananda KM, Singh I, Mondal N, Neelamegham S. von Willebrand factor self-association on platelet GpIbalpha under hydrodynamic shear: effect on shear-induced platelet activation. Blood 2010;116(19):3990–8.

[47] Yago T, Lou J, Wu T, Yang J, Miner JJ, Coburn L, et al. Platelet glycoprotein Ibα forms catch bonds with human WT vWF but not with type 2B von Willebrand disease vWF. J Clin Invest 2008;118(9):3195–207.

[48] Kumar RA, Dong J-F, Thaggard JA, Cruz MA, López JA, Mcintire LV. Kinetics of GPIbalpha-vWF-A1 tether bond under flow: effect of GPIbalpha mutations on the association and dissociation rates. Biophys J 2003;85(6):4099–109.

[49] Doggett TA, Girdhar G, Lawshé A, Schmidtke DW, Laurenzi IJ, Diamond SL, et al. Selectin-like kinetics and biomechanics promote rapid platelet adhesion in flow: the GPIb(alpha)-vWF tether bond. Biophys J 2002;83(1):194–205.

[50] Doggett TA, Girdhar G, Lawshé A, Miller JL, Laurenzi IJ, Diamond SL, et al. Alterations in the intrinsic properties of the GPIbalpha-VWF tether bond define the kinetics of the platelet-type von Willebrand disease mutation, Gly233Val. Blood 2003;102(1):152–60.

[51] Ju L, Qian J, Zhu C. Transport regulation of two-dimensional receptor-ligand association. Biophys J 2015;108(7):1773–84.

[52] Zhu C, Yago T, Lou J, Zarnitsyna VI, McEver RP. Mechanisms for flow-enhanced cell adhesion. Ann Biomed Eng 2008;36(4):604–21.

[53] Savage B, Saldívar E, Ruggeri ZM. Initiation of platelet adhesion by arrest onto fibrinogen or translocation on von Willebrand factor. Cell 1996;84(2):289–97.

[54] Springer TA. Biology and physics of von Willebrand factor concatamers. J Thromb Haemost 2011;9(Suppl. 1):130–43.

[55] De Ceunynck K, De Meyer SF, Vanhoorelbeke K. Unwinding the von Willebrand factor strings puzzle. Blood 2013;121(2):270–7.

[56] Zheng X, Chung D, Takayama TK, Majerus EM, Sadler JE, Fujikawa K. Structure of von Willebrand factor-cleaving protease (ADAMTS13), a metalloprotease involved in thrombotic thrombocytopenic purpura. J Biol Chem 2001;276(44):41059–63.

[57] Furlan M, Robles R, Lammle B. Partial purification and characterization of a protease from human plasma cleaving von Willebrand factor to fragments produced by in vivo proteolysis. Blood 1996;87(10):4223–34.

[58] Zhang Q, Zhou YF, Zhang CZ, Zhang X, Lu C, Springer TA. Structural specializations of A2, a force-sensing domain in the ultralarge vascular protein von Willebrand factor. Proc Natl Acad Sci U S A 2009;106(23):9226–31.

[59] Wu T, Lin J, Cruz MA, Dong JF, Zhu C. Force-induced cleavage of single VWFA1A2A3 tridomains by ADAMTS-13. Blood 2010;115(2):370–8.

[60] Gao W, Anderson PJ, Majerus EM, Tuley EA, Sadler JE. Exosite interactions contribute to tension-induced cleavage of von Willebrand factor by the antithrombotic ADAMTS13 metalloprotease. Proc Natl Acad Sci U S A 2006;103(50):19099–104.

[61] Akiyama M, Takeda S, Kokame K, Takagi J, Miyata T. Crystal structures of the noncatalytic domains of ADAMTS13 reveal multiple discontinuous exosites for von Willebrand factor. Proc Natl Acad Sci U S A 2009;106(46):19274–9.

[62] Gao W, Anderson PJ, Sadler JE. Extensive contacts between ADAMTS13 exosites and von Willebrand factor domain A2 contribute to substrate specificity. Blood 2008;112(5):1713–9.

[63] de Groot R, Bardhan A, Ramroop N, Lane DA, Crawley JT. Essential role of the disintegrin-like domain in ADAMTS13 function. Blood 2009;113(22):5609–16.

[64] Zanardelli S, Chion AC, Groot E, Lenting PJ, McKinnon TA, Laffan MA, et al. A novel binding site for ADAMTS13 constitutively exposed on the surface of globular VWF. Blood 2009;114(13):2819–28.

[65] Feys HB, Anderson PJ, Vanhoorelbeke K, Majerus EM, Sadler JE. Multi-step binding of ADAMTS-13 to von Willebrand factor. J Thromb Haemost 2009;7(12):2088–95.

[66] Crawley JTB, de Groot R, Xiang Y, Luken BM, Lane DA. Unraveling the scissile bond: how ADAMTS13 recognizes and cleaves von Willebrand factor. Blood 2011;118(12):3212–21.

[67] South K, Luken BM, Crawley JT, Phillips R, Thomas M, Collins RF, et al. Conformational activation of ADAMTS13. Proc Natl Acad Sci U S A 2014;111(52):18578–83.

[68] Muia J, Zhu J, Gupta G, Haberichter SL, Friedman KD, Feys HB, et al. Allosteric activation of ADAMTS13 by von Willebrand factor. Proc Natl Acad Sci U S A 2014;111(52):18584–9.

[69] Ju L, Lou J, Chen Y, Li Z, Zhu C. Force-induced unfolding of leucine-rich repeats of glycoprotein Ibα strengthens ligand interaction. Biophys J 2015;109(9):1781–4.

[70] Chen Y, Lee H, Tong H, Schwartz M, Zhu C. Force regulated conformational change of integrin alphaVbeta3. Matrix Biol 2017;60–61:70–85.

[71] Deng W, Xu Y, Chen W, Paul DS, Syed AK, Dragovich MA, et al. Platelet clearance via shear-induced unfolding of a membrane mechanoreceptor. Nat Commun 2016;7:12863.

[72] Dai K, Bodnar R, Berndt MC, Du X. A critical role for 14-3-3zeta protein in regulating the VWF binding function of platelet glycoprotein Ib-IX and its therapeutic implications. Blood 2005;106(6):1975–81.

[73] Calverley DC, Kavanagh TJ, Roth GJ. Human signaling protein 14-3-3ζ interacts with platelet glycoprotein Ib subunits Ibα and Ibβ. Blood 1998;91(4):1295–303.

[74] Kanaji T, Ware J, Okamura T, Newman PJ. GPIb regulates platelet size by controlling the subcellular localization of filamin. Blood 2012;119(12):2906–13.

[75] Nakamura F, Pudas R, Heikkinen O, Permi P, Kilpeläinen I, Munday AD, et al. The structure of the GPIb-filamin A complex. Blood 2006;107(5):1925–32.

[76] Mu F-T, Andrews RK, Arthur JF, Munday AD, Cranmer SL, Jackson SP, et al. A functional 14-3-3zeta-independent association of PI3-kinase with glycoprotein Ib alpha, the major ligand-binding subunit of the platelet glycoprotein Ib-IX-V complex. Blood 2008;111(9):4580–7.

[77] Liu J, Pestina TI, Berndt MC, Jackson CW, Gartner TK. Botrocetin/VWF-induced signaling through GPIb-IX-V produces TxA2 in an alphaIIbbetA3- and aggregation-independent manner. Blood 2005;106(8):2750–6.

[78] Cranmer S, Ashworth K, Yao Y, Berndt M, Ruggeri Z, Andrews R, et al. High shear-dependent loss of membrane integrity and defective platelet adhesion following disruption of the GPIb {alpha}-filamin interaction. Blood 2011;117(9):2718.

[79] Feghhi S, Munday AD, Tooley WW, Rajsekar S, Fura AM, Kulman JD, et al. Glycoprotein Ib-IX-V complex transmits cytoskeletal forces that enhance platelet adhesion. Biophys J 2016;111(3):601–8.

[80] Kim S, Mangin P, Dangelmaier C, Lillian R, Jackson SP, Daniel JL, et al. Role of phosphoinositide 3-kinase beta in glycoprotein VI-mediated Akt activation in platelets. J Biol Chem 2009;284(49):33763–72.

[81] Sun DS, Lo SJ, Tsai WJ, Lin CH, Yu MS, Chen YF, et al. PI3-kinase is essential for ADP-stimulated integrin alpha(IIb)beta3-mediated platelet calcium oscillation, implications for P2Y receptor pathways in integrin alpha(IIb)beta3-initiated signaling cross-talks. J Biomed Sci 2005;12(6):937–48.

[82] Lian L, Wang Y, Draznin J, Eslin D, Bennett JS, Poncz M, et al. The relative role of PLCbeta and PI3Kgamma in platelet activation. Blood 2005;106(1):110–7.

[83] Zhang S, Ye J, Zhang Y, Xu X, Liu J, Zhang SH, et al. P2Y12 protects platelets from apoptosis via PI3k-dependent Bak/Bax inactivation. J Thromb Haemost 2013;11(1):149–60.

[84] Mu FT, Cranmer SL, Andrews RK, Berndt MC. Functional association of phosphoinositide-3-kinase with platelet glycoprotein Ibalpha, the major ligand-binding subunit of the glycoprotein Ib-IX-V complex. J Thromb Haemost 2010;8(2):324–30.

[85] Jain A, Graveline A, Waterhouse A, Vernet A, Flaumenhaft R, Ingber DE. A shear gradient-activated microfluidic device for automated monitoring of whole blood haemostasis and platelet function. Nat Commun 2016;7:10176.

[86] Antithrombotic Trialists Collaboration. Baigent C, Blackwell L, Collins R, Emberson J, Godwin J, et al. Aspirin in the primary and secondary prevention of vascular disease: collaborative meta-analysis of individual participant data from randomised trials. Lancet 2009;373(9678):1849–60.

[87] Chen Y, Ju L, Rushdi MN, Ge C, Zhu C. Receptor-mediated cell mechanosensing. Mol Biol Cell 2017;28(23):3134–55. https://doi.org/10.1091/mbc.E17-04-0228.

# Mechanobiology of primary cilia in the vascular and renal systems

## 10

**Surya M. Nauli**\*,†, **Ashraf M. Mohieldin**\*, **Madhawi Alanazi**\*, **Andromeda M. Nauli**‡

*Department of Biomedical & Pharmaceutical Sciences, Chapman University, Irvine, CA, United States*\* *Department of Medicine, University of California Irvine, Irvine, CA, United States*† *Department of Pharmaceutical Sciences, Marshall B. Ketchum University, Fullerton, CA, United States*‡

## ABBREVIATION

| | |
|---|---|
| OCRL1 | oculocerebrorenal syndrome of Lowe 1 |
| PC1 | polycystin-1 |
| PC2 | polycystin-2 |
| PKD | polycystic kidney disease |

## 1 INTRODUCTION

Mechanobiology is an interdisciplinary science aiming to understand the effects of physical and mechanical forces on cells, tissues, and organ systems. The mechanobiology of the heart, blood vessels, skin, eye, and other organs has rapidly developed into an interesting topic in biomedical research. Further, cellular mechanical forces are known to exist in all organ systems in our body [1]. These mechanical forces therefore contribute to a wide spectrum of our bodily functions, from cell differentiation to organ development and from normal physiological functions to pathogeneses of various mechanodiseases.

Primary cilia have recently been recognized to be important mechanosensory organelles for sensing fluid shear force. Primary cilia are organelles that house various mechanosensory receptors, ion channels, and signaling molecules to transduce extracellular signals into intracellular responses. The study of cilia-related diseases has expanded to various medical branches, including cardiology, nephrology, neurology, and vascular biology. Cilia-targeted therapy is relatively a new concept in the field of mechanobiology. This chapter will focus on current investigations into the mechanosensory role of primary cilia in vascular endothelia and renal epithelia.

## 2 PRIMARY CILIA AS MECHANICAL SENSORS

Pinpointing the etiology of mechanodiseases *in vivo* is not an easy task. This is primarily due to two main factors. The *first factor*, described by Nauli, is that different biophysical properties can alter mechanical forces generated or experienced by one another [1a]. For example, at least five different physical forces are recognized within a blood vessel (Table 1). If pressure force was terminated through occlusion in one segment of an artery, the artery would lose the ability to sense stretch, strain, compression, or shear stress forces. The *second factor*, described by Ingber, is that a physical force applied to a biological system does not necessarily have a classic "stimulus-response" coupling [2]. Any external forces would not simply alter the cellular response, but would first impose on the preexisting forces. As such, many signaling pathways may still be better studied at a cellular and molecular level to have more precise control of surrounding mechanical forces imposed on a single cell.

Although various mechanical forces can be sensed by a single cell through multiple different sensing mechanisms, a more specific question is how cells can differentially transfer these forces to carry out a specific unique response. The primary cilium is believed to be an organelle that is particularly sensitive to fluid shear stress when compared to other mechanical forces. Once thought to be a dormant vestigial organelle, the primary cilium has garnered attention as a chemical and mechanical sensory organelle. A cilium is an antenna-like structure found on the apical surface of most mammalian cells (Fig. 1). As a sensory organelle, a cilium can be described as having six distinct compartments.

The *first* compartment is the ciliary membrane, which has a lipid composition different from that of the rest of plasma membrane [3–5]. Many sensory ion channels and receptors are localized in this domain to support the mechanosensory roles of cilium [6,7]. Ligand-activated receptors have also been observed to localize in this domain to support the chemical-sensing functions of cilia [8]. The *second* compartment is the soluble compartment or cilioplasm, also known as the matrix compartment. This compartment is made of fluid material to support various signaling proteins [9–11]. Cilioplasm also has a fairly visible bulb-like structure, whose function is still not known [12,13]. The *third* compartment is the axoneme, which is composed of nine pairs of microtubules built from α- and β-tubulin subunits that form a heterodimeric structure. The microtubules are posttranslationally modified

**Table 1** Mechanical Forces Within Blood Vessel

| Types of Forces | Differentiations of Forces |
| --- | --- |
| Stretch | Distention force by surrounding muscle |
| Cyclic strain | Pulsatile force by turbulent blood flow |
| Compression | Contractile force by differential pressure in the vessel |
| Pressure | Systolic force by blood flow |
| Shear stress | Drag force along the intima surface by blood flow |

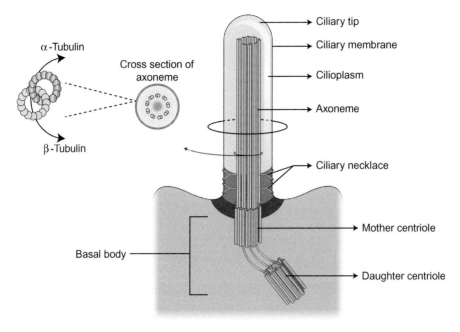

**FIG. 1**

Distinct compartments in primary cilia. The primary cilium protrudes from the cell surface into the extracellular milieu. There are at least six distinct compartments within a cilium, including the ciliary membrane, cilioplasm (soluble or matrix compartment), axoneme, ciliary tip, basal body, and ciliary necklace (transition zone). The cross section of the axoneme shows compositions of nine pairs of microtubules built from α- and β-tubulin subunits. These subunits form a total of nine heterodimer structures. The basal body is composed of two centrioles, and the primary cilium extends from mother centriole.

to support the long ciliary structure [14–16]. The axoneme plays an important role for intraflagellar transport proteins to deliver cellular components into and out of the ciliary shaft. A proper axonemal complex is needed to support assembly and maintenance of the long ciliary structure. Based on its axonemal structure, a cilium can be classified into "9+0" and "9+2" organelles [17]. While most of "9+0" cilia are considered nonmotile solitary cilia (also termed primary cilia), the "9+2" cilia are usually motile bundled cilia (also termed flagella). The *fourth* compartment is the tip or the distal part of a cilium. The ciliary tip contains specialized protein complexes whose roles still need to be further explored [18–20]. The *fifth* compartment is the basal body, which is a "mature" or "mother" centriole from which the primary cilium is projected [21–23]. Because centrosomes will become anchoring points for mitotic spindles during mitosis, many centriole proteins have thus been associated with cell division [24,25]. The *sixth* compartment is the ciliary necklace or the transition zone [23,26–29]. This compartment is thought to regulate protein and other ciliary machinery, and entry and exit to cilioplasm or the ciliary membrane. As a result, many receptors, ion channels, transporters, and sensory proteins are accumulated at this barrier to be sorted for transport to the cilium.

Much research has been carried out into how the primary cilium senses mechanical signals in the mechanotransduction process. Current theories suggest that mechanical stimulation in the form of fluid flow results in deflection of the primary cilium, causing the initiation of downstream signalling cascades. Recently, modeling techniques have been applied to investigate the mechanics of the primary cilium, predicting its behavior under in vitro fluid flow stimulation. These studies characterized the mechanical behavior of the primary cilium, with analytical and computational predictions of cilium deflection being matched to experimentally observed bending profiles of cilium deflection under fluid flow in vitro [29a,146]. This behavior was further investigated by applying multiphysics models to replicate the in vitro and in vivo mechanical environments of the primary cilia of bone cells [29b,29c].

## 3 CILIOPATHIES

A ciliopathy is defined as a disorder caused by abnormal ciliary function and structure. A ciliopathy may also include an abnormal downstream signaling pathway that blocks the signal transduction from the cilium into a normal cell function or response. An example of this is survivin; although it is not localized to the ciliary compartment, the expression of survivin is greatly dependent on ciliary function [30,31]. Given that cellular sensing is critical for a living cell to receive external clues, it is not surprising that abnormal cilia are implicated in most branches of medicine (Table 2). The mechanism by which primary cilia are involved in these pathogeneses is still largely unknown.

## 4 MECHANOBIOLOGY

Several independent studies have indicated that primary cilia function as mechanosensory compartments to house different mechanosensitive proteins, such as fibrocystin [73,74], polycystin-1 (PC1) [75,76], dopamine receptor-type 5 [77–79], TRPC1 [80–82], TRPP2 or polycystin-2 (PC2) [83], TRPP3 or polycystin-L [82,84], TRPV4 [85,86], and potential other channel receptor proteins. Studies on primary cilia in different organ systems have also confirmed that cilia are responsive to mechanical fluid shear stress. Independent studies have shown the roles of cilia in fluid sensing in cholangiocytes [87], embryonic node [88–90], left-right organizer of zebra fish [88], osteocytes [91], renal epithelia [47,74,92,93], smooth muscle cells [94], and vascular endothelia [33].

In all of these studies, calcium signaling in either cilioplasm or cytoplasm was used as ciliary readout [88,95–97]. Calcium signaling can be achieved by activating primary cilia via bending. This cilia bending can be achieved through micropipette [98], apical fluid perfusion through changing the flow rate [99,100], or twisting using

Table 2  Mechanotransduction Diseases

| Medical Branches | Examples | References |
| --- | --- | --- |
| Cardiology | Hypertension | [32–34] |
| Dermatology | Basaloid hyperplasia, hypotrichosis, pigmentation | [35–37] |
| Gastroenterology | Pancreatitis and pancreatic cysts | [38–40] |
| Endocrinology | Diabetes, obesity | [41–45] |
| Nephrology | Polycystic kidney disease | [46,47] |
| Neurology | Bardet-Biedl syndrome, epilepsy, mental retardation | [48–51] |
| Oncology | Tumorigenesis | [52,53] |
| Ophthalmology | Blindness | [54–56] |
| Orthopedics | Idiopathic scoliosis | [57–59] |
| Pediatrics | Polycystic kidney disease | [60–63] |
| Pulmonary medicine | Chronic obstructive pulmonary disease | [64,65] |
| Reproductive medicine | Infertility | [66,67] |
| Urology | Polycystic kidney disease and prostate enlargement | [68–70] |
| Vascular biology | Aneurysm and atherosclerosis | [30,71,72] |

magnetic beads [101]. Upon bending, the initial calcium influx into the cilia results in the gradual development of calcium-induced calcium-release mechanisms from intracellular stores [96,102]. Large increases in the calcium levels of a cell may activate calcium-sensitive channels or calcium-dependent processes ranging from cell proliferation to cell death.

Genetically mutated cells that do not have primary cilia show a loss in response to fluid shear stress mechanosensation. Observation of the flow-induced calcium response in cells lacking cilia demonstrated that without proper structural cilia, a cellular response to fluid flow is not detected, even though all sensory machineries are still present [34]. This indicates that having sensory proteins is insufficient to retain sensory function of cells. The sensory proteins must be properly housed in the sensory compartment within primary cilia.

## 5 CILIA LENGTH REGULATION

The notion of the cilium functioning as cellular antennae has led us to postulate that longer antennae (i.e., cilia) would provide better signal reception (i.e., function). This has prompted many scientists to examine signaling molecules that regulate cilia length, which are tissue-specific and dependent on extracellular chemical and mechanical stimuli. More specifically, different stimuli in distinct tissues could have a different impact on cilia length.

One of the many regulators of cilia length includes the orphan GPR22, a rhodopsin-like GPCR that couples to Gαi/Gαo to inhibit adenylyl cyclase. GPR22 regulates cilia length and structure and left-right symmetry in zebra fish [103]. Another regulator of cilia length is oculocerebrorenal syndrome of Lowe 1 (OCRL1), a lipid phosphatase that has been shown to modulate cilia length in renal epithelial cells. OCRL1 knockdown leads to the elongation of cilia and blunted intracellular $Ca^{2+}$ release in response to ATP [104]. There are likely multiple regulatory proteins and/or pathways that control cilium length. It has been suggested that cAMP clearance, MAPK signaling, and phosphorylation of ciliary modulator proteins are also involved in the regulation of cilia length by mechanical and chemical stimuli [105].

Of interest for research into the treatment of cilia-associated diseases is a mechanism to regulate cilia length, which is a tissue-specific phenomenon and dependent on extracellular chemical and mechanical stimuli. Renal injury [106,107] and cisplatin treatment [108,109], among other stimuli, induce the elongation of primary cilia [105,110]. Fluid shear stress also alters the mechanosensory properties of cilia [34].

## 5.1 CILIOTHERAPY IN VASCULAR HYPERTENSION

The concept of ciliotherapy has been advanced based on two-dimensional small molecular screening in determining ciliotherapeutic effect of pharmacological agents on cilia length (first dimension) and cilia function (second dimension) [79,111]. It is generally known that cilia length varies from cell to cell, as well as from condition to condition. Most importantly, changes in cilia length dictate cilia function. Cell sensitivity to fluid shear stress tends to be increased in longer cilia [112,113]. There could potentially be many pharmacological agents that could alter cilia length and function, such as dopamine [114], fenoldopam [115], and rapamycin [116]. In particular, activation of dopamine receptor-type 5 (DR5) in primary cilia increases cilia length through cofilin and actin polymerization. Furthermore, in mechanoinsensitive cells, DR5 activation has been shown to restore cilia function [79]. Because the chemosensory function of cilia via DR5 can alter the mechanosensory function through changes in sensitivity to fluid shear stress, it has been proposed that DR5 has functional chemo- and mechanosensory roles in primary cilia [79].

In clinical settings, patients with abnormal vascular endothelial cilia have presented with uncontrolled hypertension [31]. Activating DR5 can be used as a potential mechanotherapy by altering the mechanosensory function of primary cilia. This type of therapy has become known as ciliotherapy [111]. Initial drug screening has indicated that activation of DR5 increases nitric oxide (NO) biosynthesis in response to fluid shear stress in vascular endothelia. This, in turn, resulted in decreases in the overall blood pressure in the mouse model [111]. Furthermore, in clinical studies, hypertensive patients with abnormal cilia function have a significantly lower baseline level of NO compared with hypertensive-only patients. It has been demonstrated that DR5 activation decreases blood pressure in these patients [111,117,118].

**Table 3** Dopamine Receptor Overview

| Receptor Subtype | Location | Function | Effect |
| --- | --- | --- | --- |
| D1 | Central nervous system | Neuronal growth and development | Stimulatory |
| | Pulmonary artery | Vasodilation | |
| D2 | Pulmonary artery | Vasodilation | Inhibitory |
| D3 | Central nervous system | Cognition and emotion regulation | Inhibitory |
| D4 | Atria of the heart | Increased cardiac output and contractility | Inhibitory |
| | Pulmonary artery | Vasodilation | |
| D5 | Endothelial cilia of blood vessel | Blood pressure regulation Vasodilation | Stimulatory |

It is, however, worth noting that the activation of dopamine receptor is not a straightforward solution for patients with ciliopathies. Given that the dopaminergic system has a range of very diverse functions (Table 3), a more specific target of DR5 is needed if it is to be used as a ciliotherapy. In the central nervous system [119–121], dopamine receptors are involved in a number of neurological processes and are common targets of neurological drugs. They are involved in vital functions such as motion, neuroendocrine secretion, cognition and locomotion, reward, sleep, working memory, and learning. In the periphery [122–124], receptor subtypes D1, D2, D4, and D5 are expressed in pulmonary arteries, which are responsible for the vasodilator effects of dopamine in the cardiovascular system. Additionally, the pharmacological aspects of D4 receptors found in the atria of the heart indicate that dopamine signaling has a unique role of increasing cardiac output and contractility, without affecting heart rate. Similarly, dopamine receptors are also found throughout the renal system, including the renal tubules and vessels. Their main function is associated with diuresis, and they facilitate sodium excretion. Defective dopamine receptors in the kidneys have been reported to contribute to the development of hypertension.

## 5.2 MECHANOSENSORY CILIA IN VASCULAR SYSTEM

Aside from hypertension, the mechanosensory cilia function within vascular endothelia has also been associated with aneurysms and atherosclerosis formation within the cardiovascular system. Although primary cilia dysfunction is certainly related to both the formation of aneurysms and hypertension, hypertension does not necessarily cause aneurysms. Contrary to many beliefs, independent studies have shown that the development of an aneurysm involves a different process to that which causes hypertension [30,125,126]. It is therefore conceivable to understand the simple role of cilia as sensory organelles having much broader implications for vascular diseases, including hypertension, aneurysm, and atherosclerosis.

### 5.2.1 Aneurysms

An aneurysm is a localized, abnormally weak area of a blood vessel wall, with swelling and abnormal expansion in the blood vessel causing an outward bulging. Aneurysms present no observable symptoms in patients and usually go unnoticed until a catastrophic rupture occurs, regularly resulting in death of the patient due to uncontrolled internal bleeding. While aneurysms may occur in any vessel, common lethal examples include aneurysms of the thoracic and abdominal aortas, as well as the Circle of Willis in the brain.

Abnormal function or structure of primary cilia leads to abnormal calcium regulation in endothelia [33,34], and this has been associated with aneurysm formation. Further, calcium regulation by primary cilia is important to regulate PKC-Akt-NFκB-survivin pathway [30,31]. Abnormal function of mechanosensory cilia leads to survivin downregulation via PKC-Akt-NFκB-survivin pathway, which further induces aberrant ploidy formation. Disrupted survivin expression also causes abnormal cytokinesis, which results in multimitotic spindle formation and aberrant cell division orientation. The asymmetrical cell division together with abnormal planar cell polarity is believed to contribute to the expansion of tissue architecture, resulting in the expansion of arterial diameter (Fig. 2). As such, a genetic disorder that results in abnormal mechanosensory cilia function may result in loss of planar cell polarity. Direction of cell division becomes randomized in abnormal planar cell polarity, resulting in increasing vessel diameter rather than vessel elongation. If this bulging of a blood vessel continues, the vessel wall will become thin resulting in an aneurysm. Overall, the current working hypothesis indicates that the role of primary cilia may involve the following molecular and cellular pathway: primary cilia mechanosensation → calcium → PKC → Akt → NFκB → survivin → cytokinesis → polyploidy → asymmetrical cell division → planar cell polarity → vascular aneurysm (architecture expansion).

### 5.2.2 Atherosclerosis

In the simplest terms, atherosclerosis is a tissue buildup or plaque that blocks or restricts blood flow in the narrowed artery. While there are no initial symptoms as the vessel narrows, atherosclerosis can result in coronary artery disease, heart attack, stroke, peripheral artery disease, and kidney diseases. Although the exact mechanism by which it develops is still unclear, primary cilia in endothelia have also been associated with atherosclerosis [72]. Primary cilia tend to be present more in regions of atherogenesis, where they increase in number under hyperlipidemia-induced lesion formation. It has been shown that flow disturbance leads to induction of primary cilia and subsequently to atherogenesis, which suggests a role for primary cilia in endothelial activation and dysfunction. Interestingly, primary cilia actually protect against atherosclerosis. Removing endothelial cilia results in increased inflammatory gene expression and decreased eNOS activity, indicating that endothelial cilia inhibit proatherosclerotic signaling in the vascular system [71].

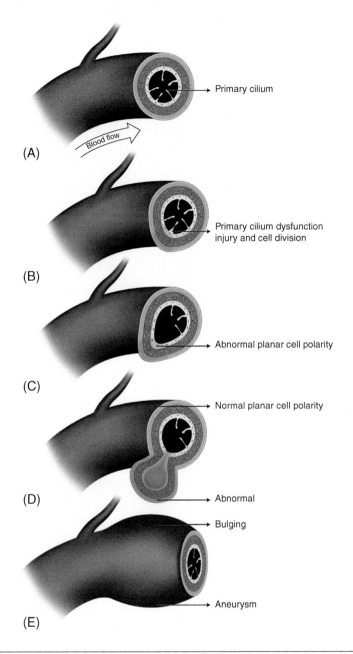

**FIG. 2**

Hypothetical model of aneurysm formation. The illustration depicts mechanosensory function of vascular endothelial cilia. (A) Each cilium plays an important role to transmit extracellular information, such as blood flow, into the cell. This message may provide critical signals to the cell regarding the direction of cell division along the blood vessel. (B) Insults, such as genetic disorder or random mutation, will result in abnormal ciliary function to sense fluid movement. (C) The functional abnormality in ciliary sensing may result in loss of planar cell polarity. (D) Direction of cell division becomes randomized, resulting in increasing vessel diameter rather than vessel elongation. (E) Bulging of the blood vessel continues resulting in vascular aneurysm. Randomized formation of vascular aneurysm along the vascular tree is illustrated on the bottom left corner. The scale of the primary cilia and bulging is exaggerated here for illustrative purposes.

## 5.3 MECHANOSENSORY CILIA IN RENAL SYSTEM

The first evidence establishing primary cilia as important mechanosensory organelles came from studies in the renal epithelial cells [47,98]. In these early studies, either primary cilia or sensory ciliary protein was ablated, resulting in the loss of mechanical sensitivity of the renal epithelial cells. Further studies indicated that cilia dysfunction causes cellular responses specific to fluid shear stress and not to other types of mechanical stimuli [34]. Encouraged by the finding that the *Caenorhabditis elegans* polycystin homologues function in cilia of sensory neurons [127,128], a study in *Chlamydomonas* further demonstrated that the formation of cilia (flagella) depends on the *ift88* gene [129]. Mutation of this orthologous gene in the mouse (*Tg737* gene) results in polycystic kidney disease (PKD), in which the kidneys reveal either an absence of cilia or abnormally shortened cilia [130]. This suggests that the cilia structure and function are required to maintain normal architecture in the kidneys, as in the vasculature.

### 5.3.1 Mechanosensory polycystins in cilia

In addition to polycystic kidney disease, mice without primary cilia exhibit liver and pancreatic cysts, hydrocephalus, skeletal defects, neural tube abnormalities, and many other malformations [129,131–133]. Many of these phenotypes have been recapitulated in PKD1 [134–139] and PKD2 [140–143] mouse models. The discovery of the mechanosensory role of ciliary polycystins was established from the early finding that both PC1 and PC2 are localized to the primary cilium [47,76,83]. This led to the hypothesis and subsequent evidence that PC1 and PC2 participate in sensing fluid shear stress [46,47,76,83]. Using genetically engineered mice, mutation in either PC1 or PC2 results in the loss of response to fluid shear stress. This result led to a proposal that PC1 and PC2 form a PC1-PC2 protein complex in the renal epithelial primary cilia to mediate mechanosensation [144,145]. Because the removal of extracellular calcium abolishes flow-induced cytosolic calcium signaling [47,98], it is believed that PC1 with its large extracellular mechanosensory domains interacts with PC2, serving as specific mechanosensitive calcium channels. In this capacity, while both PC1 and PC2 are crucial for cellular response to fluid-flow stimulation, PC2 specifically functions as a calcium gatekeeper to allow calcium ions to enter the cell in response to mechanical fluid shear stress. Since then, other laboratories have independently shown that primary cilia and polycystins play essential roles in response to flow stimulation in renal epithelia [146–148], as well as in adipose-derived stem cells [149], cholangiocytes [87], Hensen's node [89,90], osteocytes [91], and smooth muscle cells [94,150], among others.

How abnormal mechanobiology results in polycystic kidney disease is currently not fully understood. However, the process is thought to be similar to that seen in aneurysm formation, that is, abnormal cell growth and planar cell polarity. Dysfunction in cilia or ablation of ciliary proteins has been associated with abnormal planar cell polarity [151–153]. Within a single functional unit of kidney nephron, planar cell polarity is defined as cell organization where cells are arranged in a plane of tissue perpendicular to the apical-basal axis as a direction for the orientation of cell division. Cystic kidneys from the PKD patients [30,154] and different mouse models of cystic kidney disease [30,155,156] also show abnormal mitotic orientation during cell division. It is therefore postulated that abnormal mechanosensation in renal epithelia would result in abnormal planar cell polarity, which in turn triggers tubular enlargement in the kidney. The net result is abnormal architecture in the kidney.

### 5.3.2 Mechanosensory primary cilia in kidney function

Apart from calcium signaling, many ion channel and tubular transport activities are known to be induced by flow stimulation. Tubular transport of glucose [157], sodium [158], potassium [159,160], chloride [158], calcium [161,162], and magnesium [161,162] is affected by tubular flow. Tubular potassium secretion is probably one of the best-studied flow-dependent ion-transport systems that rely on the function of primary cilia [163,164]. Potassium secretion is associated with an increase in luminal flow rate along the cortical collecting tubules. Mechanical activation of primary cilia by either micropipette bending or fluid flow increases potassium secretion [165,166]. Fluid flow in cortical collecting tubules activates potassium channels only in ciliated principle cells and not in intercalated cells that are devoid of cilia [167]. Furthermore, primary cilia where PC1-PC2 complex localizes are required for flow-induced cilia activation to promote potassium secretion. Bending the primary cilium with a micropipette is also known to increase cytosolic calcium. This calcium will trigger the activation of calcium-sensitive intermediate-conductance potassium channels in renal epithelial cells [166]. Recent findings further suggest that some of these potassium channels may have new functional roles to regulate cilia formation in the kidneys [168]. It is suggested that these channels have potential implication as genetic contributors to ciliopathy pathophysiology, which may suggest future therapeutic approaches.

## 6 CONCLUSION

Primary cilia have been recognized to be mechanosensory organelles responsive toward fluid shear force. A primary cilium can be organized into six distinct domains: the ciliary membrane, cilioplasm, axoneme, ciliary tip, basal body, and ciliary necklace. Ciliopathies describe various cilia-related diseases expanding from hypertension to cancer and blindness to mental retardation. Ciliotherapy may involve a unique cilia-targeting therapy with the aim of improving mechanosensory cilia function, and thus ameliorating the associated disease.

Dopamine has the potential to activate ciliary dopamine receptors in a therapeutic manner. However, this is only the first out of many other potential cilia-targeted therapies. Improving survivin expression might also provide a potential therapeutic target for vascular complications associated with aneurysms. As a promising candidate of targeted therapy in the near future, cilia are being studied intensively, in terms of the effect of their presence, density, length, and function on different diseases.

Patient-specific information could also potentially be used to prevent, diagnose, and treat cilia-related diseases.

Primary cilia play an essential role in vascular and renal architectures. Dysfunction of these cilia will result in the formation of ball-like structures as seen in aneurysm and renal cysts. Balancing localization and function of many ciliary proteins is important to understand the complexity of ciliopathies. The balance begins by placing a weighted importance of mechanosensory primary cilia toward other potential cilia functions in chemosensing [169–171], cell division [52,53,172], gravitational sensing [173], light sensing [174,175], nodal flow sensing [89,90,176], osmosensing [177], smell sensing [178,179], and potential thermosensing [93]. Although it may be too early to envisage, cilia-targeted therapies may provide a viable path toward a personalized medicine and form an important component of precision medicine in the future.

## ACKNOWLEDGEMENTS

Authors wish to express gratitude to Elsevier artists for preparing and finalizing the figures. Authors are grateful to Chapman University School of Pharmacy for providing resources during manuscript preparation. Research conducted in the laboratory is funded by Congressionally Directed Medical Research Program PR130153.

## REFERENCES

[1] Giorgi M, Verbruggen SW, Lacroix D. In silico bone mechanobiology: modeling a multifaceted biological system. Wiley Interdiscip Rev Syst Biol Med 2016;8(6):485–505.

[1a] Nauli SM, Jin X, Hierck BP. The mechanosensory role of primary cilia in vascular hypertension. Int J Vasc Med 2011;2011:376281.

[2] Ingber DE. Mechanobiology and diseases of mechanotransduction. Ann Med 2003;35(8):564–77.

[3] Satir P. CILIA: before and after, Cilia 2017;6(1), https://ciliajournal.biomedcentral.com/articles/10.1186/s13630-017-0046-8.

[4] Lyu R, Zhou J. The multifaceted roles of primary cilia in the regulation of stem cell properties and functions. J Cell Physiol 2017;232(5):935–8.

[5] Mohieldin AM, Zubayer HS, Al Omran AJ, Saternos HC, Zarban AA, Nauli SM, et al. Vascular endothelial primary cilia: mechanosensation and hypertension. Curr Hypertens Rev 2016;12(1):57–67.

[6] Kolb RJ, Nauli SM. Ciliary dysfunction in polycystic kidney disease: an emerging model with polarizing potential. Front Biosci 2008;(13):4451–66.

[7] Nauli SM, Zhou J. Polycystins and mechanosensation in renal and nodal cilia. BioEssays 2004;26(8):844–56.

[8] Nauli SM, Pala R, Kleene SJ. Calcium channels in primary cilia. Curr Opin Nephrol Hypertens 2016;25(5):452–8.

[9] Mukherjee S, Jansen V, Jikeli JF, Hamzeh H, Alvarez L, Dombrowski M, et al. A novel biosensor to study cAMP dynamics in cilia and flagella. eLife 2016;5:pii: e14052, https://doi.org/10.7554/eLife.14052.

[10] Nguyen PA, Liou W, Hall DH, Leroux MR. Ciliopathy proteins establish a bipartite signaling compartment in a *C. elegans* thermosensory neuron. J Cell Sci 2014;127 (Pt 24):5317–30.
[11] Green JA, Mykytyn K. Neuronal primary cilia: an underappreciated signaling and sensory organelle in the brain. Neuropsychopharmacology 2014;39(1):244–5.
[12] Mohieldin AM, AbouAlaiwi WA, Gao M, Nauli SM. Chemical-free technique to study the ultrastructure of primary cilium. Sci Rep 2015;5:15982.
[13] Mohieldin AM, Haymour HS, Lo ST, AbouAlaiwi WA, Atkinson KF, Ward CJ, et al. Protein composition and movements of membrane swellings associated with primary cilia. Cell Mol Life Sci 2015;72(12):2415–29.
[14] Keeling J, Tsiokas L, Maskey D. Cellular mechanisms of ciliary length control. Cell 2016;5(1).
[15] Verhey KJ, Dishinger J, Kee HL. Kinesin motors and primary cilia. Biochem Soc Trans 2011;39(5):1120–5.
[16] Wloga D, Gaertig J. Post-translational modifications of microtubules. J Cell Sci 2010;123(Pt 20):3447–55.
[17] Abou Alaiwi WA, Lo ST, Nauli SM. Primary cilia: highly sophisticated biological sensors. Sensors (Basel) 2009;9(9):7003–20.
[18] Pedersen LB, Akhmanova A. Kif7 keeps cilia tips in shape. Nat Cell Biol 2014;16 (7):623–5.
[19] He M, Subramanian R, Bangs F, Omelchenko T, Liem Jr KF, Kapoor TM, et al. The kinesin-4 protein Kif7 regulates mammalian Hedgehog signalling by organizing the cilium tip compartment. Nat Cell Biol 2014;16(7):663–72.
[20] Marshall WF, Nonaka S. Cilia: tuning in to the cell's antenna. Curr Biol 2006;16(15): R604–14.
[21] Garcia 3rd G, Reiter JF. A primer on the mouse basal body. Cilia 2016;5:17.
[22] Moser JJ, Fritzler MJ, Ou Y, Rattner JB. The PCM-basal body/primary cilium coalition. Semin Cell Dev Biol 2010;21(2):148–55.
[23] Marshall WF. Basal bodies platforms for building cilia. Curr Top Dev Biol 2008;85:1–22.
[24] Debec A, Sullivan W, Bettencourt-Dias M. Centrioles: active players or passengers during mitosis? Cell Mol Life Sci 2010;67(13):2173–94.
[25] Mitchison T, Kirschner M. Microtubule assembly nucleated by isolated centrosomes. Nature 1984;312(5991):232–7.
[26] Takao D, Verhey KJ. Gated entry into the ciliary compartment. Cell Mol Life Sci 2016;73(1):119–27.
[27] Reiter JF, Blacque OE, Leroux MR. The base of the cilium: roles for transition fibres and the transition zone in ciliary formation, maintenance and compartmentalization. EMBO Rep 2012;13(7):608–18.
[28] Hu Q, Nelson WJ. Ciliary diffusion barrier: the gatekeeper for the primary cilium compartment. Cytoskeleton (Hoboken) 2011;68(6):313–24.
[29] Pazour GJ, Bloodgood RA. Targeting proteins to the ciliary membrane. Curr Top Dev Biol 2008;85:115–49.
[29a] Downs ME, Nguyen AM, Herzog FA, Hoey DA, Jacobs CR. An experimental and computational analysis of primary cilia deflection under fluid flow. Comput Methods Biomech Biomed Eng 2012;1–9.
[29b] Vaughan TJ, Mullen CA, Verbruggen SW, McNamara LM. Bone cell mechanosensation of fluid flow stimulation: a fluid-structure interaction model characterising the role integrin attachments and primary cilia. Biomech Model Mechanobiol 2014;2.

[29c] Khayyeri H, Barreto S, Lacroix D. Primary cilia mechanics affects cell mechanosensation: a computational study. J Theor Biol 2015;379:38–46.

[30] Aboualaiwi WA, Muntean BS, Ratnam S, Joe B, Liu L, Booth RL, et al. Survivin-induced abnormal ploidy contributes to cystic kidney and aneurysm formation. Circulation 2014;129(6):660–72.

[31] AbouAlaiwi WA, Ratnam S, Booth RL, Shah JV, Nauli SM. Endothelial cells from humans and mice with polycystic kidney disease are characterized by polyploidy and chromosome segregation defects through survivin down-regulation. Hum Mol Genet 2011;20(2):354–67.

[32] Song J, Wang L, Fan F, Wei J, Zhang J, Lu Y, et al. Role of the primary cilia on the macula densa and thick ascending limbs in regulation of sodium excretion and hemodynamics. Hypertension 2017;70(2):324–33.

[33] AbouAlaiwi WA, Takahashi M, Mell BR, Jones TJ, Ratnam S, Kolb RJ, et al. Ciliary polycystin-2 is a mechanosensitive calcium channel involved in nitric oxide signaling cascades. Circ Res 2009;104(7):860–9.

[34] Nauli SM, Kawanabe Y, Kaminski JJ, Pearce WJ, Ingber DE, Zhou J. Endothelial cilia are fluid shear sensors that regulate calcium signaling and nitric oxide production through polycystin-1. Circulation 2008;117(9):1161–71.

[35] Choi H, Shin JH, Kim ES, Park SJ, Bae IH, Jo YK, et al. Primary cilia negatively regulate melanogenesis in melanocytes and pigmentation in a human skin model. PLoS One 2016;11(12):e0168025.

[36] Croyle MJ, Lehman JM, O'Connor AK, Wong SY, Malarkey EB, Iribarne D, et al. Role of epidermal primary cilia in the homeostasis of skin and hair follicles. Development 2011;138(9):1675–85.

[37] Lehman JM, Laag E, Michaud EJ, Yoder BK. An essential role for dermal primary cilia in hair follicle morphogenesis. J Invest Dermatol 2009;129(2):438–48.

[38] Landsman L, Parent A, Hebrok M. Elevated Hedgehog/Gli signaling causes beta-cell dedifferentiation in mice. Proc Natl Acad Sci U S A 2011;108(41):17010–5.

[39] Cano DA, Sekine S, Hebrok M. Primary cilia deletion in pancreatic epithelial cells results in cyst formation and pancreatitis. Gastroenterology 2006;131(6):1856–69.

[40] Cano DA, Murcia NS, Pazour GJ, Hebrok M. Orpk mouse model of polycystic kidney disease reveals essential role of primary cilia in pancreatic tissue organization. Development 2004;131(14):3457–67.

[41] Qiu N, Fang WJ, Li HS, He ZM, Xiao ZS, Xiong Y. Impairment of primary cilia contributes to visceral adiposity of high fat diet-fed mice. J Cell Biochem 2018;119(2):1313–25.

[42] Volta F, Gerdes JM. The role of primary cilia in obesity and diabetes. Ann N Y Acad Sci 2017;1391(1):71–84.

[43] Jacobs DT, Silva LM, Allard BA, Schonfeld MP, Chatterjee A, Talbott GC, et al. Dysfunction of intraflagellar transport-A causes hyperphagia-induced obesity and metabolic syndrome. Dis Model Mech 2016;9(7):789–98.

[44] Oh EC, Vasanth S, Katsanis N. Metabolic regulation and energy homeostasis through the primary cilium. Cell Metab 2015;21(1):21–31.

[45] Shalata A, Ramirez MC, Desnick RJ, Priedigkeit N, Buettner C, Lindtner C, et al. Morbid obesity resulting from inactivation of the ciliary protein CEP19 in humans and mice. Am J Hum Genet 2013;93(6):1061–71.

[46] Nauli SM, Rossetti S, Kolb RJ, Alenghat FJ, Consugar MB, Harris PC, et al. Loss of polycystin-1 in human cyst-lining epithelia leads to ciliary dysfunction. J Am Soc Nephrol 2006;17(4):1015–25.

[47] Nauli SM, Alenghat FJ, Luo Y, Williams E, Vassilev P, Li X, et al. Polycystins 1 and 2 mediate mechanosensation in the primary cilium of kidney cells. Nat Genet 2003;33 (2):129–37.
[48] Grandone A, Torella A, Santoro C, Giugliano T, Del Vecchio Blanco F, Mutarelli M, et al. Expanding the phenotype of RTTN variations: a new family with primary microcephaly, severe growth failure, brain malformations and dermatitis. Clin Genet 2016;90 (5):445–50.
[49] Yao G, Luo C, Harvey M, Wu M, Schreiber TH, Du Y, et al. Disruption of polycystin-L causes hippocampal and thalamocortical hyperexcitability. Hum Mol Genet 2016;25 (3):448–58.
[50] Novas R, Cardenas-Rodriguez M, Irigoin F, Badano JL. Bardet-Biedl syndrome: is it only cilia dysfunction? FEBS Lett 2015;589(22):3479–91.
[51] Lee JH, Gleeson JG. The role of primary cilia in neuronal function. Neurobiol Dis 2010;38(2):167–72.
[52] Wong SY, Seol AD, So PL, Ermilov AN, Bichakjian CK, Epstein Jr EH, et al. Primary cilia can both mediate and suppress Hedgehog pathway-dependent tumorigenesis. Nat Med 2009;15(9):1055–61.
[53] Han YG, Kim HJ, Dlugosz AA, Ellison DW, Gilbertson RJ, Alvarez-Buylla A. Dual and opposing roles of primary cilia in medulloblastoma development. Nat Med 2009;15 (9):1062–5.
[54] Wheway G, Parry DA, Johnson CA. The role of primary cilia in the development and disease of the retina. Organ 2014;10(1):69–85.
[55] Rachel RA, Li T, Swaroop A. Photoreceptor sensory cilia and ciliopathies: focus on CEP290, RPGR and their interacting proteins. Cilia 2012;1(1):22.
[56] Servattalab S, Yildiz O, Khanna H. Tackling primary cilia dysfunction in photoreceptor degenerative diseases of the eye. Int J Ophthalmic Pathol 2012;1(1):pii: e101, https://doi.org/10.4172/2324-8599.1000e101.
[57] Zhang J, Dalbay MT, Luo X, Vrij E, Barbieri D, Moroni L, et al. Topography of calcium phosphate ceramics regulates primary cilia length and TGF receptor recruitment associated with osteogenesis. Acta Biomater 2017;57:487–97.
[58] Oliazadeh N, Gorman KF, Eveleigh R, Bourque G, Moreau A. Identification of elongated primary cilia with impaired mechanotransduction in idiopathic scoliosis patients. Sci Rep 2017;7:44260.
[59] Nguyen AM, Jacobs CR. Emerging role of primary cilia as mechanosensors in osteocytes. Bone 2013;54(2):196–204.
[60] Al-Bhalal L, Akhtar M. Molecular basis of autosomal recessive polycystic kidney disease (ARPKD). Adv Anat Pathol 2008;15(1):54–8.
[61] Kaimori JY, Nagasawa Y, Menezes LF, Garcia-Gonzalez MA, Deng J, Imai E, et al. Polyductin undergoes notch-like processing and regulated release from primary cilia. Hum Mol Genet 2007;16(8):942–56.
[62] Wang S, Luo Y, Wilson PD, Witman GB, Zhou J. The autosomal recessive polycystic kidney disease protein is localized to primary cilia, with concentration in the basal body area. J Am Soc Nephrol 2004;15(3):592–602.
[63] Ward CJ, Yuan D, Masyuk TV, Wang X, Punyashthiti R, Whelan S, et al. Cellular and subcellular localization of the ARPKD protein; fibrocystin is expressed on primary cilia. Hum Mol Genet 2003;12(20):2703–10.
[64] Orhon I, Dupont N, Pampliega O, Cuervo AM, Codogno P. Autophagy and regulation of cilia function and assembly. Cell Death Differ 2015;22(3):389–97.

[65] Lam HC, Cloonan SM, Bhashyam AR, Haspel JA, Singh A, Sathirapongsasuti JF, et al. Histone deacetylase 6-mediated selective autophagy regulates COPD-associated cilia dysfunction. J Clin Invest 2013;123(12):5212–30.

[66] Sha YW, Ding L, Li P. Management of primary ciliary dyskinesia/Kartagener's syndrome in infertile male patients and current progress in defining the underlying genetic mechanism. Asian J Androl 2014;16(1):101–6.

[67] Leigh MW, Pittman JE, Carson JL, Ferkol TW, Dell SD, Davis SD, et al. Clinical and genetic aspects of primary ciliary dyskinesia/Kartagener syndrome. Genet Med 2009;11(7):473–87.

[68] Ajzenberg H, Slaats GG, Stokman MF, Arts HH, Logister I, Kroes HY, et al. Non-invasive sources of cells with primary cilia from pediatric and adult patients. Cilia 2015;4:8.

[69] Kathem SH, Mohieldin AM, Nauli SM. The roles of primary cilia in polycystic kidney disease. AIMS Mol Sci 2014;1(1):27–46.

[70] Hassounah NB, Nagle R, Saboda K, Roe DJ, Dalkin BL, McDermott KM. Primary cilia are lost in preinvasive and invasive prostate cancer. PLoS One 2013;8(7):e68521.

[71] Dinsmore C, Reiter JF. Endothelial primary cilia inhibit atherosclerosis. EMBO Rep 2016;17(2):156–66.

[72] Van der Heiden K, Hierck BP, Krams R, de Crom R, Cheng C, Baiker M, et al. Endothelial primary cilia in areas of disturbed flow are at the base of atherosclerosis. Atherosclerosis 2008;196(2):542–50.

[73] Hu Q, Wu Y, Tang J, Zheng W, Wang Q, Nahirney D, et al. Expression of polycystins and fibrocystin on primary cilia of lung cells. Biochem Cell Biol 2014;92(6):547–54.

[74] Wang S, Zhang J, Nauli SM, Li X, Starremans PG, Luo Y, et al. Fibrocystin/polyductin, found in the same protein complex with polycystin-2, regulates calcium responses in kidney epithelia. Mol Cell Biol 2007;27(8):3241–52.

[75] Cai S, Bodle JC, Mathieu PS, Amos A, Hamouda M, Bernacki S, et al. Primary cilia are sensors of electrical field stimulation to induce osteogenesis of human adipose-derived stem cells. FASEB J 2017;31(1):346–55.

[76] Yoder BK, Hou X, Guay-Woodford LM. The polycystic kidney disease proteins, polycystin-1, polycystin-2, polaris, and cystin, are co-localized in renal cilia. J Am Soc Nephrol 2002;13(10):2508–16.

[77] Leaf A, Von Zastrow M. Dopamine receptors reveal an essential role of IFT-B, KIF17, and Rab23 in delivering specific receptors to primary cilia, eLife 2015;4, https://elifesciences.org/articles/06996.

[78] Omori Y, Chaya T, Yoshida S, Irie S, Tsujii T, Furukawa T. Identification of G protein-coupled receptors (GPCRs) in primary cilia and their possible involvement in body weight control. PLoS One 2015;10(6):e0128422.

[79] Abdul-Majeed S, Nauli SM. Dopamine receptor type 5 in the primary cilia has dual chemo- and mechano-sensory roles. Hypertension 2011;58(2):325–31.

[80] Bai CX, Giamarchi A, Rodat-Despoix L, Padilla F, Downs T, Tsiokas L, et al. Formation of a new receptor-operated channel by heteromeric assembly of TRPP2 and TRPC1 subunits. EMBO Rep 2008;9(5):472–9.

[81] Raychowdhury MK, McLaughlin M, Ramos AJ, Montalbetti N, Bouley R, Ausiello DA, et al. Characterization of single channel currents from primary cilia of renal epithelial cells. J Biol Chem 2005;280(41):34718–22.

[82] Maroto R, Raso A, Wood TG, Kurosky A, Martinac B, Hamill OP. TRPC1 forms the stretch-activated cation channel in vertebrate cells. Nat Cell Biol 2005;7(2):179–85.

[83] Pazour GJ, San Agustin JT, Follit JA, Rosenbaum JL, Witman GB. Polycystin-2 localizes to kidney cilia and the ciliary level is elevated in orpk mice with polycystic kidney disease. Curr Biol 2002;12(11):R378–80.

[84] Chen XZ, Vassilev PM, Basora N, Peng JB, Nomura H, Segal Y, et al. Polycystin-L is a calcium-regulated cation channel permeable to calcium ions. Nature 1999;401(6751):383–6.

[85] Luo N, Conwell MD, Chen X, Kettenhofen CI, Westlake CJ, Cantor LB, et al. Primary cilia signaling mediates intraocular pressure sensation. Proc Natl Acad Sci U S A 2014;111(35):12871–6.

[86] Ma X, Qiu S, Luo J, Ma Y, Ngai CY, Shen B, et al. Functional role of vanilloid transient receptor potential 4-canonical transient receptor potential 1 complex in flow-induced Ca2+ influx. Arterioscler Thromb Vasc Biol 2010;30(4):851–8.

[87] Masyuk AI, Masyuk TV, Splinter PL, Huang BQ, Stroope AJ, LaRusso NF. Cholangiocyte cilia detect changes in luminal fluid flow and transmit them into intracellular Ca2+ and cAMP signaling. Gastroenterology 2006;131(3):911–20.

[88] Yuan S, Zhao L, Brueckner M, Sun Z. Intraciliary calcium oscillations initiate vertebrate left-right asymmetry. Curr Biol 2015;25(5):556–67.

[89] Yoshiba S, Shiratori H, Kuo IY, Kawasumi A, Shinohara K, Nonaka S, et al. Cilia at the node of mouse embryos sense fluid flow for left-right determination via Pkd2. Science 2012;338(6104):226–31.

[90] McGrath J, Somlo S, Makova S, Tian X, Brueckner M. Two populations of node monocilia initiate left-right asymmetry in the mouse. Cell 2003;114(1):61–73.

[91] Xu H, Guan Y, Wu J, Zhang J, Duan J, An L, et al. Polycystin 2 is involved in the nitric oxide production in responding to oscillating fluid shear in MLO-Y4 cells. J Biomech 2014;47(2):387–91.

[92] Du J, Wong WY, Sun L, Huang Y, Yao X. Protein kinase G inhibits flow-induced Ca2+ entry into collecting duct cells. J Am Soc Nephrol 2012;23(7):1172–80.

[93] Kottgen M, Buchholz B, Garcia-Gonzalez MA, Kotsis F, Fu X, Doerken M, et al. TRPP2 and TRPV4 form a polymodal sensory channel complex. J Cell Biol 2008;182(3):437–47.

[94] Narayanan D, Bulley S, Leo MD, Burris SK, Gabrick KS, Boop FA, et al. Smooth muscle cell transient receptor potential polycystin-2 (TRPP2) channels contribute to the myogenic response in cerebral arteries. J Physiol 2013;591(20):5031–46.

[95] Lee KL, Guevarra MD, Nguyen AM, Chua MC, Wang Y, Jacobs CR. The primary cilium functions as a mechanical and calcium signaling nexus. Cilia 2015;4:7.

[96] Jin X, Mohieldin AM, Muntean BS, Green JA, Shah JV, Mykytyn K, et al. Cilioplasm is a cellular compartment for calcium signaling in response to mechanical and chemical stimuli. Cell Mol Life Sci 2014;71(11):2165–78.

[97] Su S, Phua SC, DeRose R, Chiba S, Narita K, Kalugin PN, et al. Genetically encoded calcium indicator illuminates calcium dynamics in primary cilia. Nat Methods 2013;10(11):1105–7.

[98] Praetorius HA, Spring KR. Bending the MDCK cell primary cilium increases intracellular calcium. J Membr Biol 2001;184(1):71–9.

[99] Prasad RM, Jin X, Nauli SM. Sensing a sensor: identifying the mechanosensory function of primary cilia. Biosensors 2014;4(1):47–62.

[100] Prasad RM, Jin X, Aboualaiwi WA, Nauli SM. Real-time vascular mechanosensation through ex vivo artery perfusion. Biol Proced Online 2014;16(1):6.

[101] Nauli SM, Jin X, AbouAlaiwi WA, El-Jouni W, Su X, Zhou J. Non-motile primary cilia as fluid shear stress mechanosensors. Methods Enzymol 2013;525:1–20.

[102] Jin X, Muntean BS, Aal-Aaboda MS, Duan Q, Zhou J, Nauli SM. L-type calcium channel modulates cystic kidney phenotype. Biochim Biophys Acta 2014;1842 (9):1518–26.
[103] Verleyen D, Luyten FP, Tylzanowski P. Orphan G-protein coupled receptor 22 (Gpr22) regulates cilia length and structure in the zebrafish Kupffer's vesicle. PLoS One 2014; 9(10):e110484.
[104] Rbaibi Y, Cui S, Mo D, Carattino M, Rohatgi R, Satlin LM, et al. OCRL1 modulates cilia length in renal epithelial cells. Traffic 2012;13(9):1295–305.
[105] Abdul-Majeed S, Moloney BC, Nauli SM. Mechanisms regulating cilia growth and cilia function in endothelial cells. Cell Mol Life Sci 2012;69(1):165–73.
[106] Wang L, Weidenfeld R, Verghese E, Ricardo SD, Deane JA. Alterations in renal cilium length during transient complete ureteral obstruction in the mouse. J Anat 2008;213 (2):79–85.
[107] Verghese E, Weidenfeld R, Bertram JF, Ricardo SD, Deane JA. Renal cilia display length alterations following tubular injury and are present early in epithelial repair. Nephrol Dial Transplant 2008;23(3):834–41.
[108] Wang S, Dong Z. Primary cilia and kidney injury: current research status and future perspectives. Am J Physiol Ren Physiol 2013;305(8):F1085–98.
[109] Wang S, Wei Q, Dong G, Dong Z. ERK-mediated suppression of cilia in cisplatin-induced tubular cell apoptosis and acute kidney injury. Biochim Biophys Acta 2013;1832(10):1582–90.
[110] Besschetnova TY, Kolpakova-Hart E, Guan Y, Zhou J, Olsen BR, Shah JV. Identification of signaling pathways regulating primary cilium length and flow-mediated adaptation. Curr Biol 2010;20(2):182–7.
[111] Kathem SH, Mohieldin AM, Abdul-Majeed S, Ismail SH, Altaei QH, Alshimmari IK, et al. Ciliotherapy: a novel intervention in polycystic kidney disease. J Geriatr Cardiol 2014;11(1):63–73.
[112] Atkinson KF, Kathem SH, Jin X, Muntean BS, Abou-Alaiwi WA, Nauli AM, et al. Dopaminergic signaling within the primary cilia in the renovascular system. Front Physiol 2015;6:103.
[113] Upadhyay VS, Muntean BS, Kathem SH, Hwang JJ, Aboualaiwi WA, Nauli SM. Roles of dopamine receptor on chemosensory and mechanosensory primary cilia in renal epithelial cells. Front Physiol 2014;5:72.
[114] Avasthi P, Marley A, Lin H, Gregori-Puigjane E, Shoichet BK, von Zastrow M, et al. A chemical screen identifies class a g-protein coupled receptors as regulators of cilia. ACS Chem Biol 2012;7(5):911–9.
[115] Spasic M, Jacobs CR. Lengthening primary cilia enhances cellular mechanosensitivity. Eur Cell Mater 2017;33:158–68.
[116] Sherpa RT, Atkinson KF, Ferreira VP, Nauli SM. Rapamycin increases length and mechanosensory function of primary cilia in renal epithelial and vascular endothelial cells. Int Educ Res J 2016;2(12):91–7.
[117] Chapman AB. Does dopamine connect the dots in ADPKD? Kidney Int 2015;87 (2):279–80.
[118] Lorthioir A, Joannides R, Remy-Jouet I, Freguin-Bouilland C, Iacob M, Roche C, et al. Polycystin deficiency induces dopamine-reversible alterations in flow-mediated dilatation and vascular nitric oxide release in humans. Kidney Int 2015;87(2):465–72.
[119] Rangel-Barajas C, Coronel I, Floran B. Dopamine receptors and neurodegeneration. Aging Dis 2015;6(5):349–68.

[120] Fuxe K, Agnati LF, Marcoli M, Borroto-Escuela DO. Volume transmission in central dopamine and noradrenaline neurons and its astroglial targets. Neurochem Res 2015;40 (12):2600–14.

[121] McHugh PC, Buckley DA. The structure and function of the dopamine transporter and its role in CNS diseases. Vitam Horm 2015;98:339–69.

[122] Hankir MK, Ashrafian H, Hesse S, Horstmann A, Fenske WK. Distinctive striatal dopamine signaling after dieting and gastric bypass. Trends Endocrinol Metab 2015;26 (5):223–30.

[123] Bialkowska M, Zajac D, Mazzatenta A, Di Giulio C, Pokorski M. Inhibition of peripheral dopamine metabolism and the ventilatory response to hypoxia in the rat. Adv Exp Med Biol 2015;837:9–17.

[124] Tayebati SK, Lokhandwala MF, Amenta F. Dopamine and vascular dynamics control: present status and future perspectives. Curr Neurovasc Res 2011;8(3):246–57.

[125] Cassis LA, Gupte M, Thayer S, Zhang X, Charnigo R, Howatt DA, et al. ANG II infusion promotes abdominal aortic aneurysms independent of increased blood pressure in hypercholesterolemic mice. Am J Physiol Heart Circ Physiol 2009;296(5):H1660–5.

[126] Vardulaki KA, Walker NM, Day NE, Duffy SW, Ashton HA, Scott RA. Quantifying the risks of hypertension, age, sex and smoking in patients with abdominal aortic aneurysm. Br J Surg 2000;87(2):195–200.

[127] Barr MM, DeModena J, Braun D, Nguyen CQ, Hall DH, Sternberg PW. The *Caenorhabditis elegans* autosomal dominant polycystic kidney disease gene homologs lov-1 and pkd-2 act in the same pathway. Curr Biol 2001;11(17):1341–6.

[128] Barr MM, Sternberg PW. A polycystic kidney-disease gene homologue required for male mating behaviour in *C. elegans*. Nature 1999;401(6751):386–9.

[129] Pazour GJ, Dickert BL, Vucica Y, Seeley ES, Rosenbaum JL, Witman GB, et al. Chlamydomonas IFT88 and its mouse homologue, polycystic kidney disease gene tg737, are required for assembly of cilia and flagella. J Cell Biol 2000;151(3):709–18.

[130] Yoder BK, Tousson A, Millican L, Wu JH, Bugg Jr CE, Schafer JA, et al. Polaris, a protein disrupted in orpk mutant mice, is required for assembly of renal cilium. Am J Physiol Ren Physiol 2002;282(3):F541–52.

[131] Taulman PD, Haycraft CJ, Balkovetz DF, Yoder BK. Polaris, a protein involved in left-right axis patterning, localizes to basal bodies and cilia. Mol Biol Cell 2001;12 (3):589–99.

[132] Murcia NS, Richards WG, Yoder BK, Mucenski ML, Dunlap JR, Woychik RP. The Oak Ridge Polycystic Kidney (orpk) disease gene is required for left-right axis determination. Development 2000;127(11):2347–55.

[133] Moyer JH, Lee-Tischler MJ, Kwon HY, Schrick JJ, Avner ED, Sweeney WE, et al. Candidate gene associated with a mutation causing recessive polycystic kidney disease in mice. Science 1994;264(5163):1329–33.

[134] Muto S, Aiba A, Saito Y, Nakao K, Nakamura K, Tomita K, et al. Pioglitazone improves the phenotype and molecular defects of a targeted Pkd1 mutant. Hum Mol Genet 2002;11(15):1731–42.

[135] Lu W, Shen X, Pavlova A, Lakkis M, Ward CJ, Pritchard L, et al. Comparison of Pkd1-targeted mutants reveals that loss of polycystin-1 causes cystogenesis and bone defects. Hum Mol Genet 2001;10(21):2385–96.

[136] Boulter C, Mulroy S, Webb S, Fleming S, Brindle K, Sandford R. Cardiovascular, skeletal, and renal defects in mice with a targeted disruption of the Pkd1 gene. Proc Natl Acad Sci U S A 2001;98(21):12174–9.

[137] Kim K, Drummond I, Ibraghimov-Beskrovnaya O, Klinger K, Arnaout MA. Polycystin 1 is required for the structural integrity of blood vessels. Proc Natl Acad Sci U S A 2000;97(4):1731–6.
[138] Lu W, Fan X, Basora N, Babakhanlou H, Law T, Rifai N, et al. Late onset of renal and hepatic cysts in Pkd1-targeted heterozygotes. Nat Genet 1999;21(2):160–1.
[139] Lu W, Peissel B, Babakhanlou H, Pavlova A, Geng L, Fan X, et al. Perinatal lethality with kidney and pancreas defects in mice with a targeted Pkd1 mutation. Nat Genet 1997;17(2):179–81.
[140] Kim I, Ding T, Fu Y, Li C, Cui L, Li A, et al. Conditional mutation of Pkd2 causes cystogenesis and upregulates beta-catenin. J Am Soc Nephrol 2009;20(12):2556–69.
[141] Pennekamp P, Karcher C, Fischer A, Schweickert A, Skryabin B, Horst J, et al. The ion channel polycystin-2 is required for left-right axis determination in mice. Curr Biol 2002;12(11):938–43.
[142] Wu G, Markowitz GS, Li L, D'Agati VD, Factor SM, Geng L, et al. Cardiac defects and renal failure in mice with targeted mutations in Pkd2. Nat Genet 2000;24(1):75–8.
[143] Wu G, D'Agati V, Cai Y, Markowitz G, Park JH, Reynolds DM, et al. Somatic inactivation of Pkd2 results in polycystic kidney disease. Cell 1998;93(2):177–88.
[144] Doerr N, Wang Y, Kipp KR, Liu G, Benza JJ, Pletnev V, et al. Regulation of polycystin-1 function by calmodulin binding. PLoS One 2016;11(8):e0161525.
[145] Cantiello HF. A tale of two tails: ciliary mechanotransduction in ADPKD. Trends Mol Med 2003;9(6):234–6.
[146] Rydholm S, Zwartz G, Kowalewski JM, Kamali-Zare P, Frisk T, Brismar H. Mechanical properties of primary cilia regulate the response to fluid flow. Am J Physiol Ren Physiol 2010;298(5):F1096–102.
[147] Xu C, Shmukler BE, Nishimura K, Kaczmarek E, Rossetti S, Harris PC, et al. Attenuated, flow-induced ATP release contributes to absence of flow-sensitive, purinergic Cai2+ signaling in human ADPKD cyst epithelial cells. Am J Physiol Ren Physiol 2009;296(6):F1464–76.
[148] Rohatgi R, Battini L, Kim P, Israeli S, Wilson PD, Gusella GL, et al. Mechanoregulation of intracellular Ca2+ in human autosomal recessive polycystic kidney disease cyst-lining renal epithelial cells. Am J Physiol Ren Physiol 2008;294(4):F890–9.
[149] Bodle JC, Rubenstein CD, Phillips ME, Bernacki SH, Qi J, Banes AJ, et al. Primary cilia: the chemical antenna regulating human adipose-derived stem cell osteogenesis. PLoS One 2013;8(5):e62554.
[150] Sharif-Naeini R, Folgering JH, Bichet D, Duprat F, Lauritzen I, Arhatte M, et al. Polycystin-1 and -2 dosage regulates pressure sensing. Cell 2009;139(3):587–96.
[151] Saburi S, Hester I, Fischer E, Pontoglio M, Eremina V, Gessler M, et al. Loss of Fat4 disrupts PCP signaling and oriented cell division and leads to cystic kidney disease. Nat Genet 2008;40(8):1010–5.
[152] Patel V, Li L, Cobo-Stark P, Shao X, Somlo S, Lin F, et al. Acute kidney injury and aberrant planar cell polarity induce cyst formation in mice lacking renal cilia. Hum Mol Genet 2008;17(11):1578–90.
[153] Fischer E, Legue E, Doyen A, Nato F, Nicolas JF, Torres V, et al. Defective planar cell polarity in polycystic kidney disease. Nat Genet 2006;38(1):21–3.
[154] Kathem SH, AbouAlaiwi WA, Zi X, Nauli SM. Capillary endothelia from two ADPKD patients are polyploidy. Ann Clin Cytol Pathol 2016;2(2):1022–6.

[155] Battini L, Macip S, Fedorova E, Dikman S, Somlo S, Montagna C, et al. Loss of polycystin-1 causes centrosome amplification and genomic instability. Hum Mol Genet 2008;17(18):2819–33.
[156] Burtey S, Riera M, Ribe E, Pennenkamp P, Rance R, Luciani J, et al. Centrosome overduplication and mitotic instability in PKD2 transgenic lines. Cell Biol Int 2008;32(10):1193–8.
[157] Garvin JL. Glucose absorption by isolated perfused rat proximal straight tubules. Am J Phys 1990;259(4 Pt 2):F580–6.
[158] Sipos A, Vargas S, Peti-Peterdi J. Direct demonstration of tubular fluid flow sensing by macula densa cells. Am J Physiol Ren Physiol 2010;299(5):F1087–93.
[159] Woda CB, Miyawaki N, Ramalakshmi S, Ramkumar M, Rojas R, Zavilowitz B, et al. Ontogeny of flow-stimulated potassium secretion in rabbit cortical collecting duct: functional and molecular aspects. Am J Physiol Ren Physiol 2003;285(4):F629–39.
[160] Taniguchi J, Imai M. Flow-dependent activation of maxi K+ channels in apical membrane of rabbit connecting tubule. J Membr Biol 1998;164(1):35–45.
[161] Dekker M, Pasch A, van der Sande F, Konings C, Bachtler M, Dionisi M, et al. Highflux hemodialysis and high-volume hemodiafiltration improve serum calcification propensity. PLoS One 2016;11(4):e0151508.
[162] Dias C, Volpini RA, Helou CM. Rosiglitazone did not induce acute kidney injury in normocholesterolemic rats despite reduction in glomerular filtration rate. Kidney Blood Press Res 2013;38(2–3):186–95.
[163] Carrisoza-Gaytan R, Carattino MD, Kleyman TR, Satlin LM. An unexpected journey: conceptual evolution of mechanoregulated potassium transport in the distal nephron. Am J Phys Cell Physiol 2016;310(4):C243–59.
[164] Praetorius HA, Spring KR. The renal cell primary cilium functions as a flow sensor. Curr Opin Nephrol Hypertens 2003;12(5):517–20.
[165] Liu W, Murcia NS, Duan Y, Weinbaum S, Yoder BK, Schwiebert E, et al. Mechanoregulation of intracellular Ca2+ concentration is attenuated in collecting duct of monocilium-impaired orpk mice. Am J Physiol Ren Physiol 2005;289(5):F978–88.
[166] Praetorius HA, Frokiaer J, Nielsen S, Spring KR. Bending the primary cilium opens Ca2+-sensitive intermediate-conductance K+ channels in MDCK cells. J Membr Biol 2003;191(3):193–200.
[167] Liu W, Wei Y, Sun P, Wang WH, Kleyman TR, Satlin LM. Mechanoregulation of BK channel activity in the mammalian cortical collecting duct: role of protein kinases A and C. Am J Physiol Ren Physiol 2009;297(4):F904–15.
[168] Slaats GG, Wheway G, Foletto V, Szymanska K, van Balkom BW, Logister I, et al. Screen-based identification and validation of four new ion channels as regulators of renal ciliogenesis. J Cell Sci 2015;128(24):4550–9.
[169] Masyuk AI, Gradilone SA, Banales JM, Huang BQ, Masyuk TV, Lee SO, et al. Cholangiocyte primary cilia are chemosensory organelles that detect biliary nucleotides via P2Y12 purinergic receptors. Am J Physiol Gastrointest Liver Physiol 2008;295(4): G725–34.
[170] Davenport JR, Watts AJ, Roper VC, Croyle MJ, van Groen T, Wyss JM, et al. Disruption of intraflagellar transport in adult mice leads to obesity and slow-onset cystic kidney disease. Curr Biol 2007;17(18):1586–94.
[171] Winkelbauer ME, Schafer JC, Haycraft CJ, Swoboda P, Yoder BK. The *C. elegans* homologs of nephrocystin-1 and nephrocystin-4 are cilia transition zone proteins involved in chemosensory perception. J Cell Sci 2005;118(Pt 23):5575–87.

[172] Christensen ST, Pedersen SF, Satir P, Veland IR, Schneider L. The primary cilium coordinates signaling pathways in cell cycle control and migration during development and tissue repair. Curr Top Dev Biol 2008;85:261–301.

[173] Moorman SJ, Shorr AZ. The primary cilium as a gravitational force transducer and a regulator of transcriptional noise. Dev Dyn 2008;237(8):1955–9.

[174] Moore A, Escudier E, Roger G, Tamalet A, Pelosse B, Marlin S, et al. RPGR is mutated in patients with a complex X linked phenotype combining primary ciliary dyskinesia and retinitis pigmentosa. J Med Genet 2006;43(4):326–33.

[175] Nishimura DY, Fath M, Mullins RF, Searby C, Andrews M, Davis R, et al. Bbs2-null mice have neurosensory deficits, a defect in social dominance, and retinopathy associated with mislocalization of rhodopsin. Proc Natl Acad Sci U S A 2004;101(47):16588–93.

[176] Field S, Riley KL, Grimes DT, Hilton H, Simon M, Powles-Glover N, et al. Pkd1l1 establishes left-right asymmetry and physically interacts with Pkd2. Development 2011;138(6):1131–42.

[177] Gradilone SA, Masyuk AI, Splinter PL, Banales JM, Huang BQ, Tietz PS, et al. Cholangiocyte cilia express TRPV4 and detect changes in luminal tonicity inducing bicarbonate secretion. Proc Natl Acad Sci U S A 2007;104(48):19138–43.

[178] Layman WS, McEwen DP, Beyer LA, Lalani SR, Fernbach SD, Oh E, et al. Defects in neural stem cell proliferation and olfaction in Chd7 deficient mice indicate a mechanism for hyposmia in human CHARGE syndrome. Hum Mol Genet 2009;18(11):1909–23.

[179] Kulaga HM, Leitch CC, Eichers ER, Badano JL, Lesemann A, Hoskins BE, et al. Loss of BBS proteins causes anosmia in humans and defects in olfactory cilia structure and function in the mouse. Nat Genet 2004;36(9):994–8.

# CHAPTER

# Neuromechanobiology

# 11

William J. Tyler

*School of Biological and Health Systems Engineering, Arizona State University, Tempe, AZ, United States*

## 1 INTRODUCTION

Neuroscience and the study of brain function have advanced tremendously over the past several decades. The primary function of neurons is to receive, process, integrate, store, retrieve, and transmit information throughout the body. This main signaling function is understood to be mediated through a variety of chemical signaling process driven by electrochemical forces. The conversion of electric signaling along axon fibers to chemical signaling at the synapse between neurons is relatively well understood. We have gained a remarkable understanding of how the electric, molecular, and genetic activity of neurons and their networks give rise to emergent behaviors and properties, such as learning and memory. Similarly, abundant amounts of data and information demonstrate how these electric and biochemical signaling mechanisms are disrupted by many different nervous system diseases and injuries. However, the functional consideration of mechanical forces as an integral feature of nervous system signaling has not been traditionally considered. In other biological systems, there are many well-characterized roles that mechanical forces play in physiology. For example, the influence of stretch on cardiac pacing is broadly implicated in normal heart function, and stretch-activated currents have been recorded in the cardiomyocytes of a variety of species. Further, the colon, bladder, and intestines also generate physiological signals through stretch-activated channels and mechanosensitive channels.

The field of mechanobiology has clearly established that cells throughout an organism are continuously subjected to stress, strain, and tension in their native environments. These mechanical forces influence cell division, gene expression, cell migration, morphogenesis, cell adhesion, fluid homeostasis, ion channel gating, and plasma membrane viscoelasticity [1–3]. Consideration of the impact of mechanical forces on neuronal function has traditionally been generated through phenomenological observations rather than being the topic of specific investigation. This can easily be attributed to a lack of available tools, but, more importantly, a theoretical framework for how micromechanical forces can play a role in endogenous brain function has been lacking.

For decades, it has been known that physical displacements occur and mechanical waves propagate along neuronal membranes during action-potential firing [4–8]. At synapses, mechanical impulses have been recorded at axon terminals during vesicle fusion [9]. Dendritic spines "twitch" [10] and experience rapid actin-mediated contractions in response to synaptic activity [11–13]. The activities of many voltage-gated channels (VGC [1,14]) and neurotransmitter receptors (e.g., NMDA receptors [15]) are regulated by mechanosensitive gating dynamics. To what extent these and other mechanical events influence nervous system function remains somewhat of a mystery [7,8]. Highlighting the importance and need for expanding our knowledge in this gap, alterations or disruptions of cellular-mechanical properties are associated with diseases and injuries, such as Alzheimer's disease [16], diffuse axonal injury, spinal cord injury, concussion, and traumatic brain injuries [17,18].

Neuroscience and the study of brain function have recently converged with mechanobiology and study of physical biology to produce a burgeoning new subdiscipline called neuromechanobiology [19]. While numerous micromechanical events have been observed and associated with neuronal activity, technology has only recently enabled the observation and modulation of mechanical forces in the nervous system. This chapter aims to discuss the importance and utility of mechanical forces in the nervous system by first highlighting the physical biology of neurons and their subcellular components. Different cellular-mechanical features of neurons and neuronal networks including lipid membranes, cytoskeletal proteins, and ion channels are discussed with reference to how physical forces may alter classical electric and molecular signaling mechanisms. Throughout the chapter, several quantitative examples of how mechanical forces can influence neuronal signaling and plasticity are provided. For example, whether flexoelectric effects or soliton-like waves in phospholipid neuronal membranes interact with conventional electric action potentials is discussed. After providing an overview of neuromechanobiology fundamentals, the chapter reviews some methods and approaches to the study of mechanical forces. Finally, the importance of cellular-mechanical activity in both normal and diseased brain function is highlighted before providing future outlooks for the study and application of neuromechanobiology.

## 2 VISCOELASTIC PLASMA MEMBRANES GOVERN NEURONAL FUNCTION

The plasma membranes of neurons are mechanically coupled to their networks through cytoskeletal filaments, extracellular matrix (ECM) proteins, and ion channels. The mechanically coupled nature of neurons to their network partners and cellular environments implies that forces affecting one layer of the architecture will in turn produce tension and strain in the other layers [1,20]. To help the reader better understand the physical biology of neurons and neuronal networks, the general properties of lipids and how their characteristics can impact neuronal signaling are described below.

Lipid membranes are non-Newtonian fluids that exhibit viscoelastic properties [21–23]. The phospholipid plasma membrane of neurons and glial cells is dynamic, undergoing changes across many time and length scales. Phospholipid bilayers are affected by mechanical forces and deform in a manner described by their elastic compression, area expansion, and bending moduli. Larger moduli exemplify greater resistance to a deformation force, whereas smaller moduli indicate lower resistances. The elastic features of cellular membranes determine their resistance and response to mechanical deformations produced by biological environmental stressors, such as osmolarity, protein inclusion, and protein conformational changes [23–25]. Briefly, rigidity can be expressed in terms of the shear modulus ($S$) or elastic modulus ($E$), given by $E=2S(1+\nu)$ where $\nu$ is the Poisson's ratio. When deflection does not change the volume, the Poisson's ratio is equal to 0.5, and $S$ will be one-third $E$ [26]. The brain has an elastic modulus depending on several factors of $\approx 0.1$–$16\,nN/\mu m^2$ (equal to $\approx 0.1$–$16\,kPa$ [27–29]).

The viscous features of the membranes are thought to determine the behavior of various dynamic and relaxation processes such as the propagation and attenuation of mechanical waves, the decay or thermal shape fluctuations, and the translational and rotational diffusion of membrane components [23]. When neurons are subjected to lateral stretching or compression, their bilayer membranes behave as a viscoelastic material with anisotropy. It has been shown that there is a close coupling between area ($A$) and thickness ($h$) variations in lipid membranes, which plays a significant role in defining their in-plane viscoelastic characteristics [23]. The compressive modulus ($K_C$) of lipid bilayers has been estimated to be between $10^9$ and $10^{10}\,N/m^2$ [25]. Bilayer tension ($t$) can influence membrane expansion as described in Eq. (1) through a simple linear relationship given by

$$t = K_A \frac{\Delta A}{A_0} \quad (1)$$

where $\Delta A$ is the increase in surface area, $A_0$ is the original surface area, and $K_A$ is the area expansion modulus [1,25]. Depending on cholesterol content, $K_A$ for lipid bilayers has been estimated to be between $10^2$ and $10^3\,mN/m$ [25]. Given that $K_C \approx 10^9\,N/m^2$, the lipid bilayer is about 10 times or more compressible in area than volume (if assuming a $K_A$ of $200\,mN$ divided by a bilayer thickness of $3\,nm$) [1]. Any change in membrane area will be accompanied by a proportional change in membrane thickness ($\Delta h$) as described by Eq. (2) given by

$$\frac{\Delta A}{A_0} = \frac{\Delta h}{h_0} \quad (2)$$

where $h_0$ is the unstressed membrane thickness, with the expansion and thickness moduli being related as $K_A = K_h \cdot h_0$ [1,25]. Expected to modulate many dynamic and relaxation processes occurring in cellular membranes, the area-thickness coupling results in large in-plane viscous moduli of membranes comparable in magnitude with their corresponding elastic moduli [23]. The expansion viscosity of lipid bilayers has been estimated to be $10^{-10}$ N s/m [25].

The stress response of a membrane is significantly larger under compression than expansion [23]. The bending modulus ($K_B$) of a bilayer estimated where $K_B \sim 10^{-19}$-Nm indicates that membrane resistance to bending is significantly less than resistance to area expansion [1,25]. Thus, the basic viscoelastic properties of a lipid membrane can be generally summarized where $K_B < K_h < K_A < K_C$. This extreme sensitivity to bending forces provides neuronal membranes with the ability to evaginate, bud, and protrude during growth, migration, and development. Further, the bending sensitivity of neuronal membranes permits the formation, release, recapture, and recycling of synaptic vesicles during neuronal communication. From a functional perspective, it is also worth considering that the dynamic viscoelastic features of neuronal membranes may mediate flexoelectric effects or influence ion channel opening and closing in manners that alter neuronal excitability. Although it has yet to be thoroughly considered, the idea that neuronal membranes experience flexoelectric effects dates to some of the first neurophysiological recordings of action potentials made from squid giant axons [30]. Attention to how membrane viscoelasticity influences brain activity and neuronal plasticity will yield new insights into nervous system function and dysfunction. As discussed later in the chapter, several neurological diseases have been associated with changes in the stiffness of neuronal circuits. Whether the global changes in the elasticity of neuronal circuits caused by a disease are the result of changes in the lipid composition of neurons or alterations in the viscoelastic properties of neuronal membranes is not yet known, but should be examined.

## 3 ELECTRODEFORMATION AND FLEXOELECTRICITY: A MECHANISM OF NEURONAL SIGNALING?

Given the physical and electric properties of lipid bilayers, neuroscience must begin to consider how flexoelectric effects contribute to neuronal membrane excitability. Studies have shown that the physical dimensions of nerves change in phase with the action potential, exerting forces normal to the membrane surface [4,5,9,31,32]. In the dominant and widely accepted electric circuit equivalent (RC) models of neurons, the lipid bilayer is characterized as a simple capacitor. The equivalent RC circuit model considers ion channels as variable resistors, and a voltage difference is generated across the insulating membrane, due to differences in electrochemical driving forces generated by leaky ion channels and different ionic concentrations across the membrane. While separate models accounting for the other nonelectric behaviors observed during the action potential have been proposed, we still lack a unifying model that wholly accounts for the electric, chemical, and mechanical phenomena observed during action-potential firing and neuronal signaling. Regardless, the coupling of mechanical and electric energy has been studied extensively (e.g., piezoelectricity), and its consideration as applied to the nervous system function is briefly considered below.

The flexoelectric effect is a liquid crystal analogue to the piezoelectric effect in solid crystals. Flexoelectricity refers specifically to the curvature-dependent polarization of the membrane. As opposed to area stretching, thickness compression, and shear deformation in solid crystals, the flexoelectric effect includes the deformation of membrane curvature. This effect is manifested in liquid crystalline membrane structures, as a curvature of membrane surface leads to a splay of lipids and proteins. The molecules would otherwise be oriented parallel to each other in the normal flat state of the local membrane. Similar to piezoelectricity of solids, flexoelectricity is also manifested as a direct and a converse effect, featuring electric field-induced curvature. This provides a basic mechanoelectric mechanism to enable nanometer-thick lipid membranes to exchange between electric and mechanical stimuli. As mentioned, consideration that neuronal membranes may have mechanoelectric properties goes back to the discussion of the possible origin of inductance in early circuit models of the nervous membrane and giant squid axons [30].

Experimentally, the generation of alternating currents by membranes subjected to oscillating gradients of hydrostatic pressure was observed in the early 1970s [33]. A quantitative description of the observed vibration response was assumed based on the change in membrane area and thus capacitance, though a detailed explanation of the mechanisms regarding transmembrane potential was not offered at the time. The oscillations of membrane curvature in these experiments can be credited as a displacement current due to the oscillating reversal of flexoelectric polarization of the curved membrane [34]. According to the Helmholtz equation, an electric potential difference appears across a polarized surface. For a membrane curvature that oscillates in time and assuming spherical curvature for simplicity, the flexopolarization leads to a transmembrane AC voltage difference with first harmonic amplitude as described by Eq. (3):

$$U_\omega = \frac{f}{\varepsilon_0} 2c_m \quad (3)$$

where $f$ is the flexoelectric coefficient, measured in coulombs; $\varepsilon_0$ is the absolute dielectric permittivity of free space; and $c_m$ is the maximal curvature. Similarly, a displacement current due to oscillating flexopolarization can also be calculated as shown in Eq. (4) by considering a membrane capacitance:

$$C_0 = \frac{\varepsilon_0 S_0}{d} \quad (4)$$

where $S_0$ is the flat membrane area and $d$ is the capacitive thickness of the membrane. The first harmonic amplitude of the membrane flexoelectric current is then described by Eq. (5):

$$I_\omega = f\frac{C_0}{\varepsilon_0} 2c_m \omega \tag{5}$$

where $\omega$ is the angular frequency of oscillations. Thus, the current through or potential across the membrane can be determined from the associated flexoelectric coefficient of the membrane and the radius of curvature. Flexoelectricity (current generation from bending) and converse flexoelectricity have been demonstrated in lipid bilayers and cell membranes [34,35]. Flexoelectricity provides a linear relationship between membrane curvature and transmembrane voltage and is likely involved in the mechanosensitivity and mechanotransduction of signals in biological cells including the nervous system. The direct and converse flexoelectric effects have been used to describe the transformation of mechanical into electric energy by stereocilia and the electromotility of outer hair cell membranes for hearing [34,35]. As discussed above, many neuronal membrane functions involve the modulation of membrane curvature (e.g., synaptic vesicle formation, exocytosis, and endocytosis), and the prospects that flexoelectricity is intricately involved in these processes to relate membrane mechanics and electrodynamics are likely. Future studies using highly sensitive membrane-based optical reporters of voltage combined with advanced microscopy methods like optical coherence tomography may be able to directly address this possibility.

Mechanical equilibrium in membranes requires that the cellular radius depends on membrane tension in order to maintain a constant pressure across the membrane, as related by the Young-Laplace equation (Eq. 6):

$$\Delta P = \gamma \left( \frac{1}{R_1} + \frac{1}{R_2} \right) \tag{6}$$

where $\Delta P$ is the pressure difference across the membrane, $\gamma$ is the membrane tension, and $R_1$ and $R_2$ are the principal radii of curvature. Considering now electric field-mediated effects on the membrane, the relation between membrane tension and an applied electrostatic potential is given by the Young-Lippmann equation (Eq. 7):

$$\gamma = \gamma^0 - \frac{CV^2}{2} \tag{7}$$

where $\gamma$ is the total chemical and electric surface tension, $\gamma^0$ is the surface tension at zero electric field, $C$ is the capacitance of the interface, and $V$ is the applied voltage. Electrowetting on dielectrics is one application concerned with membrane tension as related to an applied voltage. In electrowetting, a thin insulating layer (analogous to the cell membrane) is used to separate conductive liquid (extracellular environment) from metallic electrodes with an applied voltage (intracellular environment) to avoid electrolysis.

In the case of a bilayer membrane, differences in tension between the two interfaces will create changes in curvature, referred to earlier as converse flexoelectricity. Thus, modulation of membrane tension by transmembrane voltage in a neuron will cause movement of the membrane with magnitude and polarity governed by the cell membrane stiffness and surface potentials in order to maintain pressure across the membrane. This has been observed in real time using voltage-clamped HEK293 cells with atomic force microscopy [36]. It was observed that depolarization caused an outward movement of the membrane, with amplitude proportional to voltage. Additionally, application of the Young-Lippmann equation to both interfaces of the lipid bilayer yielded a mathematical model able to predict the membrane tension over a range of surface potentials [36]. The sum of the two interface tensions yields the total tension in the membrane ($T_t$) as shown in Eq. (8):

$$T_t = \frac{\sqrt{(2k_B T)^3 \epsilon_w \epsilon_0}}{ze_0} \left( \sqrt{n_{ex}} \left[ \sinh^{-1}\left\{ \frac{\sigma_{ex} - C_m V}{2\sqrt{n_{ex}\epsilon_w\epsilon_0 2k_B T}} \right\} \right]^2 \right.$$

$$\left. + \sqrt{n_{in}} \left[ \sinh^{-1}\left\{ \frac{\sigma_{in} + C_m V}{2\sqrt{n_{in}\epsilon_w\epsilon_0 2kT}} \right\} \right]^2 \right) + T_m \quad (8)$$

where $kT$ is the product of Boltzmann's constant and absolute temperature, $\epsilon_0$ the permittivity of free space, $\epsilon_w$ the relative permittivity of water, $z$ the solution ion valence, $e_0$ the electronic charge, $n$ the ionic strength of the solution, $\sigma$ the structural charge density at the interface, $C_m$ the specific capacitance of the membrane, $V$ the voltage, $T_m$ the voltage-independent portion of membrane tension, and the subscripts ex and in the external and internal membrane interfaces, respectively. Thus, the tension in the membrane is related to the voltage and ionic charges across the membrane. Consequently, the change in voltage with a nerve impulse is associated with a change in membrane tension, which will result in an alteration of cell radius to keep pressure constant across the membrane. This offers a mechanism and quantitative description for the observed change in the radius of nerve fibers during action-potential firing. Future investigations should aim to determine if such mechanisms exist and, if so, what their contribution is to structural changes in axons during action-potential firing as opposed to previously hypothesized mechanisms, such as cell swelling due to water transport. It is also possible that flexoelectric effects mediate some changes in the electric activity of neurons and glial cells observed, for example, during cortical spreading depression associated with headaches or some forms of epileptic seizures. Justified by several bodies of literature, future investigations into these possibilities are certainly worthy of pursuit.

## 4 THE ROLE OF MECHANOSENSITIVE ION CHANNELS IN NEURONAL FUNCTION

In addition to exerting direct flexoelectric consequences, mechanical forces acting on neuronal lipid bilayers can be transformed into consequences on the activity of membrane-bound proteins including ion channels, receptors, and enzymes. Many ion channels exhibit spring-like structures, rendering their gating kinetics sensitive to mechanical perturbations. Further, the effects of pressure, tension, stretch, and stress on cell membranes can activate and deactivate a broad range of mechanosensitive channels found in nature.

To better understand the factors influencing ion channel activity, consider the simple case of a two-state (open and closed) channel using Boltzmann statistics, where open-channel probability ($P_O$) can be described by Eq. (9) as

$$P_O = \frac{1}{1+e^{\left(\frac{\Delta G}{k_B T}\right)}} \qquad (9)$$

where $k_B$ is the Boltzmann constant, $T$ is the absolute temperature, and $\Delta G$ is an intrinsic energy difference between open and closed states ($G_{open} - G_{closed}$). $\Delta G$ dictates the likelihood of the channel occupying each state. The change in free energy can also be expressed as the sum of changes due to chemical ($\Delta G_{chem}$), electric ($\Delta G_{elec}$), and mechanical ($\Delta G_{mech}$) contributions. Various analytic models quantifying changes in free energy due to these different mechanisms are detailed in literature [37–43].

Regarding mechanical stimulation, if a force $f$ is exerted on the channel and the gating domain moves a distance $b$, then work is done, and the change in free energy is described by Eq. (10) as

$$\Delta G_{mech\ force} = -fb + \Delta u \qquad (10)$$

where $\Delta u$ is the intrinsic energy difference between states in the absence of applied force. Note that a larger movement of the gate swing ($b$) requires less force ($f$) to obtain the same $\Delta G$. Note that this expression is equivalent to the gating of voltage-dependent channels as described by Eq. (11):

$$\Delta G_{elec\ force} = f_{elec} \times \text{distance} = -Eq \times b \approx V_m \left(\frac{qb}{m}\right) \qquad (11)$$

where $E$ is the electric field strength and $q$ is the net charge on the gating region. The electric field strength is $E = V_m/m$ where $V_m$ is the transmembrane potential and $m$ is the distance over which the field drops, typically approximated as the membrane thickness. Due to difficulties in determining $b$ and $m$, the term $qb/m$ is utilized and referred to as the equivalent gating charge. The equivalent gating charge is typically 4–6 electron charges per subunit for voltage-gated channels, such that an energy difference of 1 $k_B T$ is produced by a membrane potential of ~5mV [37].

A similar expression for the free energy holds for channels influenced by tension in the membrane. Tension is the energy excess per unit area resulting from any type of stress. Most work on mechanically gated channels has used patch-clamp recording electrodes to apply suction to a patch of membrane to generate tension. A lateral stretch of a membrane generates an expanded area, producing work and lowering the free energy difference in states, and can be expressed by Eq. (12) as

$$\Delta G_{mech\ tension} = -\gamma \Delta a + \Delta u \qquad (12)$$

where $\gamma$ is lateral tension, $\Delta a$ is the change of the in-plane area of the channel after opening, and $\Delta u$ is the intrinsic energy difference between states in the absence of tension. Other forms of stress relevant to biological membranes include shear stress and bending stress, which can also contribute to changes in free energy. However, as the area elasticity modulus is much larger than the shear and bending moduli, the contributions to free energy from lateral tension will typically dominate.

The mechanosensitivity of voltage-dependent channels has been investigated for a multitude of ion channels. Mammalian cells express several families of polymodal gated ion channels, including TRP and $K_{2P}$ channels, which have been shown to activate by mechanical stimuli including membrane stretch and hydrostatic pressure (for reviews, see Refs. [14,44–46]). The TRP channel heteromers TRPC1/C3 and TRPC1/P2 are of particular interest as they have been shown to form $Ca^{2+}$-permeable MscCa channels in mammalian cells including neurons (for reviews, see [45–47]). The polymodal $K_{2P}2.1$ channel TREK-1 is mechanosensitive and is active at rest in neurons while mediating a $K^+$ leak current to regulate the membrane potential and excitability [44,48].

Neurons are also known to express a variety of pressure-sensitive channels such as transient receptor potential (TRP) or two-pore-domain $K^+$ ($K_{2p}$) channels, which both exhibit polymodal gating mechanisms. Several of the voltage-gated ion channels expressed in neurons (e.g., $Na_V1.2$, $Na_V1.5$, and $K_V1.1$) possess mechanosensitive properties that render their gating kinetics sensitive to transient changes in lipid bilayer tension [14,37]. It has been shown that $K_V$ channels exhibit sensitivity to small physiologically relevant mechanical perturbations of the cell membrane, producing shifts in the channel activation curve and an increase in the maximum open probability [49]. Tension sensitivity was accounted for in theory by having the tension act predominantly on the pore-opening transition to favor the open conformation and did not hold if membrane tension acted mainly on the voltage sensor conformational changes. This mechanically induced shift in activation kinetics could allow $K_V$ channels and likely other voltage-gated ion channels known to regulate neuronal signaling and plasticity, such as sodium and calcium channels, to play a role in neuronal mechanotransduction. While it is important to understand the influence of mechanical forces on native ion channels in the nervous system, genetic methods to render neurons mechanically sensitive in efforts to regulate their activity have also begun to emerge. For example, neurons have been genetically modified to express the bacterial large conductance mechanosensitive (MscL) channels in a manner that alters their activity [50]. Such mechanogenetic control methods represent interesting avenues for developing innovative new approaches to treating pervasive neurological and neuropsychiatric diseases as discussed further below.

## 5 CYTOSKELETAL ELEMENTS MEDIATE FORCE SENSING AND TRANSDUCTION IN NEURONS: MECHANISMS FOR PLASTICITY?

One particular approach to mechanobiology involves the application of structural analysis to the cytoskeletal and extracellular matrices of a cell and the determination of its associated effects on cellular and molecular processes. One such approach is based on the concept of tensegrity architecture as a simple mechanical model of cell structure to relate cell shape, movement, and cytoskeletal mechanics, as well as the cellular response to mechanical forces [51]. Tensegrity has allowed the mathematical formulation of the relation of tensioned and compressed parts between the extracellular and cytoskeletal matrices. A model of the intracellular cytoskeleton as a network of interconnected microfilaments, microtubules, and intermediate filaments was shown to predict dynamic mechanical properties of cells [52].

Another feature of cells that lends well to mechanical analysis is the intracellular forces of the structural matrices. Intracellular forces may be generated by the polymerization and depolymerization of cytoskeletal elements, such as actin filaments. Actin forms soft macromolecular networks of entangled and cross-linked fibers to establish part of the cellular cytoskeleton. The polymerization and depolymerization of actin filaments and microtubules generate forces that are important to many cellular processes, such as cell motility [53], to counteract plasma membrane tension and deformation changes during clathrin-mediated endocytosis [54] and to act as a molecular tension sensor regulating numerous aspects of intracellular homeostasis and function [55]. The energy for force generation by actin is provided by chemical potential differences between monomeric G-actin and its subunit incorporation into the filamentous F-actin biopolymer. When actin filaments approach a biological load (e.g., a plasma membrane), they generate pushing forces where thermal fluctuations enable the continued incorporation of G-actin monomers into the filament. This process of elongation will continue to occur until the counteracting load forces slow and stall polymerization at the thermodynamic limit [56]. This stalling (maximum) force can be estimated by Eq. (13) as

$$F_{stall} = \frac{k_B T}{\delta} \ln\left(\frac{c}{c_{crit}}\right) \qquad (13)$$

where $k_B$ is Boltzmann's constant, $T$ is the absolute temperature, $\delta$ is the elongation distance for a single G-actin monomer (2.7 nm), $c$ is the concentration of G-actin monomers in solution, and $c_{crit}$ is critical concentration for polymerization (equivalent to $k_{off}/k_{on}$) for elongation at a single filament end [57–59]. The maximum force ($F_{stall}$) a single actin filament can generate has been estimated to be $\approx 9$ pN.

In a small bundle of actin filaments, it is thought that $F_{stall}$ of the bundle is equal to the linear sum of forces generated by each fiber [58]. In addition to stalling forces, the tips of actin bundles can experience other forces that can cause filaments to buckle. Following buckling, the actin filaments can continue to elongate through the addition of G-actin. The force required to stall actin filament elongation is independent of length, whereas buckling forces vary as a function of length. The force required for filament buckling with one free end and one clamped end can be expressed by Eq. (14) as

$$F_{buckle} = \frac{\frac{\pi^2}{4} k_{mod}}{L^2} \qquad (14)$$

where $k_{mod}$ is the flexural rigidity of an actin filament (0.06 pN/μm$^2$) and $L$ is filament length [58]. To generate forces of several nanonewtons per square micrometer [60,61], actin filaments contact surface loads from a variety of different angles and are continuously undergoing nucleation and branch formation for new filament elongation near the leading edge [58].

Actin is in fact one of the best-recognized cytoskeletal contributors to synaptic function. It has been well established that the cytomechanics of axonal growth cone navigation and branching are largely mediated by actin-generated forces [62]. The actin motor protein myosin is capable of generating forces sufficient to contract muscle tissue and is known to participate in molecular cargo shuffling during cell motility and growth [63]. Recent measurements of growth cone mechanical properties have shown that growth cones have a low elastic modulus ($E = 106 \pm 21$ N/m$^2$) and that, considering its retrograde flow, actin may generate internal stress in growth cones on the order of 30 pN/μm$^2$ [64]. These results indicate growth cones are soft and weak force generators, rendering them sensitive to the mechanical properties of their environment [64], as similarly described above by $F_{stall}$.

Interestingly, dorsal root ganglion neuron cones have been shown to generate significantly greater traction forces compared with hippocampal neuron growth cones as determined using traction force microscopy [65]. Moreover, these neuronal types exhibited differential cytoskeletal adaption to substrate stiffness [65]. Such differences in cytoskeletal mechanics pose the possibility that different forces generated by actin may serve as unique mechanical scripts for synapse formation, maturation, and operation in neurons. Perhaps a mechanical environment, such as the extracellular matrix (ECM) within a given anatomical area, can change to optimize the growth dynamics of specific groups of invading axons across distinct stages of development. Any such cellular-mechanical matching for tuning patterned synapse formation is certainly a tantalizing concept. Several observations seem to provide evidence for such mechanisms.

Quantified with atomic force microscopy (AFM), different layers of the hippocampus have been shown to possess significantly different rigidities in the rodent brain (CA1 *stratum pyramidale* = 0.14 nN/μm$^2$, CA1 *stratum radiatum* = 0.20 nN/μm$^2$, CA3 *stratum pyramidale* = 0.23 nN/μm$^2$, and CA3 *stratum radiatum* = 0.31 nN/μm$^2$ [27]). Observed on substrate rigidities ranging from 0.5 to 7.5 nN/μm$^2$, hippocampal axons increase their length faster on softer substrates [66]. Neurons from embryonic spinal cord develop a fivefold higher neurite branch density on soft substrates (0.05 nN/μm$^2$) compared with more rigid ones (0.55 nN/μm$^2$ [67]). The elasticity and mechanical properties of the brain have been shown to change across different stages of development [28,68,69]. Finally, it has recently been shown that mechanical tension within axons plays an essential role in the accumulation of proteins at presynaptic terminals; biochemical signaling and recognition of synaptic partners are not sufficient [70]. Presynaptic vesicle clustering at neuromuscular synapses vanished upon severing the axon from the cell body and could be restored by applying tension to the severed end, and further stretching of intact axons could even increase vesicle clustering. Furthermore, rest tensions of approximately 1 nN in axons were restored over approximately 15 min when perturbed mechanically, implicating mechanical tension as a modulation signal of vesicle accumulation and synaptic plasticity. Increased axonal tension from the resting state may induce further actin polymerization and increased clustering via mechanical trapping or interactions between F-actin and vesicles [70]. Supported by these aforementioned primary observations, neuroscience should focus efforts on characterizing changes in growth cone traction force, elasticity, or viscosity across different anatomical and cellular regions, levels of activity, and stages of development.

Actin in dendritic spines of neurons has been shown to regulate synapse formation and spine growth [71], activity-dependent spine motility [11,13,72], and plasticity [73–75]. In gelsolin knockout mice, reduced actin depolymerization has been shown to enhance NMDA-mediated and voltage-gated (VG) $Ca^{2+}$ activity in hippocampal neurons [76]. The contribution of mechanical force changes to any of the above observations is not clearly understood. The viscoelasticity of dendritic spines was found to be critical to their function through atomic force microscopy elasticity mapping and dynamic indentation methods [77]. Through this mechanical characterization, the activity-dependent structural plasticity, metastability, and congestion in the cytoplasm of spines are all gauged by merely a few physically measurable parameters. The degree to which spines are able to remodel and retain stability is determined in large part by viscosity, where soft, malleable spines have properties likely associated with morphological plasticity for learning and the properties of rigid, stable spines are likely associated with memory retention [77]. Perhaps the stabilization or destabilization of actomyosin networks produces direct mechanical consequences on synaptic activity by increasing or decreasing plasma membrane tension to coordinate the bending or compression of presynaptic compartments and dendritic spines. Given the dynamic nature of the actin cytoskeleton in the regulation of membrane tension and channel activity as further discussed below, the aforementioned idea seems natural to investigate. Additionally, the contribution of the various other elements composing the cytoskeletal and extracellular matrices besides actin as discussed here can analogously be investigated.

Microtubules are capable of exerting tremendous molecular forces, as can be observed during mitotic spindle formation. Microtubules generate forces in cells through the polymerization of αβ-tubulin dimers or catastrophic depolymerization events due to dynamic microtubule instability [78,79]. Microtubules also generate pushing ($F_{push}$) or pulling forces ($F_{pull}$) to provide structural support to membranes and proteins or to transport molecular cargo in neurons. The maximum force a single microtubule can generate occurs when $F_{stall} \approx 5$ pN [80]. At this thermodynamic limit, microtubule elongation is thought to resemble a "Brownian ratchet" similar to actin [81]. Reinforcement from intermediate filaments, actomyosin networks, and microtubule-associated proteins (MAPs) in cells is thought to enable microtubules to handle larger compression loads and withstand buckling forces $>100$ pN [82]. Besides their classical cargo transport and structural support actions, which involve interactions with actin, microtubules have recently been shown to play previously unrealized roles in the regulation of dendritic spine morphology and synaptic plasticity [83,84].

Signaling among neighboring dendritic spines has been shown to involve the diffusion of several GTPases [85]. For instance, RhoA was recently shown to diffuse out of stimulated spines and laterally through the dendrite about 5 µm [86]. In other studies, RhoA has been shown to modulate microtubule-generated forces [87]. Diffusion of RhoA and other GTPases from an active spine is thought to signal to neighboring inactive spines, but the downstream mechanisms remain unclear. One possibility is that, in response to synaptic activity, local RhoA-induced changes in microtubule forces initiate mechanical signals that are then transmitted among small networks of microtubule-coupled spines. Indeed, microtubules can behave as elastic rods and exert forces over distances of $\approx 10$ µm [80], where the average interspine distance is $\approx 1$ µm [86]. Further, microtubules have recently been shown to invade spines and to regulate morphological changes in spines in an activity-dependent manner [83,84]. Based on the microtubule dynamics discussed above, one would expect microtubule spine invasion to be accompanied by changes in microtubule pushing ($F_{push}$) and pulling ($F_{pull}$) forces. Whether such changes in microtubule $F_{pull}$ or $F_{push}$ can trigger mechanical changes in spatially segregated spines as a means of interspine signaling is not known. Nevertheless, exploring this possibility seems justified by recent observations.

## 6 METHODS AND TOOLS FOR MECHANICAL INTERFACING WITH THE NERVOUS SYSTEM

Our early understanding of the phenomena of electric coupling with the mechanical modification of the neuronal membrane has already begun to yield innovative methods and technologies for interfacing to the nervous system. The modulation of refractive index or thickness of the cell due to transmembrane-potential-dependent deformations has allowed label-free imaging of the membrane potential. By measuring milliradian-scale phase shifts in the transmitted light, changes in individual mammalian cell's membrane potential were detected using low-coherence interferometric microscopy without the use of exogenous labels [88]. Using this technique, it was also demonstrated that propagation of electric stimuli in gap-junction-coupled cell networks could be monitored using wide-field imaging. This technique offers the advantages of simple sample preparation, low phototoxicity, and no need for photobleaching. Previous successes in label-free imaging of electric activity have been in invertebrate cells due to mammalian cells being smaller and optically transparent and scattering significantly less light compared with invertebrate nerves and neurons. While the technology still requires further refinement to be able to resolve single action potentials, the method was able to experimentally confirm that the source of light phase shifts was due to potential-mediated membrane tension, as opposed to swelling due to water transport or electrostriction of the cell membrane [88].

Regarding probing the mechanical response of mammalian cells to electric excitation, atomic force microscopy is the most commonly used tool for quantifying cellular deformation, despite its invasiveness. Recently, piezoelectric nanoribbons have been developed for electromechanical biosensing and have demonstrated that cells deflect by 1 nm when 120 mV is applied to the membrane [89]. Furthermore, these nanoribbons support the model of voltage-induced membrane tension discussed earlier and support previous investigation of cellular electromechanics using atomic force microscopy as well. These nanoribbons are made using microfabrication techniques and so can be scaled more readily than atomic force microscopy probes. Additionally, advances in microfabrication techniques could allow the manufacture of thinner nanoribbons to enhance their sensitivity and facilitate the electromechanical observation of smaller neural structures, such as axons, dendrites, and dendritic spines. The importance of observing the mechanical response of these structures is highlighted in the above discussions throughout the chapter.

The modulation and monitoring of the nervous system are pertinent to the treatment of neurological and psychiatric diseases, as well as the scientific investigation of the neural mechanisms of cognitive, sensory, and motor functions. Conventionally, interfacing with the nervous system has been conducted using electric and chemical means, such as microdialysis and deep brain stimulation. Recently, the development of devices utilizing mechanical energy to interact with the nervous system has received considerable attention. These devices include ultrasound for noninvasive neural stimulation and magnetic resonance elastography for noninvasive palpitation of the brain.

Besides its use for diagnostic imaging, ultrasound at low intensity is able to nondestructively excite nervous tissue [90,91]. The mechanisms behind ultrasound stimulation however are not well established. There are two classes of mechanisms primarily considered, thermal and mechanical. Ultrasound can heat tissue, analogous to transcranial high-intensity focused ultrasound ablation [92,93], and temperature-sensitive ion channels can be activated through tissue heating. However, negligible temperature increases have been measured during pulsed ultrasound stimulation protocols [91,94]. The premise behind the hypothesized mechanical mechanisms of ultrasound is that deformation of the cell membrane or the proteins embedded therein could affect ion channel kinetics and/or membrane capacitance to induce transmembrane currents to initiate action-potential discharge. Recently, intramembrane cavitation has been proposed as a mechanism for the effects on nervous tissue by ultrasound [95,96]. Using models of the cellular membrane, the mechanical energy from ultrasound would be absorbed and transformed by the membrane into expansions and contractions of the space between bilayer membrane leaflets [95]. Linking this model with electrodeformation, ultrasound leads to action-potential excitation via currents induced by membrane capacitance changes within the computational model [96]. The model is referred to as the bilayer sonophore and offers explanations on the requirement for long ultrasonic stimulation pulses and other experimentally observed phenomena. In addition to such direct actions of ultrasound on membranes, several different sonogenetic methods have been described. In these methods, neurons are genetically modified to express a variety of mechanosensitive channels, with ultrasound then used as a pressure source to activate them [97].

Magnetic resonance imaging is the most common imaging modality for investigating central neurological disorders as it is noninvasive and provides a number of contrast mechanisms. Magnetic resonance elastography (MRE) determines the shear modulus of tissues in vivo through the application of mechanical shear waves and the use of a phase-sensitive magnetic resonance imaging sequence to produce a map of the shear modulus of the tissue. The map is used for clinical diagnostics, as a change in cellular elasticity is associated with many diseases [98], due to their alteration of the microstructural environment of the central nervous system through neuroinflammation, neurodegeneration, and disruption of the glial matrix. This is analogous to the palpitation of tissue to identify lesions based on their differential stiffness to surrounding tissues. Application of MRE to the brain may have useful applications for characterizing brain disease based on the mechanical properties of the tissue. For example, MRE has been used to map and characterize the viscoelastic properties of the normal, aged, and diseased human brain [29,99–103]. Further, the mechanical properties of the human brain have shown a high sensitivity to neurodegeneration in initial investigations of Alzheimer's disease, multiple sclerosis, normal pressure hydrocephalus, and cancer [104]. This recent increase in the number and breadth of observations from neuroimaging studies employing MRE indicates a strong and growing interest in determining how the mechanical properties of a brain relate to its function.

Despite the growth of MRE investigation, it has yet to gain traction in clinical applications as mechanical properties are largely reported as global averages rather than local values. However, recent work in MRE has been focused on generating high-resolution, reliable, and repeatable estimates of local mechanical properties in the human brain. The mechanical properties of the corpus callosum and corona radiata were recently measured in healthy individuals using high-resolution magnetic resonance elastography and atlas-based segmentation [105]. Both structures were found to be stiffer than the overall white matter and demonstrated the feasibility of quantifying the mechanical properties of specific structures in white-matter architecture for the assessment of the localized effects of disease. The ability to reliably estimate local mechanical properties noninvasively represents a possible revolution in the clinical assessment of neurodegeneration of the human brain.

There are many other methods by which the mechanical properties of neurons and their networks can be studied [2,3,106]. As mentioned previously, modified patch-clamp methods referred to as pressure-clamp techniques can be used to study the activity of single-ion channels in response to membrane deformations [107]. Numerous fluorescence microscopy methods have also been developed for the study of mechanobiology and can be applied to neuronal cells. Fluorescence correlation spectroscopy and time-correlated single-photon counting have been useful for characterizing various mechanical forces in cells [108]. Optical methods relying on Förster resonance energy transfer (FRET) principles are particularly advantageous in that they enable the visualization of molecular mechanical actions [109].

A particularly noteworthy FRET approach involves uniquely designed strain-sensitive FRET sensors, which can be used to actively monitor the stress exerted on cytoskeletal and extracellular matrix proteins [110].

Atomic force microscopy (AFM) has proven itself a critical tool for the study of molecular and cellular mechanics [111]. AFM experiments have revealed differences in brain elasticity in different anatomical regions [27] and have been useful in describing the force-velocity relationships of growing actin networks in cells [61]. Other trapping methods using magnetic particles or optical tweezers enable additional reproducible and quantitative approaches to investigate mechanobiology [2,112] and can be useful for studying mechanical forces in neurons. Advances in microengineering have led to a recent explosion in the development of mechanobiological research devices [2]. These devices include microelectromechanical systems (MEMS) [106], integrated strain arrays [113], elastomeric micropost substrates [114], microfluidic chips [115], and others. The number of tools useful for studying neuromechanobiology has vastly increased in the past decade. New insights gained from the use of these methods will drive new innovations that enable precise control and monitoring of mechanical forces in the study and treatment of neurological diseases.

## 7 FUTURE DIRECTIONS AND OUTLOOK FOR NEUROMECHANOBIOLOGY

Many membrane functions involve the manipulation of membrane curvature (endocytosis, cell movement, etc.), and the prospects that flexoelectricity is intricately involved in these processes to relate membrane mechanics and electrodynamics are likely. In the case of a bilayer membrane, differences in tension between the two interfaces will create changes in curvature. Thus, modulation of membrane tension by transmembrane voltage in a neuron will cause movement of the membrane, with magnitude and polarity governed by the cell membrane stiffness and surface potentials in order to maintain pressure across the membrane. The field of neuroscience has not traditionally considered such mechanisms important features of neuronal signaling. The emerging discipline of neuromechanobiology has begun to draw attention to the possibility that mechanical forces and membrane dynamics work in cooperation with classical signaling mechanisms to regulate neuronal signaling and plasticity.

Features of neurons that lend themselves well to mechanical analyses include the transduction and sensing of forces by cytoskeletal elements and extracellular matrix proteins, as well as the interactions between membrane stress or tension and channel activity. Intracellular forces may be generated by the polymerization and depolymerization of cytoskeletal elements, such as actin filaments. Mechanical forces acting on cell membranes or through cytoskeletal filaments can also be transformed into consequences on membrane-bound and cytoskeletal-tethered protein activity. Membrane-bound protein activity is also influenced by the properties of the cellular phospholipid bilayer. How the properties and changes in density of phospholipid bilayers influence the propagation of mechanical waves and neuronal processes, such as action-potential initiation and propagation, is not precisely known either.

The extent to which cellular-mechanical dynamics influences neuronal activity and effectually the interfacing to the nervous system using mechanical forces remains largely unexplored. To advance neuroscience and knowledge of nervous system function, our compartmentalized analyses of electric, chemical, or mechanical energies in system characterization and manipulation of the nervous system need to be integrated into comprehensive models. By starting to consider the interplay between electric, chemical, and mechanical energy, rather than separately compartmentalizing them, a new paradigm of understanding and study of the mechanobiology of the nervous system may advance through neuromechanobiology. There is little doubt that knowledge obtained through the study of neuromechanobiology will lead to new neurodiagnostics and therapeutics for some of the most debilitating neurological and psychiatric diseases affecting human health and well-being.

## REFERENCES

[1] Hamill OP, Martinac B. Molecular basis of mechanotransduction in living cells. Physiol Rev 2001;81(2):685–740.

[2] Kim DH, Wong PK, Park J, Levchenko A, Sun Y. Microengineered platforms for cell mechanobiology. Annu Rev Biomed Eng 2009;11:203–33.

[3] Eyckmans J, Boudou T, Yu X, Chen CS. A hitchhiker's guide to mechanobiology. Dev Cell 2011;21(1):35–47.

[4] Tasaki I, Byrne PM. Volume expansion of nonmyelinated nerve-fibers during impulse conduction. Biophys J 1990;57(3):633–5.

[5] Tasaki I, Kusano K, Byrne PM. Rapid mechanical and thermal-changes in the garfish olfactory nerve associated with a propagated impulse. Biophys J 1989;55(6):1033–40.

[6] Hill DK. The volume change resulting from stimulation of a giant nerve fibre. J Physiol 1950;111(3–4):304–27.

[7] Fields RD. Signaling by neuronal swelling. Sci Signal 2011;4(155):tr1. https://doi.org/10.1126/scisignal.4155tr1.

[8] Lundstrom I. Mechanical wave propagation on nerve axons. J Theor Biol 1974;45(2):487–99.

[9] Kim GH, Kosterin P, Obaid AL, Salzberg BM. A mechanical spike accompanies the action potential in mammalian nerve terminals. Biophys J 2007;92(9):3122–9.

[10] Crick F. Do dendritic spines twitch? Trends Neurosci 1982;5:44–6.

[11] Fischer M, Kaech S, Knutti D, Matus A. Rapid actin-based plasticity in dendritic spines. Neuron 1998;20(5):847–54.

[12] Korkotian E, Segal M. Spike-associated fast contraction of dendritic spines in cultured hippocampal neurons. Neuron 2001;30:751–8.

[13] Star EN, Kwiatkowski DJ, Murthy VN. Rapid turnover of actin in dendritic spines and its regulation by activity. Nat Neurosci 2002;5(3):239–46.

[14] Morris CE, Juranka PF. Lipid stress at play: mechanosensitivity of voltage-gated channels. Curr Top Membr 2007;59:297–338.

[15] Paoletti P, Ascher P. Mechanosensitivity of Nmda receptors in cultured mouse central neurons. Neuron 1994;13(3):645–55.
[16] Puglielli L, Tanzi RE, Kovacs DM. Alzheimer's disease: the cholesterol connection. Nat Neurosci 2003;6(4):345–51.
[17] Laplaca MC, Prado GR. Neural mechanobiology and neuronal vulnerability to traumatic loading. J Biomech 2010;43(1):71–8.
[18] LaPlaca MC, Simon CM, Prado GR, Cullen DK. CNS injury biomechanics and experimental models. Prog Brain Res 2007;161:13–26.
[19] Tyler WJ. The mechanobiology of brain function. Nat Rev Neurosci 2012;13(12):867–78.
[20] Ingber DE. Tensegrity: the architectural basis of cellular mechanotransduction. Annu Rev Physiol 1997;59:575–99.
[21] Crawford GE, Earnshaw JC. Viscoelastic relaxation of bilayer lipid-membranes—frequency-dependent tension and membrane viscosity. Biophys J 1987;52(1):87–94.
[22] Harland CW, Bradley MJ, Parthasarathy R. Phospholipid bilayers are viscoelastic. Proc Natl Acad Sci U S A 2010;107(45):19146–50.
[23] Jeon J, Voth GA. The dynamic stress responses to area change in planar lipid bilayer membranes. Biophys J 2005;88(2):1104–19.
[24] Bloom M, Evans E, Mouritsen OG. Physical properties of the fluid lipid-bilayer component of cell membranes: a perspective. Q Rev Biophys 1991;24(3):293–397.
[25] Evans EA, Hochmuth RM. Mechanochemical properties of membranes. In: Bronner F, Kleinzeller A, editors. Membrane properties: mechanical aspects, receptors, energetics, and calcium-dependence of transport. Current topics in membranes and transport, vol. 10. New York: Academic Press; 1978. p. 1–64.
[26] Moore SW, Sheetz MP. Biophysics of substrate interaction: influence on neural motility, differentiation, and repair. Dev Neurobiol 2011;71(11):1090–101.
[27] Elkin BS, Azeloglu EU, Costa KD, Morrison 3rd B. Mechanical heterogeneity of the rat hippocampus measured by atomic force microscope indentation. J Neurotrauma 2007;24(5):812–22.
[28] Gefen A, Gefen N, Zhu Q, Raghupathi R, Margulies SS. Age-dependent changes in material properties of the brain and braincase of the rat. J Neurotrauma 2003;20(11):1163–77.
[29] Kruse SA, Rose GH, Glaser KJ, Manduca A, Felmlee JP, Jack CR, et al. Magnetic resonance elastography of the brain. NeuroImage 2008;39(1):231–7.
[30] Cole KS. Rectification and inductance in the squid giant axon. J Gen Physiol 1941;25(1):29–51.
[31] Iwasa K, Tasaki I. Mechanical changes in squid giant-axons associated with production of action-potentials. Biochem Biophys Res Commun 1980;95(3):1328–31.
[32] Tasaki I, Iwasa K, Gibbons RC. Mechanical changes in crab nerve-fibers during action-potentials. Jpn J Physiol 1980;30(6):897–905.
[33] Ochs AL, Burton RM. Electrical response to vibration of a lipid bilayer membrane. Biophys J 1974;14(6):473–89.
[34] Petrov AG. Electricity and mechanics of biomembrane systems: flexoelectricity in living membranes. Anal Chim Acta 2006;568(1–2):70–83.
[35] Petrov AG. Flexoelectricity of model and living membranes. BBA-Biomembranes 2002;1561(1):1–25.

[36] Zhang PC, Keleshian AM, Sachs F. Voltage-induced membrane movement. Nature 2001;413(6854):428–32.
[37] Sukharev S, Corey DP. Mechanosensitive channels: multiplicity of families and gating paradigms. Sci STKE 2004;2004(219):re4.
[38] Wiggins P, Phillips R. Analytic models for mechanotransduction: gating a mechanosensitive channel. Proc Natl Acad Sci U S A 2004;101(12):4071–6.
[39] Wiggins P, Phillips R. Membrane-protein interactions in mechanosensitive channels. Biophys J 2005;88(2):880–902.
[40] Chowdhury S, Chanda B. Estimating the voltage-dependent free energy change of ion channels using the median voltage for activation. J Gen Physiol 2012;139(1):3–17.
[41] Chowdhury S, Chanda B. Free-energy relationships in ion channels activated by voltage and ligand. J Gen Physiol 2013;141(1):11–28.
[42] Greisen P, Lum K, Ashrafuzzaman M, Greathouse DV, Andersen OS, Lundbaek JA. Linear rate-equilibrium relations arising from ion channel-bilayer energetic coupling. Proc Natl Acad Sci U S A 2011;108(31):12717–22.
[43] Sigg D. A linkage analysis toolkit for studying allosteric networks in ion channels. J Gen Physiol 2013;141(1):29–60.
[44] Dedman A, Sharif-Naeini R, Folgering JHA, Duprat F, Patel A, Honore E. The mechano-gated K-2P channel TREK-1. Eur Biophys J Biophys Lett 2009;38(3):293–303.
[45] Hamill OP. Twenty odd years of stretch-sensitive channels. Pflugers Arch - Eur J Physiol 2006;453(3):333–51.
[46] Lin SY, Corey DP. TRP channels in mechanosensation. Curr Opin Neurobiol 2005;15(3):350–7.
[47] Giamarchi A, Padilla F, Coste B, Raoux M, Crest M, Honore E, et al. The versatile nature of the calcium-permeable cation channel TRPP2. EMBO Rep 2006;7(8):787–93.
[48] Bayliss DA, Barrett PQ. Emerging roles for two-pore-domain potassium channels and their potential therapeutic impact. Trends Pharmacol Sci 2008;29(11):566–75.
[49] Schmidt D, del Marmol J, MacKinnon R. Mechanistic basis for low threshold mechanosensitivity in voltage-dependent K+ channels. Proc Natl Acad Sci U S A 2012;109(26):10352–7.
[50] Soloperto A, Boccaccio A, Contestabile A, Moroni M, Hallinan GI, Palazzolo G, et al. Mechano-sensitization of mammalian neuronal networks through expression of the bacterial mechanosensitive MscL channel. J Cell Sci 2018;131: http://jcs.biologists.org/content/131/5/jcs210393.long.
[51] Ingber DE. Tensegrity I. Cell structure and hierarchical systems biology. J Cell Sci 2003;116(7):1157–73.
[52] Sultan C, Stamenovic D, Ingber DE. A computational tensegrity model predicts dynamic rheological behaviors in living cells. Ann Biomed Eng 2004;32(4):520–30.
[53] Pollard TD, Borisy GG. Cellular motility driven by assembly and disassembly of actin filaments. Cell 2003;112(4):453–65.
[54] Boulant S, Kural C, Zeeh JC, Ubelmann F, Kirchhausen T. Actin dynamics counteract membrane tension during clathrin-mediated endocytosis. Nat Cell Biol 2011;13(9):1124–31.
[55] Galkin VE, Orlova A, Egelman EH. Actin filaments as tension sensors. Curr Biol 2012;22(3):R96–R101.

[56] Hill TL, Kirschner MW. Subunit treadmilling of microtubules or actin in the presence of cellular barriers: possible conversion of chemical free energy into mechanical work. Proc Natl Acad Sci U S A 1982;79(2):490–4.
[57] Brangbour C, du Roure O, Helfer E, Demoulin D, Mazurier A, Fermigier M, et al. Force-velocity measurements of a few growing actin filaments. PLoS Biol 2011;9(4):e1000613.
[58] Footer MJ, Kerssemakers JW, Theriot JA, Dogterom M. Direct measurement of force generation by actin filament polymerization using an optical trap. Proc Natl Acad Sci U S A 2007;104(7):2181–6.
[59] Hill TL, Kirschner MW. Bioenergetics and kinetics of microtubule and actin filament assembly-disassembly. Int Rev Cytol 1982;78:1–125.
[60] Marcy Y, Prost J, Carlier MF, Sykes C. Forces generated during actin-based propulsion: a direct measurement by micromanipulation. Proc Natl Acad Sci U S A 2004;101(16):5992–7.
[61] Parekh SH, Chaudhuri O, Theriot JA, Fletcher DA. Loading history determines the velocity of actin-network growth. Nat Cell Biol 2005;7(12):1219–23.
[62] Smith SJ. Neuronal cytomechanics: the actin-based motility of growth cones. Science 1988;242(4879):708–15.
[63] Mitchison TJ, Cramer LP. Actin-based cell motility and cell locomotion. Cell 1996;84(3):371–9.
[64] Betz T, Koch D, Lu YB, Franze K, Kas JA. Growth cones as soft and weak force generators. Proc Natl Acad Sci U S A 2011;108(33):13420–5.
[65] Koch D, Rosoff WJ, Jiang J, Geller HM, Urbach JS. Strength in the periphery: growth cone biomechanics and substrate rigidity response in peripheral and central nervous system neurons. Biophys J 2012;102(3):452–60.
[66] Kostic A, Sap J, Sheetz MP. RPTPalpha is required for rigidity-dependent inhibition of extension and differentiation of hippocampal neurons. J Cell Sci 2007;120(Pt 21):3895–904.
[67] Flanagan LA, Ju YE, Marg B, Osterfield M, Janmey PA. Neurite branching on deformable substrates. NeuroReport 2002;13(18):2411–5.
[68] Prange MT, Margulies SS. Regional, directional, and age-dependent properties of the brain undergoing large deformation. J Biomech Eng 2002;124(2):244–52.
[69] Thibault KL, Margulies SS. Age-dependent material properties of the porcine cerebrum: effect on pediatric inertial head injury criteria. J Biomech 1998;31(12):1119–26.
[70] Siechen S, Yang SY, Chiba A, Saif T. Mechanical tension contributes to clustering of neurotransmitter vesicles at presynaptic terminals. Proc Natl Acad Sci U S A 2009;106(31):12611–6.
[71] Zito K, Knott G, Shepherd GM, Shenolikar S, Svoboda K. Induction of spine growth and synapse formation by regulation of the spine actin cytoskeleton. Neuron 2004;44(2):321–34.
[72] Halpain S, Hipolito A, Saffer L. Regulation of F-actin stability in dendritic spines by glutamate receptors and calcineurin. J Neurosci 1998;18(23):9835–44.
[73] Kim CH, Lisman JE. A role of actin filament in synaptic transmission and long-term potentiation. J Neurosci 1999;19(11):4314–24.
[74] Krucker T, Siggins GR, Halpain S. Dynamic actin filaments are required for stable long-term potentiation (LTP) in area CA1 of the hippocampus. Proc Natl Acad Sci U S A 2000;97(12):6856–61.
[75] Matus A. Actin-based plasticity in dendritic spines. Science 2000;290:754–8.

[76] Furukawa K, Fu W, Li Y, Witke W, Kwiatkowski DJ, Mattson MP. The actin-severing protein gelsolin modulates calcium channel and NMDA receptor activities and vulnerability to excitotoxicity in hippocampal neurons. J Neurosci 1997;17(21):8178–86.
[77] Smith BA, Roy H, De Koninck P, Grutter P, De Koninck Y. Dendritic spine viscoelasticity and soft-glassy nature: balancing dynamic remodeling with structural stability. Biophys J 2007;92(4):1419–30.
[78] Desai A, Mitchison TJ. Microtubule polymerization dynamics. Annu Rev Cell Dev Biol 1997;13:83–117.
[79] Kueh HY, Mitchison TJ. Structural plasticity in actin and tubulin polymer dynamics. Science 2009;325(5943):960–3.
[80] Dogterom M, Yurke B. Measurement of the force-velocity relation for growing microtubules. Science 1997;278(5339):856–60.
[81] Peskin CS, Odell GM, Oster GF. Cellular motions and thermal fluctuations: the Brownian ratchet. Biophys J 1993;65:316–24.
[82] Brangwynne CP, MacKintosh FC, Kumar S, Geisse NA, Talbot J, Mahadevan L, et al. Microtubules can bear enhanced compressive loads in living cells because of lateral reinforcement. J Cell Biol 2006;173(5):733–41.
[83] Hu X, Viesselmann C, Nam S, Merriam E, Dent EW. Activity-dependent dynamic microtubule invasion of dendritic spines. J Neurosci 2008;28(49):13094–105.
[84] Jaworski J, Kapitein LC, Gouveia SM, Dortland BR, Wulf PS, Grigoriev I, et al. Dynamic microtubules regulate dendritic spine morphology and synaptic plasticity. Neuron 2009;61(1):85–100.
[85] Yasuda R, Murakoshi H. The mechanisms underlying the spatial spreading of signaling activity. Curr Opin Neurobiol 2011;21(2):313–21.
[86] Murakoshi H, Wang H, Yasuda R. Local, persistent activation of Rho GTPases during plasticity of single dendritic spines. Nature 2011;472(7341):100–4.
[87] Goldyn AM, Rioja BA, Spatz JP, Ballestrem C, Kemkemer R. Force-induced cell polarisation is linked to RhoA-driven microtubule-independent focal-adhesion sliding. J Cell Sci 2009;122(Pt 20):3644–51.
[88] Oh S, Fang-Yen C, Choi W, Yaqoob Z, Fu D, Park Y, et al. Label-free imaging of membrane potential using membrane electromotility. Biophys J 2012;103(1):11–8.
[89] Nguyen TD, Deshmukh N, Nagarah JM, Kramer T, Purohit PK, Berry MJ, et al. Piezoelectric nanoribbons for monitoring cellular deformations. Nat Nanotechnol 2012;7(9):587–93.
[90] Gavrilov LR, Tsirulnikov EM. Focused ultrasound as a tool to input sensory information to humans. Acoust Phys 2012;58(1):1–21 [Review].
[91] Tufail Y, Matyushov A, Baldwin N, Tauchmann ML, Georges J, Yoshihiro A, et al. Transcranial pulsed ultrasound stimulates intact brain circuits. Neuron 2010;66(5):681–94.
[92] Martin E, Jeanmonod D, Morel A, Zadicario E, Werner B. High-intensity focused ultrasound for noninvasive functional neurosurgery. Ann Neurol 2009;66(6):858–61.
[93] McDannold N, Clement GT, Black P, Jolesz F, Hynynen K. Transcranial magnetic resonance imaging-guided focused ultrasound surgery of brain tumors: initial findings in 3 patients. Neurosurgery 2010;66(2):323–32.
[94] Yoo SS, Bystritsky A, Lee JH, Zhang YZ, Fischer K, Min BK, et al. Focused ultrasound modulates region-specific brain activity. NeuroImage 2011;56(3):1267–75.
[95] Krasovitski B, Frenkel V, Shoham S, Kimmel E. Intramembrane cavitation as a unifying mechanism for ultrasound-induced bioeffects. Proc Natl Acad Sci U S A 2011;108(8):3258–63.

[96] Plaksin M, Shoham S, Kimmel E. Intramembrane cavitation as a predictive bio-piezoelectric mechanism for ultrasonic brain stimulation, Phys Rev X 2014;4:011004. https://journals.aps.org/prx/abstract/10.1103/PhysRevX.4.011004.
[97] Ibsen S, Tong A, Schutt C, Esener S, Chalasani SH. Sonogenetics is a non-invasive approach to activating neurons in *Caenorhabditis elegans*. Nat Commun 2015;6:8264.
[98] Mariappan YK, Glaser KJ, Ehman RL. Magnetic resonance elastography: a review. Clin Anat 2010;23(5):497–511.
[99] McCracken PJ, Manduca A, Felmlee J, Ehman RL. Mechanical transient-based magnetic resonance elastography. Magn Reson Med 2005;53(3):628–39.
[100] Zhang J, Green MA, Sinkus R, Bilston LE. Viscoelastic properties of human cerebellum using magnetic resonance elastography. J Biomech 2011;44(10):1909–13.
[101] Sack I, Streitberger KJ, Krefting D, Paul F, Braun J. The influence of physiological aging and atrophy on brain viscoelastic properties in humans. PLoS One 2011;6(9): e23451.
[102] Wuerfel J, Paul F, Beierbach B, Hamhaber U, Klatt D, Papazoglou S, et al. MR-elastography reveals degradation of tissue integrity in multiple sclerosis. NeuroImage 2010;49(3):2520–5.
[103] Murphy MC, Huston 3rd J, Jack Jr CR, Glaser KJ, Manduca A, Felmlee JP, et al. Decreased brain stiffness in Alzheimer's disease determined by magnetic resonance elastography. J Magn Reson Imaging 2011;34(3):494–8.
[104] Glaser KJ, Manduca A, Ehman RL. Review of MR elastography applications and recent developments. J Magn Reson Imaging 2012;36(4):757–74.
[105] Johnson CL, McGarry MDJ, Gharibans AA, Weaver JB, Paulsen KD, Wang H, et al. Local mechanical properties of white matter structures in the human brain. NeuroImage 2013;79:145–52.
[106] Rajagopalan J, Saif MT. MEMS sensors and microsystems for cell mechanobiology. J Micromech Microeng 2011;21(5):54002–12.
[107] DW Jr MB, Hamill OP. Pressure-clamp technique for measurement of the relaxation kinetics of mechanosensitive channels. Trends Neurosci 1993;16(9):341–5.
[108] Gullapalli RR, Tabouillot T, Mathura R, Dangaria JH, Butler PJ. Integrated multimodal microscopy, time-resolved fluorescence, and optical-trap rheometry: toward single molecule mechanobiology. J Biomed Opt 2007;12(1):014012.
[109] Wang Y, Wang N. FRET and mechanobiology. Integr Biol 2009;1(10):565–73.
[110] Meng F, Suchyna TM, Lazakovitch E, Gronostajski RM, Sachs F. Real time FRET based detection of mechanical stress in cytoskeletal and extracellular matrix proteins. Cell Mol Bioeng 2011;4(2):148–59.
[111] Azeloglu EU, Costa KD. Atomic force microscopy in mechanobiology: measuring microelastic heterogeneity of living cells. Methods Mol Biol 2011;736:303–29.
[112] Crick FHC, Hughes AFW. The physical properties of the cytoplasm. A study by means of magnetic particle method part I. Experimental. Exp Cell Res 1950;1:37–70.
[113] Simmons CS, Sim JY, Baechtold P, Gonzalez A, Chung C, Borghi N, et al. Integrated strain array for cellular mechanobiology studies. J Micromech Microeng 2011;21 (5):54016–25.
[114] Yang MT, Fu J, Wang YK, Desai RA, Chen CS. Assaying stem cell mechanobiology on microfabricated elastomeric substrates with geometrically modulated rigidity. Nat Protoc 2011;6(2):187–213.
[115] Taylor AM, Jeon NL. Micro-scale and microfluidic devices for neurobiology. Curr Opin Neurobiol 2010;20(5):640–7.

# CHAPTER

# Mechanobiology of the eye 12

Ashutosh Richhariya*, Nikhil S. Choudhari*, Ashik Mohamed*, Derek Nankivil[†],
Akshay Badakere*, Vivek P. Dave*, Sunil Punjabi[‡], Virender S. Sangwan*

*L V Prasad Eye Institute, Kallam Anji Reddy Campus, Hyderabad, India* Johnson & Johnson Vision Care Inc., Jacksonville, FL, United States[†] Department of Mechanical Engineering, Ujjain Engineering College, Ujjain, India[‡]

## ABBREVIATIONS

| | |
|---|---|
| CSF | cerebrospinal fluid |
| D | diopter |
| ICP | intracranial pressure |
| IOP | intraocular pressure |
| KC | keratoconus |
| TM | trabecular meshwork |

## 1 INTRODUCTION

The adult eye is generated from several different embryonic layers, eventually forming a unique and complex arrangement of neuronal, supportive connective tissue, optical structures, and muscular tissue. In recent years, characterization of the mechanobiology of the eye has proved to be helpful in understanding angiogenesis, function, disease progression mechanisms, and outcomes of therapies. In this chapter, the various parts of the eye, their mechanobiological equilibrium state in normal conditions, and their alterations in disease are discussed.

The primary function of the eye is to capture light and send it to the brain for interpretation. To do this, the eye can be divided into three parts: (1) the anterior segment, which transmits and focuses light; (2) the posterior segment, which converts photons into electrons and transmits signals to the brain; and (3) the extraocular segment, which governs the movement of the eye in different directions in response to neural cues (Fig. 1). Based on neural cues, the extraocular system will direct the eye to the object of interest. It also protects the eye from external damage. Post fixation, for good retinal image quality, the ray bundle of light passing through this optical

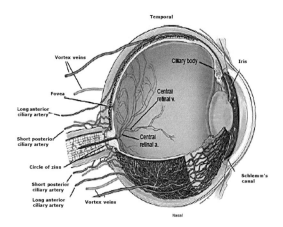

**FIG. 1**

The human eye.

*Reproduced with permission from Network CV. Developmental anatomy of the retinal and choroidal vasculature. In: The Retina and Its Disorders. Elsevier; 2011, p. 179.*

system should be focused on the retina with minimal distortion. The anterior segment is responsible for this function. Lastly, the retina converts photons into the electric signals, and while doing so, it also takes away the used up photochemical material and replenishes it to capture the next set of photons. Hence, the retina is highly metabolic.

While the ability to retain structural integrity under stress remains a common goal for all parts of the eye, other important mechanobiological parameters and how they affect function vary greatly in different parts of the eye. These stresses at all scales are severely altered in cases of diseases. The eye experiences various forces in Newtons at macro- (approximately 100 μm and larger), micro- (approximately 1–100 μm), and nanoscales (<1 μm), such as intraocular pressure (IOP), perfusion pressure, muscle stimulus on the eyeball, and eyelid forces, at a macro level. Stretch in zonules and lens for accommodation, inertial forces of the vitreous, stresses in the cornea, and papillary reaction all occur at the micro scale, while light damage, strain damage in the cells, and neural pulses occur at the nanolevel.

## 2 THE ANTERIOR SEGMENT
### 2.1 THE CORNEA AND THE TEAR FILM

The precorneal tear film is the most anterior refractive surface of the eye, and its stability plays an important role in the optical quality of the eye [1]. The tear film formation and flow is a multifacial complex fluid flow, which involves interaction of the lipids, aqueous, and mucin. The mucins are anchored to the epithelium, providing a boundary to the flow. It covers the cornea, providing almost two-thirds of the refractive power of the eye [2].

The cornea is the transparent, anterior layer of the eye that protects the eye and refracts light. The cornea is highly anisotropic in terms of mechanical and optical properties [3–5]. It has epithelial, Bowman, stromal, Descemet, and endothelial layers, and it is known that mechanobiology plays a role in the development and organization of each of these (Fig. 2).

The outermost layer of the cornea is known as epithelium, with epithelial cells originating from stem cells located in the limbus [6–10]. Post proliferation, they are known to migrate centripetally, the cause of which is the subject of current research [11,12]. It is assumed to be linked to tensile/shear strains, chemotactic factors, neuronal tracking, and biochemical and biophysical differences in the stroma [13].

The stroma consists of the collagen fibrils, which are the stress-bearing components of the cornea and are transparent to light. These fibrils form lamellae, which are embedded in a proteoglycan-rich hydrated matrix. Corneal transparency mainly depends on the uniformity of the diameters of the collagen fibrils and the distances between adjacent collagen fibrils [14]. It is suggested that the distances between these adjacent collagen fibrils are due to the equilibrium attained between two opposing forces acting on the fibrils [15]. The two opposing forces are (1) the repulsive forces between the fibrils due to the pressure exerted by water molecules attracted

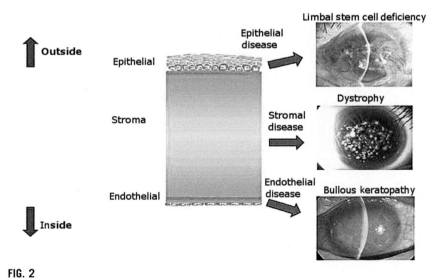

**FIG. 2**

Structure of the cornea and corneal disease. The cornea consists of three layers: epithelium, stroma, and endothelium. Vision can deteriorate due to disease in any layer.

*Reproduced with permission from Oie Y, Nishida K. Corneal regenerative medicine. Regen Ther 2016;5:40–5.*

between the fibrils, via the Donnan effect, and (2) the attractive forces due to proteoglycans and keratocytes [15,16]. This mechanism is important for corneal development and the establishment of the stromal matrix.

The fibrils are organized with specific orientations within the cornea as shown in Figs. 3 and 4 [17], thus resulting in mechanical anisotropy [18–20] and anisotropic

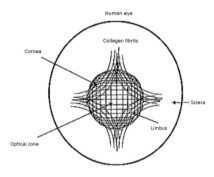

**FIG. 3**

A schematic representation of collagen fibril distribution in cornea. The central optical zone exhibits a square-like mesh of collagen fibrils with a paraboloid-like mesh near and up to the limbus.

*Reproduced with permission from Aghamohammadzadeh H, Newton RH, Meek KM. X-ray scattering used to map the preferred collagen orientation in the human cornea and limbus. Structure 2004;12(2):249–56.*

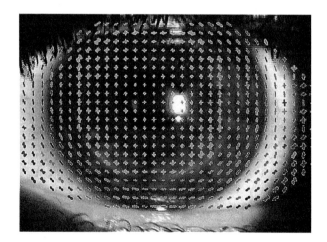

**FIG. 4**

The preferred collagen orientation at each point in the human cornea and adjacent sclera. The length of each line indicates the strength of the preferred collagen orientation, and the shade indicates the region (light gray, square-like mesh in the optic zone; gray, paraboloid-like mesh near and up to the limbus; *dark gray*, radial mesh on the conjunctiva).

*Reproduced with permission from Aghamohammadzadeh H, Newton RH, Meek KM. X-ray scattering used to map the preferred collagen orientation in the human cornea and limbus. Structure 2004;12(2):249–56.*

refractive index known as birefringence [21]. At a micro scale, these fibrils have a crimp, as in other connective tissues [22]. It has been postulated that this crimping plays an important role in the biomechanical response of the cornea [23]. In fact, collagen crimp, in combination with elastic fibers in the peripheral cornea, may explain why this region is more pliable and absorbs IOP fluctuations [24], thus preventing the deformation of the optically important central cornea.

The collagen fibrils are the stress-bearing components of the cornea and, since the fibrils are not uniformly oriented within the cornea, result in the anisotropic mechanical properties of the cornea [18–20]. With understanding of fibril direction and strength, biomechanical characterization started with the construction of constitutive equations, which is a mathematical relation between two physical quantities (e.g., stress and strain) for a specific cornea [25].

The macroscopic in vitro experiments use two common methods, namely, strip extension tests and inflation tests [26]. Strip extension tests are performed by cutting out strips of known dimensions that form the cornea and then deforming them using rheological analyzers. Strip extension tests are more reliable in terms of measurement of strain and stress because of known geometry but are different from the in vivo condition [26]. Inflation tests are performed on intact cornea; the cornea by its scleral rim is either glued or held by mechanical clamps on the test cell. The posterior side of the cornea is connected to an adjustable water column so that by changing the height of the water column, IOP can be simulated.

While in vivo biomechanical characterization was first used to study corneal mechanics, tonometry where forces are applied to the cornea and deformation or displacement gives information about the resistance offered by IOP and the cornea, various other methods [27–29] have also been proposed. All tests were nondestructive, and the deformation was measured with noninvasive imaging systems. The methods of measuring applied force and deformation with model definition used by different groups are listed in Table 1.

Noncontact methods like the ocular response analyzer (Reichert, United States) (ORA) were also used in biomechanical characterization such as corneal hysteresis [30] and corneal resistance [29]. These methods are not capable of the characterization of material properties but measure secondary parameters that are related to the mechanical properties of the cornea. Recently, Scarcelli et al. developed Brillouin optical microscopy for in vivo characterization of the cornea [35].

## 2.2 OPTICS OF THE CORNEA AND THE ROLE OF MECHANOBIOLOGY

The central part of the cornea is the optical zone, as shown in Fig. 3. Beyond the central optical region, it becomes aspheric and connects to the sclera through the limbus. It is not clear how the cornea integrates with the sclera at the limbal region, as the collagen fibrils in the limbal region lose their orientation and become opaque to light.

At the macro level, the overall optical quality of the cornea depends largely on the shape, the structure, and the stresses in the cornea [36–38]. The anisotropic properties of the cornea suggest that elongation of the cornea and resulting stress

**Table 1** Overview of Methods for Measuring the Corneal Stress and Strain In Vivo

| Research Groups | Corneal Model Definition | Modality for Exerting Stress | Modality for Measuring Strain |
|---|---|---|---|
| Tanter et al. (2009) [30] | Shear wave propagation speed is directly linked to elasticity | Remote ultrasonic palpations | Ultrafast ultrasonic imaging for shear wave propagation |
| Ford et al. (2011) [31] | Elastic and viscoelastic | Gonioscopy lens | Optical coherence tomography |
| Manapuram et al. (2012) [32] | Attenuation of sound wave is related to elasticity | Mechanical stimulus | Optical coherence tomography |
| Li et al. (2012) [33] and He and Lu (2009) [34] | Attenuation of acoustic waves with elasticity | Surface acoustic waves | Phase sensitive optical coherence tomography |

magnitudes will be different in different directions, when subjected to the forces described above [39]. It is known that there are various physiological forces, which act on the cornea. In equilibrium, these forces, together with corneal structure and anisotropic material properties, can have an impact on the geometry (topography) of the cornea and hence on the visual performance of the eye.

There are two types of birefringence in the cornea, (i) form birefringence and (ii) stress-induced birefringence, as seen in the early 19th century by Brewster [40]. Overall birefringence of the cornea is the summation of both the components of birefringence [41,42]. Hence, with changes in stresses due to IOP, surgical intervention, or topography [43], the birefringence also changes.

These properties of the cornea are used for variety of clinical assessments and disease detection. Nyquist discussed the stress-induced birefringence of the cornea and used stress-dependent dispersion of birefringence for IOP measurement [21,44]. This was further investigated by Gotzinger et al. to characterize the change in structural birefringence of the corneas with keratoconus (KC) [45]. In addition, stress-induced birefringence changes in the cornea when subjected to elevated IOP have also been demonstrated [39]. This established the links between optics and mechanobiology of the cornea.

In the clinical setting, corneal geometry is used to evaluate the effectiveness of corneal procedures [13] and disease progression [46]. The magnitude of aberrations in the normal human eye is such that vision is aberration limited. The topography of the cornea changes due to various pathologies and postintervention scenarios, and a corresponding change in the optical aberration is observed. In some disease like myopia, hyperopia, and KC, these aberrations increase significantly [37]. It is also demonstrated that in KC, the organization of the stromal lamellae change significantly. It is thought that KC involves a high displacement and slippage of lamellae, leading to thinning of the central cornea. Loss of cohesive forces and mechanical failure in regions where lamellae bifurcate is considered to be responsible for this effect [47].

Not only diseases but also surgical interventions alter forces acting on the cornea, its structure, and its material properties. Corneal biomechanics is used to predict the changes induced due to disease or clinical interventions [46]. For refractive surgeries, it has been reported that while lower order aberrations reduce, higher order aberrations increase postintervention [48,49]. Indeed, penetrating keratoplasty resulting in excessive aberrations is a known complication [50]. With the knowledge of anisotropic optical properties and stress-induced birefringence, it is clear that the modified optical properties can contribute to nonconformities in the outcome of the interventions.

At present, clinicians and researchers are endeavoring to develop lamellar grafts, especially posterior lamellar keratoplasty (DSAEK), in vitro culturing of cells from the recipient for replacement of diseased cells, manufactured tissue for replacement, and genetic replacements to prevent or reverse corneal dystrophies. Knowledge of tissue cohesion, adhesion, and shearing will be critical to the development of such technologies.

For full characterization of the cornea and use of mechanobiology in clinical forecasting or decision support systems, a method of determining mechanical properties in vivo is required. Secondary to this is the necessity to have a system for measurement of strain or stress in vivo. Thirdly, finite element or numerical modeling, where all the material properties of the cornea, stress state, and surgical interventions can be passed as an input, will expand our insight of the biomechanical environment. A fourth and final step is to integrate the model with optical parameters of the cornea. Such a system could plan or predict the effect of any disease, surgery, or condition on the vision of the patient and optimize visual rehabilitation.

### 2.3 MECHANOBIOLOGY OF THE LENS, CAPSULE, AND ZONULES

Accommodation is defined as how the eye changes focus from distance to near [51–53]. During accommodation, the eye converges, and the pupil constricts. These three movements compose the accommodative triad or near response [54,55]. The accommodative apparatus (Fig. 5) is composed of the crystalline lens, the suspensory ligament, and the ciliary body [56]. The crystalline lens is a biconvex lens situated between the aqueous and vitreous humor. The lens capsule, or capsular bag, contains the lens epithelial cells and the fiber cell mass. The epithelial cells are the active and metabolic aspect of the crystalline lens, while the fiber cell mass, composed of a cortex and nucleus, is responsible for the bulk lens phenotype [57]. The crystalline lens is suspended by a network of zonules, the zonules of Zinn, which integrate directly with the anterior hyaloid membrane and both the anterior and posterior capsule [58,59]. The anterior hyaloid membrane separates the anterior and posterior

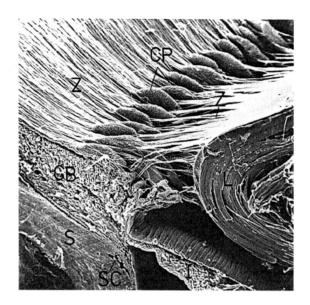

**FIG. 5**

Scanning electron microscope image of the accommodative apparatus. *CB*, ciliary body; *CP*, ciliary process; *L*, lens; *S*, sclera; *SC*, Schlemm's canal; *Z*, zonule.

Reproduced with permission from Rohen JW. Scanning electron microscopic studies of the zonular apparatus in human and monkey eyes. Invest Ophthalmol Vis Sci 1979;18(2):133–44.

chambers of the eye. The numerous zonules, grouped together in bundles only 1–6 μm in diameter, emerge from the pars plana of the ciliary body [60,61]. The ciliary body is broadly a part of the uvea, which is the metabolic engine of the eye. The ciliary body contains the ciliary epithelium and the ciliary muscle. According to the Cramer/Helmholtz theory of accommodation, accommodation is brought about by constriction of the ciliary muscle, which reduces tension on the zonules, thereby reducing the potential energy of the lens, permitting it to achieve a more rounded shape.

The human lens continues to grow throughout life by the active proliferation of epithelial cells located just in front of its equator [62]. The daughter cells elongate, differentiate into lens fibers, and are packed into lens center [63]. Therefore, the oldest fiber cells are contained in the central nucleus, with the youngest ones in the peripheral cortex. This does not cause the lens to grow excessively inside the eye, due to the activation of a process called compaction, by which inner fiber cells lose water, reduce their volume, and become compact.

In humans, lens growth is a unique process occurring in two phases: an asymptotic phase (initiated prenatally and continuing after birth into the first 3 years of life) and a linear phase of expansion thereafter. The two different regions of lens, the inner nucleus and the outer cortex, are generated by this differential growth pattern [64]. A majority of the role in lens accommodation is played by the nucleus, with a very minor contribution from the cortex [65–68].

The nucleus possesses a lower isotropic elasticity ($\sim 0.8 \times 10^3$ N/m$^2$) than the cortex, and its Young's modulus remains constant until 40 years of age, after which there is an increase to near $3.5 \times 10^3$ N/m$^2$ by 70 years [69]. The cortical elasticity increases continuously from birth ($\sim 0.8 \times 10^3$ N/m$^2$) until 50 years of age ($\sim 4 \times 10^3$ N/m$^2$). After age 50 years, the nuclear stiffness is greater by an order of magnitude than cortical stiffness, which shows a decrease [70]. In simpler terms, a young lens is soft and pliable enough to focus on near objects. As one ages, the ability to focus on near objects is lost causing a condition called presbyopia [71].

The probable causes of increasing lens stiffness, more pronounced in the nucleus, with age are (1) continuous increase in lens weight and fiber cell compaction with a net gain of solids (proteins) greater than liquid (water) [72,73], (2) increases in bonding between fiber cells and attachment of crystallin proteins to fiber cell membranes [74,75], and (3) age-related modifications such as cross-linking, deamidation, racemization, and truncation of crystallins making them more insoluble [76,77].

An elastic and transparent capsule surrounds the lens, maintaining its structural integrity. It is the thickest basement membrane in the human body, consisting of extracellular components such as collagen [57]. It also provides attachment points for the insertion of suspensory ligaments or zonules of Zinn. During accommodation, the capsule molds the lens contents and changes the lens shape. Its stress field is equibiaxial and homogeneous [78]. The stiffness of the lens capsule (Young's modulus of elasticity, 2–6 N/mm$^2$) does not change significantly with age, and senile changes do not contribute significantly to presbyopia [79,80].

The literature on the mechanical properties of the human zonule is scarce. Aside from the small dimensions of the zonular fibers, assessment of their Young's modulus is further confounded by the number and size of the fibers in each bundle. Although experimental data on isolated zonules or zonular bundles appear to be unavailable, researchers have measured the forces of the accommodative process using lens stretching devices. With this approach, estimates of the Young's modulus of the zonule of 0.35 [81] and 1.5 N/mm$^2$ [82] were obtained.

## 2.4 THE FORCES OF ACCOMMODATION

By simulating accommodation ex vivo (Fig. 6), attempts have been made to examine the response of postmortem human lenses to stretching forces [83–89]. Overall, the results of all studies showed an increasing resistance of the human crystalline lens to stretching forces, with greater shape changes in younger lenses compared with older ones. The stretching experiments also showed the ability of the lens to restore accommodation if the hardened lens material is replaced by a soft pliable material [90,91].

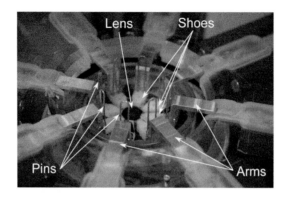

**FIG. 6**

Dissected eye containing the accommodative apparatus and mounted in the tissue chamber of the stretching apparatus used by Nankivil et al., with the lens, scleral shoes, stretching arms, and mounting pins labeled.

*Reproduced with permission from Nankivil D, Heilman BM, Durkee H, Manns F, Ehrmann K, Kelly S, et al. The zonules selectively alter the shape of the lens during accommodation based on the location of their anchorage points. Invest Ophthalmol Vis Sci 2015;56(3):1751–60.*

The load that is required to produce lens accommodation ex vivo is ~80 mN. What is more interesting is the change in lens parameters per unit applied force. The change in lens equatorial diameter decreases from ~6 μm/mN at birth to <2 μm/mN by 60 years of age. Similarly, the change in lens optical power decreases from ~0.1 D/mN at birth to 0 D/mN by 60 years of age where D stands for diopter, a unit of refractive power. The ellipsoid deformation also decreases from ~1.5 at birth to ~1.0 by 50 years of age [82].

## 2.5 SENESCENCE AND PATHOPHYSIOLOGY OF THE ACCOMMODATIVE APPARATUS

Presbyopia, as defined earlier, is the loss of the ability of the eye to focus on near objects and is one of the leading causes of vision loss globally [92]. The mean accommodation amplitude of the human eyes during the first decade of life is >12 D, where accommodation amplitude is the maximum potential of an eye to increase its optical power while focusing a near object. The potential to accommodate starts declining early in life, and the accommodation amplitude is only half a diopter or below on reaching an age range of 50–55 years [93]. There is no further decrease in life, and the residual accommodation of ~0.5 D is mainly due to depth of focus [94]. While there is no strong consensus on the exact cause of presbyopia, almost all the components of the accommodative apparatus are implicated in the etiology of presbyopia [95–98].

Although studies have shown that the ciliary muscle retains the ability to contract throughout life [99–101], forward movement of the vitreous zonule insertion zone was shown to decrease with age [102]. In senescence, the ciliary muscle becomes shorter and wider and exhibits a more inward/anterior position, and the connective tissue increases [100,103]. These age-related changes are very similar to those that occur in a young eye during ciliary muscle contraction. Together, the increase in lens thickness and the inward/anterior movement of the ciliary ring with age may, in part, be responsible for presbyopia. This could limit the potential utility of accommodative intraocular lenses designed to restore accommodation by utilizing the accommodative apparatus.

Lens stiffness increases monotonically throughout life. Lens spinning, loaded by centrifugal forces, showed that the Young's modulus of the crystalline lens ranges from roughly 0.8 kPa at birth to 3 kPa by 70 years [69,104]. In addition, concerning viscoelastic properties, compression experiments indicate that the relaxation time constants decrease by between 20% and 75% with age [105]. While presbyopia is a multifactorial phenomenon, presbyopia can largely be explained in lenticular terms [106].

Another pathology seen in the lens is cataract, which is the opacification or whitening of the lens and can be due to multiple reasons. The most common form of cataract involves the lens nucleus and occurs with aging. The causes of light scattering in senile nuclear cataracts include crystallin aggregation to form high-molecular-weight structures, fiber cell membrane damage, and formation of multilamellar bodies [107]. The stiffness of nuclear cataracts depends on the age of the patient and the stage of the disease [108]. It can vary from <40 kPa in a 40 years old with early-stage cataract to >200 kPa in an advanced cataract in an 80 years old.

## 2.6 THE AQUEOUS HUMOR AND GLAUCOMA

Filling both the anterior and posterior chambers of the eye, the aqueous humor is a transparent fluid analogous to plasma (Fig. 7). This fluid inflates the eye and thereby maintains the intraocular pressure (IOP). It contributes to the optical system of the eye by providing nutrition to the a vascular tissues like cornea and lens and has a refractive index that differs from both. It also plays a role in immune response against pathogens.

Aqueous humor is formed by the ciliary body. It enters the posterior chamber and flows to the anterior chamber through the pupil [109]. The drainage of the aqueous humor is passive and involves two distinct pathways. The conventional pathway consists of aqueous humor passing through the trabecular meshwork (TM) into the episcleral veins [110,111]. Alternatively, the aqueous humor enters the connective tissue between the muscle bundles of the ciliary body, through the suprachoroidal space, and out through the sclera (see Fig. 7) [112,113].

The balance between the production and drainage of aqueous humor determines the intraocular pressure. Fluid movement takes place down a pressure gradient by a one-way paracellular mechanism [114]. After exiting Schlemm's canal, the aqueous humor is drained into the episcleral veins. The pressure in these veins and the

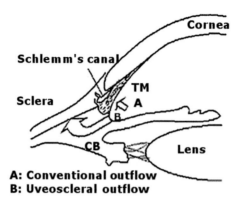

**FIG. 7**

Conventional and nonconventional pathways of aqueous humor outflow.
*Reproduced with permission from Nakajima E, Nakajima T, Minagawa Y, Shearer TR, Azuma M. Contribution of ROCK in contraction of trabecular meshwork: proposed mechanism for regulating aqueous outflow in monkey and human eyes. J Pharm Sci 2005;94(4):701–8.*

resistance of the conventional aqueous drainage tissues determine the IOP [115–118]. In humans, 75% of the resistance to the aqueous humor outflow is localized to the TM, and 25% occurs beyond Schlemm's canal [119].

The anterior tendons of the ciliary muscle insert into the outer portion of the corneoscleral junction [120]. The contraction and relaxation of the ciliary muscle can alter the aqueous outflow resistance. The muscle contraction results in widening of the inter-trabecular spaces and dilation of Schlemm's canal, thereby decreasing outflow resistance. During relaxation, the opposite effects increase outflow resistance [121]. Disruption of conventional aqueous outflow is a major risk factor in the pathogenesis of glaucoma [122]. The rate of aqueous humor turnover is estimated to be 1.0%–1.5% of the anterior chamber volume per minute [123]. The medications to reduce IOP act by either reducing the production of the aqueous humor or by enhancing its outflow by either or both outflow pathways.

The pathogenesis of glaucoma is not fully understood. However, increased IOP is the most important and modifiable risk factor. Increase in IOP is either due to obstruction of the conventional aqueous drainage pathway by the iris (angle-closure type of glaucoma) or increased resistance to aqueous outflow through the trabecular meshwork (open angle type of glaucoma). The level of intraocular pressure is related to retinal ganglion cell death. The former type of glaucoma is relatively straightforward to treat, especially if detected early, while the latter form is more complex. The cause of increased flow resistance is not completely known and likely to be multifactorial. The trabecular meshwork is a dynamic three-dimensional network that distends and compacts depending on the pressure gradient across it [124,125]. Most of the resistance to aqueous flow through the conventional outflow pathway occurs in the outermost portion of the trabecular meshwork and/or the adjacent inner wall of Schlemm's canal [126]. Because of this resistance, when the pressure gradient across the trabecular meshwork is relatively high, the outer trabecular meshwork and inner wall of Schlemm's canal are pushed out towards the outer wall of Schlemm's canal [124,125]. When the pressure gradient falls, the meshwork and inner wall move away from the outer wall. Thus, distensions of the trabecular meshwork generally correlate with intraocular pressure. Studies have demonstrated that TM cells respond in a variety of ways to mechanical loads, including increased extracellular matrix turnover, altered gene expression, cytokine release, and altered signal transduction [127]. Altered biomechanical behavior of Schlemm's canal cells is also hypothesized to contribute to aqueous outflow resistance [128].

Intraocular pressure can impart stress and strain on the structures of the eye, particularly the lamina cribrosa and surrounding tissues [129]. The sclera is perforated at the lamina where the retinal ganglion cell axons exit the eye (see Fig. 8).

This region is the frailest structure in the wall of the pressurized eye. IOP induced stress may result in compression and deformation of the lamina cribrosa, causing axonal damage and disruption of axonal transport [130,131] that interrupts retrograde delivery of essential trophic factors to retinal ganglion cells. The mechanical deformation of the lamina cribrosa is explained in Fig. 9.

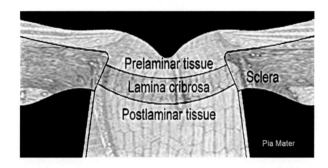

**FIG. 8**

A schematic diagram showing axial section of the eye at the level of the optic disk.

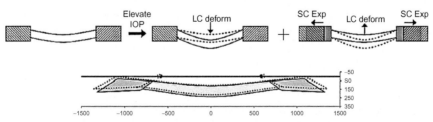

**FIG. 9**

A mechanical perspective on intraocular pressure (IOP)-induced deformation of the lamina cribrosa (LC). *SC*, scleral canal; *Exp*, expansion.

Reproduced with permission from Downs JC, Roberts MD, Sigal IA. Glaucomatous cupping of the lamina cribrosa: a review of the evidence for active progressive remodeling as a mechanism. Exp Eye Res 2011;93(2):133–40.

IOP deforms the lamina posteriorly (Fig. 9, top middle), and pushes the sclera, causing an expansion of the canal (Fig. 9, top right). The widening of the canal pulls the tissue taut thereby displacing the lamina anteriorly. These conflicting effects act simultaneously, so that typically, the two components of laminar deformation combine to produce very small (less than 10 µm) net antero-posterior laminar displacements (Fig. 9, bottom). As a result, exceedingly small antero-posterior laminar displacementmay be indicative of substantial IOL-related stresses [132].

The lamina cribrosa forms a barrier between the intraocular and the retrolaminar pressure compartments (Fig. 9). In health, there is a small, posteriorly directed pressure difference across the lamina cribrosa, the translaminar pressure difference [Intraocular pressure (IOP)–Intracranial pressure (ICP)]. Increased IOP increases posteriorly directed force at the lamina cribrosa, resulting in glaucoma. Thus, in a strict sense, IOP is the transcorneal pressure difference and might be a surrogate marker for the translaminar pressure difference (IOP–ICP). However, glaucomatous optic neuropathy can occur in individuals with intraocular pressures within the normal range. In such patients, there may be an abnormally low cerebrospinal fluid (CSF) pressure in the optic nerve subarachnoid space exaggerating the posteriorly directed pressure difference across the lamina [133,134]. While IOP is typically measured in an erect posture, CSF pressure is measured in the lateral decubitus position. However, CSF pressure in erect posture at the eye level is low and can even be measured as 0 mm Hg [135]. The subarachnoid space of the human optic nerve is not homogeneous, but divided by a complex system of arachnoid trabeculae and septa [136]. Then again, the space is a cul-de-sac and it is not yet evident how CSF circulates from this region to the general subarachnoid space [137]. To complicate the matter further, humans spend a considerable amount of time in both horizontal and vertical positions.

Unfortunately, the relationship between IOP and CSF pressure in differing positions is not known. Thus, the role of the translaminar pressure gradient in glaucoma at issue is very complex. Primary neural pathological processes may cause secondary neurodegeneration of other retinal neurons and cells in the central visual pathway by altering their environment and increasing susceptibility to damage [138].

## 3 THE POSTERIOR SEGMENT

One of the most complex neuronal tissues, the retina is constantly exposed to several different types of mechanical stresses. These stresses occur due to development, metabolism, and disease. During retinal development, cells migrate and grow along considerable distances and create a mechanically inhomogeneous environment. The retina is also exposed to light-induced damage, oxidative stress. The retina is firmly attached to the eyeball at the ora serrata and at the optic nerve (Fig. 1). The retinal pigment epithelium constantly pumps fluid and debris out from the inner retina to the outer retina or choroid, resulting in a suction force. This causes pulling on the outer retinal surface toward the interior [139]. The hydrostatic IOP exerts (positive) pressure on the inner retinal surface that pushes it toward the exterior [140].

Numerous areas of interest have been emerging from the field of mechanobiology. Mechanotransduction is one such process that centers on how cells control their mechanical properties and how physical forces regulate cellular biochemical responses [141]. Müller cells, astrocytes, and microglia are among the cells that provide structural support to the retina. It has also been shown that under stress, the calcium levels in the Muller cells increase, and transcription factors, growth factors, and neural response are elicited. It is also known that the neurons and glial cells respond to mechanical cues in their environment. Mechanical stress can alter the function of these cells in the retina and can evoke a neural response [142–149].

There are additional external random forces acting at the retina, such as inertia due to vitreous in the posterior chamber. The vitreous is a viscous material attached to both the lens and the retina; hence, with every eye or head movement, the vitreous body exerts shear or traction forces at the retinal surface [150]. These inertial forces vary with the geometry of the eye and pose a higher risk to myopes for retinal detachment. The vitreous and the retina of a highly myopic eye continuously experience shear stresses significantly higher than those of an emmetropic eye. These stresses arise from saccadic eye rotations, which are abnormally high due to the change in the vitreous chamber geometry, and thus, the shape of the vitreous chamber has a large effect on retinal stress [151]. This also holds true for shocks to the head and blunt trauma of the eye. These impacts can be classified into four stages: compression, decompression, overshooting, and oscillation. It was observed that shockwave propagation in the retina produced high strain, and the negative pressure contributed to retinal detachment. This can cause retinal break, and negative pressure with inertial motion can pull the retina away [152,153].

It has been reported that the effect of inertial forces is magnified in cases of pathology. One example is that the shape of the peripapillary basement membrane layer changes significantly in patients with papilledema or anterior ischemic optic neuropathy (AION) [154]. It is also shown that large deformations in papilledema involves temporal and nasal sides, it is argued that it is because of shifts in cerebrospinal fluid pressure against the scleral flange and hydraulic stiffening of the optic nerve sheath. The eye motion in such cases could be a factor for progression of optic neuropathies [154].

The mechanical and fluid forces affect exudative retinal detachments, where the retinal photoreceptor cells separate from the underlying retinal pigment epithelium. Parameters such as fluid production, size of the retinal blister, and location alter the amount of forces acting on a detached retina and regulate the progression of retinal detachment. Additional parameters that can affect it further are inflamed lesions, higher choroidal hydraulic conductivity, insufficient retinal pigment epithelium pump activity, and defective adhesion bonds [155].

Also in pathologies, there are different types of stresses involved in retinal degenerations, such as retinitis pigmentosa, age-related macular degeneration, and glaucoma. The mechanical properties of the retina play a crucial role in many other pathologies, such as retinoschisis, age-related idiopathic retinal detachment, and/or macular hole formation [156].

Moreover, retinal surgery also capitalizes on stresses residing inside the retina. Following a retinal detachment, scleral buckling procedures promote retinal reattachment. An eye fitted with a scleral buckle experiences large stress levels localized around the buckle [150]. Also, to fill the macular holes, retinal inner limiting membrane peeling, or epiretinal membrane removal is a standard treatment. During this surgery, significant forces are locally applied to the inner retinal surface by the surgeon. It is known that the imbalance in equilibrium of forces permit the macular holes to heal with time.

Understanding these mechanisms in the retina is important for developing new therapeutic approaches to fight blindness, and hence understanding of mechanobiology will play a crucial role.

## 4 THE EXTRA OCULAR SYSTEM

There are a total of seven extraocular muscles; six muscles that are responsible for eye movement and one muscle that controls eyelid elevation (levator palpebrae). Eye movements should be coordinated so that eyes move together in sync to ensure targets remain on the fovea, which is the part of the eye responsible for central vision. It is important to have vergence movement or convergence to make sure that the image of the object falls on the corresponding spot on both retinas. This type of movement helps in the depth perception of objects. Another important type of movement is pursuit movement (smooth pursuit) used for tracking an object's movement, so that its moving image can remain on the fovea [157]. Lastly, saccades are the rapid movement of eyes used while scanning a visual scene [158]. Governed by Hering's law of equal innervations, one of the main uses for these saccadic eye movements is to be able to scan a greater area with the high-resolution fovea of the eye.

Five of the six muscles controlling eye movement (inferior rectus, superior rectus, lateral rectus, medial rectus, and superior oblique) originate at a common tendinous ring of fibrous tissue (the Annulus of Zinn). The Annulus of Zinn surrounds the optic nerve, ophthalmic artery, and ophthalmic vein at their entrance through the apex of the orbit. The sixth muscle (inferior oblique) has a separate origin point on the orbital side of the bony maxilla at the anterior inferomedial strut [159] (Fig. 10).

### 4.1 MUSCLE FUNCTIONS

The muscles controlling eye movements act in antagonistic pairs. Horizontal motion is accomplished by the medial and lateral rectus muscles. Vertical motion is accomplished by the superior rectus, inferior rectus and oblique muscles. Torsional motion is accomplished by the oblique muscles. Adduction is controlled entirely by the

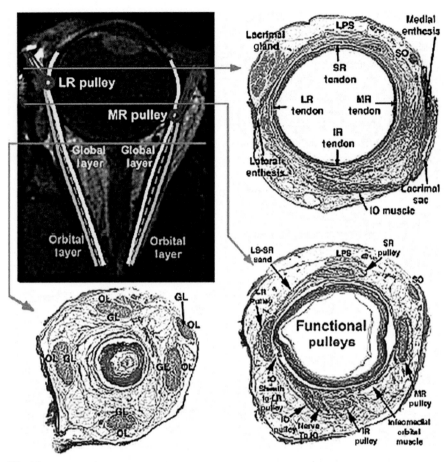

**FIG. 10**

Anatomy of the human orbit. Axial magnetic resonance imaging (MRI) scan (top left) and histopathology images of the orbital mass highlighting the muscles and insertion points on eye. Medial rectus (MR), lateral rectus (LR), orbital layers (OLs), global layers (GLs), levator palpebrae superioris (LPS) muscle. Magnetic resonance image (top left) and histology sections (top and bottom right, bottom left).

*Reproduced with permission from Wagner RS. The active pulley hypothesis. J. Pediatr. Ophthalmol. Strabismus. 2006;43(5):272.*

medial and abduction is controlled entirely by the lateral rectus muscles for horizontal movements. Vertical movements are achieved by the use of superior and inferior rectus muscles, as well as the oblique muscles. This movement is governed by Sherrington's law of reciprocal innervations [160], which states that: when a muscle contracts, its direct antagonist relaxes to an equal extent allowing smooth movement. Fig. 11 illustrates the actions of the extraocular muscles.

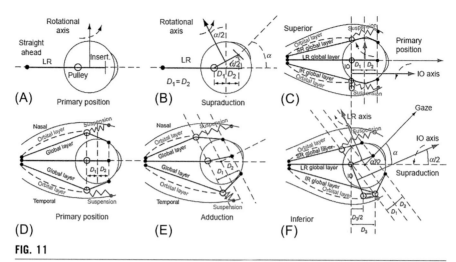

FIG. 11

Illustration of the action of the extraocular muscles (EOM). (A–C, F) are Lateral views, and (D,E) are Supra views. Lateral rectus (LR), medial rectus (MR), inferior oblique (IO), inferior rectus (IR), D1 distance from pulley to globe center, D2 distance from globe center to insertion.

*Reproduced with permission from Wagner RS. The active pulley hypothesis. J. Pediatr. Ophthalmol. Strabismus. 2006;43(5):274.*

## 4.2 RELATED DISEASE CONDITIONS

The common diseases related to extraocular muscle dysfunction are; Nystagmus, and Squint or strabismus. Nystagmus is a congenital or acquired condition where the eyes move rapidly and uncontrollably. They can move in the horizontal plane (horizontal nystagmus), vertical plane (vertical nystagmus), or torsional or rotatory nystagmus. Acquired causes are related to central nervous system dysfunction and may be related to use of drugs, secondary to a binge of alcohol, other central nervous system disorders and even vertigo. There are several medical and surgical treatments available for nystagmus [161], still there is a desire for better intervention with higher efficacy.

Squint or strabismus is a congenital or acquired condition where the eyes are misaligned. This may present in the form of: Crossed eyes or esotropia or inward deviation, wall eyed or exotropia or an outward deviation, and hyper or hypotropia where in the misalignment may be in the vertical plane.

In the acquired condition, the most commonly involved are the cranial nerves (III, IV, VI), which can weaken or palsied to cause strabismus. Third nerve palsy and superior oblique palsy are forms of paralytic strabismus.

Clinical intervention for strabismus aims to improve eye alignment which allows binocular vision. Depending on severity, treatment may involve spectacles, prisms or eye muscle surgery [158]. Strabismus surgery can be of two types weakening/ slackening or strengthening procedures [162]. Weakening procedures include

- Recession—here, a muscle is disinserted from its normal insertion point and placed slightly behind its original insertion to slacken it.
- Myotomy/myectomy—this involves cutting away a part of or the entire muscle belly.
- Tenotomy/tenectomy—this involves cutting away a part of or entire tendon of the muscle.

Strengthening procedures involve the following:

- Resection of a muscle—where in a part of muscle is cut and then the muscle is reattached to its original insertion to tighten it
- Plication—similar to resection but, instead of removing a segment of muscle, a small segment of muscle is folded on itself and sutured to the original insertion to strengthen it

The principle of strabismus surgery involves changing the moment arm and vector forces of the muscle by strengthening and weakening procedures, and hence balancing the forces responsible for eye movements.

## 5 CONCLUSION

Our understanding of the role of mechanobiology in the eye continues to develop, and reliable models truly resembling the development or disease are actively being refined. Mechanical changes in bulk ocular tissue associated with senescence and pathology are fairly well characterized, but the knowledge of cellular level mechanosensitivity, the mechanics in angiogenesis and morphogenesis, and pathogenesis of neural tissue are limited. However, this rapidly emerging field has given a whole new understanding of the functioning and developmental aspects of human body, including the eye.

## REFERENCES

[1] Rieger G. The importance of the precorneal tear film for the quality of optical imaging. Br J Ophthalmol 1992;76(3):157–8.
[2] DelMonte DW, Kim T. Anatomy and physiology of the cornea. J Cataract Refract Surg 2011;37(3):588–98.
[3] Gotzinger E, Pircher M, Sticker M, Fercher AF, Hitzenberger CK. Measurement and imaging of birefringent properties of the human cornea with phase-resolved, polarization-sensitive optical coherence tomography. J Biomed Opt 2004;9(1):94–102.
[4] Bone RA, Draper G. Optical anisotropy of the human cornea determined with a polarizing microscope. Appl Opt 2007;46(34):8351–7.
[5] Comaish IF, Lawless MA. Progressive post-LASIK keratectasia: biomechanical instability or chronic disease process? J Cataract Refract Surg 2002;28(12):2206–13.
[6] Oie Y, Nishida K. Corneal regenerative medicine. Regen Ther 2016;5:40–5.

[7] Pellegrini G, Golisano O, Paterna P, Lambiase A, Bonini S, Rama P, et al. Location and clonal analysis of stem cells and their differentiated progeny in the human ocular surface. J Cell Biol 1999;145(4):769–82.
[8] Cotsarelis G, Cheng SZ, Dong G, Sun TT, Lavker RM. Existence of slow-cycling limbal epithelial basal cells that can be preferentially stimulated to proliferate: implications on epithelial stem cells. Cell 1989;57(2):201–9.
[9] Di Girolamo N. Moving epithelia: tracking the fate of mammalian limbal epithelial stem cells. Prog Retin Eye Res 2015;48:203–25.
[10] Lavker RM, Tseng SC, Sun TT. Corneal epithelial stem cells at the limbus: looking at some old problems from a new angle. Exp Eye Res 2004;78(3):433–46.
[11] Lobo EP, Delic NC, Richardson A, Raviraj V, Halliday GM, Di Girolamo N, et al. Self-organized centripetal movement of corneal epithelium in the absence of external cues. Nat Commun 2016;7:12388.
[12] Mohammad Nejad T, Iannaccone S, Rutherford W, Iannaccone PM, Foster CD. Mechanics and spiral formation in the rat cornea. Biomech Model Mechanobiol 2015;14(1):107–22.
[13] Lavker RM, Dong G, Cheng SZ, Kudoh K, Cotsarelis G, Sun TT. Relative proliferative rates of limbal and corneal epithelia. Implications of corneal epithelial migration, circadian rhythm, and suprabasally located DNA-synthesizing keratinocytes. Invest Ophthalmol Vis Sci 1991;32(6):1864–75.
[14] Lewis PN, Pinali C, Young RD, Meek KM, Quantock AJ, Knupp C. Structural interactions between collagen and proteoglycans are elucidated by three-dimensional electron tomography of bovine cornea. Structure 2010;18(2):239–45.
[15] Meek KM, Knupp C. Corneal structure and transparency. Prog Retin Eye Res 2015;49:1–16.
[16] Cheng X, Pinsky PM. Mechanisms of self-organization for the collagen fibril lattice in the human cornea. J R Soc Interface 2013;10(87):20130512.
[17] Aghamohammadzadeh H, Newton RH, Meek KM. X-ray scattering used to map the preferred collagen orientation in the human cornea and limbus. Structure 2004;12(2):249–56.
[18] Hanna K, Jouve FE, Kaiss A, Le Tallec P. Viscoelastic model for the human cornea. High Perform Comput 1991;2:631–40.
[19] Swigert CJ, Shepherd MH, Lyon CS, McEwen WK. A comparison of tonographic-trace simulations achieved by two systems: nonlinear elastic and linear viscoelastic. IEEE Trans Biomed Eng 1971;18(2):97–103.
[20] Elsheikh A, Kassem W, Jones SW. Strain-rate sensitivity of porcine and ovine corneas. Acta Bioeng Biomech 2011;13(2):25–36.
[21] Nyquist GW. Stress-induced birefringence of the cornea. Am J Ophthalmol 1968;65(3):398–404.
[22] Grytz R, Meschke G. Constitutive modeling of crimped collagen fibrils in soft tissues. J Mech Behav Biomed Mater 2009;2(5):522–33.
[23] Liu X, Wang L, Ji J, Yao W, Wei W, Fan J, et al. A mechanical model of the cornea considering the crimping morphology of collagen fibrils. Invest Ophthalmol Vis Sci 2014;55(4):2739–46.
[24] Boyce BL, Grazier JM, Jones RE, Nguyen TD. Full-field deformation of bovine cornea under constrained inflation conditions. Biomaterials 2008;29(28):3896–904.
[25] Anderson K, El-Sheikh A, Newson T. Application of structural analysis to the mechanical behaviour of the cornea. J R Soc Interface 2004;1(1):3–15.

[26] Elsheikh A, Anderson K. Comparative study of corneal strip extensometry and inflation tests. J R Soc Interface 2005;2(3):177–85.
[27] Dorronsoro C, Pascual D, Perez-Merino P, Kling S, Marcos S. Dynamic OCT measurement of corneal deformation by an air puff in normal and cross-linked corneas. Biomed Opt Express 2012;3(3):473–87.
[28] Kaneko M, Tokuda K, Kawahara T, editors. Dynamic sensing of human eye. Proceedings of the 2005 international conference on robotics and automation {ICRA}, April 18–22Barcelona, Spain: IEEE; 2005.
[29] Elsheikh A, Wang D, Rama P, Campanelli M, Garway-Heath D. Experimental assessment of human corneal hysteresis. Curr Eye Res 2008;33(3):205–13.
[30] Tanter M, Touboul D, Gennisson JL, Bercoff J, Fink M. High-resolution quantitative imaging of cornea elasticity using supersonic shear imaging. IEEE Trans Med Imaging 2009;28(12):1881–93.
[31] Ford MR, Dupps Jr. WJ, Rollins AM, Sinha RA, Hu Z. Method for optical coherence elastography of the cornea. J Biomed Opt 2011;16(1):016005.
[32] Manapuram RK, Menodiado FM, Truong P, Aglyamov S, Emelianov S, Twa M, et al., Estimation of surface wave propagation in mouse cornea. San Francisco, CA: Society of Photo-Optical Instrumentation Engineers BiOS; 2012.
[33] Li C, Guan G, Huang Z, Johnstone M, Wang RK. Noncontact all-optical measurement of corneal elasticity. Opt Lett 2012;37(10):1625–7.
[34] He X, Liu J. A quantitative ultrasonic spectroscopy method for noninvasive determination of corneal biomechanical properties. Invest Ophthalmol Vis Sci 2009;50(11):5148–54.
[35] Scarcelli G, Pineda R, Yun SH. Brillouin optical microscopy for corneal biomechanics. Invest Ophthalmol Vis Sci 2012;53(1):185–90.
[36] Porter J, Guirao A, Cox IG, Williams DR. Monochromatic aberrations of the human eye in a large population. J Opt Soc Am A Opt Image Sci Vis 2001;18(8):1793–803.
[37] Pantanelli S, MacRae S, Jeong TM, Yoon G. Characterizing the wave aberration in eyes with keratoconus or penetrating keratoplasty using a high-dynamic range wavefront sensor. Ophthalmology 2007;114(11):2013–21.
[38] Marcos S. Aberrometry: basic science and clinical applications. Bull Soc Belge Ophtalmol 2006;302:197–213.
[39] Richhariya A, Verma Y, Rao DK, Roberts CJ, Mahmoud AM, Sangwan VS, et al. Effect of intraocular pressure and anisotropy on the optical properties of the cornea: a study using polarization sensitive optical coherence tomography. Asia Pac J Ophthalmol (Phila) 2014;3(6):348–53.
[40] Brewster D. On the communication of the structure of doubly refracting crystals to glass, muriate of soda, fluor spar, and other substances, by mechanical compression and dilatation. Philos Trans R Soc Lond 1816;106:156–78.
[41] Stanworth A, Naylor EJ. The polarization optics of the isolated cornea. Br J Ophthalmol 1950;34(4):201–11.
[42] Naylor EJ. Polarized light studies of corneal structure. Br J Ophthalmol 1953;37(2):77–84.
[43] Nicolela MT, Soares AS, Carrillo MM, Chauhan BC, LeBlanc RP, Artes PH. Effect of moderate intraocular pressure changes on topographic measurements with confocal scanning laser tomography in patients with glaucoma. Arch Ophthalmol 2006;124(5):633–40.

[44] Nyquist GW, Cloud GL. Stress-dependent dispersion of corneal birefringence: a proposed optical technique for intraocular pressure measurement. J Biomech 1970;3(3):249–53.
[45] Gotzinger E, Pircher M, Dejaco-Ruhswurm I, Kaminski S, Skorpik C, Hitzenberger CK. Imaging of birefringent properties of keratoconus corneas by polarization-sensitive optical coherence tomography. Invest Ophthalmol Vis Sci 2007;48(8):3551–8.
[46] Srodka W, Iskander DR. Optically inspired biomechanical model of the human eyeball. J Biomed Opt 2008;13(4):044034.
[47] Meek KM, Tuft SJ, Huang Y, Gill PS, Hayes S, Newton RH, et al. Changes in collagen orientation and distribution in keratoconus corneas. Invest Ophthalmol Vis Sci 2005;46(6):1948–56.
[48] Price NC, Steele AD. The correction of post-keratoplasty astigmatism. Eye (Lond) 1987;1(Pt 5):562–6.
[49] Padmanabhan P, Basuthkar SS, Joseph R. Ocular aberrations after wavefront optimized LASIK for myopia. Indian J Ophthalmol 2010;58(4):307–12.
[50] Cohen KL, Holman RE, Tripoli NK, Kupper LL. Effect of trephine tilt on corneal button dimensions. Am J Ophthalmol 1986;101(6):722–5.
[51] Young T. Observations on vision. Philos Trans R Soc Lond 1793;83:169–81.
[52] Cramer A. Het Accommodatievermogen der Oogen Physiologisch Toegelicht. Haarlem: De Erven Loosjes; 1853.
[53] Helmholtz H. Uber die akkommodation des auges. Albrecht Von Graefes Arch Ophthalmol 1855;(1):1–89.
[54] Glasser A. Accommodation: mechanism and measurement. Ophthalmol Clin N Am 2006;19(1):1–12.
[55] Kaufman PL, Levin LA, Adler FH, Alm A. Adler's physiology of the eye. 11th ed. Edinburgh: Mosby Elsevier; 2011.
[56] Lutjen-Drecoll E. Morphology and age-related changes of the accommodation apparatus. In: Guthoff R, Ludwig K, editors. Current aspects of human accommodation. Heidelberg: Kaden Verlag; 2001. p. 25–35.
[57] Danysh BP, Duncan MK. The lens capsule. Exp Eye Res 2009;88(2):151–64.
[58] Rohen JW. Scanning electron microscopic studies of the zonular apparatus in human and monkey eyes. Invest Ophthalmol Vis Sci 1979;18(2):133–44.
[59] Bernal A, Parel JM, Manns F. Evidence for posterior zonular fiber attachment on the anterior hyaloid membrane. Invest Ophthalmol Vis Sci 2006;47(11):4708–13.
[60] Bornfeld N, Spitznas M, Breipohl W, Bijvank GJ. Scanning electron microscopy of the zonule of Zinn. I. Human eyes. Albrecht Von Graefes Arch Klin Exp Ophthalmol 1974;192(2):117–29.
[61] Marshall J, Beaconsfield M, Rothery S. The anatomy and development of the human lens and zonules. Trans Ophthalmol Soc U K 1982;102(Pt 3):423–40.
[62] Augusteyn RC. On the growth and internal structure of the human lens. Exp Eye Res 2010;90(6):643–54.
[63] Duke-Elder S. Duke-Elder S, editor. System of ophthalmology: the eye in evolution. London: Kimpton; 1958.
[64] Augusteyn RC. Growth of the human eye lens. Mol Vis 2007;13:252–7.
[65] Brown N. The change in shape and internal form of the lens of the eye on accommodation. Exp Eye Res 1973;15(4):441–59.

[66] Dubbelman M, Van der Heijde GL, Weeber HA, Vrensen GF. Changes in the internal structure of the human crystalline lens with age and accommodation. Vis Res 2003;43 (22):2363–75.
[67] Koretz JF, Cook CA, Kaufman PL. Accommodation and presbyopia in the human eye. Changes in the anterior segment and crystalline lens with focus. Invest Ophthalmol Vis Sci 1997;38(3):569–78.
[68] Patnaik B. A photographic study of accommodative mechanisms: changes in the lens nucleus during accommodation. Investig Ophthalmol 1967;6(6):601–11.
[69] Fisher RF. The elastic constants of the human lens. J Physiol 1971;212(1):147–80.
[70] Heys KR, Cram SL, Truscott RJ. Massive increase in the stiffness of the human lens nucleus with age: the basis for presbyopia? Mol Vis 2004;10:956–63.
[71] Croft MA, Kaufman PL. Accommodation and presbyopia: the ciliary neuromuscular view. Ophthalmol Clin N Am 2006;19(1):13–24.
[72] Augusteyn RC. Growth of the lens: in vitro observations. Clin Exp Optom 2008;91 (3):226–39.
[73] Mohamed A, Sangwan VS, Augusteyn RC. Growth of the human lens in the Indian adult population: preliminary observations. Indian J Ophthalmol 2012;60(6):511–5.
[74] Bracchi PG, Carta F, Fasella P, Maraini G. Selective binding of aged alpha-crystallin to lens fibre ghosts. Exp Eye Res 1971;12(1):151–4.
[75] Maraini G, Fasella P. Reversible binding of soluble lens proteins to lens fibre ghosts. Exp Eye Res 1970;10(1):133–9.
[76] Coghlan SD, Augusteyn RC. Changes in the distribution of proteins in the aging human lens. Exp Eye Res 1977;25(6):603–11.
[77] Wilmarth PA, Tanner S, Dasari S, Nagalla SR, Riviere MA, Bafna V, et al. Age-related changes in human crystallins determined from comparative analysis of post-translational modifications in young and aged lens: does deamidation contribute to crystallin insolubility? J Proteome Res 2006;5(10):2554–66.
[78] Pedrigi RM, David G, Dziezyc J, Humphrey JD. Regional mechanical properties and stress analysis of the human anterior lens capsule. Vis Res 2007;47(13):1781–9.
[79] Fisher RF. Elastic constants of the human lens capsule. J Physiol 1969;201(1):1–19.
[80] Ziebarth NM, Borja D, Arrieta E, Aly M, Manns F, Dortonne I, et al. Role of the lens capsule on the mechanical accommodative response in a lens stretcher. Invest Ophthalmol Vis Sci 2008;49(10):4490–6.
[81] Fisher RF. The ciliary body in accommodation. Trans Ophthalmol Soc U K 1986; 105(Pt 2):208–19.
[82] van Alphen GW, Graebel WP. Elasticity of tissues involved in accommodation. Vis Res 1991;31(7–8):1417–38.
[83] Fisher RF. The force of contraction of the human ciliary muscle during accommodation. J Physiol 1977;270(1):51–74.
[84] Augusteyn RC, Mohamed A, Nankivil D, Veerendranath P, Arrieta E, Taneja M, et al. Age-dependence of the optomechanical responses of ex vivo human lenses from India and the USA, and the force required to produce these in a lens stretcher: the similarity to in vivo disaccommodation. Vis Res 2011;51(14):1667–78.
[85] Ehrmann K, Ho A, Parel JM. Biomechanical analysis of the accommodative apparatus in primates. Clin Exp Optom 2008;91(3):302–12.
[86] Glasser A, Campbell MC. Presbyopia and the optical changes in the human crystalline lens with age. Vis Res 1998;38(2):209–29.

[87] Manns F, Parel JM, Denham D, Billotte C, Ziebarth N, Borja D, et al. Optomechanical response of human and monkey lenses in a lens stretcher. Invest Ophthalmol Vis Sci 2007;48(7):3260–8.
[88] Pierscionek BK. Age-related response of human lenses to stretching forces. Exp Eye Res 1995;60(3):325–32.
[89] Reilly MA, Hamilton PD, Perry G, Ravi N. Comparison of the behavior of natural and refilled porcine lenses in a robotic lens stretcher. Exp Eye Res 2009;88(3):483–94.
[90] Koopmans SA, Terwee T, Barkhof J, Haitjema HJ, Kooijman AC. Polymer refilling of presbyopic human lenses in vitro restores the ability to undergo accommodative changes. Invest Ophthalmol Vis Sci 2003;44(1):250–7.
[91] Parel JM, Gelender H, Trefers WF, Norton EW. Phaco-Ersatz: cataract surgery designed to preserve accommodation. Graefes Arch Clin Exp Ophthalmol 1986;224(2):165–73.
[92] Holden BA, Fricke TR, Ho SM, Wong R, Schlenther G, Cronje S, et al. Global vision impairment due to uncorrected presbyopia. Arch Ophthalmol 2008;126(12):1731–9.
[93] Duane A. Normal values of the accommodation at all ages. JAMA 1912;59:1010–3.
[94] Duane A. Studies in monocular and binocular accommodation, with their clinical application. Trans Am Ophthalmol Soc 1922;20:132–57.
[95] Atchison DA. Accommodation and presbyopia. Ophthalmic Physiol Opt 1995;15(4):255–72.
[96] Croft MA, Kaufman PL, Crawford KS, Neider MW, Glasser A, Bito LZ. Accommodation dynamics in aging rhesus monkeys. Am J Phys 1998;275(6 Pt 2):R1885–97.
[97] Fincham EF. The mechanism of accommodation. Br J Ophthalmol 1937;8(Suppl.):5–80.
[98] Strenk SA, Strenk LM, Koretz JF. The mechanism of presbyopia. Prog Retin Eye Res 2005;24(3):379–93.
[99] Strenk SA, Semmlow JL, Strenk LM, Munoz P, Gronlund-Jacob J, DeMarco JK. Age-related changes in human ciliary muscle and lens: a magnetic resonance imaging study. Invest Ophthalmol Vis Sci 1999;40(6):1162–9.
[100] Pardue MT, Sivak JG. Age-related changes in human ciliary muscle. Optom Vis Sci 2000;77(4):204–10.
[101] Sheppard AL, Davies LN. The effect of ageing on in vivo human ciliary muscle morphology and contractility. Invest Ophthalmol Vis Sci 2011;52(3):1809–16.
[102] Croft MA, McDonald JP, Katz A, Lin TL, Lutjen-Drecoll E, Kaufman PL. Extralenticular and lenticular aspects of accommodation and presbyopia in human versus monkey eyes. Invest Ophthalmol Vis Sci 2013;54(7):5035–48.
[103] Tamm S, Tamm E, Rohen JW. Age-related changes of the human ciliary muscle. A quantitative morphometric study. Mech Ageing Dev 1992;62(2):209–21.
[104] Fisher RF. Presbyopia and the changes with age in the human crystalline lens. J Physiol 1973;228(3):765–79.
[105] Sharma PK, Busscher HJ, Terwee T, Koopmans SA, van Kooten TG. A comparative study on the viscoelastic properties of human and animal lenses. Exp Eye Res 2011;93(5):681–8.
[106] Weale RA. A biography of the eye: development, growth, age. London, UK: HK Lewis; 1982.
[107] Costello MJ, Burette A, Weber M, Metlapally S, Gilliland KO, Fowler WC, et al. Electron tomography of fiber cell cytoplasm and dense cores of multilamellar bodies from human age-related nuclear cataracts. Exp Eye Res 2012;101:72–81.

[108] Heys KR, Truscott RJ. The stiffness of human cataract lenses is a function of both age and the type of cataract. Exp Eye Res 2008;86(4):701–3.

[109] Heys JJ, Barocas VH. A boussinesq model of natural convection in the human eye and the formation of Krukenberg's spindle. Ann Biomed Eng 2002;30(3):392–401.

[110] Goldmann H. Minute volume of the aqueous in the anterior chamber of the human eye in normal state and in primary glaucoma. Ophthalmologica 1950;120(1–2):19–21.

[111] Ascher KW. Veins of the aqueous humor in glaucoma. Boll Ocul 1954;33(3):129–44.

[112] Bill A, Hellsing K. Production and drainage of aqueous humor in the cynomolgus monkey (Macaca irus). Investig Ophthalmol 1965;4(5):920–6.

[113] Johnson M, Erickson K. Mechanisms and routes of aqueous humor drainage. In: - Albert DM, Jakobiec FA, editors. Principles and practice of ophthalmology. Philadelphia: WB Saunders; 2000.

[114] Nakajima E, Nakajima T, Minagawa Y, Shearer TR, Azuma M. Contribution of ROCK in contraction of trabecular meshwork: proposed mechanism for regulating aqueous outflow in monkey and human eyes. J Pharm Sci 2005;94(4):701–8.

[115] Bill A, Svedbergh B. Scanning electron microscopic studies of the trabecular meshwork and the canal of Schlemm—an attempt to localize the main resistance to outflow of aqueous humor in man. Acta Ophthalmol 1972;50(3):295–320.

[116] Phelps CD, Armaly MF. Measurement of episcleral venous pressure. Am J Ophthalmol 1978;85(1):35–42.

[117] Brubaker RF. Determination of episcleral venous pressure in the eye. A comparison of three methods. Arch Ophthalmol 1967;77(1):110–4.

[118] Schottenstein EM. Intraocular pressure. In: Ritch R, Shields MB, Krupin T, editors. The glaucomas. St. Louis: Mosby; 1989. p. 301–17.

[119] Grant WM. Further studies on facility of flow through the trabecular meshwork. AMA Arch Ophthalmol 1958;60(4 Part 1):523–33.

[120] Rohen JW, Lutjen E, Barany E. The relation between the ciliary muscle and the trabecular meshwork and its importance for the effect of miotics on aqueous outflow resistance. A study in two contrasting monkey species, *Macaca irus* and *Cercopithecus aethiops*. Albrecht Von Graefes Arch Klin Exp Ophthalmol 1967;172(1):23–47.

[121] Barany EH. The mode of action of miotics on outflow resistance. A study of pilocarpine in the vervet monkey Cercopithecus ethiops. Trans Ophthalmol Soc U K 1966;86:539–78.

[122] Kass MA, Hart Jr. WM, Gordon M, Miller JP. Risk factors favoring the development of glaucomatous visual field loss in ocular hypertension. Surv Ophthalmol 1980;25(3):155–62.

[123] Gabelt BT, Kaufman PL. Aqueous humor hydrodynamics. In: Hart WM, editor. Adler's physiology of the eye. 9th ed. St. Louis: Mosby; 2003.

[124] Ethier CR. The inner wall of Schlemm's canal. Exp Eye Res 2002;74(2):161–72.

[125] Johnstone MA, Grant WG. Pressure-dependent changes in structures of the aqueous outflow system of human and monkey eyes. Am J Ophthalmol 1973;75(3):365–83.

[126] Johnson M. What controls aqueous humour outflow resistance? Exp Eye Res 2006;82(4):545–57.

[127] WuDunn D. Mechanobiology of trabecular meshwork cells. Exp Eye Res 2009;88(4):718–23.

[128] Overby DR, Zhou EH, Vargas-Pinto R, Pedrigi RM, Fuchshofer R, Braakman ST, et al. Altered mechanobiology of Schlemm's canal endothelial cells in glaucoma. Proc Natl Acad Sci U S A 2014;111(38):13876–81.

[129] Quigley HA, Addicks EM, Green WR, Maumenee AE. Optic nerve damage in human glaucoma. II. The site of injury and susceptibility to damage. Arch Ophthalmol 1981;99(4):635–49.
[130] Fechtner RD, Weinreb RN. Mechanisms of optic nerve damage in primary open angle glaucoma. Surv Ophthalmol 1994;39(1):23–42.
[131] Burgoyne CF, Downs JC, Bellezza AJ, Suh JK, Hart RT. The optic nerve head as a biomechanical structure: a new paradigm for understanding the role of IOP-related stress and strain in the pathophysiology of glaucomatous optic nerve head damage. Prog Retin Eye Res 2005;24(1):39–73.
[132] Downs JC, Roberts MD, Sigal IA. Glaucomatous cupping of the lamina cribrosa: a review of the evidence for active progressive remodeling as a mechanism. Exp Eye Res 2011;93(2):133–40.
[133] Wang N, Xie X, Yang D, Xian J, Li Y, Ren R, et al. Orbital cerebrospinal fluid space in glaucoma: the Beijing intracranial and intraocular pressure (iCOP) study. Ophthalmology 2012;119(10):2065–73. [e1].
[134] Ren R, Jonas JB, Tian G, Zhen Y, Ma K, Li S, et al. Cerebrospinal fluid pressure in glaucoma: a prospective study. Ophthalmology 2010;117(2):259–66.
[135] Morgan WH, Yu DY, Balaratnasingam C. The role of cerebrospinal fluid pressure in glaucoma pathophysiology: the dark side of the optic disc. J Glaucoma 2008;17(5):408–13.
[136] Killer HE, Laeng HR, Flammer J, Groscurth P. Architecture of arachnoid trabeculae, pillars, and septa in the subarachnoid space of the human optic nerve: anatomy and clinical considerations. Br J Ophthalmol 2003;87(6):777–81.
[137] Killer HE, Jaggi GP, Flammer J, Miller NR, Huber AR, Mironov A. Cerebrospinal fluid dynamics between the intracranial and the subarachnoid space of the optic nerve. Is it always bidirectional? Brain 2007;130(Pt 2):514–20.
[138] Almasieh M, Wilson AM, Morquette B, Cueva Vargas JL, Di Polo A. The molecular basis of retinal ganglion cell death in glaucoma. Prog Retin Eye Res 2012;31(2):152–81.
[139] Franze K, Francke M, Gunter K, Christ AF, Korber N, Reichenbach A, et al. Spatial mapping of the mechanical properties of the living retina using scanning force microscopy. Soft Matter 2011;7(7):3147–54.
[140] Marmor MF. Mechanisms of retinal adhesion. Prog Retin Res 1993;12:179–204.
[141] Ingber DE, Wang N, Stamenovic D. Tensegrity, cellular biophysics, and the mechanics of living systems. Rep Prog Phys 2014;77(4):046603.
[142] Franze K, Gerdelmann J, Weick M, Betz T, Pawlizak S, Lakadamyali M, et al. Neurite branch retraction is caused by a threshold-dependent mechanical impact. Biophys J 2009;97(7):1883–90.
[143] Flanagan LA, Ju YE, Marg B, Osterfield M, Janmey PA. Neurite branching on deformable substrates. Neuroreport 2002;13(18):2411–5.
[144] Georges PC, Miller WJ, Meaney DF, Sawyer ES, Janmey PA. Matrices with compliance comparable to that of brain tissue select neuronal over glial growth in mixed cortical cultures. Biophys J 2006;90(8):3012–8.
[145] Jiang X, Georges PC, Li B, Du Y, Kutzing MK, Previtera ML, et al. Cell growth in response to mechanical stiffness is affected by neuron-astroglia interactions. Open Neuroimaging J 2007;1:7–14.
[146] Moore SW, Biais N, Sheetz MP. Traction on immobilized netrin-1 is sufficient to reorient axons. Science 2009;325(5937):166.

[147] Chan CE, Odde DJ. Traction dynamics of filopodia on compliant substrates. Science 2008;322(5908):1687–91.
[148] Moshayedi P, Costa Lda F, Christ A, Lacour SP, Fawcett J, Guck J, et al. Mechanosensitivity of astrocytes on optimized polyacrylamide gels analyzed by quantitative morphometry. J Phys Condens Matter 2010;22(19):194114.
[149] Wygnanski-Jaffe T, Murphy CJ, Smith C, Kubai M, Christopherson P, Ethier CR, et al. Protective ocular mechanisms in woodpeckers. Eye (Lond) 2007;21(1):83–9.
[150] Meskauskas J, Repetto R, Siggers JH. Shape change of the vitreous chamber influences retinal detachment and reattachment processes: is mechanical stress during eye rotations a factor? Invest Ophthalmol Vis Sci 2012;53(10):6271–81.
[151] Stocchino A, Repetto R, Cafferata C. Eye rotation induced dynamics of a Newtonian fluid within the vitreous cavity: the effect of the chamber shape. Phys Med Biol 2007;52(7):2021–34.
[152] Rossi T, Boccassini B, Esposito L, Iossa M, Ruggiero A, Tamburrelli C, et al. The pathogenesis of retinal damage in blunt eye trauma: finite element modeling. Invest Ophthalmol Vis Sci 2011;52(7):3994–4002.
[153] Liu X, Wang L, Wang C, Sun G, Liu S, Fan Y. Mechanism of traumatic retinal detachment in blunt impact: a finite element study. J Biomech 2013;46(7):1321–7.
[154] Sibony P, Kupersmith MJ, Rohlf FJ. Shape analysis of the peripapillary RPE layer in papilledema and ischemic optic neuropathy. Invest Ophthalmol Vis Sci 2011;52(11):7987–95.
[155] Chou T, Siegel M. A mechanical model of retinal detachment. Phys Biol 2012;9(4):046001.
[156] Liang FQ, Godley BF. Oxidative stress-induced mitochondrial DNA damage in human retinal pigment epithelial cells: a possible mechanism for RPE aging and age-related macular degeneration. Exp Eye Res 2003;76(4):397–403.
[157] Robinson DA. The mechanics of human smooth pursuit eye movement. J Physiol 1965 Oct 1;180(3):569–91.
[158] Boghen D, Troost BT, Daroff RB, Dell'Osso LF, Birkett JE. Velocity characteristics of normal human saccades. Invest Ophthalmol Vis Sci 1974;13(8):619–23.
[159] Wright KW. Spiegel PH, editor. Pediatric ophthalmology and strabismus. New York, USA: Springer Science & Business Media; 2013.
[160] Sherrington CS. Classics. Resonance 2016;21(7):657–65.
[161] Hertle RW. Nystagmus in infancy and childhood: characteristics and evidence for treatment. Am Orthopt J 2010 Jan;60(1):48–58.
[162] Wan MJ, Hunter DG. Complications of strabismus surgery: incidence and risk factors. Semin Ophthalmol 2014;29(5–6):421–8.

## FURTHER READING

[163] Oie Y, Nishida K. Corneal regenerative medicine. Regen Ther 2016;5:40–5.
[164] Nankivil D, Heilman BM, Durkee H, Manns F, Ehrmann K, Kelly S, et al. The zonules selectively alter the shape of the lens during accommodation based on the location of their anchorage points. Invest Ophthalmol Vis Sci 2015;56(3):1751–60.
[165] Nakajima E, Nakajima T, Minagawa Y, Shearer TR, Azuma M. Contribution of ROCK in contraction of trabecular meshwork: proposed mechanism for regulating aqueous outflow in monkey and human eyes. J Pharm Sci 2005;94(4):701–8.

# CHAPTER 13

# Gastrointestinal mechanosensory function in health and disease

**Amanda J. Page**[*,†], **Hui Li**[*]

*Adelaide Medical School, University of Adelaide, Adelaide, SA, Australia* * *South Australian Health and Medical Research Institute (SAHMRI), Adelaide, SA, Australia*[†]

## ABBREVIATIONS

| | |
|---|---|
| ASIC | acid-sensing ion channels |
| CNS | central nervous system |
| COX | cyclooxygenase |
| DRG | dorsal root ganglia |
| ENS | enteric nervous system |
| GI | gastrointestinal |
| GERD | gastroesophageal reflux disease |
| HFD | high-fat diet |
| 5-HT | 5-hydroxytryptamine |
| AH | after hyperpolarization |
| IBD | inflammatory bowel disease |
| IL | interleukin |
| ICCs | interstitial cells of Cajal |
| ICC-CM | ICC-circular muscle |
| ICC-DMP | ICC-deep muscular plexus |
| ICC-LM | ICC-longitudinal muscle |
| ICC-MP | ICC-myenteric plexus |
| ICC-SEP | ICC-septa |
| ICC-SM | ICC-submucosa |
| ICC-SMP | ICC-submucosal plexus |
| ICC-SS | ICC-subserosa |
| IGVEs | intraganglionic varicose endings |
| IMAs | intramuscular arrays |
| IGLEs | intraganglionic laminar endings |
| IPANs | intrinsic primary afferent neurons |
| IBS | irritable bowel syndrome |
| BKCa | large-conductance calcium-activated potassium channels |
| BK-like | large-conductance potassium channels |
| PBMCs | peripheral blood mononuclear cells |

| | |
|---|---|
| **RAMEN** | rapidly adapting mechanosensitive enteric neurons |
| **SAMEN** | slowly adapting mechanosensitive enteric neuron |
| **SDK** | stretch-dependent potassium channel |
| **TRP** | channels transient receptor potential |
| **TRPA1** | TRP ankyrin 1 |
| **TRPV1** | TRP vanilloid 1 |
| **TNBS** | trinitrobenzenesulfonic acid |
| **USAMEN** | ultraslowly adapting mechanosensitive enteric neuron |

## 1 INTRODUCTION

The predominant roles of the gastrointestinal (GI) tract are digestion, absorption, excretion, and protection. This is achieved by the movement of food through a series of highly specialized regions along the GI tract, from the mouth to anus. The stomach and small intestine are responsible for digestion and absorption and also signaling information to central regions on the amount and nutrient content of food consumed. The primary role of the large intestine is desiccation, with reabsorption of water, and compaction of waste, prior to storage in the sigmoid colon and rectum before elimination.

Mechanosensation can occur at all levels within the gut wall, from the enterochromaffin cells to the extrinsic spinal afferents innervating the mesenteric attachment. This chapter will focus on the mechanosensory processes in the extrinsic nerves, the intrinsic enteric nervous system (ENS), interstitial cells of Cajal (ICCs), smooth muscle cells, and endocrine cells.

## 2 MECHANOSENSITIVE COMPONENTS OF THE GASTROINTESTINAL TRACT

### 2.1 GASTROINTESTINAL MECHANOSENSATION

Mechanosensation is essential for normal function of the GI tract, and abnormalities in mechanosensation have been linked to certain GI pathologies. Mechanotransduction is the conversion of mechanical stimuli into electric signals through the activation of ion channels. The key features of mechanotransduction are speed and sensitivity [1]. For speed, the mechanotransduction process must directly link the mechanical stimuli to a specific ion channel, which can subsequently open and allow the movement of ions across the membrane. The high sensitivity requires that the maximum amount of mechanical stimulus be directed at the mechanotransduction mechanism [1]. There are a number of hypotheses (Fig. 1) for how this occurs. First, mechanotransduction can occur through direct activation of an ion channel due to mechanical distortion of the cell membrane (Fig. 1A). Another direct mechanism involves the tethering of the ion channel to the extracellular matrix and/or the

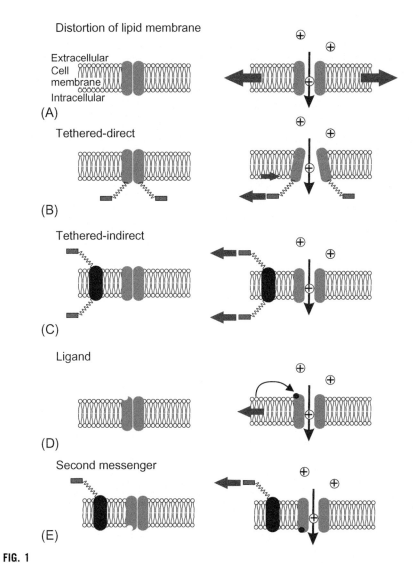

**FIG. 1**

Proposed mechanotransduction mechanisms including (A) direct activation of an ion channel due to mechanical distortion of the cell membrane; (B) direct tethering of the ion channel to the extracellular matrix and/or the intracellular cytoskeleton such that small movements can alter the tension of all elements of the system altering the opening probability; (C) indirect tethering to an accessory protein where force induces a conformational change that in turn affects the ion channel; (D) mechanical force may cause release of a ligand that activates the ion channel causing the influx of ions; and (E) the mechanosensing component may be more distant from the ion channels and, upon activation, generate a diffusible second messenger or activate a kinase that influences ion channel opening probability.

intracellular cytoskeleton [1,2], such that small movements can alter the tension of all elements of the system altering the opening probability [1] (e.g., Fig. 1B). Alternatively, the tethering can be indirect. For example, force could be conveyed to a tethered accessory protein inducing a conformational change that in turn affects the ion channel (Fig. 1C). Alternatively, mechanical force may cause release of a ligand that activates the ion channel causing the influx of ions (Fig. 1D). Finally, the mechanosensing component, of the transduction complex, may be more distant from the ion channels and, upon activation, generate a secondary signal (e.g., diffusible second messenger or activation of a kinase) that influences ion channel opening probability [1] (Fig. 1E). The intermediary ligand and second messengers have the potential to slow the mechanotransduction process. However, speed may be less critical for certain physiological responses within the GI tract, such as the release of gut hormones or possibly the responses to mechanical stimulation of high-threshold extrinsic visceral afferents [3]. In the majority of cases, the exact mechanotransduction mechanisms within the GI tract are unknown; however, advances have been made and will be discussed in this chapter.

## 2.2 EXTRINSIC INNERVATION OF THE GI TRACT
### 2.2.1 Anatomy of the extrinsic innervation
The GI tract is richly innervated with sensory nerves that convey information from the gastrointestinal tract to the central nervous system (CNS) where it is processed, and gut reflexes are coordinated with behavioral responses and sensations, such as fullness, satiety, bloating, or pain. Afferent fibers basically follow two routes to the CNS, namely, vagal and spinal pathways [4]. Vagal afferent cell bodies are located in the nodose and jugular ganglia and project centrally to the nucleus tractus solitarius. The nodose and jugular ganglia are embryologically distinct with the nodose arising from the placodal tissue and the jugular arising from the neural crest, similar to the dorsal root ganglia (DRG). The cell bodies of spinal afferents are located in the DRG. Central projections of these afferents enter the spinal cord where they make connections with second-order neurons that distribute information throughout the CNS. Spinal afferents can be further subdivided into pelvic and splanchnic afferents that have preganglionic cell bodies in the sacral and thoracolumbar spinal ganglia, respectively.

Vagal afferent endings are more prevalent in the upper GI tract, while spinal afferents innervate throughout, although pelvic afferents are predominantly located in the distal gut. About 40% of afferents innervating the stomach are vagal afferents, and 60% are spinal originating from T4 to L2 DRG [5–7]. In the mouse, duodenal afferents are less abundant than gastric afferents, although they are still equally distributed between vagal and spinal afferents [5]. Cell bodies of retrogradely labeled jejunal spinal afferents have been observed in T9-T13 DRG [8]. Vagal afferents innervating the jejunum also respond to low-threshold distension [9], in addition to chemosensory responses. Vagal afferent innervation becomes sparse at the distal end of the GI tract with the density of the vagal afferent fibers in the colon only about

10% of the density observed in the duodenum [10]. In the guinea-pig distal colon and rectum, vagal afferent fibers account for only 0.5% of cell bodies in the nodose ganglia [11]. However, in rats, the cell bodies of vagal afferents innervating the distal colon account for 10–15% of the total number of cells in the nodose ganglia [12].

The peripheral afferent endings are basically positioned in three distinct locations within the gut wall. One population has afferent endings in the serosal layer and the mesenteric attachment closely associated with blood vessels. Another population has endings in the muscularis externa either in the muscle layers (intramuscular arrays) [10,13–15] or in the myenteric plexus (intraganglionic laminar endings (IGLEs)) between the longitudinal and circular muscle layers [16]. A third population has endings in the mucosal lamina propria where they are ideally situated to detect fine tactile mechanical stimuli (e.g., food moving over the receptive field) [17] and material absorbed across the mucosal epithelium or released from specialized cells (e.g., 5-hydroxytryptamine (5-HT) from enterochromaffin cells) [18–20]. Due to the detailed studies available, this chapter will focus on the vagal innervation of the gastroesophageal region and spinal innervation of the colon.

### 2.2.2 Mechanosensory properties of vagal afferents

GI vagal afferents can be subdivided into four groups based on their mechanosensory properties that concur well with the anatomical location of their endings.

**(1)** *Mucosal afferents* are generally silent at rest but are sensitive to both mechanical and chemical stimuli. They respond to light stroking of the mucosa generating a burst of action potentials every time the stimulus passes over the receptive field [17,21] (Fig. 2). They are insensitive to distension and

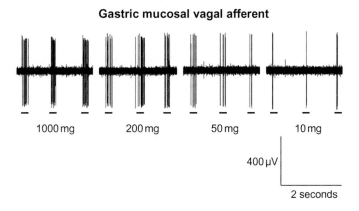

**FIG. 2**

Typical recording of a gastric mechanosensitive mucosal vagal afferent. The responses are to stroking over the receptive field with calibrated von Frey hairs ranging from 10 to 1000 mg. The afferent has no basal discharge, and the response is von Frey hair-dependent. Mucosal afferents do not respond to stretch.

contraction of the gut wall, except when distortion of the mucosa occurs as a consequence [17,21]. Although it was initially suggested that mucosal vagal afferents were a single population [18], studies have revealed the possible existence of three independent populations of mucosal afferent within the upper GI tract that may have different sensitivity to mechanical stimulation [22]. However, this requires further investigation. Hypotheses have been generated for the role of GI mechanosensory mucosal afferents; however, there is a distinct lack of evidence for any of the proposed functions. Most suggestions are based on the location and mucosal afferent responsiveness. In the gastric antrum, there are a population of mucosal afferents that are sensitive to both chemical (e.g., changes in pH) and mechanical stimuli (e.g., light probing) [23]. It has been suggested that this subtype of mucosal afferent could detect the presence and passage of luminal content and have added chemosensory ability to detect the acidity of chyme [22]. Further, there is some evidence that gastric mucosal afferents can detect particle size providing negative feedback on the control of gastric emptying [24]. In addition, an in vivo study in anesthetized ferrets demonstrated that probing the mucosa in the gastric antrum resulted in a reduction in corpus pressure and an inhibition of contractions [25]. These studies support the theory that gastric mucosal afferents are involved in the regulation of gastric motility and gastric emptying. However, to confirm, more refined experiments need to be undertaken.

(2) *Tension sensitive afferents* include the following:
- *Intraganglionic laminar endings (IGLEs)* are positioned in the myenteric plexus, between the longitudinal and circular smooth muscle layers [13,15], and often have a resting discharge of action potentials modulated in phase with ongoing contractions. They are mechanosensitive to contractions and distension, with a slowly adapting linear relationship to wall tension [17,21,26,27], and are characterized as low-threshold tension-sensitive afferents [28] (Fig. 3). It is hypothesized that IGLEs detect distortion of the tissue surrounding their endings through a mechanism likely involving ion channel(s) (see Section 2.2.4) [29]. Their stimulus-response functions saturate within the physiological range [27], unlike spinal afferents that generally signal well above the physiological range [30]. There is a dense population of IGLEs within the proximal GI tract including the esophagus, stomach, and small intestine (Fig. 3).
- *Intramuscular arrays (IMAs)* are vagal afferent endings that exist in parallel to the muscle bundles within the muscularis externa [13,15,31]. IMAs are located throughout the GI tract although the highest density is located in the wall of the stomach and the sphincters throughout the gut [10]. They have been shown to be closely associated with ICCs [32] that has led to the suggestion that they may form a functional complex with ICCs [33]. It has been hypothesized, given their morphology, that

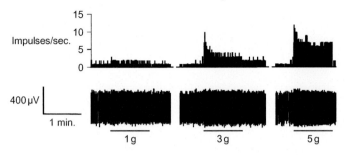

**FIG. 3**

Typical recording of a gastric tension-sensitive vagal afferent. The *upper panel* is the number of action potentials generated per second, and the *lower panel* is the raw trace. The tension-dependent responses illustrated are to circular stretch (1–5 g). Tension-sensitive vagal afferents often exhibit basal resting activity with an increased firing rate on application of stretch stimuli.

IMAs could function as tension receptors, specifically monitoring the length of the gastric muscle [34]. However, to date, there is no electrophysiological data to confirm that IMAs form a second class of tension-sensitive endings.

The properties of tension receptors appear to depend on the location, even within a single organ. For example, gastric tension-sensitive afferents within the corpus and fundus respond largely to distension of the stomach wall after consumption of a meal. In contrast, antral tension-sensitive afferents are more sensitive to contraction [35]. This is consistent with the role that each part of the stomach plays: distension of the fundus and corpus to accommodate food and contraction of the antrum to control motility. In the esophagus and maybe the small intestine, the responses of vagal tension-sensitive afferents are thought to play a regulatory role in the generation of secondary peristalsis, required to progress food through the GI tract. Esophageal vagal afferents have been further subdivided into low-threshold mechanoreceptors and high-threshold nociceptors, the latter of which are activated at noxious mechanical force and may play a role in pain signaling [36].

(3) *Tension/mucosal afferents* have to date only been observed in the striated esophagus of the ferret and not the stomach. They demonstrate mechanosensory responses to both mucosal stroking and circular stretch [21]. This subtype of vagal afferent does appear to have a spinal counterpart in the murine colonic muscular/mucosal pelvic afferents [37]. It is possible that they have a specialized role in the detection of rapidly moving liquid or food bolus along the esophagus.

### 2.2.2.1 Exaggerated or reduced vagal afferent mechanosensitivity and implications for disease

The GI tract plays an important role in sensing the arrival, amount, and chemical composition of a meal. Exaggerated or reduced GI perception of a meal can have significant implications, across the life span, for eating-related disorders including, on one end of the spectrum, obesity and, on the other, functional dyspepsia, a chronic condition characterized by enhanced GI sensitivity to luminal stimuli [38] (Fig. 4).

***2.2.2.1.1 Obesity and appetite regulation.*** In high-fat diet (HFD)-induced obese mice, the ability of gastric and small intestinal tension-sensitive afferents to respond to distension is reduced [39,40]. Such reductions in the ability to detect distension may impair the generation of satiety signals. Thus, an isovolumetric meal will exert a smaller satiety signal, which in turn could lead to the consumption of a larger meal before satiety is perceived. These observations are consistent with the increased meal size observed in obese rodent models and humans [41,42]. This dampened response to tension stimuli in HFD-induced obesity does not return to normal after an equivalent time (12 weeks) back on a standard diet [43]. Therefore, changes that occur in HFD-induced obesity are not readily reversible and could have significant implications in the maintenance of weight loss. The reduced neuronal excitability, in chronic HFD conditions [39], could be due to a disruption in the ion channel transient receptor potential vanilloid 1 (TRPV1). In TRPV1 knockout mice fed with a standard laboratory diet, there is a reduction in the mechanosensitivity of tension-sensitive gastric vagal afferents to a similar level observed in the wild-type mice fed an HFD [44,45]. These standard laboratory-diet-fed TRPV1 knockout mice also exhibit an increase in food consumption [45] compared with wild-type mice, possibly due to the reduced vagal afferent response to tension. In addition, there is no further reduction in mechanosensitivity of gastric tension-sensitive vagal afferents if the TRPV1 knockout mice are fed an HFD [45]. A similar reduction in gastric vagal afferent mechanosensitivity is observed in fasted conditions [40,46]. This suggests that a

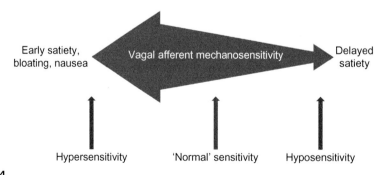

**FIG. 4**

Diagrammatic representation of the continuum of vagal afferent sensitivity to mechanical stimulation and the sensations and functional responses associated with heightened or dampened gastric vagal afferent signaling.

reduction in the plasticity of this system during the development of obesity triggers a perpetual fasted state. It has been demonstrated that under normal circumstances, gastric vagal afferents display a large degree of plasticity. Given that GI vagal afferents are involved in the control of food intake and that food intake shows strong circadian patterns, it is not surprising that gastric vagal afferents exhibit profound oscillations in mechanosensitivity, over a 24 h period, which are inversely proportional to the amount of food in the stomach [47]. It is known that there is an increase in both meal size and frequency during the dark phase, in rodents, when energy demand is high [48] due to their nocturnal nature. The reduced gastric vagal afferent mechanosensitivity, observed during the dark phase [47], would allow for the observed increase in food intake during this period before satiation is reached. In HFD-induced obese mice, this circadian variation in gastric vagal afferent mechanosensitivity is lost and associated with a grazing-like behavior in food intake [49]. Therefore, the high degree of plasticity displayed by gastric vagal afferents appears to be lost in HFD-induced obesity.

**2.2.2.1.2 Functional dyspepsia.** On the other end of the scale, functional dyspepsia is associated with enhanced GI sensitivity with no clear evidence of an organic cause [50]. Functional dyspepsia affects about 20% of the population and significantly impairs their quality of life. There are two clinically distinct syndromes:

**(1)** *Postprandial distress syndrome*, where feelings of fullness (satiety) occur early in the meal, preventing the completion of a meal and/or having persistent feelings of bloating or nausea, with symptoms occurring after eating at least several times a week
**(2)** *Epigastric pain syndrome*, where there is intermittent pain or a burning epigastrium at least once a week

Functional dyspepsia is meal-related with about 80% of patients reporting that the symptoms are aggravated by the ingestion of a meal [51]. Hypersensitivity to mechanical stimulation of the stomach is frequent in functional dyspepsia patients; however, the underlying mechanisms for this hypersensitivity are unclear. Functional dyspepsia is also associated with delayed gastric emptying and reduced gastric accommodation after a meal [52,53]. All of these processes are vagal afferent mediated, and therefore, it is highly probable that heightened gastric vagal afferent mechanosensitivity is the cause of all the symptoms.

Failure of the stomach to accommodate food is also linked to the increase in transient lower esophageal sphincter relaxations that occur in gastroesophageal reflux disease (GERD) patients [54]. This may in part explain the overlap in functional dyspepsia and GERD patients [54] and again possibly be explained by enhanced gastric vagal afferent mechanosensitivity, although this still remains to be determined. Unfortunately, unlike irritable bowel syndrome (IBS), a functional disorder of the lower GI tract, the majority of research on functional dyspepsia has occurred at the clinical level with very few basic research studies investigating the mechanisms driving hyperperception of food-related stimuli in the upper GI tract. This is mainly due to the lack of an animal model for functional dyspepsia. An animal model would allow the investigation of the molecular mechanisms driving gastric vagal afferent hypersensitivity and possibly identify new targets for the treatment of functional dyspepsia.

Esophageal vagal afferents have distinctive subtypes of vagal afferent that can respond to mechanical stimuli in the noxious range, namely, nodose and jugular C fibers but not nodose Aδ-fibers [55]. Therefore, it is possible that similar gastric nociceptive afferents, thus far unidentified, may contribute to epigastric pain syndrome along with spinal afferents, which are more commonly associated with nociception. This requires further investigation.

### 2.2.3 Mechanosensory properties of spinal afferents

In the large intestine, about one-third of the spinal afferent fibers innervate the submucosa, a quarter innervate the circular muscle, and about one-fifth innervate the myenteric ganglia [56]. Two types of spinal afferent ending have been identified in the myenteric ganglia, the intraganglionic varicose endings (IGVEs) and, in the rectum, the occasional IGLE [56]. Further, three classes of varicose ending have been identified, namely, those innervating the submucosa and circular muscle and one class innervating the longitudinal muscle, blood vessels, mucosa, intermodal strands, and crypts of Lieberkühn. Spinal afferents also innervate the upper GI tract, although there is much less diversity than in the large intestine [57]. In the stomach and esophagus, IGVEs have been identified in the myenteric ganglia with simple varicose endings in the circular muscle and mucosa [57]. Nerve endings have also been identified in the intermodal strands, blood vessels, submucosal ganglia, and longitudinal muscle, although these were observed less frequently [57]. Further, no spinal afferent IGLEs have been identified in either the esophagus or stomach [57]. Although there are numerous anatomically distinct populations, there are only five classes of spinal afferent based on mechanosensitivity [37]. Whether all anatomical subclasses of spinal afferent are mechanosensitive remains to be determined.

Mechanosensory information from the lower gut including the colon and rectum is carried to the spinal cord either in lumbar splanchnic nerves or sacral pelvic nerves. Splanchnic afferents can be subdivided into four mechanosensitive classes [37], all of which respond to probing with calibrated von Frey hairs and are distinguished by their response to other mechanical stimuli:

(1) *Mucosal afferents* respond to light stroking of the receptive field with calibrated von Frey hairs as low as 10 mg [37], but do not respond to circular stretch.
(2) *Muscular afferents* have their receptive fields in the colonic wall. They are activated by probing and maintained circular stretch; however, they are not responsive to fine tactile stimulation, such as mucosal stroking [37]. The majority of muscular afferents have low thresholds to distension; however, about 15–20% of afferents exhibit a high threshold (>30 mmHg) for response and are possibly nociceptors [30,58–62].
(3) *Serosal afferents* have their receptive field in the colonic wall. They are only activated by probing but do not respond to stretch or mucosal stroking [37].

**(4)** *Mesenteric afferents* have their receptive fields in the mesenteric attachment adjacent to the colon and respond in a graded manner to probing of the receptive field with calibrated von Frey hairs [37].

Generally, it is believed that thoracolumbar splanchnic afferents primarily convey noxious signals derived from, for example, inflammation, rapid gut distension, and/or torsion of the mesentery [37,63–67]. This pathway is probably less active under normal physiological conditions.

Pelvic afferents can also be categorized into four different classes. There are serosal, muscular, and mucosal afferents, which have similar mechanical responses to the corresponding subtypes of splanchnic afferents. In addition, a fourth subclass of pelvic afferent has been identified and classified as muscular/mucosal afferents [37]. These afferents respond to both fine mucosal stroking and circular stretch [37]. Pelvic afferents transmit physiological sensations (e.g., urgency and desire to defecate). In addition, they also form the afferent branch of extrinsic gut reflexes (e.g., cologastric inhibitory reflex) [68,69]. Further, bilateral pelvic nerve section almost entirely abolishes pain-related behavior to noxious colorectal distension in rats [37,70] and is therefore also involved in nociception.

### 2.2.3.1 Spinal afferent signalling and implications for disease

Visceral hypersensitivity refers to the increased perception to visceral stimuli (e.g., mechanical distension) and, in terms of pain, can be further subdivided into allodynia and hyperalgesia. Allodynia is when pain is perceived in response to normal physiological stimuli, whereas hyperalgesia is an increase in pain perception to stimuli that are normally perceived as pain [71]. Spinal afferents are considered the primary pathway for visceral pain signaling. Therefore, in this section, the focus will be on colorectal hypersensitivity of spinal afferents in inflammatory bowel disease (IBD) and irritable bowel syndrome (IBS). It should be noted that vagal afferents have also been implicated in the modulation of colonic pain transmission. This is supported by evidence that pain responses to colorectal distension in rats are increased after subdiaphragmatic vagotomy [72,73].

**2.2.3.1.1 Inflammatory bowel disease.** IBD, including Crohn's disease and ulcerative colitis [74], involves chronic remitting and relapsing inflammation of the intestinal mucosa due to an inappropriate activation of the immune system [75], leading to neuronal dysfunction [76–78]. The sensitization of spinal afferents, initiated by peripheral inflammation, is thought to underlie chronic pain [79]. In ulcerative colitis patients, it is assumed that colonic inflammation sensitizes primary afferent neurons to cause visceral pain. However, the findings from clinical colorectal distension studies, investigating visceral hypersensitivity in ulcerative colitis patients, are inconsistent with reports of hypersensitivity [80,81], no change [82,83] and even hyposensitivity [84]. This suggests that not all patients experience visceral hypersensitivity and, in fact, the actual reports of abdominal pain vary considerably from 40 to 90% of IBD patients [80,85,86]. On the other hand, ulcerative colitis patients have reduced stool volume and increased stool frequency suggesting that they have reduced thresholds for the urge to defecate [80]. It is possible that this is due to the sensitization of spinal afferents, although changes in muscle tone [87] could also be a contributing factor. In addition, there are a subset of Crohn's disease patients that experience urgency to defecate [88]; however, threshold pressure for discomfort is substantially higher in Crohn's patient versus healthy controls [82].

A subset of IBD patients remain hypersensitive and intolerant to rectal distension during remission [81]. Twenty to sixty percent of IBD patients experience persistent pain during remission [86,89,90], while other IBD patients report that rectal sensitivity returns to normal during remission [81,87]. In one patient study, with ulcerative colitis in remission, there was inhibition of the activity in the limbic and paralimbic brain regions, brain regions that were activated by colorectal distension in IBS patients [83]. Postinflammatory studies in mice have revealed that colonic afferents with high mechanosensory thresholds (i.e., serosal and mesenteric afferents) contribute to inflammatory hypersensitivity but not those with low threshold (i.e., muscular and mucosal afferents). Further, splanchnic afferents are implicated during both the inflammatory period and remission, whereas pelvic afferents become involved mainly during the recovery period [63]. As stated, IBD is a chronic remitting and relapsing inflammatory condition. Therefore, many of the variable outcomes in IBD research may be due to the degree of disease progression in individual patients and the number of relapse episodes experienced. It has been demonstrated that the immune response is persistently altered with successive episodes of colitis in mice [91], and therefore, it is possible that the spinal afferent response is also altered with consecutive relapse episodes. Consequently, longitudinal studies on hypersensitivity of spinal afferents are required.

*2.2.3.1.2 Irritable bowel syndrome.* IBS is a chronic functional lower GI disorder that affects about 7–14% of the population [92]. Symptoms include abdominal pain and discomfort associated with altered bowel habit. There are four main subclasses that include constipation-predominant IBS, diarrhea-predominant IBS, alternating or mixed IBS, and postinfectious IBS. Hypersensitivity of the colon and rectum is a common feature of all IBS subtypes [93,94] with 61% of IBS patients reporting altered rectal perception [95]. Current scientific evidence implicates a number of different mechanisms for the observed visceral hypersensitivity including [1] sensitization of spinal cord dorsal horn neurons; [2] altered descending excitatory and inhibitory input to the spinal cord nociceptive neurons; [3] misinterpretation of innocuous sensations as noxious due to cognitive/emotional bias; and, of significance to this chapter, [4] sensitization of peripheral visceral afferents [96,97]. Persistent hypersensitivity in IBS may occur after an acute bout of intestinal inflammation or be the consequence of low-grade inflammation not identified in routine clinical screenings [98,99]. As such, there is no clear distinction between IBS and IBD. In fact, about 40% of IBD patients in remission meet the symptom criteria for functional bowel disorders [100,101]. In addition, IBS is a risk factor for IBD [102] with the incidence of IBS higher in IBD patients in remission than the general population [103].

Postinfectious IBS patients develop symptoms after a bout of gastroenteritis [104–106]. The initial inflammation in response to the infection is a protective process necessary for wound healing. However, in postinfectious IBS patients, the peripheral hypersensitivity, in response to the infection and inflammatory response, is maintained and fails to return to normal when the healing process is complete.

Other forms of IBS are associated with low-grade inflammation within the GI tract [107,108] and altered immunologic function [109–112]. Mucosal biopsy samples from IBS patients have increased levels of interleukin (IL)-1β [113], increased numbers of mucosal mast cells [108], and increased intraepithelial lymphocytes [114]. Changes are also evident in peripheral blood mononuclear cells (PBMCs) of IBS patients [109–112] where there are data to indicate elevated T-cell activation [111,112].

There is evidence that inflammatory mediators contribute to the mechanisms driving the underlying abdominal pain in IBS patients. There is a correlation of pain with several post-inflammatory cytokines (e.g., tumor necrosis factor, IL-6, and IL-1β) found within the PBMC supernatant of IBS patients [111,112]. In addition, PBMC supernatants from IBS-D patients increase the response of colonic afferents to mechanical stimulation in mice [109–111]. Further colonic mucosal biopsy samples, from patients with IBS, activate murine extrinsic sensory nerve endings and neuronal cell bodies [115].

### 2.2.4 Contribution of ion channels to the mechanosensitivity of extrinsic afferents

The availability of genetically modified mice has enabled the identification of a number of ion channels involved in visceral afferent mechanosensitivity (Table 1). This includes a few members of the transient receptor potential (TRP) family (TRPV1, TRPV4, and TRP ankyrin 1 (TRPA1)) and the acid-sensing ion channels (ASICs, ASIC1, ASIC2, and ASIC3). The differences in mechanosensitive properties between gastroesophageal vagal afferents and colonic splanchnic and pelvic afferents and ion channels involved illustrate the complexity of the mechanotransduction process within extrinsic afferents. These differences may reflect the specific functions visceral afferent subtypes play within the gut; however, more detailed investigation is required.

#### 2.2.4.1 Transient receptor potential channels (TRPs)

The role of TRPV1 in visceral mechanosensitivity has been studied by numerous groups. TRPV1$^{-/-}$ mice exhibit deficits in gastric and jejunal vagal afferent mechanosensitivity [44,45]. For example, the response of gastric tension-sensitive afferents to circular stretch is significantly reduced in TRPV1$^{-/-}$ mice [45]. Consistent with this dampened signaling, there is also an increase in food intake in TRPV1$^{-/-}$ mice [45]. In the jejunum, TRPV1 deletion reduces the pressure-response curve of wide dynamic range fibers [116], and in the distal colon and rectum, TRPV1 deletion was shown to reduce the responses of pelvic muscular/mucosal afferents to mechanical stimulation [117]. However, the inhibitory effects are only observed in select populations of visceral afferent with no effect of TRPV1 deletion observed in gastric

**Table 1** Summary of the Effect of Ion Channel Deletion on Mechanosensitivity of Subtypes of Extrinsic Afferent Innervating the Gastrointestinal Tract and Functional Outcomes in the Whole Animal. ↓ Indicate a Reduction, ↑ Indicates in Increase, and ↔ Indicates No Change in Mechanosensitivity or Functional Outcome

| Ion channel disrupted | TRPV1 | TRPV4 | TRPA1 | ASIC1 | ASIC2 | ASIC3 |
|---|---|---|---|---|---|---|
| **Gastro-esophageal afferents** | | | | | | |
| Mucosal | ↔ | ↔ | ↓ | ↑ | ↑ | ↔ |
| Tension | ↓ | ↔ | ↔ | ↑ | ↓ | ↓ |
| **Jejunal afferents** | | | | | | |
| Low-threshold | ↔ | | | | | |
| Wide-dynamic range | ↓ | | | | | |
| High-threshold | ↔ | | | | | |
| **Colorectal splanchnic afferents** | | | | | | |
| Mucosal | | | | | | |
| Muscular | | | | | | |
| Serosal | | ↓ | ↓ | ↑ | ↑ | ↓ |
| Mesenteric | | ↓ | ↓ | ↑ | ↔ | ↓ |
| **Colorectal pelvic afferents** | | | | | | |
| Mucosal | ↔ | ↔ | ↓ | | | ↔ |
| Muscular/mucosal | ↓ | ↔ | ↔ | | | ↓ |
| Muscular | ↔ | ↔ | ↔ | | | ↔ |
| Serosal | ↔ | | | ↓ | ↓ | ↔ |
| **Functional effects** | | | | | | |
| Gastric emptying rate | ↔ | | | ↓ | ↔ | ↔ |
| Stool output | | | | ↔ | ↓ | ↔ |
| Visceromotor response to colorectal distention | ↓ | ↓ | ↓ | | | ↓ |

vagal mucosal afferents [45]; jejunal low- and high-threshold afferents [116]; and pelvic mucosal, muscular, and serosal afferents [117].

TRPV1 is not considered to be mechanically gated with effects on mechanosensation possibly due to effects on neuronal excitability or TRPV1 interactions with other ion channels/receptors. For example, the TRPV1 agonist capsaicin caused desensitization of splanchnic serosal afferents [118], an effect not observed in TRPA1$^{-/-}$ mice [119]. This suggests a link between TRPA1 and TRPV1 in the observed change in afferent function.

TRPV4 channels are localized in about 40% of gastric vagal afferent neurons [120], 60–80% of splanchnic colonic DRG neurons, and about 60% of pelvic colonic DRG neurons [120] with levels eight- and threefold higher in colonic thoracolumbar and lumbosacral neurons, respectively, compared with gastric nodose neurons [120]. Consistent with the low expression in gastric vagal neurons, deletion of TRPV4 had no effect on gastroesophageal vagal afferent mechanosensitivity [120]. In contrast, mechanosensitivity of colonic splanchnic serosal and mesenteric afferents was reduced [120]. Pelvic serosal afferents displayed similar deficits in mechanosensitivity to those in the splanchnic pathway. However, similar to the vagal pathway, mechanosensitivity of pelvic mucosal and muscular afferents was unaltered in TRPV4$^{-/-}$ mice [120]. The observed reduction in mechanosensitivity of subpopulations of colonic afferent translated to a decrease in visceromotor response to colorectal distention in TRPV4$^{-/-}$ mice. Reports suggest a strong interaction between TRPV4 and protease-activated receptor-2 (PAR$_2$) in visceral hypermechanosensitivity. TRPV4 and PAR$_2$ are anatomically and functionally expressed in neurons innervating the colon [121]. In addition, SLIGRL (PAR$_2$-activating peptide) sensitizes TRPV4-mediated currents in isolated colonic DRG neurons and activates splanchnic colonic serosal afferents in a TRPV4-dependent manner [122]. Further, it has been demonstrated that SLIGRL enhances the visceromotor response to colorectal distension in TRPV4$^{+/+}$ but not TRPV4$^{-/-}$ mice [121,122].

TRPA1 is localized in 50–60% of gastric vagal and colonic splanchnic and pelvic afferent neurons [119]. The mechanosensitivity of vagal and pelvic mucosal afferents is significantly reduced in TRPA1$^{-/-}$ mice. Deletion of TRPA1 also reduced the response to mechanical stimulation of high-threshold colonic pelvic and splanchnic mesenteric and serosal afferents [119]. In contrast, TRPA1 deletion did not alter the mechanosensitivity of gastric tension-sensitive or pelvic muscular or muscular/mucosal afferents. This suggests that TRPA1 is a peripheral mechanosensor in select afferent populations.

### 2.2.4.2 Acid-sensing ion channels (ASICs)

The potential mechanosensory properties of ASICs have also been investigated. ASIC1 is expressed in gastric and colonic neurons within the nodose and DRG neurons, respectively [123]. In ASIC1$^{-/-}$ mice, there is an increase on the mechanosensitivity of gastroesophageal vagal mucosal and tension-sensitive afferents. This corresponds with a delay in gastric emptying (vagally mediated process) in ASIC1$^{-/-}$ mice [124,125]. Similarly, the mechanosensitivity of colonic splanchnic mesenteric and serosal afferents was also demonstrated to increase in ASIC1$^{-/-}$ mice [124,125].

ASIC2 is expressed in colonic splanchnic [126] and gastric vagal [123] afferent neurons. In ASIC2$^{-/-}$ mice, there is an increase in the mechanosensitivity of gastroesophageal mucosal afferents, similar to changes observed in ASIC1$^{-/-}$ mice. In contrast, the mechanosensitivity of gastroesophageal tension receptors was significantly reduced in ASIC2$^{-/-}$ mice [123]. In addition, there was no change in the rate of gastric emptying in these mice [123]. Similarly, population differences have also been observed in the colon. Colonic splanchnic mesenteric afferent mechanosensitivity was unchanged in ASIC2$^{-/-}$ mice, whereas there was heightened sensitivity of the serosal afferents compared with wild-type controls [123].

Finally, ASIC3 is also expressed in colonic DRG and gastric vagal afferent neurons with expression about 4.5-fold higher in colonic DRG compared with gastric vagal afferents [123]. It has been demonstrated that gastric vagal afferent mucosal mechanosensitivity is unaffected by deletion of ASIC3, whereas the mechanosensitivity of the tension-sensitive afferents is dampened [44,124]. In the colon, deletion of ASIC3 caused a significant reduction in the mechanosensitivity of splanchnic mesenteric and serosal afferents [124] as well as pelvic muscular/mucosal afferents [117].

## 2.3 INTRINSIC INNERVATION OF THE GI TRACT

### 2.3.1 Anatomy of intrinsic innervation

Intrinsic nerves are located in the ENS that consists of ganglionic nerve plexuses running between the inner circular and outer longitudinal muscle layers (myenteric plexus) and in the submucosa (submucosal plexus). The ENS receives input from the CNS but can function autonomously. It contains reflex pathways that regulate a number of functions necessary for digestion and absorption of nutrients, including pathways involved in secretion, motility, and blood flow. Within the ENS, there are sensory neurons that respond to the mechanical and/or chemical environment initiating reflex pathways. In this chapter, attention is focused on the mechanosensory properties of these intrinsic sensory neurons. The vast majority of research on enteric sensory transmission in based on research using the guinea-pig ileum [127] where, based on morphology, two types of enteric neuron have been described. Dogiel type I is long filamentous neurons having a single long process, while Dogiel type II is large neurons with several long processes [128]. Based on electrophysiological properties, two main types of sensory neuron have been described [127–130]. AH neurons exhibit a prominent after hyperpolarization (AH) spike, and S-neurons receive fast excitatory synaptic potentials. S and AH neurons are predominantly Dogiel type I and II, respectively [129].

### 2.3.2 Mechanosensory properties of intrinsic neurones

Morphology and location, parallel to muscle layers, make intrinsic sensory neurons ideally positioned to detect mechanical forces generated by the gut wall. Mechanosensitive enteric neurons initiate reflex activity by responding to the constant mechanical forces (e.g., due to muscle contraction and relaxation) present in the gut wall [131,132]. Enteric mechanosensitive neurons have been subclassified into rapidly, slowly, and ultraslowly adapting neurons (RAMEN, SAMEN, and USAMEN, respectively) [133]. There are three types of mechanical stimuli: shear force (deformation by movement along a parallel plane), compressive force (compression of the neuron), and tensile stress (stretching or lengthening of a neuron). Only about 8% of enteric neurons respond to shear stress with a very low frequency of action potential generation even at high flow/shear levels [134]. Subtypes of SAMEN, RAMEN, and USAMEN all respond to compression [134]. True tension-sensitive mechanosensitive enteric neurons have been identified in the guinea-pig distal colon [135] and belong to the population of USAMEN. Tension-sensitive mechanosensitive enteric neurons also exist in the guinea-pig ileum, stomach [131,136], and esophagus [137].

Generally, AH neurons are considered mechano (or chemo)-sensory neurons, whereas S-neurons are ascending and descending interneurons and excitatory and motor neurons [128,138]. As a consequence, the AH neurons have been named intrinsic primary afferent neurons (IPANs), the first neurons in the reflex arc that respond to mechanical stimuli by firing action potentials and initiating the reflex process [129]. IPANs in the guinea-pig small intestine were initially considered to be stretch-sensitive myenteric neurons [139]. However, it is now accepted that stretch of the muscle wall causes muscle contraction, which subsequently causes deformation of IPANs resulting in action potential generation [139,140]. Other subtypes of enteric neuron have since been found to be mechanosensitive. For example, in the guinea-pig small intestine, a small number of Dogiel type I neurons [139,140] and, more recently, a large proportion of inter- and motor neurons have been shown to be mechanosensitive [131]. In addition, the ENS in the gastric corpus, although devoid of typical AH neurons, orchestrates mechanically induced motor reflexes [70]. Further, in the distal colon, stretch-sensitive S-neurons have been described [135].

When considering transduction mechanisms, it has been demonstrated that large-conductance potassium (BK-like) channels in the soma of IPANs are mechanosensitive [141]. Direct mechanical stimulation of neuronal processes evokes action potential generation, whereas direct stimulation of the soma inhibits action potential generation consistent with activation of BK-like channels. However, although some S-neurons are mechanosensitive [131], BK-like channels have not, thus far, been identified in S-neurons. Therefore, it is likely that different transduction mechanisms occur in different mechanosensory neuron populations and even in different regions of the nerve.

### 2.3.2.1 Plasticity of intrinsic mechanosensitive neurones in disease
Evidence suggests that the structure and neurochemical composition of the ENS are altered in inflammatory conditions such as IBD [142,143]. Although the number of mucosal and muscular nerve fibers increase, there is also evidence of axon degeneration in inflammatory conditions, such as Crohn's disease [144,145]. Reports on the change in myenteric neuron number vary considerably with reports of increased [146,147], decreased [148], or no change [146,149]. There are no differences in the number of ganglionic cells in the submucosal plexus of IBD patients [150]. This is

despite the close proximity of the submucosal plexus to the lamina propria, where inflammation is present in both ulcerative colitis and Crohn's disease, and the fact that inflammation of the submucosal and the myenteric plexus is common in Crohn's disease and ulcerative colitis [145,150–153].

Functionally, there is evidence of hyperexcitability of myenteric AH neurons in inflammatory conditions. In the laboratory setting, hypersensitivity of myenteric AH neurons, causing gut dysmotility, has been observed in both trinitrobenzenesulfonic acid (TNBS)- and *Trichinella spiralis*-induced inflammatory models [154,155]. The TNBS-induced effects are cyclooxygenase (COX)-2-dependent [156], whereas the *T. spiralis*-induced hyperexcitability can be suppressed using histamine, adenylate cyclase, COX, or leukotriene pathway blockers [154]. Using a guinea-pig TNBS colitis model, it has been demonstrated that inhibition of the hyperpolarization-activated cyclic nucleotide-gated channels, which contribute to AH neuron excitability, normalizes motility in the ulcerated regions [157] and increases motility at sites either side of the ulcerated region [158]. Overall, these data strongly suggest a relationship between neuronal excitability and dysmotility. This is not confined to the colon. Ileitis has also been shown to alter synaptic transmission and AH neuron excitability [159]. Inflammation in one area of the gut can affect other noninflamed regions. For example, inflammation of the guinea-pig ileum induces hypersensitivity of submucosal AH neurons in the distal colon [160]. Colitis-induced changes in the submucosal plexus of the ileum are very different with minimal effects on AH neuron hyperexcitability [161]. Conversely, colitis increases the excitability of AH neurons in the myenteric plexus of guinea-pig ileum, although the mechanisms driving this hyperexcitability are different from the increased excitability observed in ileitis [162]. Hyperexcitability of submucosal and myenteric AH neurons persists into the post-inflammatory state [162–164] and could explain the continued dysmotility observed in IBD remission patients.

## 2.4 INTERSTITIAL CELLS OF CAJAL
### 2.4.1 Structure and organisation of interstitial cells of Cajal
ICCs are found throughout the GI tract from the esophagus to the anal sphincter [165–167]. They have been classified into several subtypes based on their distribution patterns and morphology [168] (Table 2).

ICCs of the myenteric plexus (MP) are considered as pacemakers, initiating and spreading electric rhythm [169]. Intramuscular (IM) and deep muscular plexus (DMP) ICCs have a close relationship with nerve terminals and play a critical role in enteric neuromuscular transmission [182,183]. The septa isolates muscle bundles, and the ICCs (ICC-SEP) found in the septa are thought to provide a pathway to enable pacemaker activity to penetrate into the muscle [175,184].

### 2.4.2 Mechanosensory properties of interstitial cells of Cajal
A sensory role was initially proposed for ICCs based on morphology and their relationship with nerve endings [173]. ICC-IMs were suggested to be mechanoreceptors associated with both vagal and spinal afferent nerve endings in the upper GI tract and

**Table 2** Subtypes of Interstitial Cells of Cajal Based on Structure and Morphology. ICC-CM and ICC-LM Are Sometimes Grouped Together and Collectively Termed ICC-Intramuscular (IM)

| Subclass on Interstitial Cell of Cajal (ICC) | Structure/Morphology |
| --- | --- |
| ICC of the myenteric plexus (ICC-MP) [169] | • Multipolar with 3–5 primary cytoplasmic processes that project secondary, tertiary, and further branching to interact with counterparts<br>• Form cellular network around the myenteric plexus between the circular and longitudinal muscle layers<br>• Throughout GI tract but fewer in number in the gastric corpus and colon than the small intestine |
| ICC of the circular muscle (ICC-CM) [170] | • Mainly bipolar cells orientated along the long axis of surrounding smooth muscle cells<br>• Small intestine, secondary cytoplasmic branches often observed but ICCs are sparsely distributed<br>• Stomach and colon, simple elongated spindle shape densely distributed along nerve bundles<br>• Also found in the connective tissue septa (ICC-SEP) [171,172] |
| ICC of the longitudinal muscle (ICC-LM) [170] | • Similar in cell shape to ICC-CM but less numerous than ICC-CM |
| ICC of the deep muscular plexus (ICC-DMP) [173,174] | • Multipolar cells but the majority of processes show a unidirectional orientation along the circumference due to the close association with nerve bundles and circular muscle fibers |
| ICC of the submucosa and submucosal plexus (ICC-SM and ICC-SMP) [175–179] | • Cell axis parallel with adjacent circular muscle cells<br>• Multipolar cells with secondary processes forming a loose network<br>• Found at interface between the submucosal connective tissue and the innermost circular muscle layer |
| ICC of the subserosa (ICC-SS) [180,181] | • Stellate ICCs in the subserosal layer |

likely involved in extrinsic reflexes [32,33,185–187]. There is also evidence for mechanosensitive ICC-IM involvement in intrinsic motor reflexes [188]. Stretch of the gastric muscle was demonstrated to increase the basic electric rhythm, irrespective of the presence of neuronal blockers, suggesting a nonneuronal component [188]. In addition, in C-kit mutant mice that do not express ICC-IMs, gastric stretch did not affect the pacemaker activity [188]. Therefore, the ICC-IMs are likely responsible for detecting the stretch stimuli and increasing pacemaker activity.

The molecular mechanisms of mechanotransduction are less clear. Sodium channels are known to regulate the excitability of cells. A tetrodotoxin-insensitive voltage-dependent sodium channel has been detected in ICCs [189]. In addition, mechanical force (i.e., shear stress) applied to isolated ICCs has been shown to activate these sodium currents [189]. Further, mutations in the pore-forming α-subunit (encoded by SCN5A ($Na_v1.5$)) of this sodium channel have been shown to be associated with abdominal pain and a higher prevalence of the gastrointestinal symptom complexes (IBS, dyspepsia, constipation, diarrhea, abdominal pain, or reflux) [190].

### 2.4.2.1 Interstitial cells of Cajal and motility disorders

ICCs play a critical role in GI motility. They generate pacemaker activity, in the form of a slow wave, that drives propulsive rhythmic contractile activity [191]. The slow wave is a continuous, undulating change in membrane potential that propagates through the GI musculature inducing and organizing phasic contractions [192]. The mechanosensory properties of ICCs contribute to this process. It has been demonstrated that, in the presence of tetrodotoxin (to block neuronal sodium channels), the mechanosensitive sodium channel blocker QX314 hyperpolarized smooth muscle membrane potentials, decreased slow wave frequency, and slowed the rate of rise of the slow wave [189]. In contrast, stretch of human intestinal circular muscle increased slow wave frequency [189]. These data suggest a physiological role for the mechanosensitive sodium channel in ICCs in GI motility. Therefore, dysfunction of the mechanosensitive properties of ICCs could be implicated in many motility disorders (e.g., gastroparesis). Patients with gastroparesis have real or perceived gastric retention that is accompanied by symptoms of nausea, vomiting, and/or bloating. High-resolution mapping of electric activity has established abnormalities in pacemaker activity in gastroparesis, such as changes in pacemaker frequency, ectopic pacemakers, and conduction block [193]. ICCs are fundamental in the generation of gastric slow waves [194]. A decrease in ICC density has been observed in diabetic gastroparesis patients [195]. This loss of ICC-IMs in the gastric fundus could explain the low gastric tone and increased compliance. In addition, the hypomotility of the antrum could be explained by the absence of ICC slow wave generation [196]. Alterations in the mechanosensory properties of ICCs could also contribute to the changes in motility observed in gastroparesis; however, this remains to be determined.

In slow transit constipation, there is slower than normal movement of contents from the cecum to the rectum [197]. A reduction in ICCs has been observed in patients with slow transit constipation [198]. It has also been suggested that disruption of ICC electric activity may play a role in the reduced colonic motor activity observed in patients with slow transit constipation [199]. However, as with all GI motor disorders, the role if any that mechanosensitive ICCs play in this process requires further investigation.

## 2.5 SMOOTH MUSCLE CELLS

The muscular layers of the GI tract consist of an inner circular layer and an outer longitudinal layer. In addition, there is a thin layer of smooth muscle (muscularis mucosae) adjacent to the lamina propria in the mucosal layer. The main role of smooth muscle is to propel food along the GI tract by means of coordinated contractions.

Most studies on mechanosensitivity of smooth muscle cells have been performed using freshly dissociated or cultured smooth muscle cells using mechanical forces such as hypotonic cell swelling, direct pressure, and shear stress [200–202]. Each method has not only its own list of advantages but also a list of disadvantages and, therefore, should not be used in isolation. For example, the response to hypertonic swell is not equivalent to mechanosensitivity as the response may be due to, for example, changes in osmolarity. Therefore, hypotonic cell swelling cannot be used as the sole stimulus to determine mechanosensitivity. In the few heterologous studies that are available, there is the issue of whether the smooth muscle cells are mechanosensitive or whether other cells within the system form the mechanosensitive component (e.g., ICCs). Nonetheless, there are several mechanosensitive ion channels expressed on GI smooth muscle cells.

Mechanosensation at the level of the smooth muscle cells was first observed with the response to stretch in the absence of neuronal input [203]. In addition, smooth muscle swelling in response to a hypotonic solution regulates ionic conductance, including outward rectifying $Cl^-$ currents [204], a nonselective cation current [205] and a $Ca^{2+}$ current [204]. Contraction of smooth muscle cells requires the influx of calcium through voltage-sensitive L-type $Ca^{2+}$ channels [206–208]. A variety of mechanical forces, including shear stress, swelling, and membrane stretch, increase L-type $Ca^{2+}$ channel activity, and therefore, it is highly probably that these channels are mechanosensitive in GI smooth muscle cells [209–211].

Similar to ICCs, mechanosensitive sodium channels have been detected in GI smooth muscle cells [206]. Shear stress has been shown to activate a tetrodotoxin-resistant, voltage-dependent sodium current in smooth muscle cells [212]. In addition, the pore-forming α-subunit (encoded by SCN5A) has been detected in smooth muscle cells [213–215]. These mechanosensitive sodium channels are not essential for slow wave and action potential generation [206], and it has been suggested that channel effects on the membrane potential influence smooth muscle contractions through the recruitment of L-type $Ca^{2+}$ channels [212,214].

In some regions of the GI tract, smooth muscle cells do not contract and may even relax in response to stretch [216,217]. For example, the proximal stomach and colon act as storage sites and relax in response to stretch. Smooth muscle cells express a wide variety of potassium channels [207–209], and stretch-dependent $K^+$ (SDK) channels have been characterized [216]. SDK channels are thought to be TREK-1 channels [217,218]. It has been proposed that the SDK channels contribute to membrane potential with mechanical activation of this channel inducing membrane hyperpolarization and thus relaxation [217,219]. In the colon, large-conductance calcium-activated potassium channels (BKCa) are also activated by stretch [220] that produces a large outward current, hyperpolarizing the membrane and decreasing smooth muscle cell excitability. Therefore, similar to SDK channels, BKCa channels may participate in the storage function of the colon.

Dysfunction of smooth muscle cell mechanotransduction could contribute to numerous motility disorders. In animal models and humans, inflammation of the GI tract causes dysmotility possibly reflecting changes in smooth muscle function [221,222]. L-type calcium channels are crucial for the initiation of smooth muscle contraction, and therefore, any modification in this channel will impact on motility. Disturbance in L-type calcium channel activity has been reported to reduce intestinal motility [223–227]; however, the final link with mechanosensation still remains to be determined.

## 2.6 ENDOCRINE CELLS

The epithelium of the mucosa is the innermost layer of the GI tract and is in direct contact with luminal contents. The epithelium is exposed to continual mechanical stimulation through exposure to food and digestive material and also motor activity. Bulbring and colleagues were the first to put forward the idea of mucosal mechanotransducers and the role of 5-HT as an intermediate in mechanically evoked motor responses [228–230]. This has been confirmed and extended in subsequent studies. Mechanical stimulation of the mucosa has been shown to elicit motor reflexes via activation of IPANS in the enteric circuitry [231], and $5-HT_3$ antagonists block the peristaltic reflex initiated by mechanical stimulation of the mucosa [232]. Therefore, mechanical stimulation of the mucosa induces release of 5-HT from the epithelial cells, which activates IPAN endings to initiate a motor reflex [233] (Fig. 5). Within the last decade, it has been demonstrated that 5-HT release from enterochromaffin cells is important for the propulsion of colonic content [234–236]. The enterochromaffin cell has developmental [237–240] and functional [241–243] similarities to the Merkel cells (light touch sensors in the skin). The mechanosensitive ion channel Piezo2 is critical for Merkel cell mechanosensitivity [244–246]. Piezo2 is also expressed in enterochromaffin cells of the human and mouse small bowel, and stretch-induced 5-HT release was blocked by inhibition of Piezo2 [247]. Therefore, similar to Merkel cells, Piezo2 is an important enterochromaffin cell mechanosensitive ion channel (Fig. 5). However, Piezo2 likely works in concert with other mechanosensitive channels. For example, the calcium-permeable nonselective cation channel TRPA1 is highly expressed in enterochromaffin cells [248]. Further, TRPA1 agonists increase 5-HT release form enterochromaffin cells [248] and stimulate motility [249], mediated via 5-HT action at $5-HT_3$ receptors [248]. Therefore, TRPA1 likely plays a role in enterochromaffin mechanosensation (Fig. 5).

The role of 5-HT is not only limited to motor reflexes but also includes involvement in mucosal secretory reflexes and the transmission of visceral pain sensation [250–252]. Evidence suggests that alterations in 5-HT regulation and signaling mechanisms contribute to the pathogenesis of a number of diseases including IBD, IBS, vomiting, and diarrhea [155,253–261]. In addition, there are possible

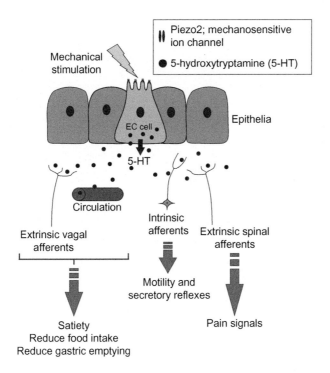

**FIG. 5**

Schematic representation of the outcomes of mechanical stimulation of the mechanosensitive Piezo2 channel on enterochromaffin cells in the gastrointestinal epithelia. Mechanical stimulation of the Piezo2 channel initiates an intracellular process that culminates in the release of 5-HT that can either enter the circulation or act on intrinsic/extrinsic neurons to elicit a physiological response.

associations with celiac disease, diverticulitis, and colorectal cancer [262]. However, the mechanosensitive properties of endocrine cells in disease conditions have not been established, and therefore, a specific role of endocrine cell mechanosensitivity in disease remains to be determined.

## 3 CONCLUSION

This chapter has outlined the huge number of mechanosensory processes that occur in the gut, from the extrinsic nerves innervating the gut to the enterochromaffin cells in the gut epithelia. Significant research has been undertaken in the quest to understand the basic fundamental mechanisms that drive extrinsic afferent mechanosensation and how this is altered in disease (i.e., IBS and obesity). However, further research is required for a more comprehensive understanding of extrinsic afferent physiology and pathophysiology before more effective treatment strategies can be established. The mechanosensory properties of other cell types in the GI tract may also play a significant role in physiology and pathophysiology; however, apart from knowledge that certain cell types (e.g., enterochromaffin cells) are mechanosensitive, the potential role of this mechanosensation in disease states remains elusive.

## REFERENCES

[1] Gillespie PG, Walker RG. Molecular basis of mechanosensory transduction. Nature 2001;413(6852):194–202.

[2] Welsh MJ, Price MP, Xie J. Biochemical basis of touch perception: mechanosensory function of degenerin/epithelial Na+ channels. J Biol Chem 2002;277(4):2369–72.

[3] Brierley SM. Molecular basis of mechanosensitivity. Auton Neurosci 2010;153(1-2):58–68.

[4] Berthoud HR, Blackshaw LA, Brookes SJH, Grundy D. Neuroanatomy of extrinsic afferents supplying the gastrointestinal tract. Neurogastroenterol Motil 2004;16:28–33.

[5] Zhong F, Christianson JA, Davis BM, Bielefeldt K. Dichotomizing axons in spinal and vagal afferents of the mouse stomach. Dig Dis Sci 2008;53(1):194–203.

[6] Sharkey KA, Williams RG, Dockray GJ. Sensory substance P innervation of the stomach and pancreas. Demonstration of capsaicin-sensitive sensory neurons in the rat by combined immunohistochemistry and retrograde tracing. Gastroenterology 1984;87(4):914–21.

[7] Green T, Dockray GJ. Calcitonin gene-related peptide and substance P in afferents to the upper gastrointestinal tract in the rat. Neurosci Lett 1987;76(2):151–6.

[8] Tan LL, Bornstein JC, Anderson CR. Distinct chemical classes of medium-sized transient receptor potential channel vanilloid 1-immunoreactive dorsal root ganglion neurons innervate the adult mouse jejunum and colon. Neuroscience 2008;156(2):334–43.

[9] Booth CE, Shaw J, Hicks GA, Kirkup AJ, Winchester W, Grundy D. Influence of the pattern of jejunal distension on mesenteric afferent sensitivity in the anaesthetized rat. Neurogastroenterol Motil 2008;20(2):149–58.

[10] Wang FB, Powley TL. Topographic inventories of vagal afferents in gastrointestinal muscle. J Comp Neurol 2000;421(3):302–24.

[11] Chen BN, Olsson C, Sharrad DF, Brookes SJ. Sensory innervation of the guinea pig colon and rectum compared using retrograde tracing and immunohistochemistry. Neurogastroenterol Motil 2016;28(9):1306–16.

[12] Herrity AN, Rau KK, Petruska JC, Stirling DP, Hubscher CH. Identification of bladder and colon afferents in the nodose ganglia of male rats. J Comp Neurol 2014;522(16):3667–82.

[13] Berthoud HR, Powley TL. Vagal afferent innervation of the rat fundic stomach: morphological characterization of the gastric tension receptor. J Comp Neurol 1992;319:261–76.

[14] Kressel M, Berthoud HR, Neuhuber WL. Vagal innervation of the rat pylorus: an anterograde tracing study using carbocyanine dyes and laser scanning confocal microscopy. Cell Tissue Res 1994;275(1):109–23.

[15] Fox EA, Phillips RJ, Martinson FA, Baronowsky EA, Powley TL. Vagal afferent innervation of smooth muscle in the stomach and duodenum of the mouse: morphology and topography. J Comp Neurol 2000;428(3):558–76.

[16] Berthoud HR, Patterson LM, Neumann F, Neuhuber WL. Distribution and structure of vagal afferent intraganglionic laminar endings (IGLEs) in the rat gastrointestinal tract. Anat Embryol 1997;195(2):183–91.

[17] Page AJ, Martin CM, Blackshaw LA. Vagal mechanoreceptors and chemoreceptors in mouse stomach and esophagus. J Neurophysiol 2002;87:2095–103.

[18] Berthoud H-R, Kressel M, Raybould H, Neuhuber W. Vagal sensors in the rat duodenal mucosa: distribution and structure as revealed by in vivo DiI-tracing. Anat Embryol 1995;191(3):203–12.

[19] Berthoud HR, Patterson LM. Anatomical relationship between vagal afferent fibers and CCK-immunoreactive entero-endocrine cells in the rat small intestinal mucosa. Acta Anat (Basel) 1996;156(2):123–31.

[20] Williams RM, Berthoud HR, Stead RH. Vagal afferent nerve fibres contact mast cells in rat small intestinal mucosa. Neuroimmunomodulation 1997;4(5-6):266–70.

[21] Page AJ, Blackshaw LA. An in vitro study of the properties of vagal afferent fibres innervating the ferret oesophagus and stomach. J Physiol 1998;512:907–16.

[22] Powley TL, Spaulding RA, Haglof SA. Vagal afferent innervation of the proximal gastrointestinal tract mucosa: chemoreceptor and mechanoreceptor architecture. J Comp Neurol 2011;519(4):644–60.

[23] Clarke GD, Davison JS. Mucosal receptors in the gastric antrum and small intestine of the rat with afferent fibres in the cervical vagus. J Physiol 1978;284(1):55–67.

[24] Becker JM, Kelly KA. Antral control of canine gastric emptying of solids. Am J Physiol Gastrointest Liver Physiol 1983;8:334–8.

[25] Andrews PL, Wood KL. Vagally mediated gastric motor and emetic reflexes evoked by stimulation of the antral mucosa in anaesthetized ferrets. J Physiol 1988;395:1–16.

[26] Blackshaw LA, Grundy D, Scratcherd T. Vagal afferent discharge from gastric mechanoreceptors during contraction and relaxation of the ferret corpus. J Auton Nerv Syst 1987;18(1):19–24.

[27] Sengupta JN, Kauvar D, Goyal RK. Characteristics of vagal esophageal tension-sensitive afferent fibers in the opossum. J Neurophysiol 1989;61(5):1001–10.

[28] Iggo A. Tension receptors in the stomach and the urinary bladder. J Physiol 1955;128 (3):593–607.

[29] Zagorodnyuk VP, Chen BN, Brookes SJH. Intraganglionic laminar endings are mechano-transduction sites of vagal tension receptors in the guinea-pig stomach. J Physiol 2001;534(1):255–68.

[30] Sengupta JN, Saha JK, Goyal RK. Stimulus-response function studies of esophageal mechanosensitive nociceptors in sympathetic afferents of opossum. J Neurophysiol 1990;64(3):796–812.

[31] Powley TL, Hudson CN, McAdams JL, Baronowsky EA, Phillips RJ. Vagal intramuscular arrays: the specialized mechanoreceptor arbors that innervate the smooth muscle layers of the stomach examined in the rat. J Comp Neurol 2016;524(4):713–37.

[32] Powley TL, Wang X, Fox EA, Phillips RJ, Liu LWC, Huizinga JD. Ultrastructural evidence for communication between intramuscular vagal mechanoreceptors and interstitial cells of Cajal in the rat fundus. Neurogastroenterol Motil 2008;20(1):69–79.

[33] Powley TL, Phillips RJ. Vagal intramuscular array afferents form complexes with interstitial cells of Cajal in gastrointestinal smooth muscle: analogues of muscle spindle organs? Neuroscience 2011;186:188–200.

[34] Phillips RJ, Powley TL. Tension and stretch receptors in gastrointestinal smooth muscle: re-evaluating vagal mechanoreceptor electrophysiology. Brain Res Rev 2000;34(1–2):1–26.
[35] Andrews PL, Grundy D, Scratcherd T. Vagal afferent discharge from mechanoreceptors in different regions of the ferret stomach. J Physiol 1980;298(1):513–24.
[36] Kollarik M, Ru F, Undem BJ. Acid-sensitive vagal sensory pathways and cough. Pulm Pharmacol Ther 2007;20(4):402–11.
[37] Brierley SM, RCW 3rd J, Gebhart GF, Blackshaw LA. Splanchnic and pelvic mechanosensory afferents signal different qualities of colonic stimuli in mice. Gastroenterology 2004;127(1):166–78.
[38] Feinle-Bisset C. Upper gastrointestinal sensitivity to meal-related signals in adult humans–relevance to appetite regulation and gut symptoms in health, obesity and functional dyspepsia. Physiol Behav 2016;162:69–82.
[39] Daly DM, Park SJ, Valinsky WC, Beyak MJ. Impaired intestinal afferent nerve satiety signalling and vagal afferent excitability in diet induced obesity in the mouse. J Physiol 2011;589(11):2857–70.
[40] Kentish S, Li H, Philp LK, O'Donnell TA, Isaacs NJ, Young RL, et al. Diet-induced adaptation of vagal afferent function. J Physiol 2012;590(1):209–21.
[41] Westerterp-Plantenga MS, Smeets A, Lejeune MP. Sensory and gastrointestinal satiety effects of capsaicin on food intake. Int J Obes 2005;29(6):682–8.
[42] Furnes M, Zhao C-M, Chen D. Development of obesity is associated with increased calories per meal rather than per day. A study of high-fat diet-induced obesity in young rats. Obes Surg 2009;19(10):1430–8.
[43] Kentish SJ, O'Donnell TA, Frisby CL, Li H, Wittert GA, Page AJ. Altered gastric vagal mechanosensitivity in diet-induced obesity persists on return to normal chow and is accompanied by increased food intake. Int J Obes 2014;38(5):636–42.
[44] Bielefeldt K, Davis BM. Differential effects of ASIC3 and TRPV1 deletion on gastroesophageal sensation in mice. Am J Physiol Gastrointest Liver Physiol 2008;294(1): G130–8.
[45] Kentish SJ, Frisby CL, Kritas S, Li H, Hatzinikolas G, O'Donnell TA, et al. TRPV1 channels and gastric vagal afferent signalling in lean and high fat diet induced obese mice. PLoS One 2015;10(8):e0135892.
[46] Kentish SJ, O'Donnell TA, Isaacs NJ, Young RL, Li H, Harrington AM, et al. Gastric vagal afferent modulation by leptin is influenced by food intake status. J Physiol 2013;591(7):1921–34.
[47] Kentish SJ, Frisby CL, Kennaway DJ, Wittert GA, Page AJ. Circadian variation in gastric vagal afferent mechanosensitivity. J Neurosci 2013;33(49):19238–42.
[48] Rosenwasser AM, Boulos Z, Terman M. Circadian organization of food intake and meal patterns in the rat. Physiol Behav 1981;27(1):33–9.
[49] Kentish SJ, Vincent AD, Kennaway DJ, Wittert GA, Page AJ. High-fat diet-induced obesity ablates gastric vagal afferent circadian rhythms. J Neurosci 2016;36(11):3199–207.
[50] Tack J, Talley NJ, Camilleri M, Holtmann G, Hu P, Malagelada JR, et al. Functional gastroduodenal disorders. Gastroenterology 2006;130(5):1466–79.
[51] Bisschops R, Karamanolis G, Arts J, Caenepeel P, Verbeke K, Janssens J, et al. Relationship between symptoms and ingestion of a meal in functional dyspepsia. Gut 2008;57(11):1495–503.
[52] Sarnelli G, Caenepeel P, Geypens B, Janssens J, Tack J. Symptoms associated with impaired gastric emptying of solids and liquids in functional dyspepsia. Am J Gastroenterol 2003;98(4):783–8.

[53] Tack J, Piessevaux H, Coulie B, Caenepeel P, Janssens J. Role of impaired gastric accommodation to a meal in functional dyspepsia. Gastroenterology 1998;115 (6):1346–52.
[54] Pauwels A, Altan E, Tack J. The gastric accommodation response to meal intake determines the occurrence of transient lower esophageal sphincter relaxations and reflux events in patients with gastro-esophageal reflux disease. Neurogastroenterol Motil 2014;26(4):581–8.
[55] Yu S, Undem BJ, Kollarik M. Vagal afferent nerves with nociceptive properties in guinea-pig oesophagus. J Physiol 2005;563(Pt 3):831–42.
[56] Spencer NJ, Kyloh M, Duffield M. Identification of different types of spinal afferent nerve endings that encode noxious and innocuous stimuli in the large intestine using a novel anterograde tracing technique. PLoS One 2014;9(11):e112466.
[57] Spencer NJ, Kyloh M, Beckett EA, Brookes S, Hibberd T. Different types of spinal afferent nerve endings in stomach and esophagus identified by anterograde tracing from dorsal root ganglia. J Comp Neurol 2016;524(15):3064–83.
[58] Janig W, Koltzenburg M. On the function of spinal primary afferent fibres supplying colon and urinary bladder. J Auton Nerv Syst 1990;30(Suppl):S89–96.
[59] Janig W, Koltzenburg M. Receptive properties of sacral primary afferent neurons supplying the colon. J Neurophysiol 1991;65(5):1067–77.
[60] Cervero F, Janig W. Visceral nociceptors: a new world order? Trends Neurosci 1992;15 (10):374–8.
[61] Sengupta JN, Gebhart GF. Characterization of mechanosensitive pelvic nerve afferent fibers innervating the colon of the rat. J Neurophysiol 1994;71(6):2046–60.
[62] Ozaki N, Gebhart GF. Characterization of mechanosensitive splanchnic nerve afferent fibers innervating the rat stomach. Am J Physiol Gastrointest Liver Physiol 2001;281 (6):G1449–59.
[63] Hughes PA, Brierley SM, Martin CM, Brookes SJ, Linden DR, Blackshaw LA. Post-inflammatory colonic afferent sensitisation: different subtypes, different pathways and different time courses. Gut 2009;58(10):1333–41.
[64] Brierley SM, Jones 3rd RC, Xu L, Gebhart GF, Blackshaw LA. Activation of splanchnic and pelvic colonic afferents by bradykinin in mice. Neurogastroenterol Motil 2005;17 (6):854–62.
[65] Lin C, Al-Chaer ED. Long-term sensitization of primary afferents in adult rats exposed to neonatal colon pain. Brain Res 2003;971(1):73–82.
[66] Lynn PA, Blackshaw LA. In vitro recordings of afferent fibres with receptive fields in the serosa, muscle and mucosa of rat colon. J Physiol 1999;518(Pt 1):271–82.
[67] Sengupta JN. Visceral pain: the neurophysiological mechanism. Handb Exp Pharmacol 2009;194:31–74.
[68] Gue M, Junien JL, Bueno L. The kappa agonist fedotozine modulates colonic distention-induced inhibition of gastric motility and emptying in dogs. Gastroenterology 1994;107 (5):1327–34.
[69] De Schepper HU, De Man JG, Van Nassauw L, Timmermans JP, Herman AG, Pelckmans PA, et al. Acute distal colitis impairs gastric emptying in rats via an extrinsic neuronal reflex pathway involving the pelvic nerve. Gut 2007;56(2):195–202.
[70] Blackshaw LA, Brookes SJ, Grundy D, Schemann M. Sensory transmission in the gastrointestinal tract. Neurogastroenterol Motil 2007;19(1 Suppl):1–19.
[71] Merskey H, Bogduk N. Classification of chronic pain. In: Merskey H, Bogduk N, editors. Part III: Pain terms, a current list with definitions and notes on usage seattle. Seattle: IASP Press; 1994. p. 209–14.

[72] Chen SL, Wu XY, Cao ZJ, Fan J, Wang M, Owyang C, et al. Subdiaphragmatic vagal afferent nerves modulate visceral pain. Am J Physiol Gastrointest Liver Physiol 2008;294(6):G1441–9.

[73] Gschossmann JM, Mayer EA, Miller JC, Raybould HE. Subdiaphragmatic vagal afferent innervation in activation of an opioidergic antinociceptive system in response to colorectal distension in rats. Neurogastroenterol Motil 2002;14(4):403–8.

[74] Podolsky DK. Inflammatory bowel disease. N Engl J Med 2002;347(6):417–29.

[75] Baumgart DC, Sandborn WJ. Crohn's disease. Lancet 2012;380(9853):1590–605.

[76] Nell S, Suerbaum S, Josenhans C. The impact of the microbiota on the pathogenesis of IBD: lessons from mouse infection models. Nat Rev Microbiol 2010;8(8):564–77.

[77] Kaser A, Zeissig S, Blumberg RS. Inflammatory bowel disease. Annu Rev Immunol 2010;28:573–621.

[78] Vermeulen W, De Man JG, Pelckmans PA, De Winter BY. Neuroanatomy of lower gastrointestinal pain disorders. World J Gastroenterol 2014;20(4):1005–20.

[79] Woolf CJ. Central sensitization: implications for the diagnosis and treatment of pain. Pain 2011;152(3 Suppl):S2–15.

[80] Rao SS, Read NW, Davison PA, Bannister JJ, Holdsworth CD. Anorectal sensitivity and responses to rectal distention in patients with ulcerative colitis. Gastroenterology 1987;93(6):1270–5.

[81] Farthing MJ, Lennard-jones JE. Sensibility of the rectum to distension and the anorectal distension reflex in ulcerative colitis. Gut 1978;19(1):64–9.

[82] Bernstein CN, Niazi N, Robert M, Mertz H, Kodner A, Munakata J, et al. Rectal afferent function in patients with inflammatory and functional intestinal disorders. Pain 1996;66(2-3):151–61.

[83] Mayer EA, Berman S, Suyenobu B, Labus J, Mandelkern MA, Naliboff BD, et al. Differences in brain responses to visceral pain between patients with irritable bowel syndrome and ulcerative colitis. Pain 2005;115(3):398–409.

[84] Chang L, Munakata J, Mayer EA, Schmulson MJ, Johnson TD, Bernstein CN, et al. Perceptual responses in patients with inflammatory and functional bowel disease. Gut 2000;47(4):497–505.

[85] Docherty MJ, Jones 3rd RC, Wallace MS. Managing pain in inflammatory bowel disease. Gastroenterol Hepatol 2011;7(9):592–601.

[86] Schirbel A, Reichert A, Roll S, Baumgart DC, Buning C, Wittig B, et al. Impact of pain on health-related quality of life in patients with inflammatory bowel disease. World J Gastroenterol 2010;16(25):3168–77.

[87] Drewes AM, Frokjaer JB, Larsen E, Reddy H, Arendt-Nielsen L, Gregersen H. Pain and mechanical properties of the rectum in patients with active ulcerative colitis. Inflamm Bowel Dis 2006;12(4):294–303.

[88] Buchmann P, Kolb E, Alexander-Williams J. Pathogenesis of urgency in defaecation in Crohn's disease. Digestion 1981;22(6):310–6.

[89] Farrokhyar F, Marshall JK, Easterbrook B, Irvine EJ. Functional gastrointestinal disorders and mood disorders in patients with inactive inflammatory bowel disease: prevalence and impact on health. Inflamm Bowel Dis 2006;12(1):38–46.

[90] Minderhoud IM, Oldenburg B, Wismeijer JA, van Berge Henegouwen GP, Smout AJ. IBS-like symptoms in patients with inflammatory bowel disease in remission; relationships with quality of life and coping behavior. Dig Dis Sci 2004;49(3):469–74.

[91] Campaniello MA, Mavrangelos C, Eade S, Harrington AM, Blackshaw LA, Brierley SM, et al. Acute colitis chronically alters immune infiltration mechanisms and sensory neuro-immune interactions. Brain Behav Immun 2017;60:319–32.

[92] Lovell RM, Ford AC. Global prevalence of and risk factors for irritable bowel syndrome: a meta-analysis. Clin Gastroenterol Hepatol 2012;10(7):712–21. e4.
[93] Mertz HR. Irritable bowel syndrome. N Engl J Med 2003;349(22):2136–46.
[94] Elsenbruch S. Abdominal pain in Irritable Bowel Syndrome: a review of putative psychological, neural and neuro-immune mechanisms. Brain Behav Immun 2011;25(3):386–94.
[95] Posserud I, Syrous A, Lindstrom L, Tack J, Abrahamsson H, Simren M. Altered rectal perception in irritable bowel syndrome is associated with symptom severity. Gastroenterology 2007;133(4):1113–23.
[96] Anand P, Aziz Q, Willert R, van Oudenhove L. Peripheral and central mechanisms of visceral sensitization in man. Neurogastroenterol Motil 2007;19(1 Suppl):29–46.
[97] Hazlett-Stevens H, Craske MG, Mayer EA, Chang L, Naliboff BD. Prevalence of irritable bowel syndrome among university students: the roles of worry, neuroticism, anxiety sensitivity and visceral anxiety. J Psychosom Res 2003;55(6):501–5.
[98] Bercik P, Verdu EF, Collins SM. Is irritable bowel syndrome a low-grade inflammatory bowel disease? Gastroenterol Clin N Am 2005;34(2):235–45. vi-vii.
[99] Spiller RC. Inflammation as a basis for functional GI disorders. Best Pract Res Clin Gastroenterol 2004;18(4):641–61.
[100] Barratt SM, Leeds JS, Robinson K, Lobo AJ, McAlindon ME, Sanders DS. Prodromal irritable bowel syndrome may be responsible for delays in diagnosis in patients presenting with unrecognized Crohn's disease and celiac disease, but not ulcerative colitis. Dig Dis Sci 2011;56(11):3270–5.
[101] Burgmann T, Clara I, Graff L, Walker J, Lix L, Rawsthorne P, et al. The Manitoba Inflammatory Bowel Disease Cohort Study: prolonged symptoms before diagnosis–how much is irritable bowel syndrome? Clin Gastroenterol Hepatol 2006;4(5):614–20.
[102] Olbe L. Concept of Crohn's disease being conditioned by four main components, and irritable bowel syndrome being an incomplete Crohn's disease. Scand J Gastroenterol 2008;43(2):234–41.
[103] Ansari R, Attari F, Razjouyan H, Etemadi A, Amjadi H, Merat S, et al. Ulcerative colitis and irritable bowel syndrome: relationships with quality of life. Eur J Gastroenterol Hepatol 2008;20(1):46–50.
[104] Barbara G, Cremon C, Pallotti F, De Giorgio R, Stanghellini V, Corinaldesi R. Postinfectious irritable bowel syndrome. J Pediatr Gastroenterol Nutr 2009;48(Suppl 2):S95–7.
[105] Spiller R, Garsed K. Postinfectious irritable bowel syndrome. Gastroenterology 2009;136(6):1979–88.
[106] Thabane M, Marshall JK. Post-infectious irritable bowel syndrome. World J Gastroenterol 2009;15(29):3591–6.
[107] Barbara G, De Giorgio R, Stanghellini V, Cremon C, Corinaldesi R. A role for inflammation in irritable bowel syndrome? Gut 2002;51(Suppl 1):i41–4.
[108] Barbara G, Stanghellini V, De Giorgio R, Cremon C, Cottrell GS, Santini D, et al. Activated mast cells in proximity to colonic nerves correlate with abdominal pain in irritable bowel syndrome. Gastroenterology 2004;126(3):693–702.
[109] Hughes PA, Harrington AM, Castro J, Liebregts T, Adam B, Grasby DJ, et al. Sensory neuro-immune interactions differ between irritable bowel syndrome subtypes. Gut 2013;62(10):1456–65.
[110] Liebregts T, Adam B, Bredack C, Gururatsakul M, Pilkington KR, Brierley SM, et al. Small bowel homing T cells are associated with symptoms and delayed gastric emptying in functional dyspepsia. Am J Gastroenterol 2011;106(6):1089–98.

[111] Liebregts T, Adam B, Bredack C, Roth A, Heinzel S, Lester S, et al. Immune activation in patients with irritable bowel syndrome. Gastroenterology 2007;132(3):913–20.

[112] Ohman L, Isaksson S, Lindmark AC, Posserud I, Stotzer PO, Strid H, et al. T-cell activation in patients with irritable bowel syndrome. Am J Gastroenterol 2009;104(5):1205–12.

[113] Gwee KA, Collins SM, Read NW, Rajnakova A, Deng Y, Graham JC, et al. Increased rectal mucosal expression of interleukin 1beta in recently acquired post-infectious irritable bowel syndrome. Gut 2003;52(4):523–6.

[114] Chadwick VS, Chen W, Shu D, Paulus B, Bethwaite P, Tie A, et al. Activation of the mucosal immune system in irritable bowel syndrome. Gastroenterology 2002;122(7):1778–83.

[115] Barbara G, Wang B, Stanghellini V, de Giorgio R, Cremon C, Di Nardo G, et al. Mast cell-dependent excitation of visceral-nociceptive sensory neurons in irritable bowel syndrome. Gastroenterology 2007;132(1):26–37.

[116] Rong W, Hillsley K, Davis JB, Hicks G, Winchester WJ, Grundy D. Jejunal afferent nerve sensitivity in wild-type and TRPV1 knockout mice. J Physiol 2004;560(3):867–81.

[117] Jones RCW, Xu L, Gebhart GF. The mechanosensitivity of mouse colon afferent fibers and their sensitization by inflammatory mediators require transient receptor potential vanilloid 1 and acid-sensing ion channel 3. J Neurosci 2005;25(47):10981–9.

[118] Brierley SM, Carter R, Jones 3rd W, Xu L, Robinson DR, Hicks GA, et al. Differential chemosensory function and receptor expression of splanchnic and pelvic colonic afferents in mice. J Physiol 2005;567(Pt 1):267–81.

[119] Brierley SM, Hughes PA, Page AJ, Kwan KY, Martin CM, O'Donnell TA, et al. The ion channel TRPA1 is required for normal mechanosensation and is modulated by algesic stimuli. Gastroenterology 2009;137(6):2084–95. e3.

[120] Brierley SM, Page AJ, Hughes PA, Adam B, Liebregts T, Cooper NJ, et al. Selective role for TRPV4 ion channels in visceral sensory pathways. Gastroenterology 2008;134(7):2059–69.

[121] Cenac N, Altier C, Chapman K, Liedtke W, Zamponi G, Vergnolle N. Transient receptor potential vanilloid-4 has a major role in visceral hypersensitivity symptoms. Gastroenterology 2008;135(3):937–46 46. e1-2.

[122] Sipe WE, Brierley SM, Martin CM, Phillis BD, Cruz FB, Grady EF, et al. Transient receptor potential vanilloid 4 mediates protease activated receptor 2-induced sensitization of colonic afferent nerves and visceral hyperalgesia. Am J Physiol Gastrointest Liver Physiol 2008;294(5):G1288–98.

[123] Page AJ, Brierley SM, Martin CM, Hughes PA, Blackshaw LA. Acid sensing ion channels 2 and 3 are required for inhibition of visceral nociceptors by benzamil. Pain 2007;133(1-3):150–60.

[124] Page AJ, Brierley SM, Martin CM, Price MP, Symonds E, Butler R, et al. Different contributions of ASIC channels 1a, 2, and 3 in gastrointestinal mechanosensory function. Gut 2005;54(10):1408–15.

[125] Page AJ, Brierley SM, Martin CM, Martinez-Salgado C, Wemmie JA, Brennan TJ, et al. The ion channel ASIC1 contributes to visceral but not cutaneous mechanoreceptor function. Gastroenterology 2004;127(6):1739–47.

[126] Hughes PA, Brierley SM, Young RL, Blackshaw LA. Localization and comparative analysis of acid-sensing ion channel (ASIC1, 2, and 3) mRNA expression in mouse colonic sensory neurons within thoracolumbar dorsal root ganglia. J Comp Neurol 2007;500(5):863–75.

[127] Furness JB, Jones C, Nurgali K, Clerc N. Intrinsic primary afferent neurons and nerve circuits within the intestine. Prog Neurobiol 2004;72(2):143–64.
[128] Furness JB. The enteric nervous system. Oxford: Blackwell; 2006.
[129] Furness JB, Kunze WA, Bertrand PP, Clerc N, Bornstein JC. Intrinsic primary afferent neurons of the intestine. Prog Neurobiol 1998;54(1):1–18.
[130] Hirst GD, Holman ME, Spence I. Two types of neurones in the myenteric plexus of duodenum in the guinea-pig. J Physiol 1974;236(2):303–26.
[131] Mazzuoli G, Schemann M. Multifunctional rapidly adapting mechanosensitive enteric neurons (RAMEN) in the myenteric plexus of the guinea pig ileum. J Physiol 2009;587 (Pt 19):4681–94.
[132] Gabella G, Trigg P. Size of neurons and glial cells in the enteric ganglia of mice, guinea-pigs, rabbits and sheep. J Neurocytol 1984;13(1):49–71.
[133] Kugler EM, Michel K, Zeller F, Demir IE, Ceyhan GO, Schemann M, et al. Mechanical stress activates neurites and somata of myenteric neurons. Front Cell Neurosci 2015;9:342.
[134] Mazzuoli-Weber G, Schemann M. Mechanosensitivity in the enteric nervous system. Front Cell Neurosci 2015;9:408.
[135] Spencer NJ, Smith TK. Mechanosensory S-neurons rather than AH-neurons appear to generate a rhythmic motor pattern in guinea-pig distal colon. J Physiol 2004;558(Pt 2):577–96.
[136] Mazzuoli-Weber G, Schemann M. Mechanosensitive enteric neurons in the guinea pig gastric corpus. Front Cell Neurosci 2015;9:430.
[137] Dong H, Jiang Y, Dong J, Mittal RK. Inhibitory motor neurons of the esophageal myenteric plexus are mechanosensitive. Am J Phys Cell Physiol 2015;308(5):C405–13.
[138] Smith TK, Spencer NJ, Hennig GW, Dickson EJ. Recent advances in enteric neurobiology: mechanosensitive interneurons. Neurogastroenterol Motil 2007;19(11):869–78.
[139] Kunze WA, Furness JB, Bertrand PP, Bornstein JC. Intracellular recording from myenteric neurons of the guinea-pig ileum that respond to stretch. J Physiol 1998;506(Pt 3):827–42.
[140] Kunze WA, Clerc N, Bertrand PP, Furness JB. Contractile activity in intestinal muscle evokes action potential discharge in guinea-pig myenteric neurons. J Physiol 1999;517 (Pt 2):547–61.
[141] Kunze WA, Clerc N, Furness JB, Gola M. The soma and neurites of primary afferent neurons in the guinea-pig intestine respond differentially to deformation. J Physiol 2000;526(Pt 2):375–85.
[142] Collins SM, Van Assche G, Hogaboam C. Alterations in enteric nerve and smooth-muscle function in inflammatory bowel diseases. Inflamm Bowel Dis 1997;3(1):38–48.
[143] Geboes K, Collins S. Structural abnormalities of the nervous system in Crohn's disease and ulcerative colitis. Neurogastroenterol Motil 1998;10(3):189–202.
[144] Dvorak AM, Osage JE, Monahan RA, Dickersin GR. Crohn's disease: transmission electron microscopic studies. III. Target tissues. Proliferation of and injury to smooth muscle and the autonomic nervous system. Hum Pathol 1980;11(6):620–34.
[145] Oehmichen M, Reifferscheid P. Intramural ganglion cell degeneration in inflammatory bowel disease. Digestion 1977;15(6):482–96.
[146] Nadorra R, Landing BH, Wells TR. Intestinal plexuses in Crohn's disease and ulcerative colitis in children: pathologic and microdissection studies. Pediatr Pathol 1986;6(2-3):267–87.
[147] Strobach RS, Ross AH, Markin RS, Zetterman RK, Linder J. Neural patterns in inflammatory bowel disease: an immunohistochemical survey. Mod Pathol 1990;3(4):488–93.

[148] Gulbransen BD, Bashashati M, Hirota SA, Gui X, Roberts JA, MacDonald JA, et al. Activation of neuronal P2X7 receptor-pannexin-1 mediates death of enteric neurons during colitis. Nat Med 2012;18(4):600–4.

[149] Neunlist M, Aubert P, Toquet C, Oreshkova T, Barouk J, Lehur PA, et al. Changes in chemical coding of myenteric neurones in ulcerative colitis. Gut 2003;52(1):84–90.

[150] Schneider J, Jehle EC, Starlinger MJ, Neunlist M, Michel K, Hoppe S, et al. Neurotransmitter coding of enteric neurones in the submucous plexus is changed in non-inflamed rectum of patients with Crohn's disease. Neurogastroenterol Motil 2001;13(3):255–64.

[151] Minderhoud IM, Smout AJ, Oldenburg B, Samsom M. A pilot study on chemospecific duodenal visceral sensitivity in inflammatory bowel disease in remission. Digestion 2006;73(2-3):151–9.

[152] Lee CM, Kumar RK, Lubowski DZ, Burcher E. Neuropeptides and nerve growth in inflammatory bowel diseases: a quantitative immunohistochemical study. Dig Dis Sci 2002;47(3):495–502.

[153] Belai A, Boulos PB, Robson T, Burnstock G. Neurochemical coding in the small intestine of patients with Crohn's disease. Gut 1997;40(6):767–74.

[154] Chen Z, Suntres Z, Palmer J, Guzman J, Javed A, Xue J, et al. Cyclic AMP signaling contributes to neural plasticity and hyperexcitability in AH sensory neurons following intestinal *Trichinella spiralis*-induced inflammation. Int J Parasitol 2007;37(7):743–61.

[155] Linden DR, Sharkey KA, Mawe GM. Enhanced excitability of myenteric AH neurons in the inflamed guinea-pig distal colon. J Physiol 2003;547(Pt 2):589–601.

[156] Linden DR, Sharkey KA, Ho W, Mawe GM. Cyclooxygenase-2 contributes to dysmotility and enhanced excitability of myenteric AH neurons in the inflamed guinea pig distal colon. J Physiol 2004;557(Pt 1):191–205.

[157] Hoffman JM, McKnight ND, Sharkey KA, Mawe GM. The relationship between inflammation-induced neuronal excitability and disrupted motor activity in the guinea pig distal colon. Neurogastroenterol Motil 2011;23(7):673-e279.

[158] Strong DS, Cornbrooks CF, Roberts JA, Hoffman JM, Sharkey KA, Mawe GM. Purinergic neuromuscular transmission is selectively attenuated in ulcerated regions of inflamed guinea pig distal colon. J Physiol 2010;588(Pt 5):847–59.

[159] Nurgali K, Qu Z, Hunne B, Thacker M, Pontell L, Furness JB. Morphological and functional changes in guinea-pig neurons projecting to the ileal mucosa at early stages after inflammatory damage. J Physiol 2011;589(Pt 2):325–39.

[160] O'Hara JR, Lomax AE, Mawe GM, Sharkey KA. Ileitis alters neuronal and enteroendocrine signalling in guinea pig distal colon. Gut 2007;56(2):186–94.

[161] Hons IM, Burda JE, Grider JR, Mawe GM, Sharkey KA. Alterations to enteric neural signaling underlie secretory abnormalities of the ileum in experimental colitis in the guinea pig. Am J Physiol Gastrointest Liver Physiol 2009;296(4):G717–26.

[162] Linden DR. Enhanced excitability of guinea pig ileum myenteric AH neurons during and following recovery from chemical colitis. Neurosci Lett 2013;545:91–5.

[163] Krauter EM, Strong DS, Brooks EM, Linden DR, Sharkey KA, Mawe GM. Changes in colonic motility and the electrophysiological properties of myenteric neurons persist following recovery from trinitrobenzene sulfonic acid colitis in the guinea pig. Neurogastroenterol Motil 2007;19(12):990–1000.

[164] Lomax AE, O'Hara JR, Hyland NP, Mawe GM, Sharkey KA. Persistent alterations to enteric neural signaling in the guinea pig colon following the resolution of colitis. Am J Physiol Gastrointest Liver Physiol 2007;292(2):G482–91.

[165] Faussone-Pellegrini MS, Cortesini C. Ultrastructural features and localization of the interstitial cells of Cajal in the smooth muscle coat of human esophagus. J Submicrosc Cytol 1985;17(2):187–97.
[166] Torihashi S, Horisawa M, Watanabe Y. c-Kit immunoreactive interstitial cells in the human gastrointestinal tract. J Auton Nerv Syst 1999;75(1):38–50.
[167] Hagger R, Gharaie S, Finlayson C, Kumar D. Distribution of the interstitial cells of Cajal in the human anorectum. J Auton Nerv Syst 1998;73(2-3):75–9.
[168] Hanani M, Farrugia G, Komuro T. Intercellular coupling of interstitial cells of cajal in the digestive tract. Int Rev Cytol 2005;242:249–82.
[169] Sanders KM, Koh SD, Ward SM. Interstitial cells of cajal as pacemakers in the gastrointestinal tract. Annu Rev Physiol 2006;68:307–43.
[170] Komuro T. Structure and organization of interstitial cells of Cajal in the gastrointestinal tract. J Physiol 2006;576(Pt 3):653–8.
[171] Horiguchi K, Komuro T. Ultrastructural characterization of interstitial cells of Cajal in the rat small intestine using control and Ws/Ws mutant rats. Cell Tissue Res 1998;293(2):277–84.
[172] Mazet B, Raynier C. Interstitial cells of Cajal in the guinea pig gastric antrum: distribution and regional density. Cell Tissue Res 2004;316(1):23–34.
[173] Rumessen JJ, Thuneberg L, Mikkelsen HB. Plexus muscularis profundus and associated interstitial cells. II. Ultrastructural studies of mouse small intestine. Anat Rec 1982;203(1):129–46.
[174] Zhou DS, Komuro T. Interstitial cells associated with the deep muscular plexus of the guinea-pig small intestine, with special reference to the interstitial cells of Cajal. Cell Tissue Res 1992;268(2):205–16.
[175] Horiguchi K, Semple GS, Sanders KM, Ward SM. Distribution of pacemaker function through the tunica muscularis of the canine gastric antrum. J Physiol 2001;537(Pt 1):237–50.
[176] Seki K, Komuro T. Distribution of interstitial cells of Cajal and gap junction protein, Cx 43 in the stomach of wild-type and W/Wv mutant mice. Anat Embryol (Berl) 2002;206(1-2):57–65.
[177] Mitsui R, Komuro T. Distribution and ultrastructure of interstitial cells of Cajal in the gastric antrum of wild-type and Ws/Ws rats. Anat Embryol (Berl) 2003;206(6):453–60.
[178] Berezin I, Huizinga JD, Daniel EE. Interstitial cells of Cajal in the canine colon: a special communication network at the inner border of the circular muscle. J Comp Neurol 1988;273(1):42–51.
[179] Ishikawa K, Komuro T. Characterization of the interstitial cells associated with the submuscular plexus of the guinea-pig colon. Anat Embryol (Berl) 1996;194(1):49–55.
[180] Thuneberg L. Interstitial cells of Cajal: intestinal pacemaker cells? Adv Anat Embryol Cell Biol 1982;71:1–130.
[181] Vanderwinden JM, Rumessen JJ, Bernex F, Schiffmann SN, Panthier JJ. Distribution and ultrastructure of interstitial cells of Cajal in the mouse colon, using antibodies to Kit and Kit(W-lacZ) mice. Cell Tissue Res 2000;302(2):155–70.
[182] Ward SM, Sanders KM. Involvement of intramuscular interstitial cells of Cajal in neuroeffector transmission in the gastrointestinal tract. J Physiol 2006;576(Pt 3):675–82.
[183] Klein S, Seidler B, Kettenberger A, Sibaev A, Rohn M, Feil R, et al. Interstitial cells of Cajal integrate excitatory and inhibitory neurotransmission with intestinal slow-wave activity. Nat Commun 2013;4:1630.

[184] Lee HT, Hennig GW, Fleming NW, Keef KD, Spencer NJ, Ward SM, et al. Septal interstitial cells of Cajal conduct pacemaker activity to excite muscle bundles in human jejunum. Gastroenterology 2007;133(3):907–17.

[185] Huizinga JD, Reed DE, Berezin I, Wang XY, Valdez DT, Liu LW, et al. Survival dependency of intramuscular ICC on vagal afferent nerves in the cat esophagus. Am J Phys Regul Integr Comp Phys 2008;294(2):R302–10.

[186] Ward SM, Bayguinov J, Won KJ, Grundy D, Berthoud HR. Distribution of the vanilloid receptor (VR1) in the gastrointestinal tract. J Comp Neurol 2003;465(1):121–35.

[187] Ward SM, Sanders KM, Hirst GD. Role of interstitial cells of Cajal in neural control of gastrointestinal smooth muscles. Neurogastroenterol Motil 2004;16(Suppl 1):112–7.

[188] Won KJ, Sanders KM, Ward SM. Interstitial cells of Cajal mediate mechanosensitive responses in the stomach. Proc Natl Acad Sci U S A 2005;102(41):14913–8.

[189] Strege PR, Ou Y, Sha L, Rich A, Gibbons SJ, Szurszewski JH, et al. Sodium current in human intestinal interstitial cells of Cajal. Am J Physiol Gastrointest Liver Physiol 2003;285(6):G1111–21.

[190] Locke 3rd GR, Ackerman MJ, Zinsmeister AR, Thapa P, Farrugia G. Gastrointestinal symptoms in families of patients with an SCN5A-encoded cardiac channelopathy: evidence of an intestinal channelopathy. Am J Gastroenterol 2006;101(6):1299–304.

[191] Sanders KM, Koh SD, Ro S, Ward SM. Regulation of gastrointestinal motility–insights from smooth muscle biology. Nat Rev Gastroenterol Hepatol 2012;9(11):633–45.

[192] Kelly KA, La Force RC. Role of the gastric pacesetter potential defined by electrical pacing. Can J Physiol Pharmacol 1972;50(10):1017–9.

[193] O'Grady G, Angeli TR, Du P, Lahr C, Lammers WJ, Windsor JA, et al. Abnormal initiation and conduction of slow-wave activity in gastroparesis, defined by high-resolution electrical mapping. Gastroenterology 2012;143(3):589–98. e1-3.

[194] Forster J, Damjanov I, Lin Z, Sarosiek I, Wetzel P, McCallum RW. Absence of the interstitial cells of Cajal in patients with gastroparesis and correlation with clinical findings. J Gastrointest Surg 2005;9(1):102–8.

[195] Long QL, Fang DC, Shi HT, Luo YH. Gastro-electric dysrhythm and lack of gastric interstitial cells of cajal. World J Gastroenterol 2004;10(8):1227–30.

[196] Ordog T, Redelman D, Horvath VJ, Miller LJ, Horowitz B, Sanders KM. Quantitative analysis by flow cytometry of interstitial cells of Cajal, pacemakers, and mediators of neurotransmission in the gastrointestinal tract. Cytometry A 2004;62(2):139–49.

[197] Nyam DC, Pemberton JH. Long-term results of lateral internal sphincterotomy for chronic anal fissure with particular reference to incidence of fecal incontinence. Dis Colon Rectum 1999;42(10):1306–10.

[198] Lyford GL, He CL, Soffer E, Hull TL, Strong SA, Senagore AJ, et al. Pan-colonic decrease in interstitial cells of Cajal in patients with slow transit constipation. Gut 2002;51(4):496–501.

[199] Shafik A, Shafik AA, El-Sibai O, Mostafa RM. Electric activity of the colon in subjects with constipation due to total colonic inertia: an electrophysiologic study. Arch Surg 2003;138(9):1007–11. discussion 11.

[200] Sachs F. Mechanical transduction in biological systems. Crit Rev Biomed Eng 1988;16(2):141–69.

[201] Markin VS, Martinac B. Mechanosensitive ion channels as reporters of bilayer expansion. A theoretical model. Biophys J 1991;60(5):1120–7.

[202] Morris CE. Mechanosensitive ion channels. J Membr Biol 1990;113(2):93–107.
[203] Bulbring E. Correlation between membrane potential, spike discharge and tension in smooth muscle. J Physiol 1955;128(1):200–21.
[204] Xu WX, Kim SJ, So I, Kang TM, Rhee JC, et al. Effect of stretch on calcium channel currents recorded from the antral circular myocytes of guinea-pig stomach. Pflugers Arch 1996;432(2):159–64.
[205] Waniishi Y, Inoue R, Ito Y. Preferential potentiation by hypotonic cell swelling of muscarinic cation current in guinea pig ileum. Am J Phys 1997;272(1 Pt 1):C240–53.
[206] Farrugia G. Ionic conductances in gastrointestinal smooth muscles and interstitial cells of Cajal. Annu Rev Physiol 1999;61:45–84.
[207] Horowitz B, Ward SM, Sanders KM. Cellular and molecular basis for electrical rhythmicity in gastrointestinal muscles. Annu Rev Physiol 1999;61:19–43.
[208] Sanders KM. Regulation of smooth muscle excitation and contraction. Neurogastroenterol Motil 2008;20(Suppl 1):39–53.
[209] Farrugia G, Holm AN, Rich A, Sarr MG, Szurszewski JH, Rae JL. A mechanosensitive calcium channel in human intestinal smooth muscle cells. Gastroenterology 1999;117(4):900–5.
[210] Holm AN, Rich A, Sarr MG, Farrugia G. Whole cell current and membrane potential regulation by a human smooth muscle mechanosensitive calcium channel. Am J Physiol Gastrointest Liver Physiol 2000;279(6):G1155–61.
[211] Lyford GL, Strege PR, Shepard A, Ou Y, Ermilov L, Miller SM, et al. alpha(1C)(Ca(V)1.2) L-type calcium channel mediates mechanosensitive calcium regulation. Am J Phys Cell Physiol 2002;283(3):C1001–8.
[212] Strege PR, Holm AN, Rich A, Miller SM, Ou Y, Sarr MG, et al. Cytoskeletal modulation of sodium current in human jejunal circular smooth muscle cells. Am J Phys Cell Physiol 2003;284(1):C60–6.
[213] Ou Y, Strege P, Miller SM, Makielski J, Ackerman M, Gibbons SJ, et al. Syntrophin gamma 2 regulates SCN5A gating by a PDZ domain-mediated interaction. J Biol Chem 2003;278(3):1915–23.
[214] Holm AN, Rich A, Miller SM, Strege P, Ou Y, Gibbons S, et al. Sodium current in human jejunal circular smooth muscle cells. Gastroenterology 2002;122(1):178–87.
[215] Ou Y, Gibbons SJ, Miller SM, Strege PR, Rich A, Distad MA, et al. SCN5A is expressed in human jejunal circular smooth muscle cells. Neurogastroenterol Motil 2002;14(5):477–86.
[216] Koh SD, Monaghan K, Ro S, Mason HS, Kenyon JL, Sanders KM. Novel voltage-dependent non-selective cation conductance in murine colonic myocytes. J Physiol 2001;533(Pt 2):341–55.
[217] Sanders KM, Koh SD. Two-pore-domain potassium channels in smooth muscles: new components of myogenic regulation. J Physiol 2006;570(Pt 1):37–43.
[218] Hwang SJ, O'Kane N, Singer C, Ward SM, Sanders KM, Koh SD. Block of inhibitory junction potentials and TREK-1 channels in murine colon by Ca2+ store-active drugs. J Physiol 2008;586(4):1169–84.
[219] Koh SD, Sanders KM. Stretch-dependent potassium channels in murine colonic smooth muscle cells. J Physiol 2001;533(Pt 1):155–63.
[220] Wang W, Huang H, Hou D, Liu P, Wei H, Fu X, et al. Mechanosensitivity of STREX-lacking BKCa channels in the colonic smooth muscle of the mouse. Am J Physiol Gastrointest Liver Physiol 2010;299(6):G1231–40.

[221] Vermillion DL, Huizinga JD, Riddell RH, Collins SM. Altered small intestinal smooth muscle function in Crohn's disease. Gastroenterology 1993;104(6):1692–9.
[222] Vrees MD, Pricolo VE, Potenti FM, Cao W. Abnormal motility in patients with ulcerative colitis: the role of inflammatory cytokines. Arch Surg 2002;137(4):439–45. discussion 45-6.
[223] Akbarali HI, Pothoulakis C, Castagliuolo I. Altered ion channel activity in murine colonic smooth muscle myocytes in an experimental colitis model. Biochem Biophys Res Commun 2000;275(2):637–42.
[224] Kinoshita K, Sato K, Hori M, Ozaki H, Karaki H. Decrease in activity of smooth muscle L-type Ca2+ channels and its reversal by NF-kappaB inhibitors in Crohn's colitis model. Am J Physiol Gastrointest Liver Physiol 2003;285(3):G483–93.
[225] Liu X, Rusch NJ, Striessnig J, Sarna SK. Down-regulation of L-type calcium channels in inflamed circular smooth muscle cells of the canine colon. Gastroenterology 2001;120(2):480–9.
[226] Shi XZ, Pazdrak K, Saada N, Dai B, Palade P, Sarna SK. Negative transcriptional regulation of human colonic smooth muscle Cav1.2 channels by p50 and p65 subunits of nuclear factor-kappaB. Gastroenterology 2005;129(5):1518–32.
[227] Shi XZ, Sarna SK. Impairment of Ca(2+) mobilization in circular muscle cells of the inflamed colon. Am J Physiol Gastrointest Liver Physiol 2000;278(2):G234–42.
[228] Bulbring E, Crema A. Observations concerning the action of 5-hydroxytryptamine on the peristaltic reflex. Br J Pharmacol Chemother 1958;13(4):444–57.
[229] Bulbring E, Crema A. The action of 5-hydroxytryptamine, 5-hydroxytryptophan and reserpine on intestinal peristalsis in anaesthetized guinea-pigs. J Physiol 1959;146(1):29–53.
[230] Bulbring E, Lin RC. The effect of intraluminal application of 5-hydroxytryptamine and 5-hydroxytryptophan on peristalsis; the local production of 5-HT and its release in relation to intraluminal pressure and propulsive activity. J Physiol 1958;140(3):381–407.
[231] Kunze WA, Furness JB. The enteric nervous system and regulation of intestinal motility. Annu Rev Physiol 1999;61:117–42.
[232] Neya T, Mizutani M, Yamasato T. Role of 5-HT3 receptors in peristaltic reflex elicited by stroking the mucosa in the canine jejunum. J Physiol 1993;471:159–73.
[233] Bertrand PP. Real-time measurement of serotonin release and motility in guinea pig ileum. J Physiol 2006;577(Pt 2):689–704.
[234] Heredia DJ, Dickson EJ, Bayguinov PO, Hennig GW, Smith TK. Localized release of serotonin (5-hydroxytryptamine) by a fecal pellet regulates migrating motor complexes in murine colon. Gastroenterology 2009;136(4):1328–38.
[235] Heredia DJ, Gershon MD, Koh SD, Corrigan RD, Okamoto T, Smith TK. Important role of mucosal serotonin in colonic propulsion and peristaltic reflexes: in vitro analyses in mice lacking tryptophan hydroxylase 1. J Physiol 2013;591(23):5939–57.
[236] Keating DJ, Spencer NJ. Release of 5-hydroxytryptamine from the mucosa is not required for the generation or propagation of colonic migrating motor complexes. Gastroenterology 2010;138(2):659–70 70. e1-2.
[237] Yang Q, Bermingham NA, Finegold MJ, Zoghbi HY. Requirement of Math1 for secretory cell lineage commitment in the mouse intestine. Science 2001;294(5549):2155–8.
[238] Li HJ, Ray SK, Singh NK, Johnston B, Leiter AB. Basic helix-loop-helix transcription factors and enteroendocrine cell differentiation. Diabetes Obes Metab 2011;13(Suppl 1):5–12.

[239] Roach G, Heath Wallace R, Cameron A, Emrah Ozel R, Hongay CF, Baral R, et al. Loss of ascl1a prevents secretory cell differentiation within the zebrafish intestinal epithelium resulting in a loss of distal intestinal motility. Dev Biol 2013;376(2):171–86.
[240] Wright MC, Reed-Geaghan EG, Bolock AM, Fujiyama T, Hoshino M, Maricich SM. Unipotent, Atoh1+ progenitors maintain the Merkel cell population in embryonic and adult mice. J Cell Biol 2015;208(3):367–79.
[241] Raybould HE, Cooke HJ, Christofi FL. Sensory mechanisms: transmitters, modulators and reflexes. Neurogastroenterol Motil 2004;16(Suppl 1):60–3.
[242] Nakatani M, Maksimovic S, Baba Y, Lumpkin EA. Mechanotransduction in epidermal Merkel cells. Pflugers Arch 2015;467(1):101–8.
[243] Chang W, Kanda H, Ikeda R, Ling J, DeBerry JJ, Gu JG. Merkel disc is a serotonergic synapse in the epidermis for transmitting tactile signals in mammals. Proc Natl Acad Sci U S A 2016;113(37):E5491–500.
[244] Woo SH, Ranade S, Weyer AD, Dubin AE, Baba Y, Qiu Z, et al. Piezo2 is required for Merkel-cell mechanotransduction. Nature 2014;509(7502):622–6.
[245] Ranade SS, Woo SH, Dubin AE, Moshourab RA, Wetzel C, Petrus M, et al. Piezo2 is the major transducer of mechanical forces for touch sensation in mice. Nature 2014;516(7529):121–5.
[246] Ikeda R, Gu JG. Piezo2 channel conductance and localization domains in Merkel cells of rat whisker hair follicles. Neurosci Lett 2014;583:210–5.
[247] Wang F, Knutson K, Alcaino C, Linden DR, Gibbons SJ, Kashyap P, et al. Mechanosensitive ion channel Piezo2 is important for enterochromaffin cell response to mechanical forces. J Physiol 2017;595(1):79–91.
[248] Nozawa K, Kawabata-Shoda E, Doihara H, Kojima R, Okada H, Mochizuki S, et al. TRPA1 regulates gastrointestinal motility through serotonin release from enterochromaffin cells. Proc Natl Acad Sci U S A 2009;106(9):3408–13.
[249] Doihara H, Nozawa K, Kawabata-Shoda E, Kojima R, Yokoyama T, Ito H. Molecular cloning and characterization of dog TRPA1 and AITC stimulate the gastrointestinal motility through TRPA1 in conscious dogs. Eur J Pharmacol 2009;617(1-3):124–9.
[250] Cooke HJ, Christofi FL. Enteric neural regulation of mucosal secretion. In: Johnson LR, editor. Physiology of the gastrointestinal tract. 4th ed. New York: Academic Press; 2006. p. 737–62.
[251] Christofi FL. Purinergic receptors and gastrointestinal secretomotor function. Purinergic Signal 2008;4(3):213–36.
[252] Mawe GM, Hoffman JM. Serotonin signalling in the gut–functions, dysfunctions and therapeutic targets. Nat Rev Gastroenterol Hepatol 2013;10(8):473–86.
[253] Fujimiya M, Okumiya K, Kuwahara A. Immunoelectron microscopic study of the luminal release of serotonin from rat enterochromaffin cells induced by high intraluminal pressure. Histochem Cell Biol 1997;108(2):105–13.
[254] Linan-Rico A, Wunderlich JE, Grants IS, Frankel WL, Xue J, Williams KC, et al. Purinergic autocrine regulation of mechanosensitivity and serotonin release in a human EC model: ATP-gated P2X3 channels in EC are downregulated in ulcerative colitis. Inflamm Bowel Dis 2013;19(11):2366–79.
[255] Kordasti S, Sjovall H, Lundgren O, Svensson L. Serotonin and vasoactive intestinal peptide antagonists attenuate rotavirus diarrhoea. Gut 2004;53(7):952–7.
[256] Gershon MD. Review article: serotonin receptors and transporters–roles in normal and abnormal gastrointestinal motility. Aliment Pharmacol Ther 2004;20(Suppl 7):3–14.

[257] Galligan JJ. 5-hydroxytryptamine, ulcerative colitis, and irritable bowel syndrome: molecular connections. Gastroenterology 2004;126(7):1897–9.
[258] Coates MD, Mahoney CR, Linden DR, Sampson JE, Chen J, Blaszyk H, et al. Molecular defects in mucosal serotonin content and decreased serotonin reuptake transporter in ulcerative colitis and irritable bowel syndrome. Gastroenterology 2004;126(7):1657–64.
[259] Crowell MD. Role of serotonin in the pathophysiology of the irritable bowel syndrome. Br J Pharmacol 2004;141(8):1285–93.
[260] Crowell MD, Shetzline MA, Moses PL, Mawe GM, Talley NJ. Enterochromaffin cells and 5-HT signaling in the pathophysiology of disorders of gastrointestinal function. Curr Opin Investig Drugs 2004;5(1):55–60.
[261] O'Hara JR, Ho W, Linden DR, Mawe GM, Sharkey KA. Enteroendocrine cells and 5-HT availability are altered in mucosa of guinea pigs with TNBS ileitis. Am J Physiol Gastrointest Liver Physiol 2004;287(5):G998–1007.
[262] Manocha M, Shajib MS, Rahman MM, Wang H, Rengasamy P, Bogunovic M, et al. IL-13-mediated immunological control of enterochromaffin cell hyperplasia and serotonin production in the gut. Mucosal Immunol 2013;6(1):146–55.

CHAPTER

# Mechanobiology of skin diseases and wound healing

# 14

Sun Hyung Kwon, Jagannath Padmanabhan, Geoffrey C. Gurtner

*Hagey Laboratory, Division of Plastic Surgery, Department of Surgery, Stanford University School of Medicine, Stanford, CA, United States*

## ABBREVIATIONS

| | |
|---|---|
| **AADM** | age-associated dermal microenvironment |
| **ADAM** | a disintegrin and metalloproteinase domain |
| **AGE** | advanced glycation end products |
| **CCN** | cysteine-rich protein |
| **ECM** | extracellular matrix |
| **ERK** | extracellular regulated kinases |
| **FAK** | focal adhesion kinase |
| **FEA** | finite element analysis |
| **FKBP14** | FK506-binding protein |
| **GPCR** | G-protein-coupled receptor |
| **HTS** | hypertrophic scar |
| **MAPK** | mitogen-activated protein kinases |
| **MCP** | monocyte chemoattractant protein |
| **MMP** | matrix metalloproteinase |
| **ROS** | reactive oxygen species |
| **siRNA** | small interfering ribonucleic acid |
| **TGF-β** | transforming growth factor beta |
| **TIMP** | tissue inhibitors of metalloproteinase |
| **UV** | ultraviolet |
| **VEGF** | vascular endothelial growth factor |

## 1 INTRODUCTION

Mechanical forces influence the growth and homeostasis in virtually all tissues and organ systems in the human body [1–3]. In particular, the skin is vulnerable to mechanical forces as it is exposed to several environmental insults as the outermost layer of the body [3,4]. In the past decade, it has become increasingly clear that skin cells integrate biochemical signals with mechanical information from their microenvironment to respond to a number of insults including connective tissue diseases, wounds, and trauma [3,4]. The primary cell types that constitute the skin such as fibroblasts and keratinocytes are both known to respond to mechanical forces [3,5]. Fibroblasts have been studied extensively for modulating the extracellular matrix (ECM) in response to changes in their mechanical microenvironment [6–12]. Keratinocytes respond to mechanical forces by modulating proliferation and matrix-remodeling proteins [13–17]. The epidermis contains many other cells such as melanocytes, Langerhans cells and T cells, and resident progenitor cells that could potentially respond to mechanical cues [4,18]. The dermis, which is more vascularized, contains endothelial cells; pericytes; myofibroblasts; dermal adipocytes; and resident inflammatory cells such as dendritic cells, T cells, and macrophages [19–23]. In addition, circulating cells that are recruited to the skin such as neutrophils, monocytes, and fibrocytes also display mechanoresponsiveness [24,25].

An understanding of the mechanobiology of skin cells is important for studying normal skin homeostasis as well as skin diseases and wound healing processes. In particular, skin mechanobiology is critical in the context of wound healing because breaks in cutaneous continuity alter the cellular mechanical environment and could potentially signal the beginning of the wound healing cascade. When cutaneous integrity is reestablished, mechanical forces may guide skin remodeling, leading to intact and durable repair. Alterations in skin mechanobiology lead to abnormal wound repair, which occurs across a spectrum ranging from overhealing wounds with excessive fibrosis and scarring to underhealing wounds clinically typified by nonhealing chronic ulcers. Similarly, defects in mechanoresponsive skin cells have been observed in several skin diseases such as keloids, scleroderma, and psoriasis.

Skin diseases and dysfunctional wound healing pose a significant health-care challenge (Fig. 1). For example, overhealing occurring in burns and trauma leads

**FIG. 1**

Spectrum of skin wound healing. Abnormal skin wound healing occurs across a spectrum ranging from overhealing wounds such as hypertrophic scars (left) to underhealing wounds such as chronic diabetic ulcers (right).

to severe functional disabilities costing the economy over $4 billion per year [4,26–28]. Keloids, which are characterized by excessive fibrosis in the areas of the skin with high strain [29], represent another major fibrotic skin disorder [27,30]. Dupuytren's disease represents another category of skin disease characterized by excessive fibrosis of the subcutaneous fascia and is closely associated with dermal tissue [31–36]. Finally, over 40 million patients undergo surgical procedures [37,38], in which the formation of hypertrophic scars (HTS) is a common complication leading to substantial morbidity and disfigurement [3,39]. At the other end of the spectrum, nonhealing wounds associated with diabetes and aging cost the US economy over $25 billion annually [40–43]. Diabetic foot ulcers develop in about 15% of patients and are the leading cause of nontraumatic lower-limb amputations in the United States [40,44–47].

Recent evidence has revealed that a range of skin and soft tissue pathologies including HTS, keloids, Dupuytren's contracture, and chronic diabetic wounds are largely mediated by alterations in skin mechanobiology [4,16,30,36,41,48,49]. The importance of mechanical forces in skin diseases is further underscored by the recently developed clinical wound treatment paradigms that reduce scar formation in wounds by redistributing mechanical forces around the wound [50,51]. Furthermore, recent animal studies have implicated specific skin cells and molecular pathways that are activated in skin diseases such as HTS, keloids, and diabetic wounds [30,41,49]. Identification of specific molecular pathways underlying altered mechanobiology has also led to the development of pharmacological strategies to control or reverse skin diseases [52]. Additionally, advancements in sequencing technologies and single-cell transcriptional analyses have revealed previously uncharacterized heterogeneity in skin cell populations, with implications in skin diseases and wound healing [6,53–55]. Here, we will consolidate recent advancements in our understanding of skin mechanobiology and its role in skin disease and wound healing disorders, and identify key challenges and outstanding issues and opportunities for further research in the field.

## 2 MECHANOBIOLOGY OF THE SKIN
### 2.1 MECHANOBIOLOGY IN SKIN HOMEOSTASIS

Mechanical cues are quantified using three major parameters, namely, stress, strain, and stiffness [56]. Stress is a measure of the intensity of force or force per unit area and is reported in units of newton per square meter. Strain is a dimensionless measure of the deformation of a material in response to stress, which is expressed as the ratio of deformed length to the original length. Stiffness of material is defined as a measure of resistance to deformation and is dependent on inherent mechanical stiffness and material geometry. All three parameters of mechanical cues have been shown to be key regulators of skin homeostasis. For example, the human skin has been shown to have static lines of maximal stress called the Langer's lines, which surgeons use to

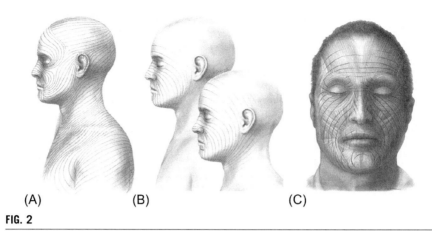

**FIG. 2**

Various skin anatomical lines of the face and neck. (A) Langer's lines, (B) Borges' relaxed skin tension lines (left), and Straith's lines (right). Straith's lines follow Borges' lines very closely, differing only at the glabella, columella, and mentolabial fold. (C) Wrinkle lines, as described by Kraissl, superimposed on facial musculature [58].

orient incisions to minimize scar formation (Fig. 2) [57,58]. These naturally occurring anatomical lines of skin tension constitute the earliest evidence of the importance of mechanobiology in skin homeostasis. Other topological lines of the skin have also emerged since Langer's lines, which are adopted by many surgeons to date when planning surgical incisions. In addition, joint movement, muscle activity, and gravity lead to dynamic stress in the skin. Moreover, mechanical strain such as introduced by pregnancy, weight gain, or subcutaneously implanted devices (e.g., tissue expanders) can induce an increase in mass, volume, and area of the skin [59]. Furthermore, Young's modulus for the skin has been reported to have a range between 5 kPa and 140 MPa, depending on the methodology used and the source of the skin being tested. It is believed that once the skin matrix stiffness is established, it is maintained throughout the lifetime of an individual in the absence of disease, increased age, or trauma [56,60,61]. Since the skin, as the outermost layer of the body, is exposed to a variety of mechanical insults and deformations, cutaneous cell populations have to be able to sense mechanical cues and respond accordingly to guarantee homeostasis and proper mechanical function. An elaborated system of mechanoresponsive cells, ECM proteins, and proteins that govern cell-ECM interactions are involved in maintaining skin homeostasis as discussed below [56].

## 2.2 MECHANORESPONSIVE SKIN CELLS

The skin is composed of two layers: the epidermis and the dermis separated by the basement membrane made of collagen IV and laminin [62]. The outer layer epidermis is a highly cellular layer consisting mostly of keratinocytes, melanocytes, Merkel cells, and Langerhans cells. The dermis is predominantly composed of collagen I, collagen III, elastins, and proteoglycans and is populated by capillaries and fibroblasts. The subcutaneous layer beneath the dermis is made of adipose tissue.

Cutaneous cells including fibroblasts, keratinocytes, and endothelial cells have been shown to be mechanosensitive. Fibroblasts isolated from the human skin are capable of pulling on collagen with 800–1000 μN per one million cells in vitro. In vivo, fibroblasts interact with extracellular components such as collagen via integrins, which are connected to the actin cytoskeleton and transmit extracellular mechanical forces to intracellular components [63]. In addition to integrin-associated signaling, mechanotransduction is also mediated by G-protein-coupled receptors (GPCRs) and stretch-activated ion channels [3,62]. Similarly, epidermal keratinocytes have also been shown to be mechanosensitive. For example, keratinocytes respond to mechanical strain with increased mitosis and proliferation [59]. Numerous reports have shown that endothelial cells that line capillaries in the dermis are also mechanosensitive and respond to mechanical cues by changing the expression of adhesion molecules, inflammatory proteins, and reactive oxygen species (ROS) production [64,65]. Similarly, mechanical strain has been shown to reduce proliferation of adipose cells and results in overall thinning of adipose tissue layer in the subcutaneous tissue [66]. Other cells such as resident macrophages and other inflammatory cells in the skin may also be mechanosensitive [3,67]. While mechanosensitive cells regulate the maintenance of skin mechanical homeostasis in response to environmental insults, aberrant activation of mechanosensitive skin cells underlies abnormal wound healing and skin diseases.

## 2.3 MOLECULAR PATHWAYS MEDIATING SKIN MECHANOBIOLOGY

Advancements in our understanding of skin mechanobiology have identified several molecular pathways that mediate mechanotransduction in skin cells during homeostasis and pathology. The most well-studied mechanosensitive cells are the fibroblasts that populate the dermis and contribute to the production and maintenance of ECM production in the skin. Stimulation of fibroblasts by transforming growth factor beta (TGF-β) and upregulation of extracellular regulated kinases (ERK) induce differentiation into myofibroblasts, which exhibit increased ECM production and contractile capacity [56,63]. Apart from TGF-β receptors, the integrins have been widely studied for their role in cellular mechanotransduction. Integrins are heterodimeric transmembrane receptor proteins, which are composed of two components: one intracellular component in contact with the actin cytoskeleton and another extracellular component that binds to extracellular components [3,56,68]. Integrins are composed of different combinations of 18 α-subunits and 8 β-subunits and mediate ligand-binding-induced signal transduction and signal transduction in response to mechanical cues. Specific integrins such as α1β1, α5β1, and αvβ3 mediate adhesion of fibroblasts to ECM components such as collagens, vitronectin, and fibronectin [69]. Integrins α2β1, α3β1, and α5β1 mediate keratinocyte binding to collagen, laminin, and fibronectin, respectively. Integrin-ECM binding sites are characterized by focal adhesion complexes, which are composed of linker proteins such as vinculin, paxillin, and talin. These linker proteins link the integrin molecules to the actin cytoskeleton of the cell and mediate mechanotransduction. Focal adhesion protein talin binds to β-integrin and F-actin and contains several binding sites for vinculin, which in turn binds to many cytoplasmic proteins [70–72]. Stretching talin induces increased binding of vinculin to talin, which in turn reinforces the ECM-integrin-cytoskeleton mechanotransduction axis [56]. Mechanical activation of integrins and focal adhesion complexes has been associated with consequent activation of intracellular molecular pathways that alter transcription and cell function.

One of the key molecular mediators of this intracellular signaling is focal adhesion kinase (FAK). Mechanical activation of focal adhesions leads to phosphorylation of intracellular FAK via the mechanistic target of rapamycin (mTOR)-phosphoinositide 3-kinase (PI3K)-Akt-dependent pathway leading to the phosphorylation of FAK, the active form of FAK, and can catalyze its substrates [73]. Modulation of p-FAK levels has been linked to inflammatory pathways in various cell types [49,74]. A second set of important mediators of mechanotransduction is the Rho-family GTPases. Rho GTPases act in concert with myosin II and polymerized actin to mediate cell contraction. Conversely, mechanical stress can induce the activation of Rho GTPases or Ras and subsequent cytoplasmic responses including the activation of the NF-kB pathway [3,56,75]. Activation of Rho GTPases is also modulated by mechanical cues such as nanoscale topography, which leads to changes in cell morphology and ECM production in fibroblasts [76]. Similarly, mechanical stress has been shown to activate mitogen-activated protein kinase (MAPK) in keratinocytes resulting in increased proliferation [59,77]. It has also been shown that blocking of β1 integrin leads to the attenuation of MAPK in keratinocytes [78].

There are several other mechanotransduction pathways that can act independent of the ECM-integrin-cytoskeleton axis (Fig. 3) [79]. For example, mechanical perturbation of fibroblasts activates N-cadherin-associated stress-sensitive calcium channels, which increases actin polymerization in fibroblasts [75]. Stress-induced activation of calcium channel-mediated signaling can be propagated to other fibroblasts and keratinocytes in the skin. Activation of stress-sensitive ion channels leads to the activation of protein kinase C (PKC), which in turn upregulates the expression of MAPK and subsequent transcriptional changes. Another category of mechanotransduction mediators is GPCRs. Mechanical activation of GPCRs upregulates collagen synthesis, induces contraction in fibroblasts, and mediates fibrosis [5].

Activation of intracellular mechanotransduction pathways during homeostasis as well as wound healing and skin pathology is mediated in response to mechanical cues such as stress, strain, and stiffness, which are relayed to the cells via the ECM. The role of ECM components in skin mechanobiology will be discussed in detail.

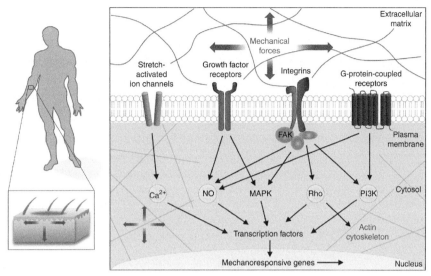

**FIG. 3**

Intracellular mechanotransduction of the skin. Important mediators of transduction signals from the biomechanical environment include integrin-matrix interactions, growth factor receptors (e.g., TGF receptors), G-protein-coupled receptors (GPCRs), ion channels (calcium), and cytoskeletal strain responses. Once transmitted over the cell membrane, mechanical force activates multiple interrelated signaling pathways [79].

## 2.4 SKIN MECHANOBIOLOGY AND THE ECM

Mechanical stress acts through multiple extracellular and intracellular mediators to convert mechanical signals to biochemically relevant signals (Fig. 3). In addition to a variety of cell surface receptors and their effector proteins, the ECM components also play pivotal roles in this mechanotransduction. The ECM provides the physical scaffolding and chemical ligands for cells to reside in and bind to and has proved to be critical to the form and function of the skin. While skin cells produce, maintain, and remodel the ECM during homeostasis, the ECM modulates cell function and phenotype, especially in response to mechanical cues [56]. The ECM is composed of proteins, glycosaminoglycans, and proteoglycans. Collagens, elastins, fibrillins, fibronectin, and hyaluronan are the predominant constituents of the skin ECM. Type IV collagen is the major component of the basement membrane at the epidermis-dermis junction. The dermis contains fibrillar form collagen types I, III, and V; fibril-associated collagens XII and XIV; and nonfibril-forming collagens IV, VI, and VII [75]. Fibrillar collagen provides the skin with mechanical stiffness, and its remodeling and turnover are critical to the mechanical homeostasis of the skin [56]. Nonfibrillar collagens and glycoproteins such as tenascin C and proteoglycans such as decorin enhance the mechanical stability of collagen fibrils and the overall ECM.

Elastic fibers made of an elastin core and a surrounding sheath of microfibrils and fibrillin contribute to the extensibility and elastic nature of the skin [56]. Collagen fibers and elastic fibers are arranged in parallel, and both contribute to the mechanical homeostasis of the skin [56]. Elastic stress-strain curves of the skin reflect the energies stored via elastic and collagen fiber stretching [75]. At low levels of mechanical strain, the collagen fibrils offer little resistance, and the slope of the elastic stress–strain curve reflects the stretching of elastic fibers. At high levels of elastic strain (30%–60%), collagen fibers contribute to a higher elastic modulus range, yet higher levels of uniaxial strain on the skin lead to the defibrillation of collagen fibers.

In addition to mechanical cues, the ECM provides ligands for binding of integrins for cell-ECM adhesion. Ligand binding to integrins induces intracellular signaling that influences cytoskeletal reorganization, gene expression, and cellular differentiation as discussed above [75]. Conformational changes in ECM can lead to altered spatiotemporal distribution of binding sites, hidden domains, and soluble growth factors such as TGF-β, which further alter cell behavior [1,2,56]. ECM mechanical cues in the form of stiffness and topography, biochemical cues in the form of integrin-binding ligands, growth factors and soluble effectors, and the physical reciprocal interactions with cells are all important for skin homeostasis. Disruption of these ECM components underlies a variety of skin pathologies and aberrant wound healing [3].

## 2.5 ALTERED MECHANOBIOLOGY IN SKIN DISEASES AND WOUND HEALING

Disruption of skin homeostasis by wounds, genetic alterations, or underlying diseases can lead to aberrant mechanobiology in the skin [3]. Breaks in cutaneous continuity can alter the cellular mechanical environment and signal the mediation of the wound healing cascade (Fig. 4) [4]. When cutaneous integrity is reestablished, mechanical forces guide skin remodeling, leading to intact and durable repair. Altered mechanotransduction in the skin leads to abnormal wound repair, which occurs across a spectrum ranging from overhealing wounds with excessive fibrosis and scarring to underhealing wounds clinically typified by nonhealing chronic ulcers. Both ends of the dysfunctional wound healing and repair spectrum pose a significant health-care challenge. The rest of this chapter will discuss the role of altered skin mechanobiology in the spectrum of wound healing disorders and skin pathologies.

## 3 MECHANOTRANSDUCTION IN SKIN FIBROSIS AND FIBROTIC WOUND HEALING

Skin fibrosis and fibrotic wound healing represent a significant source of morbidity in the United States. The formation of cutaneous scar following injury, trauma, or disease varies considerably in humans based on the mechanism of injury, degree

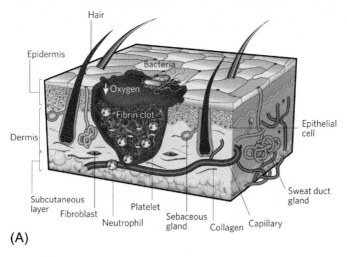

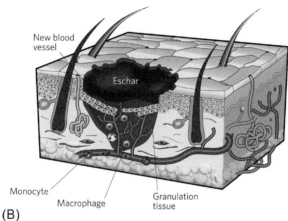

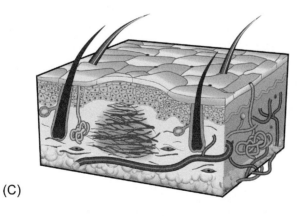

**FIG. 4**

Classic stages of wound repair. There are three classic stages of wound repair: inflammation (A), new tissue formation (B), and remodeling (C). (A) Inflammation. This stage lasts about 48 h after injury. The wound becomes a hypoxic environment, and inflammatory cells are abundant in the wound bed. (B) New tissue formation. This stage occurs about 2–10 days after injury. An eschar (scab) has formed on the surface of the wound, and new blood vessels have populated the area. Epithelial cells have migrated under the eschar. (C) Remodeling. This stage lasts for a year or longer. Collagen deposition by fibroblasts can be disorganized. Reepithelialized wound can be slightly higher than surrounding surface [4].

of damage, anatomical location, and genetic predispositions [5]. It has become increasingly clear that altered mechanotransduction in the skin leads to aberrant cutaneous wound repair and scar formation. Scar formation such as occurs in burns and trauma leads to severe functional disabilities costing the economy over $4 billion per year [4,5,28].

Following trauma and wounding, components of the coagulation cascade, inflammatory pathways, and the immune system are activated to prevent ongoing blood and fluid losses, to remove dead tissue debris, and to prevent infection [4]. A fibrin matrix is formed at the wound site, which becomes the scaffold for infiltrating neutrophils. Subsequently, monocytes from the bloodstream are recruited to the wound, which differentiate into macrophages. Macrophages coordinate the inflammatory phase of the progression of wound healing response. At the end of the inflammatory phase (2–3 days), new tissue forms at the wound site. Keratinocytes from neighboring tissue migrate to reepithelialize the wound area. Sprouts of capillaries associated with fibroblasts and macrophages replace the fibrin matrix with granulation tissue, which forms a new substrate for keratinocyte migration. Fibroblasts, which are attracted from the edge of the wound or from the bone marrow, differentiate into myofibroblasts, which contract to bring the edges of the wound together. Fibroblasts produce ECM, mainly in the form of collagen, which ultimately forms the scar tissue. Over the next few weeks, tissue remodeling occurs through the action of matrix metalloproteinase (MMP) secreted by fibroblasts and other cell types. At the end of remodeling, the remaining tissue consists of mostly collagen type I and a few cells. All steps of this wound repair process are modulated by skin mechanobiology, and aberrant mechanotransduction leads to fibrosis and scar formation [3,28]. Here, we will discuss the role of mechanotransduction in specific examples of skin fibrosis and fibrotic healing.

### 3.1 HYPERTROPHIC SCARS

Annually, in the United States, over 34 million patients undergo surgical procedures, 2 million patients are treated for injury related to motor vehicle accidents, and over 1 million are treated for burn-related injuries [37]. Formation of HTS following these events is a common complication following both injury and repair, and treatments are estimated to cost over $4 billion per year. HTS formation is a fibroproliferative disorder characterized by an exaggerated wound repair response, scar contracture, and overdeposition of ECM [48,80]. Clinically, overhealed HTS formation is worse in the regions of high physical stress, and mechanical forces are known to play a critical role in a variety of fibroproliferative pathologies including HTS.

We have previously reported the reproduction of humanlike HTS in murine models using mechanical straining of healing wounds [8,48]. The resultant scars were raised, lacked adnexal structures and displayed collagen whorls, increased vascularity, and increased mast cell density, similar to human HTS. We found that in fibroblasts, FAK acts to mechanically trigger the secretion of a potent chemokine, monocyte chemoattractant protein-1 (MCP-1) that is linked to inflammatory cell recruitment and human fibrotic disorders. In contrast, fibroblasts derived from transgenic FAK KO exhibited diminished fibrogenic responses and attenuated MCP-1 secretion [49]. To further interrogate the role of FAK activation in fibroblasts, we developed fibroblast-restricted FAK KO transgenic mice and compared HTS formation in the transgenic mice with wild-type controls [49]. Mechanical load-induced formation of HTS was significantly abrogated in the FAK KO mice. Fibroblast-specific deletion of FAK resulted in significantly smaller scar area and reduced fibrosis as confirmed by trichrome staining of scar tissues. Another recent study has shown that the activation of TRPC3 and NF-kB also mediates the formation of HTS [80]. Additionally, the role of endothelin-1 secreted from endothelial cells in inducing differentiation of fibroblasts into contractile myofibroblasts on HTS is also under investigation [81].

Similarly, the manipulation of mechanical strain has been shown to alter the extent of fibrosis and scar formation in porcine models as well [50]. Microarray profiling revealed that mechanical strain induced altered gene expression of key connective tissue remodeling and mechanotransduction pathways. Based on our findings in small- and large-animal studies, we designed a clinical trial to confirm the importance of mechanical force in human scar formation [50]. Patients undergoing abdominal procedures were recruited, and an external polymer device was used to off-load mechanical strain on the surgical site. At 12 months postsurgery, scars were dramatically reduced with diminished strain, compared with the within-patient standard treatment incisions. Similarly, a rehabilitation massage therapy to reduce mechanical strain on burn-induced HTS has shown promising results [82]. These clinical data strongly demonstrate that mechanical strain can be modulated to produce different outcomes in human wounds, further validating the translational potential of targeting skin mechanotransduction [50,51,83–85].

## 3.2 KELOIDS

Keloid formation is another major fibroproliferative skin disorder, which is caused by prolonged inflammation after cutaneous injury [86]. Keloids are characterized by collagen accumulation and blood vessel proliferation in the reticular layer of the dermis. Occurrence of keloids is significantly higher in people with skin pigments as opposed to Caucasians [87,88]. Such observations and familial keloid occurrence indicate that keloids might have an associated genetic predisposition.

Skin sites that are subject to high mechanical stress exhibit high susceptibility to develop keloid disease. For example, recurrence rates of keloids have been shown to be higher in high-mechanical-stress regions such as the anterior chest wall and the scapular and suprapubic regions as compared with the earlobes [89,90]. Custom-made ear clips that reduce mechanical stress around keloid tissues have been shown to decrease keloid reoccurrence by 70% [91].

Altered structure of the ECM plays a key role in modulating fibroblast function in keloid pathogenesis and progression. As mentioned previously, collagen fibers are arranged in parallel to the elastic fibers and the epidermis in the normal skin.

In contrast, keloid tissues display highly irregular distribution and nodular structures in the reticular dermis [89,92]. Another unique aspect of keloid progression is that keloids grow and spread beyond the boundaries of the original wound, especially in the direction of increased mechanical stress [29]. Finite element analysis (FEA) of keloid tissue has shown that the boundaries of a keloid exhibit high mechanical stress, which contributes to keloid progression. Atomic force microscopy and experiments with micropost array detectors have demonstrated that fibroblasts that mediate keloid progression exhibit a FAK-dependent polarization of filament elasticity and force generation [93].

Several mechanoresponsive genes have been linked to keloid pathogenesis and progression including FAK, ERK, TGF-β, platelet-derived growth factor (PDGF), vascular endothelial growth factor (VEGF), and heat-shock protein (HSP)47 [30,87,93–95]. Expression of endothelin-1 (which enhances myofibroblast differentiation) has been found to be higher in keloid tissue as compared with controls, indicating that endothelial cells from the microvasculature might contribute to increased keloid formation [81,96]. It has been hypothesized that the current modes of treatment for keloids, which include radiation, steroid administration, and laser treatment, work in part by targeting endothelial cells [96]. Another recent study has shown that keloid tissues exhibit increased epithelial-to-mesenchymal transition, where epidermal cells express mesenchymal markers and could differentiate into fibroblasts and contribute to keloid progression. Further research into cell-specific mechanotransduction pathways mediating keloid pathogenesis and progression will help identify targets and develop targeted drug therapies to control and treat keloids.

### 3.3 SCLERODERMA (SYSTEMIC SCLEROSIS)

Scleroderma is a complex fibrotic disorder that is characterized by extensive fibrosis of the skin, a vascular dysfunction, and an autoimmune component [97]. The manifestation of scleroderma in patients can be either (i) limited, where it affects the arms, face, and hands, or (ii) diffuse, where a rapidly progressive disease affects a large part of the skin and several internal organs. A third of the patients with scleroderma can develop pulmonary hypertension [97]. Localized or limited scleroderma is linked directly to fibrosis of the skin and subcutaneous tissue [98,99]. Scleroderma is present in patients with atrophic epidermis and increased collagen production in the dermis, which leads to stiffening of the skin. It is often accompanied by increased infiltration of the dermal tissue by lymphocytes and plasma, indicating exacerbated and prolonged inflammation.

The key effectors of scleroderma have been identified to be the fibroblasts [97]. Atomic force microscopy-aided studies have revealed that fibroblasts from scleroderma tissue are significantly stiffer than control fibroblasts. Altered mechanobiological properties of scleroderma fibroblasts induce altered response to mechanical stimuli such as biomechanical forces via integrins and lead to increased ECM production [97,100]. Inflammatory factors such as cytokines, toll-like receptors, and oxidative stress also contribute to increased ECM production in scleroderma fibroblasts [101]. Endothelial cell-derived endothelin-1 and thrombin are also thought to contribute to the process [97,101]. A recent study has implicated major histocompatibility complex (MHC)-compatible bone-marrow-derived stromal cells (PDGFRa+ Sca-1+ cells), which activate T cells and increase the production of inflammatory cytokines such as IL-6, contributing to fibrosis in a scleroderma animal model [102]. Increased production of inflammatory cytokines such as MCP-1 and interferon-inducible protein 10 and decreased production of overall VEGF have also been reported in patients with scleroderma [103,104].

Scleroderma is a poorly understood fibrotic disorder. Radiation-induced decrease in collagen production and increase in matrix metalloproteases have shown some success in limiting scleroderma in animal models [105]. Similarly, targeting adiponectin, a potent regulator of fibroblasts, has also shown to be effective in limiting scleroderma in animal models [106]. These results need further validation since mouse models and other preclinical models can only reflect part of the scleroderma disease and need to be optimized for the specific interventions being assessed [99]. Further development of reliable animal models and the development of multimodal drug therapies targeting autoantibodies, profibrotic factors, and mechanotransduction pathways could lead to effective therapies to control scleroderma.

### 3.4 DUPUYTREN'S CONTRACTURE

Dupuytren's disease is a fibroproliferative disorder characterized by excess fibrosis of the subcutaneous fascia and is commonly thought to be triggered by mechanical stress or vibrations in patients with a genetic predisposition to fibrosis [31–35]. Although Dupuytren's disease is the result of excessive fibrosis of the subcutaneous fascia, it has been shown to have a dermal component in many patients [36]. Since this type of fibrotic response does not normally occur in patients with no known family history, the affected patients likely have an innate dysfunction of the cells and molecules responsible for normal tissue homeostasis.

Due to the mechanical trigger, these abnormalities likely occur within mechanotransduction pathways and mechanoresponsive cells. For example, tissue from Dupuytren's disease exhibits increased levels of β-catenin, fibronectin, periostin, and insulin-like growth factor (IGF)-II [107]. While Dupuytren's disease is not malignant, the expression of these markers indicates that the tissue shows similarities to tumors. Fibroblasts isolated from Dupuytren's disease also display the upregulation of Wilms' tumor 1 (WT1), another gene associated with cancer [107]. A separate study showed that the subcutaneous fat tissue from Dupuytren's disease exhibits significant upregulation of a disintegrin and metalloproteinase domain (ADAM12), aldehyde dehydrogenase 1 (ALDH1A1), and iroquois-class homeobox protein 6 (IRX6) [108]. All these genes could serve as targets for new therapies. Further research into specific mechanotransduction pathways upstream of these markers could also serve as therapeutic targets.

Unlike other fibrotic skin disorders including keloids and scleroderma, one of the primary treatments of choice is surgical excision [109–112]. Newer modes of treatment are under investigation; for example, collagenase injections are being explored as a potential treatment for Dupuytren's disease. While collagenase treatment does reverse the effects of Dupuytren's disease by 70%, there might be an increased risk for skin rupture as compared with needle fasciectomies [113]. Similarly, radiotherapy at the early stages has also been shown to result in the regression of symptoms/lack of progression in 40% of patients with low recurrence rates [114]. Local silencing of mechanotransduction pathways along with radiotherapy and collagenase treatments could prove to be effective in controlling Dupuytren's disease.

### 3.5 OTHER FIBROTIC SKIN DISORDERS

There are several other fibrotic skin disorders where altered mechanobiology of skin components has been implicated in the pathogenesis and progression of the disease. For example, cutis hyperelastica is an inherited disorder with genetic defects that lead to improper synthesis of collagen types I, III, and V [97]. Abnormal collagen production leads to significantly increased elasticity and consequently abnormal wound healing. This leads to the formation of broad and atrophic scars and increased propensity for bruising and bleeding in the skin [97]. Cutis hyperelastica is often accompanied by other mutations and defects that can manifest in other tissues. For example, some patients display a loss of FK506-binding protein (FKBP14), a protein involved in collagen folding [115]. Loss of FKBP14 leads to progressive kyphoscoliosis, myopathy, and hearing loss. Similarly, a recent report also identified cutis hyperelastica affecting the vasculature [116]. Another fibrotic disorder that affects the mechanical properties of the skin is the cutis laxa, where the elastic fibers are not properly formed leading to inelastic skin and increased folds and wrinkles [97]. It has been postulated that altered mechanobiology in the skin of patients with cutis laxa leads to abnormal wound healing response and scarring. Other skin disorders that are potentially regulated by altered skin mechanobiology are reviewed by Ogawa and colleagues [97].

The prevalence of fibrotic skin disorders and fibrotic wound healing necessitates focused investigation of the mechanotransduction pathways in normal skin homeostasis and how they are altered during pathologies. Identification of specific intracellular targets or altering the ECM components could provide a means to restore skin mechanical homeostasis and cure fibrotic skin disorders.

## 4 MECHANOBIOLOGY OF DEGENERATIVE CUTANEOUS WOUND HEALING

At the opposite end of the cutaneous tissue healing and repair spectrum stand pathologically underhealing chronic wounds. While overhealing fibrotic skin disorders involve excessive and disorganized production of collagen and related ECM proteins leading to benign scars, underhealing wounds largely occur as a result of insufficient or delayed production of connective tissue proteins and other ECM components.

Such poor wound healing in the form of either chronic or nonhealing wounds imposes a significant health burden on our society.

In the United States alone, the estimated socioeconomic cost of chronic wound management exceeds $25 billion annually, but clinical outcomes have been largely suboptimal with currently available therapies [40,117,118]. Underlying pathophysiological factors that can cause chronic and/or nonhealing wounds include degenerative conditions such as diabetes and aging, although underhealing wounds can also result from sedentary lifestyle with very little or almost no physical activity [40]. Increasing evidence supports the importance of mechanotransduction in the pathogenesis of underhealing wounds associated with disease states; however, precise implications of its biochemistry and pathological outcomes remain poorly defined [41]. In the following sections, delayed wound healing in two principal degenerative conditions is reviewed. Recent evidence on how mechanical equilibrium and mechanotransduction pathway components might be dysregulated in these processes is particularly highlighted.

## 4.1 DIABETIC SKIN ULCERS

Diabetic ulcers are major complications of diabetes mellitus and markedly impair the quality of life of patients irrespective of its severity [119,120]. Open sores most frequently develop in the legs and feet in approximately up to 25% of patients with diabetes [40,121,122]. Foot ulcers account for 20% of all hospital admissions in diabetic patients, and it is estimated that 12%–24% of diabetic foot ulcers require amputation, the leading cause of nontraumatic lower-limb amputations in the United States [40,121,122]. Poorly controlled diabetes is usually the starting point. The lack of healthy blood circulation due to uncontrolled hyperglycemia can cause wounds to heal abnormally [123]. Diabetes-induced peripheral neuropathy and nerve damage also contribute to developing ulcers as sensitivity to pain or discomfort reduces under these conditions. Impaired blood circulation obstructs the flow of inflammatory cells (i.e., neutrophils and macrophages) that migrate and populate the wound bed [123,124]. Local inflammatory cells release a variety of cellular growth factors, cytokines, and chemokines that can recruit keratinocytes, fibroblasts, and circulating endothelial progenitor cells that are responsible for proper wound repair and remodeling, which may be prevented in diabetic wound healing and resulting in delayed tissue reconstruction [123,125]. On the other hand, the early inflammatory phases can be abnormally sustained beyond the normal time line, and persistence of neutrophils and macrophages can interfere with wound progression into subsequent remodeling phases [123]. Diminished cellular responses to hypoxic and ischemic conditions can lead to reduced angiogenesis and necrotic cell death that has been linked to abnormal inflammation and infection, further complicating wound healing under diabetic conditions [123].

Prolonged systemic high-glucose conditions in diabetes can alter the biochemical and biophysical homeostasis of various cells that constitute the epidermis and dermis of the skin [41,126]. To date, the complex and intricate cascade of events leading to the malfunction of cells is far from completely understood. One component that has been shown to be the major cause of many diabetic complications including neuropathy, nephropathy, and retinopathy is the formation of advanced glycation end products (AGE) that form via biochemical reaction of proteins and lipids with sugar [127,128]. Glycation-mediated modification of cellular components can impose deleterious effects on the normal functions of cell and organs by denaturation and functional decline of proteins, generation of oxidative stress, and accumulation of AGE in tissues (organopathy) [127,128]. Impaired wound healing in diabetes, therefore, is associated with physiological disruption of the skin resulting from persistent systemic hyperglycemia [123].

Diabetic skin is thought to be intrinsically inferior to the normal skin. Tissue integrity deteriorates with decreased collagen synthesis and elevated levels of glycated collagens and ECM proteins that disturb proper cross-linking of structural components [41,129]. In the context of wound healing, such defects become more prominent as epidermal and dermal cells should be actively producing a wide array of skeletal proteins to repair the breach. During wound healing, keratinocytes can play both regulatory and secretory roles and are primarily responsible for reepithelialization of wounds [130]. At the cellular level, exposure to high glucose is associated with changes in keratinocyte morphology, proliferation and differentiation rates (presumably via reduced sensitivity to growth factors), and expression of various surface proteins that are important for glucose transport and utilization [131]. As their cellular biochemistry and metabolic profile change, epidermal mechanics can also undergo pathological dysregulation during wound remodeling [59,131]. Dermal fibroblasts that provide support for keratinocytes and endothelial cells that constitute small and large blood vessels are affected with hyperglycemia, and their behavior during wound healing can be defective [41,132]. High glucose impacts proliferation, differentiation, senescence, and apoptotic capacity of fibroblasts derived from many different organs [132–135]. Fibroblast production of collagen and other ECM components is also altered under hyperglycemia [133,136]. Endothelial cells also respond to changes in extracellular glucose concentration. Endothelial cell permeability, cell adhesiveness, coagulative activity, and endothelium-dependent relaxation properties are impaired, all of which further impacts the normal functions of blood microvasculature in the wound bed [137]. Not only these structural cells but also circulating progenitor and immune cells present in the granulation tissues may further display abnormalities [138–140]. As granulation tissues undergo structural remodeling to reoptimize the mechanical properties of the healing skin, cellular and molecular components important for these processes may be defective and dysfunctional.

Although diabetic ulcers often successfully heal with appropriate therapies and proper clinical management, healed skin can be substantially inferior to normally healed skin in terms of architecture and skin functionality [41,129,141]. In animal studies, healed diabetic ulcers are overall fragile and brittle, characterized by thinner dermis and significantly reduced tensile strength (the ability of the skin to withstand external forces) compared with that of euglycemic animals after injury [41].

Such abnormal architecture of healed diabetic skin leads to increased susceptibility to future insult, and slow-healing wounds reoccur due to pressure, shear, and friction. The cellular and molecular basis of inferior biomechanical characteristics of diabetic skin and healed diabetic skin is still unclear. In the current literature, unwounded diabetic skin from mouse models and human cadaveric skin both display decrease in cross-sectional area, tensile strength, and Young's modulus (the modulus of longitudinal elasticity) [129]. Protein content of mature collagen I and III, main components of ECM responsible for structural integrity, is lower in diabetic skin [129]. Collagen content is regulated by the balancing act of synthesizing new collagen and degradation of existing collagen via MMP [142]. Expression and activity of these enzymes are regulated by another group of peptidases involved in the degradation of ECM known as tissue inhibitors of metalloproteinase (TIMP) [142]. Inflammatory cytokines, such as tumor necrosis factor-α (TNF-α) and TGF-β1, are also implicated in these processes and can mediate the activation of other pathological events, such as fibrosis, as we addressed in earlier sections. This balance between MMP and TIMP enzymes is critical for ECM composition and homeostasis [142,143]. Clinical data have demonstrated that diabetic wounds that have a greater propensity for healing have lower MMP enzyme activities but higher levels of TIMP1 [144]. Perhaps, this at least partially explains the mechanistic basis of poor inherent mechanical properties found in diabetic skin and defective underhealing of diabetic wounds.

There is emerging evidence that mechanotransduction factors other than MMP/TIMP imbalance can play an important role in producing abnormal architecture of healed diabetic wounds. We previously reviewed that an array of different mechanotransduction signaling pathways are involved in wound healing and remodeling of the matrix. Often, exaggerated activation of these mechanotransduction pathways can lead to fibrogenic activities and overhealing skin disorders. On the other hand, deactivation or insufficient activation of these pathways can result in opposing outcomes in underhealing wounds [41,145,146]. Animal research has found that under high-glucose conditions, degradation of a key mechanotransduction pathway mediator, FAK, can be abnormally enhanced due to increased activity of the proteolytic enzymes (Fig. 5) [41]. FAK is a central player in skin mechanosensing, and recent publications have shown that overactivation of the FAK pathway mediates skin fibrosis via the activation of inflammatory responses [49]. Suppression of FAK in fibroblasts, using either genetically engineered animals or pharmacological inhibitors of FAK, results in reduced collagen deposition, reduced skin inflammation, and attenuated scar formation during wound healing and remodeling [49]. Keratinocyte-specific inhibition of FAK, however, exhibits delayed reepithelialization and reduced mechanical strength with increased MMP activities [16]. The mechanisms by which FAK activation/deactivation and synthesis/degradation are regulated are incompletely understood in the context of skin mechanobiology. A group of proteolytic enzymes that target FAK degradation is calpains [147]. Calpains play a vital role in adhesion dynamics, hence, regulating cell migration, and their target substrates include various proteins that are present in the focal adhesions [147]. Recent evidence suggests that calpain-mediated FAK breakdown is

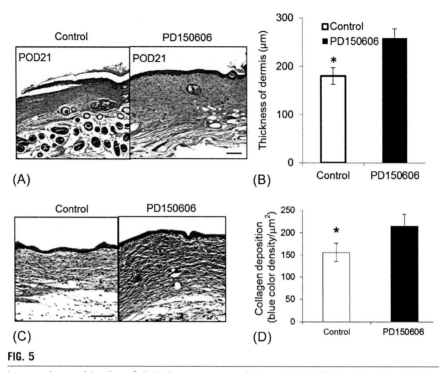

**FIG. 5**

Improved wound healing of diabetic mouse wounds by calpain inhibition. Healed diabetic wounds displayed significantly thicker dermis and increased collagen expression with calpain inhibition. (A) Hematoxylin and eosin histology staining and (B) dermal thickness with calpain inhibitor (PD150606) treatment. Scale bar = 100 μm (B) of wound sites on day 21 postwounding. *$P=0.005$ versus PD150606, $n=14$. (C) Masson's trichrome staining of diabetic control and PD150606-treated wounds in db/db on day 21 postwounding. Scale bar = 100 μm. *Blue*, collagen; *red*, cytoplasm and muscle; and *black*, cell nuclei. PD150606 significantly increased dermal thickness, FAK expression in the dermis, and collagen deposition. (D) Blue color density per unit area in the dermal area representing collagen expression in Masson's trichrome staining images. Error bars: mean ± SD. *$P=0.0014$, $n=14$ [41].

accelerated in diabetic wounds due to abnormally elevated calpain activity under hyperglycemia [41]. It is postulated that increased calpain action is the result of AGE-mediated activation of myriad cellular pathways through the receptors of advanced glycation end products (RAGE) [41,148]. Given that physiological levels of FAK are important for collagen and ECM formation and normalizing skin mechanical properties, the loss of FAK in underhealing may underlie dermal defects. Whether other elements of the mechanotransduction machinery are dysregulated in the process is an ongoing research.

It is clear that pathological pathways exist in diabetic ulcers and poor wound healing. Studies to understand the inherent dermal defects and inferior healing properties of diabetic wounds point toward the disruption of key mechanotransduction pathway components and aberrantly regulated ECM-degrading events. Therapeutic strategies to correct these aberrations, therefore, can improve mechanical parameters of diabetic skin and wound healing outcomes and potentially lead to significant health and socioeconomic benefits related to diabetes complications.

## 4.2 WOUND HEALING IN AGING

Robustness of skin physical properties diminishes as the human body ages. Aging is another degenerative factor that perturbs the physiology of normal wound healing. Age-associated differences in wound healing have been well documented in the gerontology literature; however, the importance of understanding the detailed mechanisms of poor healing and development of treatment strategies is increasing with the growing aging population worldwide. Many of the processes involved in wound healing may be either delayed or defective with aging, but wound healing in the elderly is not always impaired [149,150]. Studies show that once wounds completely heal in healthy elderly patients, healed skin can be qualitatively similar to that of young subjects [150]. However, concomitant medical conditions such as cardiovascular disease, diabetes, cancer, and other degenerative problems in elderly can largely confound the wound healing processes [151].

Wound healing in the human fetus and infants is surprisingly accelerated often in the absence of scar formation [152]. Despite extensive investigation, the mechanisms of scarless wound regeneration especially in early gestation fetal days are still unclear. Numerous studies have demonstrated that multiple features in the healing process are different between fetal, pediatric, and adult wound healing [152–154]. Fetal and postnatal wounds differ in terms of ECM composition, inflammatory response, extra- and intracellular mediators (e.g., growth factors), and global gene expression profiles, all of which may lead to differential healing properties [152–155]. Clearly, there are inherent age-related differences of intact postnatal skin. As the skin structure and also composition change dramatically with development and time, so do the mechanical parameters [155]. Skin viscoelasticity is a well-characterized parameter that shows distinct age-dependent differences. On average, skin tension and Young's modulus of pediatric skin are both greater than the tension and modulus of an elderly adult [156]. Moreover, the mean ultimate skin deformation before breaking is typically lower for the elderly [156]. Changes in these parameters indicate that the overall skin becomes stiffer, less tense, and less flexible with aging, suggesting that its protective functionality against external forces and injuries is reduced. The anisotropy of the skin increases as its directional characteristics change with aging and is more prominently observed in anatomical locations where the skin is tender such as the inner region of the upper arm and the neck [156,157]. Studies have also shown that the directions of dermal tension and cleavage lines (e.g., Langer's lines, Kraissl's lines, and Borges' lines) can change with age progression [156,157].

Age-dependent changes of skin mechanical properties perhaps fundamentally arise from skin structure and composition changes over the course of time. In some literature, this is collectively referred to as age-associated dermal microenvironment (AADM) alterations [155]. There are phenotypic differences in major connective tissue protein contents and organization. For example, marked changes in collagen fibrils responsible for skin tensile strength can cause stiffer skin with natural aging [155]. It is thought that the dysregulation of ECM-remodeling proteins and signaling pathways governing the physiology of such proteins underlies these structural alterations of the skin [155]. Like other organs, the skin has been exposed to oxidative stress caused by ROS and different external insults such as UV irradiation and physical and chemical stress over the course of natural aging [155,158–160]. Chronic exposure to these assaults may damage or interfere with the cells' normal metabolism. Prolonged exposure to ROS in dermal fibroblasts upregulates cysteine-rich protein 61 (CCN1), a member of the family of proteins that mediates aberrant collagen homeostasis, leading to decreased TGF-$\beta$ signaling and increased multiple MMP expression and pro-inflammatory cytokines [155,161–163]. These important changes result in exaggerated inflammation and decreased new collagen synthesis, while collagen breakdown and fragmentation are increased, overall leading to thinner and damaged collagenous ECM. Although there are numerous factors implicated, most likely, no one factor is thought to be the sole cause of age-related skin deterioration.

In the context of impaired wound healing in the elderly, age and other underlying pathologies such as degenerative diseases can work as serious risk factors. Studies have shown that CCN1-mediated AADM alterations are rapidly induced with skin wounding even in young tissues with healthy architecture, affecting both the inflammatory phase and the remodeling phase of repair [155]. Constitutive elevation of CCN1 in aged skin, therefore, can predispose aged skin to impairment in wound healing presumably via extended inflammatory actions and/or defective remodeling of ECM. Clear understanding of age-related changes in skin integrity and how such mechanisms contribute to chronic wounds in aged skin will improve our insight into age-related treatment strategies.

## 5 THERAPEUTIC STRATEGIES FOR MODULATION OF DYSFUNCTIONAL MECHANOBIOLOGY

As current research suggests that scars following aberrant wound healing and other fibrous disorders of the skin may be improved via the manipulation of the skin mechanoenvironment, the management of such diseases utilizing new medical devices that modulate skin biomechanics has produced a fast-growing market, in addition to conventional operational revisions. Noninvasive therapeutic strategies that pharmacologically target novel key biomolecular mechanotransduction cascade components have also received highlight in recent studies, although these may still be at the experimental stage with translational potential. Various mechanomodulatory methods to mitigate dysfunctional skin mechanobiology are discussed in the context of skin scar management.

## 5.1 PREOPERATIVE SURGICAL TECHNIQUES

Historically in cutaneous surgery, surgical incisions on the skin are made parallel to Langer's lines to produce least tension as incisions made this way generally heal better with less scarring and fibrosis [164]. These lines naturally run parallel to major collagen bundles but do not always coincide with the lines of wrinkles [164,165]. Relaxed skin tension lines, also called Borges' lines, follow furrows when the skin is relaxed and are produced by pinching the skin [166]. Generally, Langer's and Borges' lines run similar over many parts of the body but are different in mechanically complex areas such as the mouth corner, temple, and lateral canthal area [167]. There are multiple other guidelines of minimal tension in the surgical literature that surgeons have been employing in practice as preventive measures of scarring complications. While conventional skin lines largely provide guidance for incision design, the skin dynamics and tension can change with excisional wounds (e.g., excision after skin cancer) [168]. In recent literature, the importance of understanding the difference between incisional skin lines and biodynamics of excisional skin tension lines has received attention, as surgical design based on conventional lines may not be necessarily suitable for excision surgeries. Mechanical parameters such as closing tension, extensibility, and midline and terminal Young's modulus are particularly important considerations for excisional lines, and mechanical responses can be more asymmetrical for excisions [168]. In silico and computer-based simulation models are now popular methods to map and predict such dynamic processes in skin biology including wound healing, inflammatory responses, and drug permeability through the skin [169–171]. For example, FEA methods have been applied to study the role of mechanical forces in wound contracture and HTS formation [169,172,173]. More comprehensive understanding of the skin and wound biomechanics will improve our insight into adopting preventive measures for cutaneous surgeries to limit mechanically induced healing complications.

## 5.2 MECHANOMODULATORY DEVICES

Exuberant cutaneous scar formation, following either surgical intervention or traumatic injury, can be surgically resected. Surgical scar revision does not erase but modifies scars to appear less noticeable and minimized. Nonsurgical techniques utilizing laser therapies and nonoperative modalities using topical application of steroid- or silicone-based creams can also work effectively for scar management. Although numerous treatment modalities have been tested and used in the clinics for decades, inconsistent clinical outcome has become disappointing [50]. In earlier sections, we addressed that exaggerated scar formation and possibly other fibrous skin disorders may arise from mechanical stress that acts through mechanotransduction and inflammatory pathways during the course of healing. An emerging trend in excessive scar management, therefore, would be to postsurgically modulate skin mechanics to reduce the effects of mechanical stress on healing wounds and to reestablish mechanical homeostasis (Fig. 6). For high-tension incisions, mechanomodulatory devices or often called stress-shielding or off-loading devices and applicators that lower wound tension have been developed to actively control the wound mechanoenvironment [84]. Elastomeric polymer technology allows the production of conveniently used silicone applicator sheets designed to transfer tension from the lateral edges of the wound bed to the polymer device [173]. Preclinical animal studies and human clinical trials have been performed using these devices and demonstrated scar-reducing efficacy by stress-shielding without causing problematic dehiscence or delayed healing [84,174]. These topical devices, however, may work best with linear incisional wounds on noncomplex body regions and may not be amenable to treat wounds affecting large body areas as seen in massive contact burn wounds and blast injuries. Alternative mechanomodulatory approaches using biochemical agents are discussed in the section below.

## 5.3 BIOCHEMICAL INHIBITORS

In addition to physically blocking mechanical stress and high tension with topical devices, recent research has demonstrated that biochemical and pharmacological inhibitors that modulate the mechanotransduction pathway components may also be effective tools for fibroproliferative scar management. Nonoperational and nondevice strategies are attractive in that this approach would be more relevant to treating large-area wounds and/or irregularly contoured wounds (e.g., hand and face). Inhibitory molecules such as small interfering RNAs (siRNA) delivered via gene therapy technologies and small-molecule inhibitors of mechanotransduction signaling pathway components and ECM-remodeling factors that play important roles in cellular fibrosis have been of particular interest to researchers. Some of these agents have been proved to be effective in treating fibrosis of internal organs in animal models. For example, inhibition of a key histone transcriptional regulator using a specific siRNA has recently been shown to attenuate the progression of myocardial fibrosis by suppressing TGF-$\beta$1-mediated fibrogenic events in cardiac myofibroblasts [175]. In idiopathic lung fibrosis, inhibition of canonical Wnt/$\beta$-catenin pathway via overexpression of a Wnt antagonist, Dickkopf-1, potently reduced profibrotic effects of TGF-$\beta$1 and ameliorated lung fibrosis [176]. In lung and skin fibrosis, small-molecule-mediated inhibition of the mechanosensor FAK in the integrin mechanotransduction pathway has been successfully used to prevent experimental lung fibrosis and mechanically induced skin scar formation [49,177,178]. FAK inhibition significantly reduced fibroblast migration, production of $\alpha$-smooth muscle actin, and aberrant collagen deposition in these studies [49,177,178]. Collectively, biochemically targeted inhibition of elements that are critical in fibrosis and scar mechanotransduction pathways is highly translational and may have major

# 5 Therapeutic strategies for modulation of dysfunctional mechanobiology 437

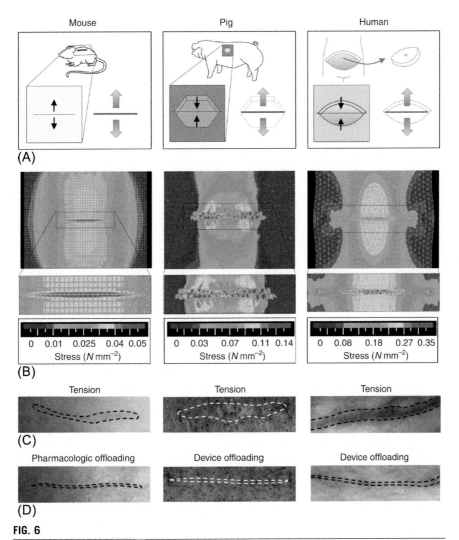

**FIG. 6**

Scar mechanotransduction in animal models and humans and off-loading of tension on incisions. (A) Schematics of mouse and pig models of overhealing based on mechanical forces. An analogous situation occurs following closure of abdominoplasty wounds in humans. (B) Linear elastic finite element analysis predicts that linear incisions experience increased tension in these models. (C) Incisions exposed to high tension in mouse (21 days postinjury), pig (8 weeks postinjury), and human (8 months postinjury). (D) High-tension incisions treated with either pharmacologic (FAK inhibitor) or device (stress-shielding polymer) approaches to lower tension and attenuate fibrosis [79].

advantages over other approaches. However, the development of relevant drug delivery systems and extensive safety studies on potential adverse effects of each agent will also be imperative to successful pharmaceutical therapies against fibrous scar formation.

## 6 FUTURE RESEARCH AND DIRECTIONS

A better understanding of the skin mechanobiology and how homeostasis of the skin's mechanoenvironment can be disrupted under pathological conditions will be of paramount importance in designing effective new techniques to treat associated skin disorders. Current mechanomodulatory strategies largely aim to mitigate over-healing wounds and aberrant fibrotic disorders; however, recent evidence has begun to suggest that underhealing wounds are also associated with disrupted skin mechanoenvironment and devising targeted therapies against underhealing wounds is anticipated. Furthermore, the development of novel drug delivery systems that are topically applied to target aberrant mechanotransduction pathways could lead to the development of effective therapies.

## 7 CONCLUSION

The cellular pathways governing skin mechanotransduction are complex and are linked to other pathophysiological events such as inflammation and fibrosis, all of which may be dysregulated in over- and underhealing wounds. As the skin is the most frequently injured organ, the importance of understanding the mechanobiology of wound healing deserves ample attention. Our improved knowledge of this system will complement current strategies targeted at modulating the mechanobiology of the skin for the treatment of skin diseases.

## REFERENCES

[1] Ingber DE. Cellular mechanotransduction: putting all the pieces together again. FASEB J 2006;20:811–27.
[2] Wang N, Tytell JD, Ingber DE. Mechanotransduction at a distance: mechanically coupling the extracellular matrix with the nucleus. Nat Rev Mol Cell Biol 2009;10:75–82.
[3] Duscher D, Maan ZN, Wong VW, Rennert RC, Januszyk M, Rodrigues M, et al. Mechanotransduction and fibrosis. J Biomech 2014;47:1997–2005.
[4] Gurtner GC, Werner S, Barrandon Y, Longaker MT. Wound repair and regeneration. Nature 2008;453:314–21.
[5] Wong VW, Longaker MT, Gurtner GC. Soft tissue mechanotransduction in wound healing and fibrosis. Semin Cell Dev Biol 2012;23:981–6.

[6] Rinkevich Y, Walmsley GG, Hu MS, Maan ZN, Newman AM, Drukker M, et al. Skin fibrosis. Identification and isolation of a dermal lineage with intrinsic fibrogenic potential. Science 2015;348:aaa2151.

[7] Driskell RR, Lichtenberger BM, Hoste E, Kretzschmar K, Simons BD, Charalambous M, et al. Distinct fibroblast lineages determine dermal architecture in skin development and repair. Nature 2013;504:277–81.

[8] Derderian CA, Bastidas N, Lerman OZ, Bhatt KA, Lin SE, Voss J, et al. Mechanical strain alters gene expression in an in vitro model of hypertrophic scarring. Ann Plast Surg 2005;55:69–75.

[9] He Y, Macarak EJ, Korostoff JM, Howard PS. Compression and tension: differential effects on matrix accumulation by periodontal ligament fibroblasts in vitro. Connect Tissue Res 2004;45:28–39.

[10] Junker JP, Kratz C, Tollback A, Kratz G. Mechanical tension stimulates the transdifferentiation of fibroblasts into myofibroblasts in human burn scars. Burns 2008;34:942–6.

[11] Wang Z, Fong KD, Phan TT, Lim IJ, Longaker MT, Yang GP. Increased transcriptional response to mechanical strain in keloid fibroblasts due to increased focal adhesion complex formation. J Cell Physiol 2006;206:510–7.

[12] Paterno J, Vial IN, Wong VW, Rustad KC, Sorkin M, Shi Y, et al. Akt-mediated mechanotransduction in murine fibroblasts during hypertrophic scar formation. Wound Repair Regen 2011;19:49–58.

[13] Reno F, Traina V, Cannas M. Mechanical stretching modulates growth direction and MMP-9 release in human keratinocyte monolayer. Cell Adhes Migr 2009;3:239–42.

[14] Yano S, Komine M, Fujimoto M, Okochi H, Tamaki K. Mechanical stretching in vitro regulates signal transduction pathways and cellular proliferation in human epidermal keratinocytes. J Invest Dermatol 2004;122:783–90.

[15] Yano S, Komine M, Fujimoto M, Okochi H, Tamaki K. Activation of Akt by mechanical stretching in human epidermal keratinocytes. Exp Dermatol 2006;15:356–61.

[16] Wong VW, Garg RK, Sorkin M, Rustad KC, Akaishi S, Levi K, et al. Loss of keratinocyte focal adhesion kinase stimulates dermal proteolysis through upregulation of MMP9 in wound healing. Ann Surg 2014;260:1138–46.

[17] Januszyk M, Kwon SH, Wong VW, Padmanabhan J, Maan ZN, Whittam AJ, et al. The role of focal adhesion kinase in fibrogenic gene expression. Int J Mol Sci 2017;18:1915.

[18] Fuchs E. Epithelial skin biology: three decades of developmental biology, a hundred questions answered and a thousand new ones to address. Curr Top Dev Biol 2016;116:357–74.

[19] Dupasquier M, Stoitzner P, van Oudenaren A, Romani N, Leenen PJ. Macrophages and dendritic cells constitute a major subpopulation of cells in the mouse dermis. J Invest Dermatol 2004;123:876–9.

[20] Mastrogiannaki M, Lichtenberger BM, Reimer A, Collins CA, Driskell RR, Watt FM, et al. Beta-catenin stabilization in skin fibroblasts causes fibrotic lesions by preventing adipocyte differentiation of the reticular dermis. J Invest Dermatol 2016;136:1130–42.

[21] Driskell RR, Jahoda CA, Chuong CM, Watt FM, Horsley V. Defining dermal adipose tissue. Exp Dermatol 2014;23:629–31.

[22] Chien S. Mechanotransduction and endothelial cell homeostasis: the wisdom of the cell. Am J Physiol Heart Circ Physiol 2007;292:H1209–24.

[23] Liu HB, Zhang J, Xin SY, Liu C, Wang CY, Zhao D, et al. Mechanosensitive properties in the endothelium and their roles in the regulation of endothelial function. J Cardiovasc Pharmacol 2013;61:461–70.

[24] Nauseef WM, Borregaard N. Neutrophils at work. Nat Immunol 2014;15:602–11.
[25] Auffray C, Fogg D, Garfa M, Elain G, Join-Lambert O, Kayal S, et al. Monitoring of blood vessels and tissues by a population of monocytes with patrolling behavior. Science 2007;317:666–70.
[26] Bloemen MC, van der Veer WM, Ulrich MM, Zuijlen PP, Niessen FB, Middelkoop E. Prevention and curative management of hypertrophic scar formation. Burns 2009;35: 463–75.
[27] Bock O, Schmid-Ott G, Malewski P, Mrowietz U. Quality of life of patients with keloid and hypertrophic scarring. Arch Dermatol Res 2006;297:433–8.
[28] Aarabi S, Longaker MT, Gurtner GC. Hypertrophic scar formation following burns and trauma: new approaches to treatment. PLoS Med 2007;4:e234.
[29] Ogawa R. Mechanobiology of scarring. Wound Repair Regen 2011;19(Suppl 1):s2–9.
[30] Wong VW, You F, Januszyk M, Gurtner GC, Kuang AA. Transcriptional profiling of rapamycin-treated fibroblasts from hypertrophic and keloid scars. Ann Plast Surg 2014;72:711–9.
[31] Hindocha S, McGrouther DA, Bayat A. Epidemiological evaluation of Dupuytren's disease incidence and prevalence rates in relation to etiology. Hand 2009;4:256–69.
[32] Palmer KT, Bovenzi M. Rheumatic effects of vibration at work. Best Pract Res Clin Rheumatol 2015;29:424–39.
[33] Palmer KT, D'Angelo S, Syddall S, Griffin MJ, Cooper C, Coggon D. Dupuytren's contracture and occupational exposure to hand-transmitted vibration. Occup Environ Med 2014;71:241–5.
[34] Descatha A, Bodin J, Ha C, Goubault P, Lebreton M, Chastang JF, et al. Heavy manual work, exposure to vibration and Dupuytren's disease? Results of a surveillance program for musculoskeletal disorders. Occup Environ Med 2012;69:296–9.
[35] Elliot D, Ragoowansi R. Dupuytren's disease secondary to acute injury, infection or operation distal to the elbow in the ipsilateral upper limb—a historical review. J Hand Surg (Br) 2005;30:148–56.
[36] Wade R, Igali L, Figus A. Skin involvement in Dupuytren's disease. J Hand Surg Eur 2016;41:600–8.
[37] American Burn Association. Burn incidence fact sheet. ABA; 2016, http://ameriburn.org/who-we-are/media/burn-incidence-fact-sheet/.
[38] Improving America's Hospitals. The joint commission's annual report on quality and safety; 2007, https://www.jointcommission.org/annualreport.aspx.
[39] Walmsley GG, Maan ZN, Wong VW, Duscher D, Hu MS, Zielins ER, et al. Scarless wound healing: chasing the holy grail. Plast Reconstr Surg 2015;135:907–17.
[40] Sen CK, Gordillo GM, Roy S, Kirsner R, Lambert L, Hunt TK, et al. Human skin wounds: a major and snowballing threat to public health and the economy. Wound Repair Regen 2009;17:763–71.
[41] Liu W, Kun M, Kwon SH, Garg R, Patta YR, Fujiwara T, et al. The abnormal architecture of healed diabetic ulcers is the result of FAK degradation by calpain 1. J Invest Dermatol 2017;137:1155–65.
[42] Stefano GB, Challenger S, Kream RM. Hyperglycemia-associated alterations in cellular signaling and dysregulated mitochondrial bioenergetics in human metabolic disorders. Eur J Nutr 2016;55:2339–45.
[43] Graves DT, Kayal RA. Diabetic complications and dysregulated innate immunity. Front Biosci 2008;13:1227–39.

[44] Yazdanpanah L, Nasiri M, Adarvishi S. Literature review on the management of diabetic foot ulcer. World J Diabetes 2015;6:37–53.

[45] Alavi A, Sibbald RG, Mayer D, Goodman L, Botros M, Armstrong DG, et al. Diabetic foot ulcers: part II. Management. J Am Acad Dermatol 2014;70:21. e1–4; quiz 45–6.

[46] Cavanagh PR, Lipsky BA, Bradbury AW, Botek G. Treatment for diabetic foot ulcers. Lancet 2005;366:1725–35.

[47] Pecoraro RE, Reiber GE, Burgess EM. Pathways to diabetic limb amputation. Basis for prevention. Diabetes Care 1990;13:513–21.

[48] Aarabi S, Bhatt KA, Shi Y, Paterno J, Chang EI, Loh SA, et al. Mechanical load initiates hypertrophic scar formation through decreased cellular apoptosis. FASEB J 2007;21: 3250–61.

[49] Wong VW, Rustad KC, Akaishi S, Sorkin M, Glotzbach JP, Januszyk M, et al. Focal adhesion kinase links mechanical force to skin fibrosis via inflammatory signaling. Nat Med 2011;18:148–52.

[50] Gurtner GC, Dauskardt RH, Wong VW, Bhatt KA, Wu K, Vial IN, et al. Improving cutaneous scar formation by controlling the mechanical environment: large animal and phase I studies. Ann Surg 2011;254:217–25.

[51] Januszyk M, Wong VW, Bhatt KA, Vial IN, Paterno J, Longaker MT, et al. Mechanical offloading of incisional wounds is associated with transcriptional downregulation of inflammatory pathways in a large animal model. Organ 2014;10:186–93.

[52] Kwon SH, Ma K, Duscher D, Padmanabhan J, Dong Y, Inayathullah M, et al. In: Targeted delivery of small molecule focal adhesion kinase inhibitor via pullulan-collagen hydrogel scaffold for scarless wound healing. Abstract Presented at Annual Meeting of the 2017 Wound Healing Society, April 5, San Diego, CA; 2017.

[53] Rennert RC, Januszyk M, Sorkin M, Rodrigues M, Maan ZN, Duscher D, et al. Microfluidic single-cell transcriptional analysis rationally identifies novel surface marker profiles to enhance cell-based therapies. Nat Commun 2016;7:11945.

[54] Rennert RC, Sorkin M, Januszyk M, Duscher D, Kosaraju R, Chung MT, et al. Diabetes impairs the angiogenic potential of adipose-derived stem cells by selectively depleting cellular subpopulations. Stem Cell Res Ther 2014;5:79.

[55] Suga H, Rennert RC, Rodrigues M, Sorkin M, Glotzbach JP, Januszyk M, et al. Tracking the elusive fibrocyte: identification and characterization of collagen-producing hematopoietic lineage cells during murine wound healing. Stem Cells 2014;32:1347–60.

[56] Humphrey JD, Dufresne ER, Schwartz MA. Mechanotransduction and extracellular matrix homeostasis. Nat Rev Mol Cell Biol 2014;15(12):802.

[57] Edlich RF, Carl BA. Predicting scar formation: from ritual practice (Langer's lines) to scientific discipline (static and dynamic skin tensions). J Emerg Med 1998;16:759–60.

[58] Waldorf JC, Perdikis G, Terkonda SP. Planning incisions. Oper Tech Gen Surg 2002;4:199–206.

[59] Evans ND, Oreffo RO, Healy E, Thurner PJ, Man YH. Epithelial mechanobiology, skin wound healing, and the stem cell niche. J Mech Behav Biomed Mater 2013;28:397–409.

[60] Liang X, Boppart SA. Biomechanical properties of in vivo human skin from dynamic optical coherence elastography. IEEE Trans Biomed Eng 2010;57:953–9.

[61] Kalra A, Lowe A, Al-Jumaily AM. Mechanical behaviour of skin: a review. J Mater Sci Eng 2016;5:254.

[62] Limbert G. Mathematical and computational modelling of skin biophysics: a review. Proc Math Phys Eng Sci 2017;473(2203):20170257.
[63] Eckes B, Krieg T. Regulation of connective tissue homeostasis in the skin by mechanical forces. Clin Exp Rheumatol 2004;22:S73–6.
[64] Chatterjee S, Fisher AB. Mechanotransduction in the endothelium: role of membrane proteins and reactive oxygen species in sensing, transduction, and transmission of the signal with altered blood flow. Antioxid Redox Signal 2014;20:899–913.
[65] Koller A. Signaling pathways of mechanotransduction in arteriolar endothelium and smooth muscle cells in hypertension. Microcirculation 2002;9:277–94.
[66] Svedman P. Mechanical homeostasis regulating adipose tissue volume. Head Face Med 2007;3:34.
[67] Padmanabhan J, Augelli MJ, Cheung B, Kinser ER, Cleary B, Kumar P, et al. Regulation of cell-cell fusion by nanotopography. Sci Rep 2016;6:33277.
[68] Schwarz US, Gardel ML. United we stand: integrating the actin cytoskeleton and cell-matrix adhesions in cellular mechanotransduction. J Cell Sci 2012;125:3051–60.
[69] Kenny FN, Connelly JT. Integrin-mediated adhesion and mechano-sensing in cutaneous wound healing. Cell Tissue Res 2015;360:571–82.
[70] Hemmings L, Rees DJ, Ohanian V, Bolton SJ, Gilmore AP, Patel B, et al. Talin contains three actin-binding sites each of which is adjacent to a vinculin-binding site. J Cell Sci 1996;109(Pt 11):2715–26.
[71] Bakolitsa C, de Pereda JM, Bagshaw CR, Critchley DR, Liddington RC. Crystal structure of the vinculin tail suggests a pathway for activation. Cell 1999;99:603–13.
[72] Lee SE, Kamm RD, Mofrad MR. Force-induced activation of talin and its possible role in focal adhesion mechanotransduction. J Biomech 2007;40:2096–106.
[73] Lee FY, Zhen YY, Yuen CM, Fan R, Chen YT, Sheu JJ, et al. The mTOR-FAK mechanotransduction signaling axis for focal adhesion maturation and cell proliferation. Am J Transl Res 2017;9:1603–17.
[74] Lee HJ, Diaz MF, Ewere A, Olson SD, Cox Jr. CS, Wenzel PL, et al. Focal adhesion kinase signaling regulates anti-inflammatory function of bone marrow mesenchymal stromal cells induced by biomechanical force. Cell Signal 2017;38:1–9.
[75] Silver FH, Siperko LM, Seehra GP. Mechanobiology of force transduction in dermal tissue. Skin Res Technol 2003;9:3–23.
[76] Padmanabhan J, Kinser ER, Stalter MA, Duncan-Lewis C, Balestrini JL, Sawyer AJ, et al. Engineering cellular response using nanopatterned bulk metallic glass. ACS Nano 2014;8:4366–75.
[77] Takei T, Han O, Ikeda M, Male P, Mills I, Sumpio BE. Cyclic strain stimulates isoform-specific PKC activation and translocation in cultured human keratinocytes. J Cell Biochem 1997;67:327–37.
[78] Kippenberger S, Bernd A, Loitsch S, Guschel M, Muller J, Bereiter-Hahn J, et al. Signaling of mechanical stretch in human keratinocytes via MAP kinases. J Invest Dermatol 2000;114:408–12.
[79] Wong VW, Akaishi S, Longaker MT, Gurtner GC. Pushing back: wound mechanotransduction in repair and regeneration. J Invest Dermatol 2011;131:2186–96.
[80] Ishise H, Larson B, Hirata Y, Fujiwara T, Nishimoto S, Kubo T, et al. Hypertrophic scar contracture is mediated by the TRPC3 mechanical force transducer via NFkB activation. Sci Rep 2015;5:11620.

[81] Kiya K, Kubo T, Kawai K, Matsuzaki S, Maeda D, Fujiwara T, et al. Endothelial cell-derived endothelin-1 is involved in abnormal scar formation by dermal fibroblasts through RhoA/Rho-kinase pathway. Exp Dermatol 2017;26:705–12.

[82] Cho YS, Jeon JH, Hong A, Yang HT, Yim H, Cho YS, et al. The effect of burn rehabilitation massage therapy on hypertrophic scar after burn: a randomized controlled trial. Burns 2014;40:1513–20.

[83] Longaker MT, Rohrich RJ, Greenberg L, Furnas H, Wald R, Bansal V, et al. A randomized controlled trial of the embrace advanced scar therapy device to reduce incisional scar formation. Plast Reconstr Surg 2014;134:536–46.

[84] Wong VW, Beasley B, Zepeda J, Dauskardt RH, Yock PG, Longaker MT. A mechanomodulatory device to minimize incisional scar formation. Adv Wound Care 2013;2:185–94.

[85] Lim AF, Weintraub J, Kaplan EN, Januszyk M, Cowley C, McLaughlin P, et al. The embrace device significantly decreases scarring following scar revision surgery in a randomized controlled trial. Plast Reconstr Surg 2014;133:398–405.

[86] Kuwahara H, Tosa M, Egawa S, Murakami M, Mohammad G, Ogawa R. Examination of epithelial mesenchymal transition in keloid tissues and possibility of keloid therapy target. Plast Reconstr Surg Glob Open 2016;4:e1138.

[87] Ogawa R. Keloid and hypertrophic scarring may result from a mechanoreceptor or mechanosensitive nociceptor disorder. Med Hypotheses 2008;71:493–500.

[88] Marneros AG, Norris JE, Olsen BR, Reichenberger E. Clinical genetics of familial keloids. Arch Dermatol 2001;137:1429–34.

[89] Ogawa R, Miyashita T, Hyakusoku H, Akaishi S, Kuribayashi S, Tateno A. Postoperative radiation protocol for keloids and hypertrophic scars: statistical analysis of 370 sites followed for over 18 months. Ann Plast Surg 2007;59:688–91.

[90] Ogawa R, Okai K, Tokumura F, Mori K, Ohmori Y, Huang C, et al. The relationship between skin stretching/contraction and pathologic scarring: the important role of mechanical forces in keloid generation. Wound Repair Regen 2012;20:149–57.

[91] Tanaydin V, Beugels S, Piatkowski A, Colla C, van den Kerckhove E, Hugenholtz GC, et al. Efficacy of custom-made pressure clips for ear keloid treatment after surgical excision. J Plast Reconstr Aesthet Surg 2016;69:115–21.

[92] Schneider D, Wickstrom SA. Force generation and transmission in keloid fibroblasts: dissecting the role of mechanosensitive molecules in cell function. Exp Dermatol 2015;24:574–5.

[93] Harn HI, Wang YK, Hsu CK, Ho YT, Huang YW, Chiu WT, et al. Mechanical coupling of cytoskeletal elasticity and force generation is crucial for understanding the migrating nature of keloid fibroblasts. Exp Dermatol 2015;24:579–84.

[94] Suarez E, Syed F, Alonso-Rasgado T, Bayat A. Identification of biomarkers involved in differential profiling of hypertrophic and keloid scars versus normal skin. Arch Dermatol Res 2015;307:115–33.

[95] Suarez E, Syed F, Rasgado TA, Walmsley A, Mandal P, Bayat A. Skin equivalent tensional force alters keloid fibroblast behavior and phenotype. Wound Repair Regen 2014;22:557–68.

[96] Ogawa R, Akaishi S. Endothelial dysfunction may play a key role in keloid and hypertrophic scar pathogenesis—keloids and hypertrophic scars may be vascular disorders. Med Hypotheses 2016;96:51–60.

[97] Ogawa R, Hsu CK. Mechanobiological dysregulation of the epidermis and dermis in skin disorders and in degeneration. J Cell Mol Med 2013;17:817–22.
[98] Laga AC, Larson A, Granter SR. Histopathologic spectrum of connective tissue diseases commonly affecting the skin. Surg Pathol Clin 2017;10:477–503.
[99] Denton CP, Khanna D. Systemic sclerosis. Lancet 2017. https://doi.org/10.1016/S0140-6736(17)30933-9.
[100] Reich A, Meurer M, Eckes B, Friedrichs J, Muller DJ. Surface morphology and mechanical properties of fibroblasts from scleroderma patients. J Cell Mol Med 2009;13:1644–52.
[101] Bhattacharyya S, Wei J, Varga J. Understanding fibrosis in systemic sclerosis: shifting paradigms, emerging opportunities. Nat Rev Rheumatol 2011;8:42–54.
[102] Ogawa Y, Morikawa S, Okano H, Mabuchi Y, Suzuki S, Yaguchi T, et al. MHC-compatible bone marrow stromal/stem cells trigger fibrosis by activating host T cells in a scleroderma mouse model. eLife 2016;5:e09394.
[103] Rentka A, Harsfalvi J, Szucs G, Szekanecz Z, Szodoray P, Koroskenyi K, et al. Membrane array and multiplex bead analysis of tear cytokines in systemic sclerosis. Immunol Res 2016;64:619–26.
[104] Rentka A, Harsfalvi J, Berta A, Koroskenyi K, Szekanecz Z, Szucs G, et al. Vascular endothelial growth factor in tear samples of patients with systemic sclerosis. Mediat Inflamm 2015;2015:573681.
[105] Karpec D, Rudys R, Leonavicinene L, Mackiewicz Z, Bradunaite R, Kirdaite G, et al. The impact of high-dose narrowband ultraviolet A1 on dermal thickness, collagen and matrix-metalloproteinases in animal model of scleroderma. J Photochem Photobiol B 2017;173:448–55.
[106] Marangoni RG, Masui Y, Fang F, Korman B, Lord G, Lee J, et al. Adiponectin is an endogenous anti-fibrotic mediator and therapeutic target. Sci Rep 2017;7:4397.
[107] Crawford J, Raykha C, Charles D, Gan BS, O'Gorman DB. WT1 expression is increased in primary fibroblasts derived from Dupuytren's disease tissues. J Cell Commun Signal 2015;9:347–52.
[108] Shih B, Brown JJ, Armstrong DJ, Lindau T, Bayat A. Differential gene expression analysis of subcutaneous fat, fascia, and skin overlying a Dupuytren's disease nodule in comparison to control tissue. Hand 2009;4:294–301.
[109] Dibenedetti DB, Nguyen D, Zografos L, Ziemiecki R, Zhou X. Prevalence, incidence, and treatments of Dupuytren's disease in the United States: results from a population-based study. Hand 2011;6:149–58.
[110] Derk CT, Jimenez SA. Systemic sclerosis: current views of its pathogenesis. Autoimmun Rev 2003;2:181–91.
[111] Akaishi S, Ogawa R, Hyakusoku H. Keloid and hypertrophic scar: neurogenic inflammation hypotheses. Med Hypotheses 2008;71:32–8.
[112] Citron N, Hearnden A. Skin tension in the aetiology of Dupuytren's disease; a prospective trial. J Hand Surg (Br) 2003;28:528–30.
[113] Scherman P, Jenmalm P, Dahlin LB. One-year results of needle fasciotomy and collagenase injection in treatment of Dupuytren's contracture: a two-centre prospective randomized clinical trial. J Hand Surg Eur Vol 2016;41:577–82.
[114] Eberlein B, Biedermann T. To remember: radiotherapy—a successful treatment for early Dupuytren's disease. J Eur Acad Dermatol Venereol 2016;30:1694–9.
[115] Giunta C, Baumann M, Fauth C, Lindert U, Abdalla EM, Brady AF, et al. A cohort of 17 patients with kyphoscoliotic Ehlers-Danlos syndrome caused by biallelic mutations

in FKBP14: expansion of the clinical and mutational spectrum and description of the natural history. Genet Med 2017. https://doi.org/10.1038/gim.2017.70.

[116] Cereda AF, Canova PA, Soriano FS. Spontaneous coronary artery dissection after pregnancy as first manifestation of a vascular Ehlers-Danlos syndrome. J Invasive Cardiol 2017;29:E67–8.

[117] Stojadinovic A, Carlson JW, Schultz GS, Davis TA, Elster EA. Topical advances in wound care. Gynecol Oncol 2008;111:S70–80.

[118] Brem H, Stojadinovic O, Diegelmann RF, Entero H, Lee B, Pastar I, et al. Molecular markers in patients with chronic wounds to guide surgical debridement. Mol Med 2007;13:30–9.

[119] American Diabetes Association. Direct and indirect costs of diabetes in the United States, ADA; 2007. http://www.diabetes.org/diabetes-statistics/cost-of-diabetes-in-us.jsp.

[120] IDF. Diabetes atlas, 7th ed. International Diabetes Federation; 2017. http://www.diabetesatlas.org.

[121] Singh N, Armstrong DG, Lipsky BA. Preventing foot ulcers in patients with diabetes. JAMA 2005;293:217–28.

[122] Amputee Coalition of America. National limb loss information fact sheet; diabetes and lower extremity amputations, Amputee Coalition of America; 2008. http://amputee-coalition.org/fact_sheets/diabetes_leamp.html.

[123] Baltzis D, Eleftheriadou I, Veves A. Pathogenesis and treatment of impaired wound healing in diabetes mellitus: new insights. Adv Ther 2014;31:817–36.

[124] Alexiadou K, Doupis J. Management of diabetic foot ulcers. Diab Ther 2012;3:4.

[125] Koh TJ, DiPietro LA. Inflammation and wound healing: the role of the macrophage. Expert Rev Mol Med 2011;13:e23.

[126] Kawahito S, Kitahata H, Oshita S. Problems associated with glucose toxicity: role of hyperglycemia-induced oxidative stress. World J Gastroenterol 2009;15:4137–42.

[127] Ahmed N. Advanced glycation endproducts—role in pathology of diabetic complications. Diabetes Res Clin Pract 2005;67:3–21.

[128] Meerwaldt R, Links T, Zeebregts C, Tio R, Hillebrands JL, Smit A, et al. The clinical relevance of assessing advanced glycation endproducts accumulation in diabetes. Cardiovasc Diabetol 2008;7:29.

[129] Bermudez DM, Herdrich BJ, Xu J, Lind R, Beason DP, Mitchell ME, et al. Impaired biomechanical properties of diabetic skin: implications in pathogenesis of diabetic wound complications. Am J Pathol 2011;178:2215–23.

[130] Werner S, Krieg T, Smola H. Keratinocyte–fibroblast interactions in wound healing. J Invest Dermatol 2007;127:998–1008.

[131] Spravchikov N, Sizyakov G, Gartsbein M, Accili D, Tennenbaum T, Wertheimer E. Glucose effects on skin keratinocytes: implications for diabetes skin complications. Diabetes 2001;50:1627–35.

[132] Xuan YH, Huang BB, Tian HS, Chi LS, Duan YM, Wang X, et al. High-glucose inhibits human fibroblast cell migration in wound healing via repression of bFGF-regulating JNK phosphorylation. PLoS One 2014;9:e108182.

[133] Andreea SI, Marieta C, Anca D. AGEs and glucose levels modulate type I and III procollagen mRNA synthesis in dermal fibroblasts cells culture. Exp Diabetes Res 2008;2008:473603.

[134] Asbun J, Manso AM, Villarreal FJ. Profibrotic influence of high glucose concentration on cardiac fibroblast functions: effects of losartan and vitamin E. Am J Physiol Heart Circ Physiol 2005;288:H227–34.

[135] Zhang B, Cui S, Bai X, Zhuo L, Sun X, Hong Q, et al. SIRT3 overexpression antagonizes high glucose accelerated cellular senescence in human diploid fibroblasts via the SIRT3-FOXO1 signaling pathway. Age (Dordr) 2013;35:2237–53.

[136] Yevdokimova NY. High glucose-induced alterations of extracellular matrix of human skin fibroblasts are not dependent on TSP-1-TGFbeta1 pathway. J Diabetes Complications 2003;17:355–64.

[137] Popov D. Endothelial cell dysfunction in hyperglycemia: phenotypic change, intracellular signaling modification, ultrastructural alteration, and potential clinical outcomes. Int J Diabetes Mellitus 2010;2:189–95.

[138] Khanna, S, Biswas S, Shang Y, Collard E, Azad A, Kauh C, et al. Macrophage dysfunction impairs resolution of inflammation in the wounds of diabetic mice. Neeraj V, editor. PLoS One 2010;5:e9539.

[139] Liu BF, Miyata S, Kojima H, Uriuhara A, Kusunoki H, Suzuki K, et al. Low phagocytic activity of resident peritoneal macrophages in diabetic mice: relevance to the formation of advanced glycation end products. Diabetes 1999;48:2074–82.

[140] Georgescu A. Vascular dysfunction in diabetes: the endothelial progenitor cells as new therapeutic strategy. World J Diabetes 2011;2:92–7.

[141] Niu Y, Cao X, Song F, Xie T, Ji X, Miao M, et al. Reduced dermis thickness and AGE accumulation in diabetic abdominal skin. Int J Low Extrem Wounds 2012;11:224–30.

[142] Arpino V, Brock M, Gill SE. The role of TIMPs in regulation of extracellular matrix proteolysis. Matrix Biol 2015;44–46:247–54.

[143] Lu P, Takai K, Weaver VM, Werb Z, et al. Extracellular matrix degradation and remodeling in development and disease. Cold Spring Harb Perspect Biol 2011;3: pii: a005058.

[144] Ladwig GP, Robson MC, Liu R, Kuhn MA, Muir DF, Schultz GS. Ratios of activated matrix metalloproteinase-9 to tissue inhibitor of matrix metalloproteinase-1 in wound fluids are inversely correlated with healing of pressure ulcers. Wound Repair Regen 2002;10:26–37.

[145] Bielefeld KA, Amini-Nik S, Alman BA. Cutaneous wound healing: recruiting developmental pathways for regeneration. Cell Mol Life Sci 2013;70:2059–81.

[146] Stojadinovic O, Brem H, Vouthounis C, Lee B, Fallon J, Stallcup M, et al. Molecular pathogenesis of chronic wounds: the role of beta-catenin and c-myc in the inhibition of epithelialization and wound healing. Am J Pathol 2005;167:59–69.

[147] Chan KT, Bennin DA, Huttenlocher A. Regulation of adhesion dynamics by calpain-mediated proteolysis of focal adhesion kinase (FAK). J Biol Chem 2010;285:11418–26.

[148] Chen SC, Guh JY, Hwang CC, Chiou SJ, Lin TD, Ko YM, et al. Advanced glycation end-products activate extracellular signal-regulated kinase via the oxidative stress-EGF receptor pathway in renal fibroblasts. J Cell Biochem 2010;109:38–48.

[149] Gerstein AD, Phillips TJ, Rogers GS, Gilchrest BA. Wound healing and aging. Dermatol Clin 1993;11:749–57.

[150] Gosain A, DiPietro LA. Aging and wound healing. World J Surg 2004;28:321–6.

[151] Sgonc R, Gruber J. Age-related aspects of cutaneous wound healing: a mini-review. Gerontology 2013;59:159–64.

[152] Larson BJ, Longaker MT, Lorenz HP. Scarless fetal wound healing: a basic science review. Plast Reconstr Surg 2010;126:1172–80.

[153] Balaji S, Watson CL, Ranjan R, King A, Bollyky P, et al. Chemokine involvement in fetal and adult wound healing. Adv Wound Care 2015;4:660–72.

[154] Adzick NS, Harrison MR, Glick PL, Beckstead JH, Villa RL, Scheuenstuhl H, et al. Comparison of fetal, newborn, and adult wound healing by histologic, enzyme-histochemical, and hydroxyproline determinations. J Pediatr Surg 1985;20:315–9.
[155] Quan T, Fisher GJ. Role of age-associated alterations of the dermal extracellular matrix microenvironment in human skin aging: a mini-review. Gerontology 2015;61: 427–34.
[156] Pawlaczyk M, Lelonkiewicz M, Wieczorowski M. Age-dependent biomechanical properties of the skin. Postepy Dermatol Alergol 2013;30:302–6.
[157] Ruvolo EC, Stamatas GN, Kollias N. Skin viscoelasticity displays site- and age-dependent angular anisotropy. Skin Pharmacol Physiol 2007;20:313–21.
[158] Wlaschek M, Tantcheva-Poor I, Naderi L, Ma W, Schneider LA, Razi-Wolf Z, et al. Solar UV irradiation and dermal photoaging. J Photochem Photobiol B 2001;63:41–51.
[159] Yaar M, Eller MS, Gilchrest BA. Fifty years of skin aging. J Investig Dermatol Symp Proc 2002;7:51–8.
[160] Rinnerthaler M, Bischof J, Streubel MK, Trost A, Richter K, et al. Oxidative stress in aging human skin. Biomol Ther 2015;5:545–89.
[161] Quan T, He T, Shao Y, Lin L, Kang S, Voorhees JJ, et al. Elevated cysteine-rich 61 mediates aberrant collagen homeostasis in chronologically aged and photoaged human skin. Am J Pathol 2006;169:482–90.
[162] Quan T, Qin Z, Robichaud P, Voorhees JJ, Fisher GJ. CCN1 contributes to skin connective tissue aging by inducing age-associated secretory phenotype in human skin dermal fibroblasts. J Cell Commun Signal 2011;5:201–7.
[163] Qin Z, Fisher GJ, Quan T. Cysteine-rich protein 61 (CCN1) domain-specific stimulation of matrix metalloproteinase-1 expression through alphaVbeta3 integrin in human skin fibroblasts. J Biol Chem 2013;288:12386–94.
[164] Wilhelmi BJ, Blackwell SJ, Phillips LG. Langer's lines: to use or not to use. Plast Reconstr Surg 1999;104:208–14.
[165] Bush J, Ferguson MW, Mason T, McGrouther G. The dynamic rotation of Langer's lines on facial expression. J Plast Reconstr Aesthet Surg 2007;60:393–9.
[166] Borges AF. Relaxed skin tension lines (RSTL) versus other skin lines. Plast Reconstr Surg 1984;73:144–50.
[167] Son D, Harijan A. Overview of surgical scar prevention and management. J Korean Med Sci 2014;29:751–7.
[168] Paul SP, Matulich J, Charlton N. A new skin tensiometer device: computational analyses to understand biodynamic excisional skin tension lines. Sci Rep 2016;6: 30117.
[169] Wong VW, Sorkin M, Glotzbach JP, Longaker MT, Gurtner GC, et al. Surgical approaches to create murine models of human wound healing. J Biomed Biotechnol 2011;2011:969618.
[170] Menke NB, Cain JW, Reynolds A, Chan DM, Segal RA, Witten TM, et al. An in silico approach to the analysis of acute wound healing. Wound Repair Regen 2010;18: 105–13.
[171] Li NY, Verdolini K, Clermont G, Mi Q, Rubinstein EN, Hebda PA, et al. A patient-specific in silico model of inflammation and healing tested in acute vocal fold injury. PLoS One 2008;3:e2789.

[172] Akaishi S, Akimoto M, Ogawa R, Hyakusoku H. The relationship between keloid growth pattern and stretching tension: visual analysis using the finite element method. Ann Plast Surg 2008;60:445–51.
[173] Vermolen FJ, Javierre E. Computer simulations from a finite-element model for wound contraction and closure. J Tissue Viability 2010;19:43–53.
[174] Mustoe TA, Cooter RD, Gold MH, Hobbs FD, Ramelet AA, Shakespeare PG, et al. International clinical recommendations on scar management. Plast Reconstr Surg 2002;110:560–71.
[175] Tao H, Yang JJ, Hu W, Shi KH, Li J. HDAC6 promotes cardiac fibrosis progression through suppressing RASSF1A expression. Cardiology 2016;133:18–26.
[176] Akhmetshina A, Palumbo K, Dees C, Bergmann C, Venalis P, Zerr P, et al. Activation of canonical Wnt signalling is required for TGF-beta-mediated fibrosis. Nat Commun 2012;3:735.
[177] Lagares D, Busnadiego O, Garcia-Fernandez RA, Kapoor M, Liu S, Carter DE, et al. Inhibition of focal adhesion kinase prevents experimental lung fibrosis and myofibroblast formation. Arthritis Rheum 2012;64:1653–64.
[178] Kinoshita K, Aono Y, Azuma M, Kishi J, Takezaki A, Kishi M, et al. Antifibrotic effects of focal adhesion kinase inhibitor in bleomycin-induced pulmonary fibrosis in mice. Am J Respir Cell Mol Biol 2013;49:536–43.

# CHAPTER 15

# Mechanobiology of metastatic cancer

**Martha B. Alvarez-Elizondo, Rakefet Rozen, Daphne Weihs**
*Faculty of Biomedical Engineering, Technion—Israel Institute of Technology, Haifa, Israel*

## ABBREVIATIONS

| | |
|---|---|
| **AFM** | atomic force microscopy |
| **CTCs** | circulating tumor cells |
| **E-cadherin** | epithelial cadherin |
| **ECM** | extracellular matrix |
| **EMT** | epithelial-mesenchymal transition |
| **F-actin** | long actin filaments |
| **FE** | finite elements |
| **G-actin** | globular actin monomers |
| **Lp** | persistence length |
| **MMP** | matrix metalloproteinase |
| **mRNA** | messenger RNA |
| **N-cadherin** | Neural cadherin |
| **PAM** | polyacrylamide |
| **PDE** | partial differential equation |
| **PDMS** | polydimethylsiloxane |
| **PEG** | polyethylene glycol |
| **RT-PCR** | reverse transcriptase polymerase chain reaction |
| **TFM** | traction force microscopy |
| **VEGF** | vascular endothelial growth factor |

## 1 INTRODUCTION

The major cause for cancer-associated deaths is the formation of metastases, causing 90% of patient deaths. The term metastasis was coined in 1829 by Jean Claude Recamier [1], and the importance of mechanics in the metastasis progression has been recognized since that time. Metastasis describes the spread of the tumor cells from the original site to distant regions and new organs. The spread of invasive cells includes various biochemical and mechanical interactions, requiring changes in cell morphology, dynamics, internal structure, and mechanics, as well as cell-cell and cell-microenvironment interactions. These changes affect the mechanics of the interactions of invading cells and have been termed the mechanobiology of the cells. While the role of mechanobiology is identified as important, confirming the mechanisms leading to the formation of metastasis has proved to be as tortuous as the process itself.

In 1858, Rudolf Virchow [2] suggested that metastatic tumor dissemination was caused by trapping of circulating tumor cells (CTCs) at mechanical constriction sites in narrowing vasculature, causing tumor cell embolisms, which are the start of migrating tumors. This theory was refuted in 1889, when Stephen Paget [3] recognized that metastatic cells/tumors do not follow vasculature structure and are not commonly found within the vasculature. Instead, he identified the important role of cell-cell interactions at the metastatic site. Paget proposed the "seed and soil" hypothesis, explaining that the way that cancer cell "seeds" disseminate in the body depends on the "soil" that they encounter, that is, the properties of the microenvironment at the secondary organ/site [3]. Forty years later, Ewing [4] challenged Paget's hypothesis, again proposing that metastasis was determined by the anatomy of the vascular and the lymphatic channels connected to the primary tumor. Only in the 1970s did a series of studies conclusively demonstrate the actuality of the "seed and soil" principle [5–9]. These studies showed that although tumor cells traversed the vasculature throughout the body, metastases develop selectively in certain organs [7].

The "seed and soil" metastasis principle is a process that includes a complex series of sequential and interlinked steps. The stages of metastasis include (i) detachment of invasive cells from the primary tumor, followed by (ii) invasion of the stroma (tumor microenvironment) and basement membrane, referred to as intravasation, then (iii) dissemination into the blood stream or lymph nodes, referred to as extravasation; and finally (iv) the outgrowth of a secondary tumor. These stages are marked by significant changes in gene expression of the cells and tumor markers, which have been studied extensively. In parallel, cells dynamically modify their mechanobiology including cell mechanics and cell-cell and cell-microenvironment interactions to facilitate responses to the changing environments.

The physical properties of cells, extracellular matrix (ECM), and their interactions have garnered recent interest. The mechanical properties of cancer cells and their interactions with the surrounding environment are implicated in many steps of metastasis, such as cancer migration through different microenvironments and structures, including basal membrane, blood vessels with an endothelial barrier, through the vascular system and tissues to reach a suitable secondary site to grow. The primary tumor is composed of different subpopulations of cells with distinct genotypes or phenotypes [10] that may be exposed to diverse biochemical conditions (e.g., different oxygen levels), due to the tumor topography. The different cell subpopulations within the tumor are usually characterized by their proliferation rates, genotype, and phenotypes, including invasive phenotypes. Invasive cells are those that are capable of metastasizing and can flexibly modify their dynamics and mechanics depending on their interactions with other cells and the microenvironment. The ability of metastatic cells to overcome physical barriers, that is, their dynamic response to changing environments, determines the probability of metastasis formation. This probability is typically termed the metastatic potential, or metastatic capacity of the cells. As will be shown in this chapter, the metastatic potential of cancer cells is directly determined by their mechanobiology, which is uniquely suited to facilitate their invasiveness.

Mechanobiological measurements rely on a combination of physical and biological tools to characterize, control, and manipulate intracellular, cell-cell and cell-microenvironment dynamics and mechanics. The goal of mechanobiology studies in cancer research is to develop different model systems, experimental or simulated, to reveal important aspects and mechanisms related to invasion of cells through varying environments. Such models, mainly in vitro, have already contributed to the understanding of the stages of invasion and how the microenvironment affects cancer cell mechanostructural characteristics and response, for example, cytoskeletal dynamics, cell adhesion, migration, and cell force generation. Expanding our knowledge of how mechanobiological factors are involved in cancer progression and metastasis will help us to obtain a more integrated understanding of the disease, enabling us to develop more effective diagnosis/prognosis tools, preventive treatments, and novel therapies.

In this chapter, we review the mechanobiology of cancer cells, with specific focus on migration, invasion, and metastasis focusing on approaches to study, determine, and predict these features. Initially, we describe the mechanical structure of the cell, the role of the microenvironment, and the mechanical interaction between the cells and their microenvironment, which affect migration and invasion of cells. In the second section, we describe the general types of migration that may be utilized by cells during metastasis, in 2D and 3D. We explain how the cells utilize their dynamic cytoskeletons to generate different motility modes, depending on the encountered microenvironment and mechanical stimuli. In the third section, we focus on the metastatic cascade, where we describe the steps of metastasis from a mechanical perspective, including the factors involved and their mechanical contributions to this process. In the subsequent sections, we present the most commonly used in vitro methods and techniques to perform mechanobiology studies. We highlight recent advances in techniques for measurement of cell mechanics, in vitro assays to evaluate the migration capabilities of cells, methods to evaluate cell strength, and approaches to determine the metastatic potential. Determining the metastatic potential and the subsequent metastatic risk of a subset of cells from a tumor is crucial for providing accurate prognosis and for selecting the most effective treatment; treatments not only are typically designed to reduce tumor growth but also are being directed at preventing metastases. Finally, we briefly review different mathematical models that have been used as a powerful tool to study cell motility. Those play an important role in the understanding of cancer progression and metastasis, and allow expansion to conditions that are difficult to test experimentally.

## 2 CELL MECHANICS, STRUCTURE, AND INTERACTIONS

In this section, we provide an overview of the mechanical structure of the cell and how the different elements of the cytoskeleton are involved in cell migration and invasion, and highlight the role of the surrounding matrix and the mechanotransduction process between the cell and its microenvironment.

### 2.1 MECHANICAL STRUCTURE OF THE CELL—THE CYTOSKELETON

The main cellular structures that produce both the mechanical and dynamic responses of the cells are the elements of the cell cytoskeleton and the cytoskeleton-associated molecular motors [11]. The cytoskeleton is a dynamic and adaptive 3D scaffold, composed of three different types of 3D, biopolymer protein networks: actin filaments, microtubules, and intermediate filaments. Each of these biopolymer elements are distinguished by their specific function and different (dynamic) remodeling abilities and rigidity [11]; rigidity is described by the persistence length (Lp) of the filament, where a larger value indicates a more rigid structure. The cytoskeleton has three main roles: (i) spatially organize the cell content by maintaining a cell-wide structure that facilitates intracellular transport, (ii) connect the cell physically and biochemically to the extracellular microenvironment, and (iii) generate coordinated forces for cell motility and morphological changes [12]. While actin and microtubules are more dynamic and rapidly responsive, the function of the intermediate filaments is to provide structural support.

The actin filament network is the most dynamic element of the cytoskeleton. Actin filaments are composed of globular actin monomers (G-actin) that polymerize to form long actin filaments (F-actin). These actin filaments are semiflexible polymers with $Lp \sim 17\,\mu m$ and $\sim 7\,nm$ in diameter [13]. Actin filaments can remain as single filaments but typically either form mechanically stable bundles, called stress fibers, or may be cross-linked by mobile myosin molecular motors to form a dynamic network. The simple and polar structure of actin filaments enables their rapid dynamics, and thus, the main role of actin is to rapidly interact with the cells' changing environment. This includes responses to mechanical, environmental cues or actively interacting with the microenvironment, that is, with the ECM or neighboring cells. Actin is typically located close to the plasma membrane of the cell, where it provides mechanostructural support to the membrane against external forces, for example, shear and compression. At the cell membrane, actin filaments can form bundles, or stress fibers, which provide structural support and have a mechanosensory role [13,14]. Concurrently, the actin connects to the extracellular microenvironment through transmembrane receptors called integrins.

Integrins connect the actin with the ECM form stable adhesion through clusters called focal adhesions. Those adhesion sites allow cells to generate forces required to maintain cell adhesion, to generate cell motion, or to rapidly change shape [15]. Through focal adhesions and cell-cell junctions, the actin mechanical networks allow external forces to be transmitted directly to the nucleus [16–18]; the nucleus may induce a mechanical response in the cell. The combination of integrin connection sites with a cross-linked actomyosin network facilitates application of shear forces by the cells, allowing transport, cell motion, and force generation.

The microtubules, the stiffer of the polymers, are hollow, rodlike structures with a ~25 μm diameter and an Lp of 100–5000 μm [13]. They are dynamic, yet their more complex structure and α/β tubulin heterodimer composition results in slower remodeling. The role of the microtubules is in cell organization, cell division, and organelle transport within the cells, as it connects the nucleus with the membrane. Microtubules are polarized and directional with respect to the nucleus, having plus and minus ends with different structure and kinetics [19,20]; directional polarization is achieved as a result of the heterodimer orientation. The positive end is located in the periphery close to cell membrane, while the negative end is located at the centrosome close to the cell nucleus. The microtubules' polarity allows specific motors, dynein and kinesin, to move, respectively, to and from the nucleus moving cargo across the cell.

The third element of the cytoskeleton is the intermediate filaments, which have a dense, stiff, and complex structure of biofilaments, resulting in more limited dynamic remodeling capabilities. Thus, intermediate filaments are less involved in cell motility and migration as compared with the actin and microtubules and mostly provide structural support. Their main role is to be a binding factor for all cell elements, and they also are assumed to absorb mechanical stress [21,22].

## 2.2 THE CYTOSKELETON IN MOTILITY AND FORCE APPLICATION

The cytoskeleton and associated molecular motors allow cells to generate forces, to change shape, to move, and to divide. Specifically, in metastasis, the forces generated by the cytoskeleton play a role in enabling cells to migrate, invade, and reach a site where a secondary tumor may be initiated. Cancer cells have been shown by our lab and others to have a reduced external [23,24] and internal stiffness [25–27], due to changes in organization of the cytoskeleton network [26–28]. Cancer cells exhibit a reduction in well-defined F-actin filaments and stress fibers, particularly the submembrane actin [29], while the microtubule network is sparse with large voids [26]. These changes result in a softer, more pliable structure that facilitates rapid morphological changes required for cells to squeeze through narrow regions during metastasis-related migration and invasion. Actin has typically been associated with both membrane flexibility and force application. Cells can use actin to apply forces in two ways: using the actomyosin network [30] or using only the actin cytoskeleton [31]. These are used to generate, respectively, forces that are normal to the cell membrane or shear forces that are parallel to the membrane. The actomyosin shear forces are applied when myosin motors that cross-link two actin filaments induce relative motion of the filaments. When one of the filaments is anchored to the ECM (via integrins) the relative motion of the filaments induces a shear. This process leads to the generation of actomyosin contractile forces that are essential, for example, in pulling the cell forward during crawling [32] or in anchoring adhered cells

[33,34]. Specifically, in cancer, a marked elevation in actomyosin-generated forces has been correlated with increased invasiveness in metastatic cells [35].

Actin filament treadmilling, including remodeling typically mediated by Rho GTPases [36], is a crucial mechanism for cell invasion during metastasis. Treadmilling forces are induced when one end of the filament (the minus end) is broken down and the other (the plus end, e.g., at the membrane) is built up, thus pushing the membrane forward. The actin filaments can grow (by polymerization) in length or shorten (by depolymerization) depending on the concentration of monomeric actin. The actin monomer concentration at the minus end is six times higher than at the plus end, respectively, ~0.6 and 0.1 μM [14]; when the actin monomer concentration is between these two values, the actin filaments will grow in one direction.

To apply and provide force, actin filaments assemble into different protrusive and contractile structures such as stress fibers and filopodia. Stress fibers are composed of bundles of 10–30 actin filaments, which are cross-linked together by α-actinin [37]. These stress fibers are usually anchored to focal adhesions that connect the ECM with the actin cytoskeleton allowing tension to be applied and propagated across the cell; this enables the cell to apply force and move. When these actin bundles extend out beyond the lamellar edge, they form structures known as filopodia; the filopodia membrane is linked to the actin filaments by ezrin-moesin-radixin family proteins [38]. The main function of filopodia is as cell sensors (chemical and mechanical) to direct cell migration through tissue. Actin-containing membrane protrusions appear to transmit extracellular forces across the membrane and to the mechanically connected nucleus [17,18].

The integrins (connected to actin) also play an important role in force application. To move the cells forward, cell extensions need to adhere to the substrate via integrins and exert traction force [39]. Integrins are connected to the intracellular actin filaments through other adaptor proteins. Integrins are also an important part of the bidirectional signaling network between the ECM and cytoskeleton, that controls mechanotransduction [40–42]. In the cell membrane, integrins are coupled to the ECM cluster and interact with proteins that modify the cytoskeleton, like talin and vinculin [43,44]. The sites of these integrin clusters are points of matrix metalloproteinase (MMP) release, specialized actin-rich protrusions that are well known as invadopodia. Invadopodia formation is very important for cell invasion [45], since invadopodia perforate the basement membrane, elongate, and then guide the cell to the stroma. These exploratory protrusions rely not only on proteolytic degradation but also on actin, filament, and microtubule extension [46]. By blocking integrins, Weaver et al. observed the reversion of the malignant phenotype of human breast cells in three-dimensional culture showing the crucial role integrins have in cancer progression [47]. When disassembly occurs, integrins are internalized by endocytosis to be reused [48]. Motility is assessed with the coordination between assembly and disassembly of adhesions, along with matrix degradation through MMPs. Additionally, adhesions must be disassembled to allow protrusions and cell motility.

The main role of microtubules is typically considered to be intracellular organization and transport, yet they are also likely have a role in force application, even as support for the actin. The microtubules exhibit dynamics that are similar to actin: they are polarized, and they use treadmilling polymerization, which can be used to apply force [49]. Intracellularly, microtubule pushing and pulling forces mainly support the correct spatial distribution of chromosomes, mitotic spindles, and nuclei in cells [49]. Extracellularly, microtubules also regulate focal adhesion size [50], which is correlated with traction forces applied by actin [51]. The mechanical roles of the actin and microtubules in force application have been studied independently, but there is evidence of direct mechanical coupling between the actin and microtubule networks [52]. Besides the actin and microtubules, the nucleus, the largest and stiffest organelle in the cell, also affects motility; its deformation rate and capacity limits the ability of cells to squeeze through constrictions. It has been shown that nuclear deformation is a response to changes in cell adhesion, but its role in force application is not well defined [53,54].

Cell motility, invasion, and force application depend on cell dynamics and the rapid ability of cells to respond to stimuli. Results from our lab provide insights on the differences in intracellular dynamics of cancer cells with different metastatic potential and compared with benign cells. By tracking intracellular particles in low (MDA-MB-468) and high (MDA-MB-231) metastatic potential (MP) breast cancer cells (both derived from pleural effusion) and in benign cells (MCF-10A), we have observed differences in dynamics of the cells [26,27,55]; highly invasive cells were most dynamic and structurally sparse and soft. Concurrently, we demonstrated the role of the cell elements in intracellular dynamics, by targeting the cytoskeleton, molecular motors, and ATP energy [27]. In the low MP cells, particles are mostly transported by motors on single microtubules. In contrast, in high MP particle, motion is driven by indirect motor transport, highly impacted by microtubule fluctuations.

We have also observed differences in structural localization of the cytoskeleton between metastatic and benign cells, specifically as metastatic cells apply invasive forces to soft, impenetrable, elastic gels [28], with cells indenting the gels in attempted invasion [56–58]. On soft, elastic gels, all of the benign cells retain a rounded morphology and exhibit a condensed cytoskeleton that is not distinctly fibrous (Fig. 1). This is in contrast to the same cells (MDA-MB-231) [28] and other cells from lines (MCF-7, MCF-10, and 3T3) [29,30] that spread out on glass. Differences in spatial organization of the cytoskeleton on the gels are apparent between the benign and the invasive cancer cells. We have observed that the actin and the microtubules are nearly uniformly distributed in benign cells [28], which adhere to the gel substrate without attempting to invade it [33,56]. In contrast, in the metastatic cancer cells that indent the gels, actin accumulates at grip handles on the gel and at the cell's leading edge, located at the bottom of the indentation dimple, followed by the nucleus [28]. The actin at the leading edge likely relates to force application during cell migration, for example, within membrane protrusions. We have shown that the nucleus moves and deforms (Fig. 2), likely to facilitate invasion and force application [28]. Concurrently, microtubules may localize above and/or below the nucleus, indicating their lesser participation in the force application in the gel setup.

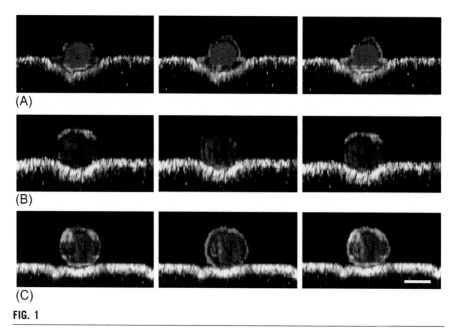

**FIG. 1**

Confocal side view of cells seeded on a 2.3 kPa gel with 200 nm particles marking the surface of the gel (*bottom surface*) and stained nuclei (*middle blob*). Image columns, from left, show the actin, the microtubules, and their merge. In rows, both the (A) high MP cell and (B) low MP cell indent the gel, as apparent in the dimple in the gel. (C) The benign cells mostly remain rounded on the gel surface and do not significantly indent it. Scale bar is 10 μm.

*Adapted from Dvir L, Nissim R, Alvarez-Elizondo MB, Weihs D. Quantitative measures to reveal coordinated cytoskeleton-nucleus reorganization during in vitro invasion of cancer cells. New J Phys 2015;17(4):43010. Available from: http://stacks.iop.org/1367-2630/17/i=4/a=043010.*

## 2.3 ROLE OF THE TUMOR MICROENVIRONMENT

The tumor microenvironment includes the neighboring cells and the mechanobiologically responsive ECM, the latter being the focus of the current section. The ECM provides the framework for intercellular crosstalk, adhesion, and migration [59]. The ECM stiffness induces signal transduction into cells by activating integrin and focal adhesion sites, resulting in a change in cell morphology. ECM stiffness influences cell adhesion [33], cell motility mode [60], and also the cells' ability to invade [61,62]. The ability of different cancer cells to cross through the tumor tissue, the surrounding tissue, and even enter the blood vessels depend on the ECM structure and its mechanics [61,63]. Specifically, ECM stiffness, which may be remodeled by cells in its vicinity, plays a crucial role in cancer progression. Furthermore, the stiffness of the entire tumor tissue has been shown to be mainly due to the ECM mechanics. Consequently, the stiffness of tissue has been used for centuries to detect tumors by simple manual palpation, either through the skin or directly on resected tissues. Solid tumors have generally been shown to be stiffer than healthy tissue [64].

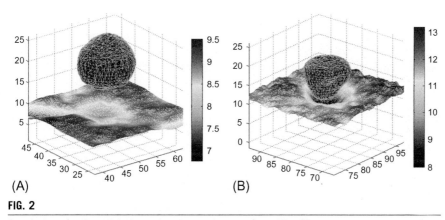

**FIG. 2**

Rendering of cell nuclei above the detected gel surface. The 3D scales are consistent (in microns), and differences in cell indentation depth, nucleus shape, and nucleus depth are apparent. Depth in the gel indicates cell lower edge, where cytoskeleton and membrane stains are not shown for clarity. Note that the depths differ between panels showing the following: (A) benign cells tend to localize at inherent valleys (scratches) in the gel; (B) highly metastatic cells indent the gel deeply.

*Adapted from Dvir L, Nissim R, Alvarez-Elizondo MB, Weihs D. Quantitative measures to reveal coordinated cytoskeleton-nucleus reorganization during in vitro invasion of cancer cells. New J Phys 2015;17(4):43010. Available from: http://stacks.iop.org/1367-2630/17/i=4/a=043010.*

The ECM of solid tumors exhibits increased stiffness and cross-linking that can promote invasive phenotypes in cancer cells [65]. ECM cross-linking has been shown to result in aggregation and clustering of integrins, as well as enhanced signals via phosphoinositide 3-kinase (PI3K) that induce invasion [64,66]. The stiffness of the surroundings also regulates the number of MMP release tunnels, called invadopodia, through signaling molecules FAK and P130Cas [67], all of which greatly affect invasion. The increment of cell adhesion, along with the secretion level of proteolytic enzymes, have also been correlated to metastatic potential [68]. Recently, we and others have shown that cancer cells change morphology and apply different traction forces depending on the stiffness of the surroundings [33,69]. This is a result of interactions between the substrate stiffness (extracellular force) and cytoskeleton tension (intracellular force) and indicates the role of the increased ECM stiffness in promoting the malignant and invasive phenotypes [64].

## 2.4 MECHANOTRANSDUCTION IN CANCER CELLS

Metastasis requires physical interactions between cancer cells and the microenvironment, involving processes known as mechanosensing and mechanotransduction. In mechanotransduction, cells translate mechanical cues from their environment into biochemical signals, in activated signaling cascades. This process has an important role in cancer, as cells respond rapidly to changes in the environments that they encounter. Mechanotransduction occurs through mechanosensitive proteins that activate different signaling pathways through force-induced conformational changes [70]. One example is cofilin, a mechanosensitive protein required to sense the mechanical stimuli that enhance invasion [71]. Another mechanotransduction example involves gene expression in the nucleus.

During tumor growth, the ECM is remodeled, and increased mechanical pressures are exerted by the surrounding tissue on the tumor and vice versa. Forces sensed or generated by the actin-myosin apparatus in the cell are transmitted to the nucleus of the tumor cells via the cytoskeleton and can alter its structure and that of the cells [48,71–73]. Deformation of the nucleus mediates changes in gene expression that regulate cancer progression [74,75]. Changes in extracellular mechanical forces can also cause dysregulation of signaling pathways inside the cell and lead to malignancy. For example, it has been shown that mechanical compression can promote invasion and metastasis that also enhances cell-substrate adhesion in 2D cell culture [76] and that facilitates invasion by increasing the release of MMPs.

## 3 CELL MOTILITY AND MIGRATION
### 3.1 MOTILITY MODES

Cell migration is a natural process, occurring, for example, in wound healing. It is also intrinsic to metastasis, being part of every step of the metastatic cascade. Depending on the cell type and the cross talk of cells with their microenvironment, metastatic cells use different motility modes. The basic modes are amoeboid or mesenchymal, and under different conditions metastatic cells migrate in single mode or in groups (collective migration). Fig. 3 shows the different modes of motility typically observed in metastasizing cells.

#### 3.1.1 Single cell migration

Single-cell migration can be classified as mesenchymal or amoeboid, based on the cell morphology and interaction with the microenvironment. During mesenchymal migration, cells typically exhibit a spindle shape with a leading edge (Fig. 3A), which is RAC-induced actin-rich cell protrusions (i.e., lamellipodia) with active actin polymerization to maintain polarity. These cells are able to remodel the microenvironment by proteolysis, using MMPs to generate free pathways for invasion. In contrast, amoeboid cells (Fig. 3A) utilize changes in morphology and force application to traverse the ECM without biochemically remodeling it. The amoeboid cells present a round morphology with bleb-like protrusions that form due to intracellular pressure rather than actin polymerization. Those protrusions move parts of the cell body through existing pores in the ECM and actively apply forces to the ECM through RhoA-/ROCK-mediated actomyosin contractions and nonintegrin adhesion mechanisms [77].

Metastatic cells can easily switch between these two modes of motility, where interactions of the cells with the encountered microenvironment trigger dynamic adaptive responses of the cells [78,79]. For example, fibrosarcoma cells have been

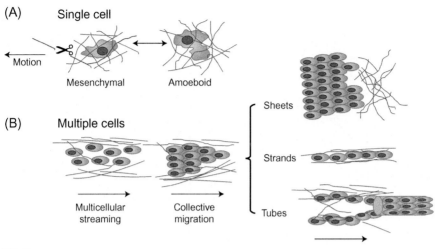

**FIG. 3**

Motility modes during metastasis. (A) Single-cell motion, mesenchymal and amoeboid motility modes; (B) multiple cells, multicellular streaming and collective migration; collective migration can be shown in different morphological organizations such as sheets, strands and tubes.

observed to switch to an amoeboid migration mode when MMPs were inhibited [80]. Conversely, the inhibition of cell contractility through the ROCK pathway induced transition from amoeboid migration in 3D to mesenchymal migration [79]. Consequently, various cancer treatments that target adhesion receptors or proteases (e.g., MMPs) have not been effective in stopping cancer progression in clinical trials, mainly because cancer cells can easily respond by modifying their migration modes. Thus, the most effective way to reduce cell migration is likely by combined inhibition of MMPs and actomyosin contractility, targeting both basic modes of cell migration and invasion.

### 3.1.2 Collective cell migration

Collective cell migration is the motion of a multicellular group, in which the cell-cell junctions are maintained for long times (Fig. 3B). Motion is attained through traction forces that are generated at the leading edge. Actomyosin protrusion and contractility is generated by the cells on the lateral edge (the periphery) of the group that are in contact with the microenvironment; cells within the group are only in contact with neighboring cells and any intercellular matrix between cell-cell junctions. The cell-cell junction is coupled by cadherins, tight junction proteins, adhesion receptors, and gap junctions that maintain the cell-cell mechanical cohesion, polarity, and cell signaling.

Collective migration can display different morphological organizations, especially at the leading edge, as shown in Fig. 3B. Morphologies vary from cell strands that are 1–2 cells wide to large clusters that form 3D structures; the latter have frequently been observed in breast [81], pancreatic [82], and prostate carcinomas [83]. This type of migration integrates mechanisms used by single cells [84], such as mesenchymal migration. Mesenchymal mode is mainly observed at the leading edge, where the cells exhibit actin-rich protrusions that generate sufficient traction force to move forward. The leading cells also generate a proteolyzed zone on the boundaries to guide the group of cells behind them, and the following cells align and enlarge the invading strands [85].

A number of recent studies have shown that carcinoma and melanoma cells and other cancers can move and invade as sheets, as is shown in Fig. 3, using the edges to extend and pull the cell sheet ahead, while maintaining the cell-to-cell contacts [86]. Similar patterns of cell migration, collective and multicellular streaming, have been observed previously, during wound healing and embryogenesis [87,88]. Collective migration has been observed in different cell lines like carcinoma cells, melanoma cells, and other cancers that move and invade by forming a sheet. Recently, we have shown that physical proximity of cells promotes cell invasiveness, allowing cells to synergistically apply force and invade [57]. Forces may be applied by each neighboring cell independently, as was observed in multicellular streaming of orthotopic breast cancer and melanoma models [89,90], which generate traction forces on the surrounding environment and low-adhesion forces between the cells. Such migration is usually guided by extracellular structures. Motility guidance cues, stimuli, and general mechanisms are discussed in the next section.

### 3.2 STIMULI FOR DIRECTIONAL MOTILITY OR MIGRATION OF CELLS

Metastatic cells can migrate randomly in the presence of a homogeneous concentration of soluble growth factors/chemokines and a homogeneous substrate, which is a very unlikely situation in vivo. Typically, inhomogeneities of factors in the microenvironment will induce directional, directed migration; such stimuli are encountered as metastatic cells move in the body. The different types of possible stimuli are usually categorized through the causing factor or the biochemical or biomechanical gradient involved. The Latin word "taxis" (directional movement in response to an external stimulus) is connected to a word specifying the nature of the taxis stimulus. In cells, the taxis process can roughly be separated into four categories:

1. Chemical/biochemical cues
   - *Haptotaxis*—Hapto in Latin means attachment or binding. It indicates migration of cells along a gradient of ECM-bound chemoattractant [91].
   - *Chemotaxis*—Migration of cells directed by the direction of changing concentration of a specific soluble or suspended substances [92,93].

2. Substrate mechanics or mechanotaxis
   - *Tensotaxis*—Tenso borrowed from the Latin word *tēnsus* is an adjective for tension, that is, stretching or strain. It denotes migration along a gradient of mechanical tension or toward an area that is mechanically strained.
   - *Durotaxis*—Duro in Latin means hard or rigid. Migration is directed by stiffness gradients in the substrate or microenvironment [94].
3. Environmental topology
   Migration preference toward a topological structure in the microenvironment. The surface topology can also result in different stiffness or tension, intersecting with the motion by durotaxis or tensotaxis.
4. Intercellular mechanics
   *Cohesotaxis or plithotaxis*—To migrate when cells cohere, to stick together, or are a united part of the same mass. Plithos, in Greek, means a crowd or swarm. Migration of each cell within a monolayer of cells in the direction of minimal intercellular tension.

These types of directed migration can be difficult to distinguish due to the presence of multiple in vivo signals, causing simultaneous biochemical and mechanical stimuli.

## 4 STEPS OF METASTASIS

DNA mutations occur constantly, yet affected cells are typically removed by the body's immune system. Occasionally, a mutant cell evades removal, potentially causing a malignancy. One of the typical cancerous mutations causes uncontrolled cell proliferation, thus generating a tumor mass. As primary tumor grows, oxygen requirements increase leading to angiogenesis (blood vessel formation) surrounding the tumor, and more mutations accumulate. The metastatic cascade is activated by a complex of signaling pathways, leading to various changes in the cells including their mechanics and interactions with the surrounding microenvironment. In this section, we present the steps of metastasis from a mechanical perspective, focusing on the mechanical factors that are involved in each step. The steps and factors are summarized in Fig. 4.

### 4.1 LOCAL INVASION THROUGH TISSUE

The first step of the metastatic cascade is the acquisition of invasive capabilities in the cancer cells, as a result of changes in gene expression and function. After these transformations, cancer cells become metastatic cells and may dissociate from the primary tumor. This dissociation from the epithelium and subsequent invasion of the surrounding basement membrane and ECM has been well characterized as the epithelial-mesenchymal transition (EMT, Fig. 5) [95,96]. During EMT, through the downregulation of epithelial cadherin (E-cadherin), a molecule responsible for

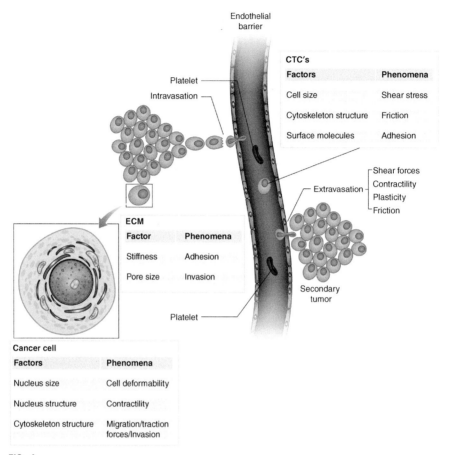

**FIG. 4**

Mechanical properties of cancer cells and tumor microenvironment during metastasis. The different steps of metastasis: EMT transition, intravasation, cells in blood stream, and extravasation are shown with the contributing factors and physical phenomena that can be observed or measured during each of the steps.

adherence junctions between epithelial cells, cancer cells reduce their adhesion capacity, contributing to malignant phenotype [97]. The downregulation of E-cadherin also generates drastic changes in cell mechanics and transforms cell morphology from epithelial to mesenchymal, activating mechanisms that facilitate migration. In addition, during EMT, Neural cadherin (N-cadherin), the cell-cell adhesion molecule typically expressed by mesenchymal cells, is increased in a process known as cadherin switching [98]. Once the cadherin switch is activated, metastatic cells lose their affinity for epithelial neighbors and gain affinity for stromal cells, triggering cell migration [99]. Cadherin switch has been clinically correlated with poor prognosis [100].

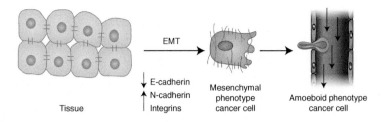

**FIG. 5**

Epithelial-mesenchymal transition (EMT) and resulting migratory cells. In this transition, cancer cells change their morphology from cube-like (epithelial) to elongated (mesenchymal) by the downregulation of E-cadherin junctions and upregulation of N-cadherin junctions. Using the mesenchymal phenotype, the cells are able to invade the ECM. Following that, cells transition to the amoeboid phenotype that allows them to cross the endothelial barrier and reach the blood stream.

Once the EMT has occurred, metastatic cells can utilize one of the two modes of migration/invasion: mesenchymal or amoeboid. These require, respectively, a degradable matrix and cell-degradation capabilities or, for amoeboid migration, preexistence of pores in the matrix through which the cells may invade. Cells may switch between the modes depending on their microenvironment [78,101]. Mesenchymal cells overexpress MMPs on their membrane, promoting digestion of the collagen and laminin of the surrounding basement membrane. The overexpression of MMPs also helps metastatic cells to overcome the ECM, which contains collagen and fibronectin. It is intuitive that metastatic cell motility and invasion of the surrounding tissue is modulated by the physiochemical properties of the ECM. The ability of metastatic cells to proteolytically degrade the basement membrane has been considered as an indicator of metastatic potential [102]. However, MMP-mediated digestion is not the only way metastatic cells invade; cancer cells can also migrate amoeboidally, by squeezing through subcellular-sized pores in the matrix. Amoeboid migration is the chosen approach if the matrix is nondegradable or if, for example, MMPs are inhibited.

## 4.2 CELL MECHANICS IN THE INTRAVASATION PHASE

Once metastatic cells cross the local tissue, including the basement membrane and ECM, they invade into the blood stream by traversing the endothelial barrier in a process called intravasation. It has been shown that while the most metastatic cells survive this process, the majority of the less-metastatic cells will fragment and perish when attempting to enter a blood vessel, indicating that intravasation constitutes an important barrier in metastasis [103].

The mechanical properties of the cell facilitate both intravasation and extravasation (exiting) through a blood vessel wall; extravasation also includes (blood) flow effects and is discussed in a separate section below. During intravasation and extravasation, metastatic cells typically use amoeboid motility. Cells squeeze between endothelial cells and undergo large deformations to penetrate the endothelial cell-cell junctions. These changes are driven by cytoskeletal remodeling, and require rapid adaptation to appropriate structures and mechanics. Cells must be able to rapidly deform and also apply forces, that is, if they are too soft or too stiff, they cannot deform to migrate efficiently. The rate limiting factor in the invasion or migration of cancer cells across the endothelial barrier is the deformation of the nucleus, which is the largest and stiffer organelle of the cell [104,105].

Cells within the tumor are heterogeneous and display a wide range of mechanical properties that can contribute to successful negotiation of the endothelial barrier. However, different optimal mechanics are required for each step of the metastatic process. Hence, cells that are unable to dynamically adapt and change to survive different microenvironments may be destroyed at later stages.

## 4.3 METASTATIC CELLS IN THE BLOOD STREAM

Once in the blood stream, most metastatic cells are subjected to various types of physical stress such as hemodynamic shear forces, collision with solid elements in the blood (e.g., blood cells), and interactions with the immune cells and system, all of which compromise their survival [106]. Thus, to survive, they are required to be highly elastic and resistant to high deformation rates. The motion of CTCs in the blood stream is influenced by physical and mechanical parameters, such as the pattern of the blood flow, the blood vessel diameter, and the interplay between shear flow and adhesion that leads to the arrest of movement in large vessels. Shear flow induces deformation and changes the motion of the CTC toward the vessel walls. At the vessel walls, the shear stress is maximal, varying between 0.1 and 0.4 Pa (or 1–4 dyne/cm$^2$) in venous circulation and 0.4–3 Pa in the arteries; the shear stress is a product of the viscosity of the blood (4–5 mPa s) [107] and the shear rate. Shear flow has a crucial role since it determines the orientation and time constants associated with receptor-ligand interactions that lead to adhesion. Computational and experimental models of the response of cancer cell to shear flow are reviewed in detail elsewhere [108].

Paradoxically, exposure to shear stress can activate specific signaling pathways in cancer cells. This can result in a dramatic reorganization of the cytoskeleton and adhesive system, which facilitates attachment to the vascular wall [47]; attachment is the first step toward extravasation. It has been shown that shear flow affects the binding kinetics between neutrophils and melanoma cells, which may contribute to melanoma extravasation [109]. Additionally, shear flow alters the viability and proliferation of CTCs [110] as well as their mechanics [111]. Different methods have been designed to capture and manipulate CTCs. The most common are reverse transcriptase polymerase chain reaction (RT-PCR) and techniques to identify cancer-related genes through messenger RNA (mRNA) [112].

## 4.4 EXTRAVASATION

To exit the blood stream, a CTC must first bind to a blood vessel wall arresting its flow-induced motion. This can occur by cell adhesion that is enabled by physical occlusion, for example, dependent on the blood vessel diameter. When the cell enters a vessel with a diameter smaller than its own (~10 μm), the cell may become mechanically trapped. This phenomenon has been observed in brain tissue and blood vessels and was followed by extravasation and metastasis formation in mouse models [113]. Tumor cells with smaller diameters will follow the trajectory of blood vessels and, while in the flow, will experience collisions with other cells. These collisions can direct the cancer cell to collide with the vessel wall. Extravasation of cancer cells in large diameter blood vessels requires specific strong bonds or ligands to overcome the shear flow and stress at the blood vessel wall; these may be formed as the cell moves next to the vessel wall and interacts with it. A cell moving along a vessel has translational and angular velocity, with the translational velocity always being larger. That results in a slipping motion that increases the chances of bonding a single receptor on the cell surface with the ligands on the vessel wall; cell rotation exposes several receptors to contact with the vessel wall. The adhesion will depend on the tensile strength of the individual receptor-ligand bonds and also on their number [114].

When cells adhere to the endothelium, they undergo a process termed diapedesis, where the cells extend pseudopods to penetrate cell-cell junctions in the endothelium [115]. This requires local and dynamic changes in cellular mechanics driven by cytoskeletal remodeling. This cytoskeletal remodeling is combined with changes in adhesion molecules from cell-cell to cell-ECM, similar to the cadherin switch in EMT that include the activation of various new and existing integrins. The ability of cells to transmigrate an endothelial coculture is directly correlated with the expression of chemokine receptor CXCR2 [116].

In addition to interactions with cells in the blood stream, animal studies have shown that platelets can help cancer cells to avoid immune elimination and promote their arrest at the endothelium [117]. Metastasis was successfully inhibited by inducing genetic or antibody-induced platelet depletion and restored when platelets where infused. Additionally, platelets facilitate cancer cell adhesion and then extravasation by releasing growth factors such as vascular endothelial growth factor (VEGF) at the endothelium, promoting vascular permeability.

Cancer cells withstand and exert mechanical forces on their environment in each of the metastatic cascade steps; the cytomechanical methods used to evaluate mechanical properties of the cells, migration, and cell strength will be reviewed in the next section.

## 5 METHODS FOR CYTOMECHANOBIOLOGY IN CANCER CELLS

To overcome the different microenvironments encountered during metastasis, cells must dynamically adapt by undergoing profound mechanostructural changes. Adaptive changes in the cell structure, mechanics, and its ability to interact and apply force require a highly dynamic and responsive cell cytoskeleton. In this section, we present different in vitro cytomechanobiology techniques that reveal dynamic changes in the various aspects of the mechanobiology of the cells; we emphasize the application of these techniques in the study of cancer metastasis. We present in vitro techniques to evaluate cell mechanics, assays for migration, and approaches to measure forces applied by the cells. We then specifically focus on different techniques for determining the metastatic potential of cells, based on the cell-substrate mechanical interactions.

## 5.1 CELL MECHANICS ASSAYS

Here, we briefly summarize several categories of experimental methods used to obtain the mechanical properties of cancer cells (Fig. 6); the approaches are generally applicable to noncancerous cells as well. We focus on whole cell (external) measurements such as micropipette aspiration [61–65], optical stretching [57–60], traction force microscopy (TFM) [66–69], and atomic force microscopy (AFM) [70–73] and also discuss intracellular measurements in the form of intracellular particle tracking [26,27,55,118–121]. Cells are characterized by quantifying their structure and viscoelasticity, where external measures can be reduced to one criterion: deformability. This can be used to distinguish between malignant and benign cancer cells. Indeed, deformability of cancer cells has recently been recognized as a label-free phenotype biomarker of cancer invasiveness. All of the following techniques have shown that cancer cells are more dynamic and deformable than their benign counterparts.

*Micropipette aspiration* uses suction pressure to deform cells and derive their mechanical properties (Fig. 6A). A typical micropipette aspiration system consists of a pressure generator, a pressure transducer, a glass micropipette, an *x-y-z* micropositioner, and an optical microscope for imaging. By measuring geometry changes from the images obtained, the viscoelastic properties of the cell can be derived. This technique has been used to distinguish cancerous from benign cells in both breast

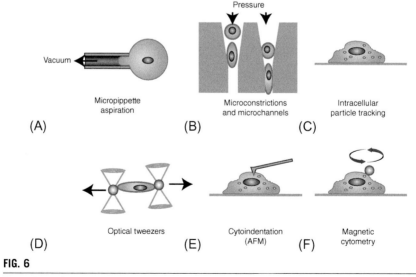

**FIG. 6**

Techniques for evaluating cell mechanics. (A) Micropipette aspiration, (B) microconstriction and microchannels, (C) intracellular particle tracking, (D) optical tweezers, (E) cytoindentation (AFM), and (F) magnetic cytometry.

[122–124] and liver tissue [125]. Alterations in mechanical properties of prostate cancer cells due to shear flow has been also measured with this technique [111].

*Microconstrictions and microchannels* are approaches that apply microchannels of various shapes that are designed to act as barriers to natural cell motion or to test the effects of shear flow (Fig. 6B). Recently, it has been shown that subnucleus-sized barriers induce different invasion phases and morphologies in cancer cells [126]. These barriers, in combination with different cytoskeleton inhibitors/stabilizers, can help to reveal the mechanisms of nonproteolytic invasion. For example, it was shown that the stabilization of microtubules by Taxol affects the invasion capacity of breast cancer cells [126]. This approach has also been used to measure the mechanical properties of cells in suspension. Cell elasticity and cell fluidity have been measured by microconstriction arrays by recording entry time, cell size, and applied pressure [127].

Similarly, shear flow can be applied, either within a parallel-plate flow chamber or with microchannels in which cells are subjected to different various flows. Channel dimensions or a viscometer (a stationary flat plate and a rotating inverted cone) can be used to generate laminar and turbulent flows. In both cases, the shear stress applied to cells is known or can be quantified. With this approach, breast cancer cells have been distinguished from benign cells by two parameters: deformation while passing through the channel and by the time taken to enter a microchannel while in flow [128]. Likewise, it has also been shown, based on migration under flow through a microchannel, that breast cancer and osteosarcoma cells are softer than benign cells [129]. The same approach has been used to characterize leukocytes and malignant cells in pleural effusions, and the deformability has been correlated to disease state in cancer patients [130]. Additionally, microfluidics has been used to sort and analyze CTCs of breast cancer using a microchannel-based device [131]. Mak et al. [123] have created an automated micropipette system with spatial, temporal, and force scales that match physiological and biomechanical processes during metastasis. This system can measure the deformation and relaxation dynamics of single cells serially and in a parallel manner. Using MDA-MB-231 metastatic breast cancer cells, they evaluate the effect that paclitaxel has on cell deformation. Additionally, their system allows evaluation of the contribution of the cell nucleus.

*Intracellular particle tracking* has been used to quantify the viscoelasticity and other rheological parameters of small, microscale samples through motion of particles embedded within a material sample [132]. The dynamics of particle motion reveal the sample response, to either externally applied stresses/deformations (active microrheology) or driven by thermal fluctuations (passive microrheology). Evaluation of the particle dynamics provides quantitative information about the local microenvironment of the sample, revealing its rheology [133]; the viscoelastic properties of a fluid can be determined by the Stokes-Einstein-Sutherland equation given the particle size, temperature, and viscosity, $\eta$ [55].

Inside living cells, however, motion is not only thermally driven, as (unknown) energetic processes continuously occur (e.g., molecular motor motion and treadmilling). Thus, microrheological calculations may not be used, albeit much can be learned about cell dynamics and structure through advanced particle motion analysis approaches [55,134]. Using intracellular particle tracking, we have shown that metastatic breast cancer cells are softer, with a sparse cytoskeleton, and exhibit enhanced internal dynamics, as compared with benign cells [25–27].

*Optical (laser) tweezers* is a technique by which a laser beam is focused through a microscope objective creating a strong dielectric field gradient. This gradient generates traction force between a dielectric bead of high refractive index and a laser beam, pulling the bead toward the focal point of the trap. One of the applications is the optical stretcher [135], where a cell in a capillary tube is stretched simultaneously by two trapped microbeads (typically one to several micrometers in diameter) attached to the ends of the cell. A quantitative index of deformability can be derived from the known forces applied by the trapped beads (200–500 pN). The optical stretcher has been used to distinguish cancer from benign cells in breast [24], gastrointestinal [136], and oral [137] tissues. Indeed, even different breast cancer cells of varying metastatic potential can be distinguished by stretching using optical tweezers. Optical tweezers have also been used for micromanipulation of particles inside the cells. They have also been used in active microrheology to define the pericellular changes due to contractility and degradation of the ECM [138]. Other applications of laser tweezers in mechanobiology can be found elsewhere [139].

*Cytoindentation by AFM* technique incorporates a cantilever with an ultrafine probe used to apply force on the cell surface. The magnitude of this force is derived by the deflection of the cantilever. The force is used to determine mechanical properties of the cell surface and to stimulate the cell mechanically [140]. This technique was successfully used by Cross et al. [23,141] to distinguish between cancerous and benign tissue from the lung, breast, and pancreas. Similarly, Li et al. [29], Lekka et al. [142,143], and Xu et al. [144] were able to discriminate between breast, bladder, and ovarian cancer cells and their benign counterparts through their mechanics. Besides single-cell analysis, AFM has also been used to distinguish cancer cells within tissue slices [145]. Scanning force by a cantilever has been also used to compare cancer cells with their benign counterparts in five cell lines from different tissues, finding that cancer cells are more deformable than benign [146]. A different approach to microrheology using AFM has provided extracellular mechanics measurements of thyroidal and breast cancer cells as compared with benign [147,148], showing reduction of membrane-area stiffness after contractility inhibition by ML7 [147].

Cytoindentation is a similar technique used to determine the mechanics of the cell by poking the cell surface with a probe [149]. In this technique, a thick glass beam (75 μm diameter) is perpendicular to the tip probe that is attached to a piezoelectric translator. The probe, which can have different adherent coatings, is moved against the cell, with the resulting beam deflection imaged by a microscope. This beam deflection, based on previous calibration, can be used to calculate force of reaction exerted by the cell surface [150].

*Magnetic rotational cytometry* is a mechanical characterization that is performed by applying torque to a magnetic bead (4.5 μm) on the surface of an adherent cell. Low frequencies (0.05–5 Hz) are used to twist the bead. Differences in stiffness have been observed on cancer cells from skin, kidney, prostate, and bladder using this technique [151]. Magnetic beads have also been used as probes for intracellular dynamics, in active particle-tracking microrheology experiments [152].

## 5.2 CELL MIGRATION ASSAYS

Several in vitro assays have been developed to investigate the mechanisms regulating cancer cell migration. These include the following assays: transmembrane migration assays (Boyden chamber), scratch, gap closure or exclusion zone, and migration using microfluidic devices. Fig. 7 shows these in vitro migration assays.

### 5.2.1 Experimental methods for cell migration

*Gap closure scratching/crushing assays* are a straightforward method that quantifies migration by time-lapse microscopy. A cell monolayer is damaged by crushing or scratching (in a round area or a straight line, respectively), and the 2D motion of cells into a cell-free area is quantified by image processing or different quantitation software. This gap closure is always a combination of migration and proliferation—unless the experiment is very short (relative to proliferation or doubling time of the cells) or proliferation is otherwise inhibited. Cells are often serum starved for 8–24h prior to scratching to avoid cell proliferation altering the rate of migration. The disadvantages of this assay are the cell damage during scratching and different wound-area sizes. These types of assays have been used to study breast cancer cell migration [153], and using this setup, we have demonstrated that stretching can promote faster migration [154].

*Gap closure exclusion zone assay* is an assay in which a monolayer grows around an insert, blocking part of the substrate and excluding cells from growing at that location. The insert is removed once the monolayer is confluent. The advantage of this assay is that there is no cell damage (i.e., wound-free migration), and the shape and

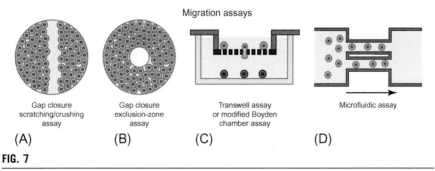

**FIG. 7**

In vitro migration assays. (A) Gap closure scratching/crushing assays, (B) gap closure exclusion zone assay, (C) transwell or modified Boyden chamber assay, and (D) microfluidic assay.

size of the initial cell-free area are well controlled, increasing reproducibility. Quantification is done in the same methods as in scratch cell migration assay.

*Transwell or modified Boyden chamber assay* is a cell migration assay that involves seeding serum-starved cells into a chamber with a membrane with subcellular size micropores (usually ECM coated) separating upper and lower compartments. This assay can evaluate migration and invasion (when coated with thick layer of ECM [155]), and it has been used to determine the in vitro metastatic potential of cells based on their ability to pass the subcellular size pores on the membrane [153,156]. Days after seeding, the number of cells that successfully transmigrate from the upper to the lower compartment is quantified. This movement is induced by a chemotactic gradient that is created by the addition of serum or other chemoattractant in the lower compartment. The smaller the pore that a cell can cross, the more deformable it is and presents an invasive/migratory phenotype. The advantage of this approach is that it can be used with adherent or nonadherent cells, and cells can also be collected for further analysis. We have recently combined this approach with a gel-based mechanical invasion assay for a combined evaluation of migration and invasion [58]. The only disadvantage of this assay is that it has a set endpoint, at a long time after seeding, preventing kinetic analysis of cell migration.

*Microfluidic assays* are usually composed of two chambers connected by one or more microchannels. Cells are seeded in one of the chambers, and their movement through the channel can be monitored, to obtain chemokinetic measurements. Different geometric designs are now available commercially. This assay is suitable for drug testing of primary cells, since it requires small volume of cells.

### 5.2.2 Mathematical modeling in cell motility

Cell motility, especially in the context of cancer, is a complex process involving many different elements of the cells, their environment, and interactions. As controlled experiments that may isolate only specific aspects are extremely difficult to design, perform, and analyze, mathematical modeling has become an indispensable tool. The output of a good predictive model can be and have been successfully used to describe biological and medical processes [157]. In this section, we provide an overview of the different models and simulations that have been used to study cell motility. These models are used to describe migration in the context of cancer and metastasis, as well as in cell sorting, cell aggregation, wound healing, and more [158,159].

Cell motility models can be roughly separated into two main models: the cell lattice models and the continuous approach. In each study, a different function or functions are used for the simulations and analysis. The most widely used functions are population density, energy, and flux. A model includes a set of functions and an iteration rule, and the step of each iteration can be applied in a computer to give simulation results of the model. The following are the main models, simulations (steps), and elements that are used for cell motility research, either modeling cells or cell motion.

### 5.2.3 Cell lattice models

These logical and easy-to-compute models were used in the past when computational abilities were limited. They are still in use today [159], although most studies employ more complex methods that provide more in-depth understanding. Specifically, such models can include multiple facets of the mechanical attributes (cell and environment) involved in migration. The concept involves spreading cells on a grid, where the grid can have different shapes. A function or a set of algorithm steps are used to decide how the cells will move in each successive step. In this way, through many iterations, a long-scale process can be simulated. Running several such simulations can reveal the model's convergence and be used for predicting the results of a process under the model's conditions.

Algorithms of cell lattice models differ in how the steps of the objects (cells) are determined. The most common algorithms are the cellular automata [160], the Ising model, and the Potts model [159]. The latest progress in cellular automata was of Stephen Wolfram. In his study, a one-dimensional (1D) rule can explain how from three cells a cell will iterate in the next step, by changing the middle cell value according to his two left and right neighbors and itself, for example, from [010] to [000]. Ernest Ising added the polarity of elements. In the Ising model, an energy function that includes the polarity of an element is taken into account, and a probability function on performing the step was added. The use of such function is called the Metropolis algorithm. In 1952, Renfrey Potts introduced the Potts model [161], a 2D model that is like the Ising model, but now, each object can spin in Q states and not only two. An XY model [162] generalizes the above to a 3D model, by a spherical angle.

### 5.2.4 Cell lattice models with subcellular attributes

In the previously described lattice models, cells are a black box that can only live, die, or proliferate. Taking into account the real attributes of cells that can influence their ability to move is more computationally complex but may give better simulations of a process. The following cell lattice models include a cell morphology characterization. The iterating method employed is the same as in the Ising model, but the cells are distributed differently. Cell-centric models [163] give points as the center of cells, and a Dirichlet or Voronoli tessellation is created from each point. Cells are distributed in physical initial stage; the edges of the cells are moved in each iteration. The energy change is calculated by the change of all neighboring cells influenced from the edge change. In subcellular lattice [162], on a regular grid, an initial cell distribution is determined, but as cells can occupy more than one location, they can be spread in several continuous locations. Finite elements (FEs) are the most recent and commonly used modeling method at present. It best describes the cell, as each cell is meshed in to a triangular grid, and each part of the cell may have a different set of parameters. This better describes a cell that contains different organelles in different parts of the cell. For example, the viscosity of the nuclear is different from the cytoplasm. The mesh can be as small as computationally possible [164–167].

### 5.2.5 Continuous functions

Another approach of cell motility is the continuous function approach [166,168]. This approach deals with the motility process as a bulk of cells, by one or more partial differential equation (PDE) that describe one or more attributes. The attributes and process are continuous over time, space, or others and not as discrete points, for example, energy, flux, ECM, and cell density with the attributes distributed on the bulk. These models are very different from one another and are harder to package in a specific technique. Computational PDE proximity algorithms are used to solve the PDEs and give a simulated prediction of the given setup [169]. The advantage of this method that is widely used these days is that the steps can be calculated to any degree of accuracy. The values are not related to a given mesh. They are commonly used for that reason; nevertheless, FE models are also very common; in both models, you can iterate or mesh as fine as computationally possible, but in FE, the mesh size is determined before the model is complete, and in the continuous model, you can decide the size of your steps or attribute after the model is built.

### 5.2.6 Considerations in choice of motility model

Deciding what are the attributes that influence a specific process is the most important part of modeling and simulating a process. The different attributes, as stated before, can be separated into four main categories: mechanical, chemical, intercellular mechanics, and topology. Other commonly assessed attributes are the use of a single cell or a population bulk. Sometimes the population is built from different phenotypes, polarization of the cell, and probability function, depending on the model approach. For example, looking at a growing tumor in a specific direction might fit better a bulk described in a PDE set of functions or a lattice. The inner cell activities are treated as a black box attribute [170]. On the other hand, adhesion or invasion will probably fit attributes that include inner cell fibers moving in and out, affecting one another and the substrate. This can also be modeled in a PDE [171] or a finite element model [164]. A game theory researcher would prefer looking at the cells as a black box, having the process modeled in strategy function that converges into an equilibrium, most probably simulating his results by a lattice model [172]. Depending on the specific process to be modeled and knowledge from in vivo or in vitro experimental results, the different attributes should be considered or neglected. To sum up, there is no wrong or right model; the decision will most probably be chosen by the history and prior knowledge of the researchers and the attributes that they want to emphasize and learn more about.

## 5.3 CELL STRENGTH EVALUATION

Cell strength can be quantified through forces that can be directly measured or through the results of force application. Specifically, when cells apply forces to their microenvironment, we can evaluate the strain energy that is accumulated in the surrounding gel/ECM or the deformations induced in them. Fig. 8 shows the most used techniques for measuring cell strength. Many of these methods utilize variations of

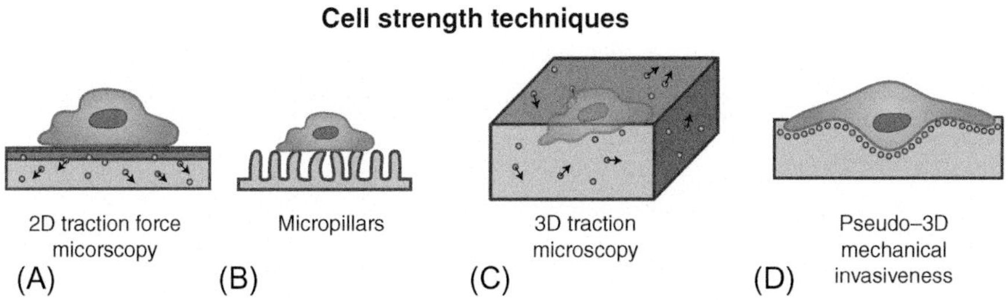

**Cell strength techniques**

**FIG. 8**

Techniques for measuring cell strength. (A) 2D traction force microscopy, (B) micropillars, (C) 3D traction microscopy, and (D) pseudo-3D mechanical invasiveness.

traction microscopy, which involve tracking the deformation of elastic, nondegradable, typically polymeric substrates, due to cell-applied forces. Forces can be applied by cells laterally, normally, or as a combination of both. Cells are seeded on a nondegradable hydrogel, such that the cell forces can be decoupled from the mechanical properties of the substrate. Due to linear elastic behavior and well-characterized mechanical properties, polyacrylamide (PAM) (1–100 kPa) [94,173] and silicone (10–100 kPa) embedded with fluorescent beads (~1 μm diameter) are used as a substrate in TFM. To facilitate and promote cell adhesion, gels are coated by a thin layer of physiologically relevant ECM or molecules such as collagen or fibronectin.

### *5.3.1 2D traction force microscopy*

A typical TFM experiment requires two microscopy images to be analyzed, one with the cells attached firmly to the substrate and a second image of the same field of view after cell detachment. Image processing is used to determine the displacement of the beads embedded on the gel from the stressed state (when cell was attached) to the relaxed state (after cell detachment). Due to the size of the beads, much smaller than the cells, forces can be derived from the displacement maps with subcellular resolution. The forces are converted into traction stress fields [174,175]. By solving the inverse problem, the possible traction stress field can be obtained from the displacement maps and the measured deformation [175–178]. An example of the process is shown in Fig. 9 from Ref. [33]. These algorithms can calculate subpiconewton forces from subnanometric displacements. Due to the high sensitivity of this technique, stronger traction forces can be measured on cells expressing type I cadherins than on those expressing type II cadherins [179].

The force generated by the cell can be decomposed into surface-parallel components of the substrate and a normal component, which is perpendicular to the substrate surface. Lateral forces, whose components are parallel to the surface, induce deformation on the surface that can be observed and quantified by the analysis of conventional microscopy images. Since it measures only lateral cell tractions, this

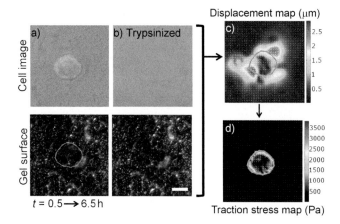

**FIG. 9**

Traction force microscopy protocol. (A) Cells are seeded on a PAM gel (*top*) that have fluorescent beads embedded in its surface (*bottom*); the *drawn circular line* is the cell boundary. Cell is imaged every 1 h after seeding, and gel deformations due to cell-applied forces are identified through bead displacement. (B) The cell is removed from the gel with trypsin, providing the undeformed, reference state. (C) The displacement map of the beads between the cell deformed and relaxed gel is calculated for each time point; (D) traction stress map is calculated from the displacement map by using the Young's modulus and Poisson's ratio of the gel. Scale bar is 20 μm.

*Adapted from Massalha S, Weihs D. Metastatic breast cancer cells adhere strongly on varying stiffness substrates, initially without adjusting their morphology. Biomech Model Mechanobiol 2017;16(3):961–70. Available from: http://link.springer.com/10.1007/s10237-016-0864-4 [cited 2016 December 24].*

is considered 2D method. This technique has been used to characterize force dynamics in different cell processes, such as adhesion [33], migration [69], and differentiation [32].

### 5.3.2 Micropillars

In a similar approach to TFM, posts of known stiffness and structure are built. Force applied by tissue or cell contracting can be derived using beam theory, from the cantilever bending imaged by optical microscopy. These vertical cantilevers are fabricated from polydimethylsiloxane (PDMS) by soft lithography and are built as constructs that can measure forces of several cells. Specifically, they have been very useful for measuring contraction forces in cardiomyocytes [180]. The main advantage is that several cantilevers can be measured simultaneously. By combining this technique with optical tweezer-based active microrheology, the local intracellular mechanics can be measured [181]. This technique has recently used to measure mechanical properties of multicellular tumor spheroids [182], to measure nuclear deformability and cell mechanical properties [183].

### 5.3.3 Three-dimensional traction microscopy

In 3D gels, the elastic strain energy stored in the matrix due to cell traction-induced deformations are calculated; traction forces are not available due to theoretical and algorithmic limitations. In this technique, cells are usually either embedded within collagen prior to polymerization or cultured on a collagen gel surface and then allowed to invade. In both conditions, the cells spread into the collagen exerting traction forces that deform the gel. The expended energy is stored as elastic strain energy in the gels. If the elastic properties of the collagen are known, the strain energy can be derived by measuring the deformation of the collagen gels. These gel deformations are determined from the positions of fluorescent microbeads embedded throughout the gel. Cells of different types or under varying conditions are compared with a scalar measure for the total cellular contractility.

In vivo, cells exist within a 3D environment, and it has been shown in many reviews how three-dimensionality induces changes in morphology, phenotype, and cell behavior when comparing with 2D. Collagen I is the most commonly used ECM material for 3D culture; however, the nonlinear fibrillar nature of collagen along with the high degradation rate prevents proper calculation of traction forces from ECM deformation. Three-dimensional traction requires the use of a linearly elastic material such as the synthetic MMP-cleavable polyethylene glycol (PEG). However, displacement measurements in 3D are cumbersome and challenging, and confocal microscopy is required [174,184–186].

### 5.3.4 Psuedo 3D mechanical invasiveness

Using a soft, nondegradable PAM gel with submicron pores, we have effectively allowed only mechanical interactions of a single cell with the gel; the gel is coated with collagen for cell attachment. Our gel system allows cells to indent gels by applying force, yet cells are not able to penetrate the gel. Gel indentation is obtained by monitoring deformations in the initially flat gel [56]. The gel surface is embedded with fluorescent microparticles (200 nm diameter) within the top layer of the gel. Hence, using microscopy, we can detect changes in focal depth of the particles that indicate indentation. Using this gel-based invasion assay, we have successfully distinguished between highly metastatic, low-metastatic potential, and nonmetastatic breast cell lines, based only on the deformation induced by the normal forces applied by metastatic cells when attempting invasion. We have observed that metastatic breast cancer cells deform and indent a soft gel while benign breast cells do not (Fig. 10). Furthermore, we have also observed that cells exhibit preference for gels of specific stiffness, soft enough to indent and stiff enough to effectively grasp [56].

In the previous sections, we have reviewed different methodologies and assays to quantify forces and motility abilities. In the next section, we will show and discuss how those may be used to evaluate the metastatic potential of invasive cancer cells.

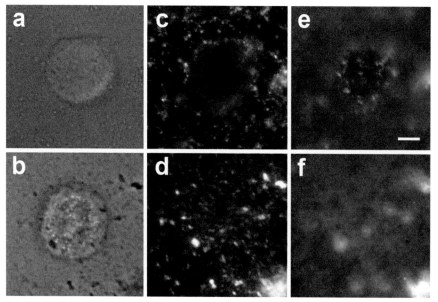

**FIG. 10**

Single, metastatic, breast cancer cells (MDA-MB-231) seeded on 2.4 kPa stiffness PAM gels either (top row) indent the gels or (bottom row) adhere but do not indent. (A, B) DIC image of indenting and nonindenting cells showing that both maintain a rounded morphology; (C, D) fluorescence image of 200 nm-diameter fluorescent beads embedded at the gel surface. In (C), beads directly under the cell are displaced to a lower plane due to cell-induced indentation; (E) beads are in focus at 6.28 μm below gel surface, indicating the indentation depth generated in the gel; (F) for the nonindenting cell, the gel remains flat, and beads are out of focus below the gel surface. Scale bar is 10 μm. DIC, differential interference contrast.

*Adapted from Alvarez-Elizondo MB, Weihs D. Cell-gel mechanical interactions as an approach to rapidly and quantitatively reveal invasive subpopulations of metastatic cancer cells. Tissue Eng Part C Methods 2017;23(3):180–7. Available from: http://online.liebertpub.com/doi/10.1089/ten.tec.2016.0424 [cited 2017 February 14].*

## 6 EVALUATING METASTATIC POTENTIAL THROUGH MECHANICAL INTERACTIONS

Cell-microenvironment mechanical interactions are present in all stages of metastasis. The capacity of cells to rapidly change morphology and reorganize the cytoskeleton to apply forces is what drives and facilitates their invasiveness—their metastatic potential. The metastatic potential of cancer cells is indicative of their ability to reach later stages of cancer progression. The ability to distinguish metastatic potential provides a basis for developing more efficient treatments for metastasis. Here, we present methods for distinguishing metastatic potential of cancer cells based on their mechanical interactions with an elastic nondegradable hydrogel embedded with microparticles.

## 6.1 ADHESION AND ADHESIVE FORCES

Cell adhesion is a time- and microenvironment-dependent process that can be divided in two steps: attachment and spreading. Together, these lead to stably adhered cells. Both steps result from a combination of biochemical and biomechanical processes, albeit attachment is mostly biochemical while spreading is more biomechanical. Interactions of a cell with its microenvironment occur through adhesion receptors [187] and focal adhesions [188] by utilizing the cytoskeletal networks. Following initial attachment through biochemical connections, the cytoskeleton and its associated molecular motors undergo dynamic rearrangement that allows active cell migration and force generation. Adherent and migrating cells typically apply lateral traction forces to the underlying substrate.

Our lab has quantified and evaluated the strength of adherence through the 2D lateral traction forces applied by metastatic breast cancer cell lines during the short ($<7$ h) initial stages of cell adhesion [33], which differ from the forces applied during stable, long-scale ($>24$ h) adherence [69]. We identify differences in time-dependent morphology and strength of adherence in these initial interactions, just 6.5 h following seeding, which are likely correlated to the metastatic potential of the cells. Specifically, low-metastatic-potential cells (MDA-MB-468) apply different typically lower 2D traction forces than high MP cells (MDA-MB-231), with both being stronger than forces applied by benign cells (MCF-10A). In addition, the stiffness of a substrate directly affects the morphology of all cells and the forces that metastatic cells apply, with a smaller effect on the benign cells. Metastatic cancer cells remain mainly rounded, while benign cells elongate with time, especially on the stiffer gels, as shown in Fig. 11A.

We determined the lateral traction forces applied by cells on the surface of PAM gels with stiffness 2.4–10.6 kPa, between 0.5 and 6.5 h after seeding, as shown in Fig. 11B–E. We note that the evaluated cells exhibit a wide distribution of time-dependent traction forces, depending on each cell's specific interaction times. We have developed a novel approach to effectively normalize the different dynamic timescales of the single cells. Using the local force maxima approach [33,56], we observed statistically significant ($P<0.05$) variation between the forces applied by each evaluated cell type, allowing us to distinguish the cell lines with different metastatic potentials. On gels with Young's moduli larger than 2.4 kPa, both the high and low MP cells apply significantly larger traction forces during adhesion, as compared with the benign cells. In contrast, on the 2.4 kPa gels, the traction forces applied by the cell lines were indistinguishable. Many of the metastatic cells interact differently with those soft gels, indenting them, as will be discussed in the next section. Thus, the evaluated metastatic, breast cancer cell lines exhibited gel-stiffness-dependent differences in traction forces, strain energies, and morphologies during the initial stages of adhesion, which likely relate to their metastatic capacity; the cells also significantly differ from benign cells.

After cells are well adhered, traction forces can be measured; these forces or their output as strain energy has also been proposed as a biophysical marker of invasiveness. Mechanical interactions of cancer cells with their microenvironment measured

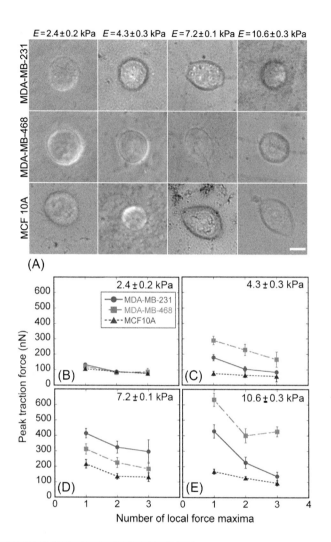

**FIG. 11**

(A) Cell morphologies on 2.4–10.6 kPa PAM gels at $t=6.5$ h after seeding. Cell images and areas indicate that high MP (MDA-MB-231) and low MP (MDA-MB-468) cells are mainly round on the gels. Benign (MCF-10A) cells are round on soft gels ($E<7$ kPa) and elongate on stiffer gels ($E>7$ kPa). Scale bar is 10 μm. (B–E) Average of peak, total traction force applied at local force maxima, correlated with time after seeding. The force amplitudes on all gel stiffness decrease with time. Only cells that applied measurable forces during at least five time points out of the total of seven experimental time points were included, and all cells exhibited three local force maxima during the experiment time. (B) The cells apply the same amplitude of forces on the soft gel (2.4 kPa). (C, E) Low MP cells apply larger forces than high MP, and both apply more force than benign cells; (D) high MP cells apply the highest forces. Error bars are standard errors.

*Adapted from Massalha S, Weihs D. Metastatic breast cancer cells adhere strongly on varying stiffness substrates, initially without adjusting their morphology. Biomech Model Mechanobiol 2017;16(3):961–70. Available from: http://link.springer.com/10.1007/s10237-016-0864-4 [cited 2016 December 24].*

by TFM have revealed differences between cells with varying metastatic potential during cell processes, such as cell migration [69,189]. However, conflicting studies with different cell lines have shown either reduced or increased traction forces produced in cells following malignant transformation (EMT) [176]. Using 2D traction force microscopy, significantly different contractile forces have been shown across varying matrix mechanostructures (varying stiffness and collagen density) in highly metastatic breast, prostate, and lung cancer cell lines as compared with benign cells [69]. Higher traction forces have also been measured in highly metastatic bladder cancer cell lines as compared with benign [189]. In contrast, an inverse relationship between traction forces and metastatic capacity was observed in four murine breast cancer cell lines derived from a single primary tumor, and each of these cell lines was capable of completing different stages of metastasis [190]. This study showed a reduction in the number of focal adhesions and in the 2D migration of both single and collectively migrating cells as the metastatic capacity increased. A gradual shift from migration modes was expected, and as predicted, higher migration speeds were observed in 3D for more metastatic cells. Furthermore, 3D traction forces measured on several cell lines of different cancer types showed higher strain energy (resulting from the 3D traction forces) and more elongated spindle-like morphologies in higher metastatic potential cells [191]; fully 3D traction forces are experimentally unmeasurable. Thus, the mechanical interactions of the cells with their substrate, including morphology and microenvironment deformations induced by cell-applied forces, may serve as a measure for the metastatic potential and capacity of the cells.

## 6.2 MECHANICAL INVASIVENESS

We have previously observed that forces applied by invasive, metastatic cells may result in a combination of normal and lateral forces; this is in contrast to the lateral adhesive and migratory forces described in the previous section. We have shown that on soft, initially 2D PAM gels (e.g., 2.4 kPa), metastatic cells rapidly (within 2 h of seeding) induce indentations, likely in the initial stages of attempted invasion [11,28,56–58]. This phenomenon was observed both in single cells and in groups of adjacent metastatic cells, while benign cells do not indent. Evaluating these indentations provided a measure to successfully differentiate between high metastatic potential, low MP, and nonmetastatic breast and also pancreatic cell lines [56,57]. We term this measure the "mechanical invasiveness" of the cells, which correlates directly with their metastatic potential and migratory capacity [58].

The combined lateral and normal forces applied by the cells to their substrate (i.e., the PAM gel) require a different approach to evaluate the force, as traction force microscopy (2D/3D) only considers lateral forces. In this more complex case, a model for the force application mechanism is required, entailing assumptions on the cell morphology, interactions, and force application mechanisms. Thus, for simplification, we determine the mechanical invasiveness of the cancer cells through the experimentally measurable percent of indenting cells and indentation depths of each cell. Small indentations (<1 μm) have been observed previously in noncancerous cells [192,193]. In contrast, the normal forces applied by metastatic cells result in indentations in the range of 10–20 µm comparable with the cell diameter. By evaluating the percentage of intending cells and the indentation depth, induced either by single cells or by groups of cells, we have determined an invasive subpopulation of metastatic cancer cells within the cell line, able to indent a soft gel. Interestingly, the distribution of indentation depths of all indenting cells differs between single cells and closely adjacent cells.

When the metastatic cells are seeded in high density on the gel and are adjacent, we have observed synergistic interactions of indenting cells, likely related to the efficiency of collective invasion. Specifically, groups of metastatic cells are able to reach larger indentation depths than single cells [57]; benign cells do not significantly indent in either scenario. The indentation depths induced by the single, well-spaced, high and low MP cells on 2.4 kPa gels were single peak Gaussian-like distribution and always <10 µm, with statistically indistinguishable means. In contrast, when the metastatic cells are seeded in high density on PAM gels (still in 2D), we observe a two-peak distribution of indentation depths. The lower peak, <10 µm, overlaps perfectly with that observed in the single cells. The adjacent cells further exhibit a subset of the indenting subpopulation of cells that is able to synergistically reach indentations >10 µm. The subset of high and low MP cells able to synergistically indent were different, being, respectively, 65% versus 15% [57], providing a differentiation between the more and less invasive cells. This synergistic capability of the cells to attain larger indentation depths is easily disrupted, for example, by chemotherapy or by mechanical perturbations (unpublished work).

The in vivo implications of the synergistic mechanical invasiveness phenomenon are clear—collective migration/invasion is more efficient. Accordingly, mice injected with clusters of tumor cells develop more metastases compared with mice injected with an equal total number of single cells [194]. Similarly, collective cell streaming was observed in cancer cell migration [90]. Thus, proximity and interactions of the metastatic cells promote mechanical invasiveness and show why collective cancer cell invasion could be more efficient.

Further, correlating our mechanical invasiveness measure to commonly used cell migration/invasion assays shows its direct link to determination of metastatic potential. We have correlated the gel-indentation assay with the Boyden chamber transmigration/invasion assay [58]; the combination of these two assays may be used to evaluate in vitro metastatic potential and the ensuing mechanical invasiveness of cancer cells on a single-cell level. Both the Boyden chamber and our gel mechanobiology assay reveal an invasive subpopulation of cancer cells that is capable of both indentation and invasion. We have observed that the Boyden chamber effectively concentrates the more invasive cells. That is, the subpopulation of cells that successfully transmigrate through an 8 µm Boyden chamber membrane is also the subpopulation that will deeply indent the gels. A larger subpopulation of indenting cells was observed in the high MP cells as compared with the low MP cells, under all conditions (Fig. 12). Importantly, the gel-indentation or mechanical invasiveness measure is rapid (<2 h) and quantitative, allowing focus on specific cells as they indent; the response of specific cells to treatment may also be monitored [56].

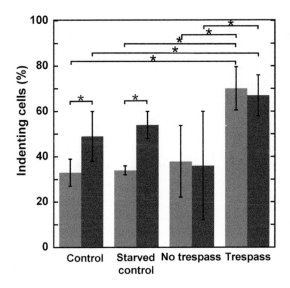

**FIG. 12**

Percentage of MDA-MB-468 *(light gray)* and MDA-MB-231 *(dark gray)* metastatic breast cancer cells that indent the 2.4 kPa PAM gels. Controls include all cells plated on the gels under either serum-full or serum-starved conditions; control cells were dissociated with trypsin. We compare the whole population controls with two cell populations that are plated on the gels following Boyden chamber migration for 72 h being, respectively, the trespassing subpopulation (collected from the lower chamber) and the nontrespassing subpopulation (upper chamber). Results are presented as the mean ± standard deviation. Asterisks indicate statistically significant differences ($P < 0.05$).

*Adapted from Alvarez-Elizondo MB, Weihs D. Cell-gel mechanical interactions as an approach to rapidly and quantitatively reveal invasive subpopulations of metastatic cancer cells. Tissue Eng Part C Methods 2017;23(3):180–7. Available from: http://online.liebertpub.com/doi/10.1089/ten.tec.2016.0424 [cited 2017 February 14].*

# 7 SUMMARY

Tumor staging (risk level) is currently based on a qualitative, typically manual characterization of a biopsy; a biopsy being a sample of the tissue is composed of malignant and healthy cells within the surrounding microenvironment (basement membrane and stromal ECM). While this approach can reveal malignancy, it relies on statistics, genetics, and molecular markers for a prognosis of metastasis; this can result in inaccurate prognosis. Mechanobiology is a novel approach to obtain quantitative measures that could reveal likelihood for metastasis. By evaluating cell mechanics and cell-applied forces, we have shown that inherent differences are observed between benign, malignant, and invasive cells. Further, the understanding of cell movement, the underlying mechanisms, and the forces induced during invasion will enable more accurate prognosis and can lead to new therapeutic strategies for metastasis.

To predict metastasis more effectively and reveal controlling aspects, we should identify the "invasive phenotype," that is, the set of characteristics that allow a subpopulation of cells to successfully traverse through the body, including all the discussed obstacles. Such characteristics already include genetic, hormonal, and molecular markers but as we have shown should include mechanobiological measures; those aspects are available experimentally and/or through computational modeling, which expands available conditions. In this chapter, we have thus provided a description of cell mechanics, the mechanical interactions, and the different cell motility modes during the steps of metastasis, which are all important factors that contribute to the invasive phenotype. In addition, we provided an up-to-date review of current cytomechanobiology approaches and methods to determine features of the in vitro mechanobiology of cancer cells.

# REFERENCES

[1] Talmadge JE, Fidler IJ. AACR centennial series: the biology of cancer metastasis: historical perspective. Cancer Res 2010;70(14):5649–69.

[2] Virchow R. Cellular pathology. As based upon physiological and pathological histology. Lecture XVI—atheromatous affection of arteries. 1858. Nutr Rev 1989;47(1):23–5.

[3] Paget S. The distribution of secondary growths in cancer of the breast. Lancet 1889;133 (3421):571–3.

[4] Neoplastic diseases, a treatise on tumors. Can Med Assoc J 1924;14(5):466. [cited 2017 June 25]. Available from: https://www.ncbi.nlm.nih.gov/pmc/articles/PMC1707671/.

[5] Fidler IJ. Tumor heterogeneity and the biology of cancer invasion and metastasis. Cancer Res 1978;38(9):2651–60. [cited 2017 June 25]. Available from: http://www.ncbi.nlm.nih.gov/pubmed/354778.

[6] Fidler IJ. Metastasis: quantitative analysis of distribution and fate of tumor emboli labeled with 125 I-5-iodo-2′-deoxyuridine, J Natl Cancer Inst 1970;45(4):773–82. [cited 2017 June 25]. Available from: http://www.ncbi.nlm.nih.gov/pubmed/5513503.

[7] Talmadge JE, Fidler IJ. Cancer metastasis is selective or random depending on the parent tumour population. Nature 1982;297(5867):593–4. https://doi.org/10.1038/297593a0.

[8] Raz A, Hanna N, Fidler IJ. In vivo isolation of a metastatic tumor cell variant involving selective and nonadaptive processes. J Natl Cancer Inst 1981;66(1):183–9. [cited 2017 June 25]. Available from: http://www.ncbi.nlm.nih.gov/pubmed/6935459.

[9] Fidler IJ, Talmadge JE. Evidence that intravenously derived murine pulmonary melanoma metastases can originate from the expansion of a single tumor cell. Cancer Res 1986;46(10):5167–71. [cited 2017 June 25]. Available from: http://www.ncbi.nlm.nih.gov/pubmed/3756870.

[10] Nicolson GL, Brunson KW, Fidler IJ. Specificity of arrest, survival, and growth of selected metastatic variant cell lines. Cancer Res 1978;38(11 Pt 2):4105–11. [cited 2017 June 25]. Available from: http://www.ncbi.nlm.nih.gov/pubmed/359132.

[11] Kristal-Muscal R, Dvir L, Schvartzer M, Weihs D. Mechanical interaction of metastatic cancer cells with a soft gel. Procedia IUTAM 2015;12:211–9.

[12] Fletcher DA, Mullins R. Cell mechanics and the cytoskeleton. Nature 2010;463 (7280):485–92.

[13] Gittes F, Mickey B, Nettleton J, Howard J. Flexural rigidity of microtubules and actin filaments measured from thermal fluctuations in shape. J Cell Biol 1993;120(4):923–34.
[14] Ananthakrishnan R, Ehrlicher A. The forces behind cell movement. Int J Biol Sci 2007;3:303–17.
[15] Pollard TD, Cooper JA. Actin, a central player in cell shape and movement. Science 2009;326(5957):1208–12.
[16] Dupin I, Etienne-Manneville S. Nuclear positioning: mechanisms and functions. Int J Biochem Cell Biol 2011;43(12):1698–707.
[17] Lombardi ML, Lammerding J. Keeping the LINC: the importance of nucleocytoskeletal coupling in intracellular force transmission and cellular function. Biochem Soc Trans 2011;39:1729–34.
[18] Maniotis AJ, Chen CS, Ingber DE. Demonstration of mechanical connections between integrins cytoskeletal filaments, and nucleoplasm that stabilize nuclear structure. Proc Natl Acad Sci U S A 1997;94(3):849–54.
[19] Valiron O, Caudron N, Job D. Microtubule dynamics. Cell Mol Life Sci 2001;58(14):2069–84.
[20] Strobl JS, Nikkhah M, Agah M. Actions of the anti-cancer drug suberoylanilide hydroxamic acid (SAHA) on human breast cancer cytoarchitecture in silicon microstructures. Biomaterials 2010;31(27):7043–50. Available from: http://www.ncbi.nlm.nih.gov/pubmed/20579727.
[21] Herrmann H, Bar H, Kreplak L, Strelkov SV, Aebi U. Intermediate filaments: from cell architecture to nanomechanics. Nat Rev Mol Cell Biol 2007;8(7):562–73.
[22] Nagle RB. A review of intermediate filament biology and their use in pathological diagnosis. Mol Biol Rep 1994;19(1):3–21.
[23] Cross SE, Jin YS, Rao J, Gimzewski JK. Nanomechanical analysis of cells from cancer patients. Nat Nanotechnol 2007;2(12):780–3.
[24] Guck J, Schinkinger S, Lincoln B, Wottawah F, Ebert S, Romeyke M, et al. Optical deformability as an inherent cell marker for testing malignant transformation and metastatic competence. Biophys J 2005;88(5):3689–98.
[25] Gal N, Weihs D. Experimental evidence of strong anomalous diffusion in living cells. Phys Rev E 2010;81(2):020903(R).
[26] Gal N, Weihs D. Intracellular mechanics and activity of breast cancer cells correlate with metastatic potential. Cell Biochem Biophys 2012;63(3):199–209.
[27] Goldstein D, Elhanan T, Aronovitch M, Weihs D. Origin of active transport in breast-cancer cells. Soft Matter 2013;9(29):7167–73. Available from: http://xlink.rsc.org/?DOI=c3sm50172h.
[28] Dvir L, Nissim R, Alvarez-Elizondo MB, Weihs D. Quantitative measures to reveal coordinated cytoskeleton-nucleus reorganization during in vitro invasion of cancer cells. New J Phys 2015;17(4):43010. Available from: http://stacks.iop.org/1367-2630/17/i=4/a=043010.
[29] Li QS, Lee GYH, Ong CN, Lim CT. AFM indentation study of breast cancer cells. Biochem Biophys Res Commun 2008;374(4):609–13.
[30] Ghibaudo M, Saez A, Trichet L, Xayaphoummine A, Browaeys J, Silberzan P, et al. Traction forces and rigidity sensing regulate cell functions. Soft Matter 2008;4(9):1836–43.
[31] Mogilner A, Oster G. Force generation by actin polymerization II: the elastic ratchet and tethered filaments. Biophys J 2003;84(3):1591–605. [cited 2017 August 30]. Available from: http://linkinghub.elsevier.com/retrieve/pii/S0006349503749698.

[32] Abuhattum S, Weihs D. Asymmetry in traction forces produced by migrating preadipocytes is bounded to 33%. Med Eng Phys 2016;38(9):834–8.

[33] Massalha S, Weihs D. Metastatic breast cancer cells adhere strongly on varying stiffness substrates, initially without adjusting their morphology. Biomech Model Mechanobiol 2017;16(3):961–70. [cited 2016 December 24], https://doi.org/10.1007/s10237-016-0864-4.

[34] Teo A, Lim M, Weihs D. Embryonic stem cells growing in 3-dimensions shift from reliance on the substrate to each other for mechanical support. J Biomech 2015;48 (10):1777–81. https://doi.org/10.1016/j.jbiomech.2015.05.009.

[35] Wyckoff JB, Pinner SE, Gschmeissner S, Condeelis JS, Sahai E. ROCK- and myosin-dependent matrix deformation enables protease-independent tumor-cell invasion in vivo. Curr Biol 2006;16(15):1515–23. Available from: http://www.ncbi.nlm.nih.gov/pubmed/16890527.

[36] Ridley AJ. Rho GTPase signalling in cell migration. Curr Opin Cell Biol 2015;36:103–12. [cited 2017 August 30]. Available from: http://linkinghub.elsevier.com/retrieve/pii/S0955067415001106.

[37] Tojkander S, Gateva G, Lappalainen P. Actin stress fibers—assembly, dynamics and biological roles. J Cell Sci 2012;125:1855–64. [cited 2017 August 29]. Available from:http://jcs.biologists.org/content/joces/125/8/1855.full.pdf.

[38] Ponuwei GA. A glimpse of the ERM proteins [cited 2017 August 30]. Available from: https://jbiomedsci.biomedcentral.com/track/pdf/10.1186/s12929-016-0246-3?site=jbiomedsci.biomedcentral.com

[39] Mierke CT, Rosel D, Fabry B, Brabek J. Contractile forces in tumor cell migration. Eur J Cell Biol 2008;87(8–9):669–76.

[40] Albelda SM. Role of integrins and other cell adhesion molecules in tumor progression and metastasis. Lab Investig 1993;68(1):4–17. Available from: http://www.ncbi.nlm.nih.gov/pubmed/8423675.

[41] Ganguly KK, Pal S, Moulik S, Chatterjee A. Integrins and metastasis, Cell Adhes Migr 2013;7(3):251–61. Available from: http://www.ncbi.nlm.nih.gov/pubmed/23563505.

[42] Hart IR. Role of integrins in tumor invasion and metastasis. Exp Dermatol 2004;13 (10):663. Available from: http://www.ncbi.nlm.nih.gov/pubmed/15447732.

[43] Dumbauld DW, Lee TT, Singh A, Scrimgeour J, Gersbach CA, Zamir EA, et al. How vinculin regulates force transmission. Proc Natl Acad Sci U S A 2013;110 (24):9788–93.

[44] Huang H, Kamm RD, Lee RT. Cell mechanics and mechanotransduction: pathways, probes, and physiology. Am J Phys Cell Physiol 2004;287(1):C1–11. Available from:http://www.ncbi.nlm.nih.gov/pubmed/15189819.

[45] Weaver AM. Invadopodia: specialized cell structures for cancer invasion. Clin Exp Metastasis 2006;23(2):97–105.

[46] Schoumacher M, Goldman RD, Louvard D, Vignjevic DM. Actin, microtubules, and vimentin intermediate filaments cooperate for elongation of invadopodia. J Cell Biol 2010;189(3) [cited 2017 June 26]. Available from: http://jcb.rupress.org/content/189/3/541.short.

[47] Weaver VM, Petersen OW, Wang F, Larabell CA, Briand P, Damsky C, et al. Reversion of the malignant phenotype of human breast cells in three-dimensional culture and in vivo by integrin blocking antibodies. J Cell Biol 1997;137(1):231–45. [cited 2017 June 27]. Available from: http://www.ncbi.nlm.nih.gov/pubmed/9105051.

[48] Huang S, Ingber DE. Cell tension, matrix mechanics, and cancer development. Cancer Cell 2005;8(3):175–6. [cited 2017 June 26]. Available from: http://www.ncbi.nlm.nih.gov/pubmed/16169461.

[49] Dogterom M, Kerssemakers JW, Romet-Lemonne G, Janson ME. Force generation by dynamic microtubules. Curr Opin Cell Biol 2005;17(1):67–74. [cited 2017. August 29]. Available from: http://linkinghub.elsevier.com/retrieve/pii/S0955067404001796.

[50] Ezratty EJ, Partridge MA, Gundersen GG. Microtubule-induced focal adhesion disassembly is mediated by dynamin and focal adhesion kinase. Nat Cell Biol 2005;7(6):581–90. [cited 2017 August 8]. Available from: http://www.ncbi.nlm.nih.gov/pubmed/15895076.

[51] Balaban NQ, Schwarz US, Riveline D, Goichberg P, Tzur G, Sabanay I, et al. Force and focal adhesion assembly: a close relationship studied using elastic micropatterned substrates. Nat Cell Biol 2001;3(5):466–72.

[52] Wang N, Naruse K, Stamenović D, Fredberg JJ, Mijailovich SM, Tolić-Nørrelykke IM, et al. Mechanical behavior in living cells consistent with the tensegrity model. Proc Natl Acad Sci U S A 2001;98(14):7765–70. [cited 2017 August 8]. Available from: http://www.ncbi.nlm.nih.gov/pubmed/11438729.

[53] Jean RP, Gray DS, Spector AA, Chen CS. Characterization of the nuclear deformation caused by changes in endothelial cell shape. J Biomech Eng 2004;126(5):552–8. Available from: http://www.ncbi.nlm.nih.gov/pubmed/15648807.

[54] Zwerger M, Ho CY, Lammerding J. Nuclear mechanics in disease. Annu Rev Biomed Eng 2011;13:397–428. Available from: http://www.ncbi.nlm.nih.gov/pubmed/21756143.

[55] Gal N, Lechtman-Goldstein D, Weihs D. Particle tracking in living cells: a review of the mean square displacement method and beyond. Rheol Acta 2013;52(5):425–43.

[56] Kristal-Muscal R, Dvir L, Weihs D. Metastatic cancer cells tenaciously indent impenetrable, soft substrates. New J Phys 2013;15:35022.

[57] Merkher Y, Weihs D. Proximity of metastatic cells enhances their mechanobiological invasiveness. Ann Biomed Eng 2017;45(6):1399–406. [cited 2017 February 26], https://doi.org/10.1007/s10439-017-1814-8.

[58] Alvarez-Elizondo MB, Weihs D. Cell-gel mechanical interactions as an approach to rapidly and quantitatively reveal invasive subpopulations of metastatic cancer cells. Tissue Eng Part C Methods 2017;23(3):180–7. [cited 2017 February 14], https://doi.org/10.1089/ten.tec.2016.0424.

[59] Frantz C, Stewart KM, Weaver VM. The extracellular matrix at a glance. J Cell Sci 2010;123(Pt 24):4195–200. [cited 2017 August 29]. Available from: http://www.ncbi.nlm.nih.gov/pubmed/21123617.

[60] Petrie RJ, Yamada KM. At the leading edge of three-dimensional cell migration. J Cell Sci 2012;125(24):5917–26. [cited 2017 June 27]. Available from: http://www.ncbi.nlm.nih.gov/pubmed/23378019.

[61] Pickup MW, Mouw JK, Weaver VM. The extracellular matrix modulates the hallmarks of cancer. EMBO Rep 2014;15(12):1243–53. [cited 2017 June 26]. Available from: http://www.ncbi.nlm.nih.gov/pubmed/25381661.

[62] Min-Cheol Kim JW, Silberberg YR, Kamm RD, Harry Asada H. Cell invasion dynamics into a three dimensional extracellular matrix fibre network. PLoS Comput Biol 2015;5(1004535):1–29. Available from: http://journals.plos.org/ploscompbiol/article/asset?id=10.1371/journal.pcbi.1004535.PDF.

[63] Nelson CM, Bissell MJ. Of extracellular matrix, scaffolds, and signaling: tissue architecture regulates development, homeostasis, and cancer. Annu Rev Cell Dev Biol

2006;22(1):287–309. [cited 2017 June 27], https://doi.org/10.1146/annurev.cellbio.22.010305.104315.

[64] Paszek MJ, Zahir N, Johnson KR, Lakins JN, Rozenberg GI, Gefen A, et al. Tensional homeostasis and the malignant phenotype. Cancer Cell 2005;8(3):241–54.

[65] Giussani M, Merlino G, Cappelletti V, Tagliabue E, Daidone MG. Tumor-extracellular matrix interactions: identification of tools associated with breast cancer progression. Semin Cancer Biol 2015;35:3–10. [cited 2017 August 29]. Available from: http://linkinghub.elsevier.com/retrieve/pii/S1044579X15000930.

[66] Levental KR, Yu H, Kass L, Lakins JN, Egeblad M, Erler JT, et al. Matrix crosslinking forces tumor progression by enhancing integrin signaling. Cell 2009;139(5):891–906. [cited 2017 June 27]. Available from: http://www.ncbi.nlm.nih.gov/pubmed/19931152.

[67] Alexander N, Branch K, Parekh A, Clark E. Extracellular matrix rigidity promotes invadopodia activity. Curr Biol 2008; [cited 2017 June 26]; Available from: http://www.sciencedirect.com/science/article/pii/S096098220801035X.

[68] Rolli M, Fransvea E, Pilch J, Saven A, Felding-Habermann B. Activated integrin alphavbeta33 cooperates with metalloproteinase MMP-9 in regulating migration of metastatic breast cancer cells. Proc Natl Acad Sci U S A 2003;100(16):9482–7. [cited 2017 June 26]. Available from: http://www.ncbi.nlm.nih.gov/pubmed/12874388.

[69] Kraning-Rush CM, Califano JP, Reinhart-King CA. Cellular traction stresses increase with increasing metastatic potential. PLoS One 2012;7(2):e32572. Available from: http://www.ncbi.nlm.nih.gov/pubmed/22389710.

[70] Jaalouk DE, Lammerding J. Mechanotransduction gone awry. Nat Rev Mol Cell Biol 2009;10(1):63–73. [cited 2017 June 27], https://doi.org/10.1038/nrm2597.

[71] Menon S, Beningo KA, Lock P, Sang Q, Howard L, Alexander N, et al. Cancer cell invasion is enhanced by applied mechanical stimulation. Gullberg D, editor. PLoS One 2011;6(2):e17277. https://doi.org/10.1371/journal.pone.0017277 [cited 2016 November 27].

[72] Vogel V, Sheetz M. Local force and geometry sensing regulate cell functions. Nat Rev Mol Cell Biol 2006;7(4):265–75. [cited 2017 June 27], https://doi.org/10.1038/nrm1890.

[73] Paszek MJ, Weaver VM. The tension mounts: mechanics meets morphogenesis and malignancy. J Mammary Gland Biol Neoplasia 2004;9(4):325–42. Available from: http://www.ncbi.nlm.nih.gov/pubmed/15838603.

[74] Bissell MJ, Hall HG, Parry G. How does the extracellular matrix direct gene expression? J Theor Biol 1982;99(1):31–68. [cited 2017 June 26]. Available from: http://linkinghub.elsevier.com/retrieve/pii/0022519382903885.

[75] Ingber DE, Huang S, Ingber DE. The structural and mechanical complexity of cell-growth control. Nat Cell Biol 1999;1(5):E131–8. [cited 2017 June 26], https://doi.org/10.1038/13043.

[76] Tse JM, Cheng G, Tyrrell JA, Wilcox-Adelman SA, Boucher Y, Jain RK, et al. Mechanical compression drives cancer cells toward invasive phenotype. Proc Natl Acad Sci U S A 2012;109(3):911–6. [cited 2017 July 10]. Available from: http://www.ncbi.nlm.nih.gov/pubmed/22203958.

[77] Yilmaz M, Christofori G. Mechanisms of motility in metastasizing cells. Mol Cancer Res 2010;8(5):629–42. Available from: http://www.ncbi.nlm.nih.gov/pubmed/20460404.

[78] Friedl P. Prespecification and plasticity: shifting mechanisms of cell migration. Curr Opin Cell Biol 2004;16(1):14–23.

[79] Sahai E, Marshall CJ. Differing modes of tumour cell invasion have distinct requirements for Rho/ROCK signalling and extracellular proteolysis. Nat Cell Biol 2003;5 (8):711–9. [cited 2017 June 27]. Available from: http://www.ncbi.nlm.nih.gov/pubmed/12844144.

[80] Polte TR, Eichler GS, Wang N, Ingber DE. Extracellular matrix controls myosin light chain phosphorylation and cell contractility through modulation of cell shape and cytoskeletal prestress. Am J Physiol Cell Physiol 2004;286(3):518C–528. [cited 2017 June 27]. Available from: http://www.ncbi.nlm.nih.gov/pubmed/14761883.

[81] Cheung KJ, Padmanaban V, Silvestri V, Schipper K, Cohen JD, Fairchild AN, et al. Polyclonal breast cancer metastases arise from collective dissemination of keratin 14-expressing tumor cell clusters. Proc Natl Acad Sci U S A 2016;113(7):E854–63. [cited 2016 October 9]. Available from: http://www.ncbi.nlm.nih.gov/pubmed/26831077.

[82] Hardikar AA, Marcus-Samuels B, Geras-Raaka E, Raaka BM, Gershengorn MC. Human pancreatic precursor cells secrete FGF2 to stimulate clustering into hormone-expressing islet-like cell aggregates. Proc Natl Acad Sci U S A 2003;100 (12):7117–22. [cited 2017 August 30]. Available from: http://www.ncbi.nlm.nih.gov/pubmed/12799459.

[83] Harryman WL, Hinton JP, Rubenstein CP, Singh P, Nagle RB, Parker SJ, et al. The cohesive metastasis phenotype in human prostate cancer HHS public access. Biochim Biophys Acta 2016;1866(2):221–31. [cited 2017 August 30]. Available from: https://www.ncbi.nlm.nih.gov/pmc/articles/PMC5534328/pdf/nihms886024.pdf.

[84] Friedl P, Wolf K. Tumour-cell invasion and migration: diversity and escape mechanisms. Nat Rev Cancer 2003;3(5):362–74. [cited 2017 June 27], https://doi.org/10.1038/nrc1075.

[85] Friedl P, Locker J, Sahai E, Segall JE. Classifying collective cancer cell invasion. Nat Cell Biol 2012;14(8):777–83. Available from: http://www.ncbi.nlm.nih.gov/pubmed/22854810.

[86] Friedl P, Hegerfeldt Y, Tusch M. Collective cell migration in morphogenesis and cancer. Int J Dev Biol 2004;48(5–6):441–9.

[87] Quaranta V. Motility cues in the tumor microenvironment. Differentiation 2002;70 (9–10):590–8. [cited 2017 June 27]. Available from: http://www.ncbi.nlm.nih.gov/pubmed/12492500.

[88] Schock F, Perrimon N. Molecular mechanisms of epithelial morphogenesis. Annu Rev Cell Dev Biol 2002;18(1):463–93. [cited 2017 June 27]. Available from: http://www.ncbi.nlm.nih.gov/pubmed/12142280.

[89] Roussos ET, Balsamo M, Alford SK, Wyckoff JB, Gligorijevic B, Wang Y, et al. Mena invasive (MenaINV) promotes multicellular streaming motility and transendothelial migration in a mouse model of breast cancer. J Cell Sci 2011;124(13):2120–31. [cited 2017 June 27]. Available from: http://www.ncbi.nlm.nih.gov/pubmed/21670198.

[90] Patsialou A, Bravo-Cordero JJ, Wang Y, Entenberg D, Liu H, Clarke M, et al. Intravital multiphoton imaging reveals multicellular streaming as a crucial component of in vivo cell migration in human breast tumors. Intravital 2013;2(2):e25294. [cited 2016 September 26]. Available from: http://www.ncbi.nlm.nih.gov/pubmed/25013744.

[91] Oudin MJ, Jonas O, Kosciuk T, Broye LC, Guido BC, Wyckoff J, et al. Tumor cell-driven extracellular matrix remodeling drives haptotaxis during metastatic progression.

Cancer Discov 2016;6(5):516–31. [cited 2017 July 10]. Available from: http://www.ncbi.nlm.nih.gov/pubmed/26811325.

[92] Roussos ET, Condeelis JS, Patsialou A. Chemotaxis in cancer. Nat Rev Cancer 2011;11(8):573–87.

[93] Ye Q, Kantonen S, Gomez-Cambronero J. Serum deprivation confers the MDA-MB-231 breast cancer line with an EGFR/JAK3/PLD2 system that maximizes cancer cell invasion. J Mol Biol 2013;425(4):755–66. Available from: http://www.ncbi.nlm.nih.gov/pubmed/23238254.

[94] Lo CM, Wang HB, Dembo M, Wang YL. Cell movement is guided by the rigidity of the substrate. Biophys J 2000;79(1):144–52.

[95] Polyak K, Weinberg RA. Transitions between epithelial and mesenchymal states: acquisition of malignant and stem cell traits. Nat Rev Cancer 2009;9(4):265–73. [cited 2017 June 25]. Available from: http://www.ncbi.nlm.nih.gov/pubmed/19262571.

[96] Kang Y, Massagué J. Epithelial-mesenchymal transitions. Cell 2004;118(3):277–9. [cited 2017 June 25]. Available from: http://linkinghub.elsevier.com/retrieve/pii/S0092867404007020.

[97] Yilmaz M, Christofori G. EMT, the cytoskeleton, and cancer cell invasion. Cancer Metastasis Rev 2009;28(1–2):15–33. [cited 2017 June 25]. Available from: http://www.ncbi.nlm.nih.gov/pubmed/19169796.

[98] Wheelock MJ, Shintani Y, Maeda M, Fukumoto Y, Johnson KR. Cadherin switching. J Cell Sci 2008;121(6) [cited 2017 June 25]. Available from: http://jcs.biologists.org/content/121/6/727.

[99] Nieman MT, Prudoff RS, Johnson KR, Wheelock MJ. N-cadherin promotes motility in human breast cancer cells regardless of their E-cadherin expression. J Cell Biol 1999;147(3):631–44. [cited 2017 June 25]. Available from: http://www.ncbi.nlm.nih.gov/pubmed/10545506.

[100] Gravdal K, Halvorsen OJ, Haukaas SA, Akslen LA. A switch from E-cadherin to N-cadherin expression indicates epithelial to mesenchymal transition and is of strong and independent importance for the progress of prostate cancer. Clin Cancer Res 2007;13(23):7003–11. [cited 2017 June 25]. Available from: http://www.ncbi.nlm.nih.gov/pubmed/18056176.

[101] Friedl P, Wolf K. Plasticity of cell migration: a multiscale tuning model. J Cell Biol 2010;188(1):11–9.

[102] Liotta LA, Tryggvason K, Garbisa S, Hart I, Foltz CM, Shafie S. Metastatic potential correlates with enzymatic degradation of basement membrane collagen. Nature 1980;284(5751):67–8. [cited 2017 June 26], https://doi.org/10.1038/284067a0.

[103] Wyckoff JB, Jones JG, Condeelis JS, Segall JE. A critical step in metastasis: in vivo analysis of intravasation at the primary tumor. Cancer Res 2000;60(9):2504–11. [cited 2017 June 26]. Available from: http://www.ncbi.nlm.nih.gov/pubmed/10811132.

[104] Yamazaki D, Kurisu S, Takenawa T. Involvement of Rac and Rho signaling in cancer cell motility in 3D substrates. Oncogene 2009;28(13):1570–83. [cited 2017 June 27], https://doi.org/10.1038/onc.2009.2.

[105] Friedl P, Wolf K, Lammerding J. Nuclear mechanics during cell migration. Curr Opin Cell Biol 2011;23(1):55–64. Available from: http://www.ncbi.nlm.nih.gov/pubmed/21109415.

[106] Mehlen P, Puisieux A. Metastasis: a question of life or death. Nat Rev Cancer 2006;6(6):449–58. [cited 2017 April 3], https://doi.org/10.1038/nrc1886.

[107] Baskurt OK, Meiselman HJ. Blood rheology and hemodynamics. Semin Thromb Hemost 2003;29(5):435–50. [cited 2017 June 27], https://doi.org/10.1055/s-2003-44551.

[108] Mitchell MJ, King MR. Computational and experimental models of cancer cell response to fluid shear stress. Front Oncol 2013;3:44. [cited 2017 June 28]. Available from: http://www.ncbi.nlm.nih.gov/pubmed/23467856.

[109] Liang S, Hoskins M, Dong C. Tumor cell extravasation mediated by leukocyte adhesion is shear rate dependent on IL-8 signaling. Mol Cell Biomech 2010;7(2):77–91. [cited 2017 June 28]. Available from: http://www.ncbi.nlm.nih.gov/pubmed/20379392.

[110] Fan R, Emery T, Zhang Y, Xia Y, Sun J, Wan J. Circulatory shear flow alters the viability and proliferation of circulating colon cancer cells. [cited 2017 June 28]; Available from: https://www.nature.com/articles/srep27073.pdf; 2016.

[111] Vigmostad S, Krog B, Nauseef J, Henry M, Keshav V. Alterations in cancer cell mechanical properties after fluid shear stress exposure: a micropipette aspiration study. Cell Health Cytoskelet 2015;7:25. [cited 2017 June 28]. Available from: http://www.ncbi.nlm.nih.gov/pubmed/25908902.

[112] Hughes AD, King MR. Nanobiotechnology for the capture and manipulation of circulating tumor cells. Wiley Interdiscip Rev Nanomed Nanobiotechnol 2012;4(3):291–309. [cited 2017 June 28]. Available from: http://www.ncbi.nlm.nih.gov/pubmed/22162415.

[113] Reymond N, d'Água BB, Ridley AJ. Crossing the endothelial barrier during metastasis. Nat Rev Cancer 2013;13(12):858–70. [cited 2017 September 3]. Available from: http://www.ncbi.nlm.nih.gov/pubmed/24263189.

[114] Wirtz D, Konstantopoulos K, Searson PC. The physics of cancer: the role of physical interactions and mechanical forces in metastasis. Nat Rev Cancer 2011;11(7):512–22. [cited 2017 June 27], https://doi.org/10.1038/nrc3080.

[115] Kumar S, Weaver VM. Mechanics, malignancy, and metastasis: the force journey of a tumor cell. Cancer Metastasis Rev 2009;28(1–2):113–27. Available from: http://www.ncbi.nlm.nih.gov/pubmed/19153673.

[116] Mierke CT, Zitterbart DP, Kollmannsberger P, Raupach C, Schlotzer-Schrehardt U, Goecke TW, et al. Breakdown of the endothelial barrier function in tumor cell transmigration. Biophys J 2008;94(7):2832–46. [cited 2017 June 28]. Available from: http://www.ncbi.nlm.nih.gov/pubmed/18096634.

[117] Gay LJ, Felding-Habermann B. Contribution of platelets to tumour metastasis. Nat Publ Gr 2011;11(2):123–34. [cited 2017 June 27]. Available from: https://www.nature.com/nrc/journal/v11/n2/pdf/nrc3004.pdf.

[118] Yamada S, Wirtz D, Kuo SC. Mechanics of living cells measured by laser tracking microrheology. Biophys J 2000;78(4):1736–47.

[119] Xu JY, Viasnoff V, Wirtz D. Compliance of actin filament networks measured by particle-tracking microrheology and diffusing wave spectroscopy. Rheol Acta 1998;37(4):387–98.

[120] Wei MT, Zaorski A, Yalcin HC, Wang J, Hallow M, Ghadiali SN, et al. A comparative study of living cell micromechanical properties by oscillatory optical tweezers. Opt Express 2008;16(12):8594–603.

[121] Abuhattoum S, Weihs D. Location-dependent intracellular particle tracking using a cell-based coordinate system. Comput Methods Biomech Biomed Eng 2013;16(10):1042–9. https://doi.org/10.1080/10255842.2012.761694.

[122] Lee LM, Liu AP. A microfluidic pipette array for mechanophenotyping of cancer cells and mechanical gating of mechanosensitive channels. Lab Chip 2015;15 (1):264–73. [cited 2017 June 28]. Available from: http://www.ncbi.nlm.nih.gov/pubmed/25361042.

[123] Mak M, Erickson D. A serial micropipette microfluidic device with applications to cancer cell repeated deformation studies. Integr Biol 2013;5(11):1374. [cited 2017 June 28]. Available from: http://www.ncbi.nlm.nih.gov/pubmed/24056324.

[124] Mohammadalipour A, Choi YE, Benencia F, Burdick MM, Tees DFJ. Investigation of mechanical properties of breast cancer cells using micropipette aspiration technique. FASEB J 2012;26(1 Suppl.):905.9. [cited 2016 September 1]. Available from: http://www.fasebj.org/content/26/1_Supplement/905.9.short?related-urls=yes&legid=fasebj;26/1_Supplement/905.9.

[125] Wu ZZ, Zhang G, Long M, Wang HB, Song GB, Cai SX. Comparison of the viscoelastic properties of normal hepatocytes and hepatocellular carcinoma cells under cytoskeletal perturbation. Biorheology 2000;37(4):279–90. [cited 2017 June 28]. Available from: http://www.ncbi.nlm.nih.gov/pubmed/11145074.

[126] Mak M, Reinhart-King CA, Erickson D. Elucidating mechanical transition effects of invading cancer cells with a subnucleus-scaled microfluidic serial dimensional modulation device. Lab Chip 2013;13(3):340–8. Available from: http://www.ncbi.nlm.nih.gov/pubmed/23212313.

[127] Lange JR, Steinwachs J, Kolb T, Lautscham LA, Harder I, Whyte G, et al. Microconstriction arrays for high-throughput quantitative measurements of cell mechanical properties. Biophys J 2015;109(1):26–34. [cited 2017 August 29]. Available from: http://www.ncbi.nlm.nih.gov/pubmed/26153699.

[128] Hou HW, Li QS, Lee GYH, Kumar AP, Ong CN, Lim CT. Deformability study of breast cancer cells using microfluidics. Biomed Microdevices 2009;11(3):557–64.

[129] Hur SC, Henderson-MacLennan NK, McCabe ERB, Di Carlo D, Ishida S, Spang R, et al. Deformability-based cell classification and enrichment using inertial microfluidics. Lab Chip 2011;11(5):912. [cited 2017 July 4]. Available from: http://xlink.rsc.org/?DOI=c0lc00595a.

[130] Gossett DR, Tse HTK, Lee SA, Ying Y, Lindgren AG, Yang OO, et al. Hydrodynamic stretching of single cells for large population mechanical phenotyping. Proc Natl Acad Sci U S A 2012;109(20):7630–5. [cited 2017 July 4]. Available from: http://www.ncbi.nlm.nih.gov/pubmed/22547795.

[131] Riahi R, Gogoi P, Sepehri S, Zhou Y, Handique K, Godsey J, et al. A novel microchannel-based device to capture and analyze circulating tumor cells (CTCs) of breast cancer. Int J Oncol 2014;44(6):1870–8. [cited 2017 July 4]. Available from: http://www.ncbi.nlm.nih.gov/pubmed/24676558.

[132] MacKintosh FC, Schmidt CF. Microrheology. Curr Opin Colloid Interface Sci 1999;4 (4):300–7.

[133] Mason TG, Ganesan K, vanZanten JH, Wirtz D, Kuo SC. Particle tracking microrheology of complex fluids. Phys Rev Lett 1997;79(17):3282–5.

[134] Weihs D, Mason TG, Teitell MA. Bio-microrheology: a frontier in microrheology. Biophys J 2006;91(11):4296–305. Available from: http://www.sciencedirect.com/science/article/pii/S0006349506721439%5Cnhttp://pdn.sciencedirect.com/science?_ob=MiamiImageURL&_cid=277708&_user=606230&_pii=S0006349506721439&_check=y&_origin=article&_zone=toolbar&_coverDate=2006-01&view=c&originContent Famil.

[135] Guck J, Ananthakrishnan R, Cunningham CC, Kas J. Stretching biological cells with light. J Phys: Condens Matter 2002;14(19):4843–56.

[136] Suresh S, Spatz J, Mills JP, Micoulet A, Dao M, Lim CT, et al. Connections between single-cell biomechanics and human disease states: gastrointestinal cancer and malaria. Acta Biomater 2005;1(1):15–30.

[137] Remmerbach TW, Wottawah F, Dietrich J, Lincoln B, Wittekind C, Guck J. Oral cancer diagnosis by mechanical phenotyping. Cancer Res 2009;69(5):1728–32.

[138] Keating M, Kurup A, Alvarez-Elizondo M, Levine AJ, Botvinick E. Spatial distributions of pericellular stiffness in natural extracellular matrices are dependent on cell-mediated proteolysis and contractility. Acta Biomater 2017;57:304–12. [cited 2017 June 28]. Available from: http://www.sciencedirect.com/science/article/pii/S174270611730291X.

[139] Kniazeva E, Weidling JW, Singh R, Botvinick EL, Digman MA, Gratton E, et al. Quantification of local matrix deformations and mechanical properties during capillary morphogenesis in 3D. Integr Biol 2012;4(4):431–9. Available from: http://www.ncbi.nlm.nih.gov/pubmed/22281872.

[140] Kuznetsova TG, Starodubtseva MN, Yegorenkov NI, Chizhik SA, Zhdanov RI. Atomic force microscopy probing of cell elasticity. Micron 2007;38(8):824–33. [cited 2017 June 28]. Available from: http://linkinghub.elsevier.com/retrieve/pii/S0968432807001047.

[141] Cross SE, Jin YS, Tondre J, Wong R, Rao J, Gimzewski JK. AFM-based analysis of human metastatic cancer cells. Nanotechnology 2008;19(38)384003.

[142] Lekka M, Laidler P, Gil D, Lekki J, Stachura Z, Hrynkiewicz AZ. Elasticity of normal and cancerous human bladder cells studied by scanning force microscopy. Eur Biophys J Biophys Lett 1999;28(4):312–6.

[143] Lekka M. Discrimination between normal and cancerous cells using AFM. Bionanoscience 2016;6:65–80. [cited 2017 June 28]. Available from: http://www.ncbi.nlm.nih.gov/pubmed/27014560.

[144] Xu, WW., Mezencev, R., Kim, B., Wang, LJ., McDonald, J., Sulchek, T. Cell stiffness is a biomarker of the metastatic potential of ovarian cancer cells. Batra SK, editor. PLoS One. 2012;7(10):e46609. doi:https://doi.org/10.1371/journal.pone.0046609 [cited 2017 June 28]

[145] Lekka M, Gil D, Pogoda K, Dulińska-Litewka J, Jach R, Gostek J, et al. Cancer cell detection in tissue sections using AFM. Arch Biochem Biophys 2012;518(2):151–6. [cited 2017 June 28]. Available from: http://linkinghub.elsevier.com/retrieve/pii/S0003986111004255.

[146] Jonas O, Mierke CT, Kas JA. Invasive cancer cell lines exhibit biomechanical properties that are distinct from their noninvasive counterparts. Soft Matter 2011;7(24):11488–95.

[147] Lyapunova E, Nikituk A, Bayandin Y, Naimark O, Rianna C, Radmacher M. Passive microrheology of normal and cancer cells after ML7 treatment by atomic force microscopy; 2016. p. 101. [cited 2017 June 28], https://doi.org/10.1063/1.4960265.

[148] Rother J, Noding H, Mey I, Janshoff A. Atomic force microscopy-based microrheology reveals significant differences in the viscoelastic response between malign and benign cell lines. Open Biol 2014;4(5):140046.

[149] Yao W, Yoshida K, Fernandez M, Vink J, Wapner RJ, Ananth CV, et al. Measuring the compressive viscoelastic mechanical properties of human cervical tissue using

indentation. J Mech Behav Biomed Mater 2014;34:18–26. Available from: http://www.ncbi.nlm.nih.gov/pubmed/24548950.

[150] Shin D, Athanasiou K. Cytoindentation for obtaining cell biomechanical properties. J Orthop Res 1999;17(6):880–90. [cited 2017 July 16]. Available from: http://www.ncbi.nlm.nih.gov/pubmed/10632455.

[151] Coughlin MF, Bielenberg DR, Lenormand G, Marinkovic M, Waghorne CG, Zetter BR, et al. Cytoskeletal stiffness, friction, and fluidity of cancer cell lines with different metastatic potential. Clin Exp Metastasis 2013;30(3):237–50.

[152] Wilhelm C. Out-of-equilibrium microrheology inside living cells. Phys Rev Lett 2008;101(2):28101.

[153] McSherry EA, Brennan K, Hudson L, Hill AD, Hopkins AM. Breast cancer cell migration is regulated through junctional adhesion molecule-A-mediated activation of Rap1 GTPase. Breast Cancer Res 2011;13(2):R31. [cited 2017 July 16]. Available from: https://breast-cancer-research.biomedcentral.com/track/pdf/10.1186/bcr2853?site=breast-cancer-research.biomedcentral.com.

[154] Toume S, Gefen A, Weihs D. Low-level stretching accelerates cell migration into a gap. Int Wound J 2017;14(4):698–703. [cited 2016 October 31], https://doi.org/10.1111/iwj.12679.

[155] Justus CR, Leffler N, Ruiz-Echevarria M, Yang LV. In vitro cell migration and invasion assays. J Vis Exp 2014;(88) [cited 2017 April 24]. Available from: http://www.ncbi.nlm.nih.gov/pubmed/24962652.

[156] Nelson MT, Short A, Cole SL, Gross AC, Winter J, Eubank TD, et al. Preferential, enhanced breast cancer cell migration on biomimetic electrospun nanofiber "cell highways". BMC Cancer 2014;14:825. [cited 2017 July 16]. Available from: http://www.ncbi.nlm.nih.gov/pubmed/25385001.

[157] Sun E, Cohen FE. Computer-assisted drug discovery—a review. Gene 1993;137(1):127–32. [cited 2017 July 15]. Available from: http://www.sciencedirect.com/science/article/pii/037811199390260A.

[158] Weihs D, Gefen A, Vermolen FJ. Review on experiment-based two- and three-dimensional models for wound healing. Interface Focus 2016;6(5):20160038. [cited 2016 October 9]. Available from: http://www.ncbi.nlm.nih.gov/pubmed/27708762.

[159] Dudaie M, Weihs D, Vermolen FJ, Gefen A. Modeling migration in cell colonies in two and three dimensional substrates with varying stiffnesses. Silico Cell Tissue Sci 2015;2(1):1–14.

[160] Lee Y, Kouvroukoglou S, McIntire LV, Zygourakis K. A cellular automaton model for the proliferation of migrating contact-inhibited cells. Biophys J 1995;69(4):1284–98. Available from: http://www.ncbi.nlm.nih.gov/pmc/articles/PMC1236359/%5Cnhttp://linkinghub.elsevier.com/retrieve/pii/S0006349595799969.

[161] Potts RB, Domb C. Some generalized order-disorder transformations. Math Proc Camb Philos Soc 1952;48(1):106. [cited 2017 July 15]. Available from: http://www.journals.cambridge.org/abstract_S0305004100027419.

[162] Glazier JA, Graner F. Simulation of the differential adhesion driven rearrangement of biological cells. Phys Rev E 1993;47(3):2128–54.

[163] Honda H. Description of cellular patterns by Dirichlet domains: the two-dimensional case. J Theor Biol 1978;72(3):523–43. [cited 2017 July 15]. Available from: http://linkinghub.elsevier.com/retrieve/pii/0022519378903156.

[164] Brodland GW, Veldhuis JH. The mechanics of metastasis: insights from a computational model. PLoS One 2012;7(9):e44281.
[165] Borau C, Kamm RD, García-Aznar JM. Mechano-sensing and cell migration: a 3D model approach. Phys Biol 2011;8(6):66008.
[166] Chaplain MAJ, Lolas G. Mathematical modelling of cancer invasion of tissue: dynamic heterogeneity. Netw Heterog Media 2006;1(3):399–439. Available from: http://aimsciences.org.
[167] Moreo P, Garcia-Aznar JM, Doblare M, García-Aznar JM, Doblaré M. Modeling mechanosensing and its effect on the migration and proliferation of adherent cells. Acta Biomater 2008;4(3):613–21.
[168] Marchant BP, Norbury J, Sherratt JA. Travelling wave solutions to a haptotaxis-dominated. Nonlinearity 2001;14:1653–71.
[169] Anderson ARA, Chaplain MAJ, Newman EL, Steele RJC, Thompson AM. Mathematical modelling of tumour invasion and metastasis. J Theor Med 2000;2(2):129–54. Available from: http://www.hindawi.com/journals/cmmm/2000/490902/abs/.
[170] Vermolen FJ, Van Der MRP, van Es M, Gefen A, Weihs D. Towards a mathematical formalism for semi-stochastic cell-level computational modeling of tumor initiation. Ann Biomed Eng 2015;43(7):1680–94. Available from: http://www.ncbi.nlm.nih.gov/pubmed/25670322.
[171] Chalovich JM, Eisenberg E. NIH public access. Biophys Chem 2012;257(5):2432–7.
[172] Basanta D, Deutsch A. A game theoretical perspective on the somatic evolution of cancer. Evol Immune Compet Ther 2008;97–112.
[173] Kandow CE, Georges PC, Janmey PA, Beningo KA. Polyacrylamide hydrogels for cell mechanics: steps toward optimization and alternative uses. Methods Cell Biol 2007;83:29–46. [cited 2016 November 27]. Available from: http://www.ncbi.nlm.nih.gov/pubmed/17613303.
[174] Wang JHC, Lin JS. Cell traction force and measurement methods. Biomech Model Mechanobiol 2007;6(6):361–71.
[175] Butler JP, Tolic-Norrelykke IM, Fabry B, Fredberg JJ. Traction fields, moments, and strain energy that cells exert on their surroundings. Am J Physiol 2002;282(3):C595–605.
[176] Munevar S, Wang YL, Dembo M. Traction force microscopy of migrating normal and H-ras transformed 3T3 fibroblasts. Biophys J 2001;80(4):1744–57.
[177] Sabass B, Gardel ML, Waterman CM, Schwarz US. High resolution traction force microscopy based on experimental and computational advances. Biophys J 2008;94(1):207–20.
[178] Physical Sciences-Oncology Centers N, Agus DB, Alexander JF, Arap W, Ashili S, Aslan JE, et al. A physical sciences network characterization of non-tumorigenic and metastatic cells. Sci Rep 2013;3:1449. Available from: http://www.ncbi.nlm.nih.gov/pubmed/23618955.
[179] Jasaitis A, Estevez M, Heysch J, Ladoux B, Dufour S. E-cadherin-dependent stimulation of traction force at focal adhesions via the Src and PI3K signaling pathways. Biophys J 2012;103(2):175–84. [cited 2017 July 4]. Available from: http://www.ncbi.nlm.nih.gov/pubmed/22853894.
[180] Ramade A, Legant WR, Picart C, Chen CS, Boudou T. Microfabrication of a platform to measure and manipulate the mechanics of engineered microtissues. In: Methods in cell biology. 2014. p. 191–211. [cited 2017 July 16]. Available from: http://www.ncbi.nlm.nih.gov/pubmed/24560511.

[181] Mandal K, Asnacios A, Goud B, Manneville J-B. Mapping intracellular mechanics on micropatterned substrates. Proc Natl Acad Sci U S A 2016;113(46):E7159–68. [cited 2016 November 16]. Available from: http://www.ncbi.nlm.nih.gov/pubmed/27799529.

[182] Aoun L, Weiss P, Laborde A, Ducommun B, Lobjois V, Vieu C, et al. Microdevice arrays of high aspect ratio poly(dimethylsiloxane) pillars for the investigation of multicellular tumour spheroid mechanical properties. Lab Chip 2014;14(13):2344–53. [cited 2017 July 16]. Available from: http://xlink.rsc.org/?DOI=C4LC00197D.

[183] Ermis M, Akkaynak D, Chen P, Demirci U, Hasirci V. A high throughput approach for analysis of cell nuclear deformability at single cell level. Nat Publ Gr 2016; [cited 2017 July 16]. Available from: https://www.nature.com/articles/srep36917.pdf.

[184] Maskarinec SA, Franck C, Tirrell DA, Ravichandran G, Boxer SG. Quantifying cellular traction forces in three dimensions. Proc Natl Acad Sci U S A 2009;106(52):22108–13.

[185] Franck C, Maskarinec SA, Tirrell DA, Ravichandran G. Three-dimensional traction force microscopy: a new tool for quantifying cell-matrix interactions. PLoS One 2011;6(3):e17833. Available from: http://www.ncbi.nlm.nih.gov/pubmed/21468318.

[186] Style RW, Boltyanskiy R, German GK, Hyland C, MacMinn CW, Mertz AF, et al. Traction force microscopy in physics and biology. Soft Matter 2014;10(23):4047–55. 2014/04/18. Available from: http://www.ncbi.nlm.nih.gov/pubmed/24740485.

[187] Bershadsky AD, Balaban NQ, Geiger B. Adhesion-dependent cell mechanosensitivity. Annu Rev Cell Dev Biol 2003;19:677–95. Available from: http://www.ncbi.nlm.nih.gov/pubmed/14570586.

[188] Ingber DE. Tensegrity and mechanotransduction. J Bodyw Mov Ther 2008;12(3):198–200. Available from: http://www.ncbi.nlm.nih.gov/pubmed/19083675.

[189] Peschetola V, Laurent VM, Duperray A, Michel R, Ambrosi D, Preziosi L, et al. Time-dependent traction force microscopy for cancer cells as a measure of invasiveness. Cytoskeleton 2013;70(4):201–14. Available from: http://www.ncbi.nlm.nih.gov/pubmed/23444002.

[190] Indra I, Undyala V, Kandow C, Thirumurthi U, Dembo M, Beningo KA. An in vitro correlation of mechanical forces and metastatic capacity. Phys Biol 2011;8(1):15015. [cited 2017 July 4]. Available from: http://www.ncbi.nlm.nih.gov/pubmed/21301068.

[191] Koch TM, Munster S, Bonakdar N, Butler JP, Fabry B. 3D traction forces in cancer cell invasion. PLoS One 2012;7(3):e33476. Available from: http://www.ncbi.nlm.nih.gov/pubmed/22479403.

[192] Hur SS, Zhao YH, Li YS, Botvinick E, Chien S. Live cells exert 3-dimensional traction forces on their substrata. Cell Mol Bioeng 2009;2(3):425–36.

[193] Delanoe-Ayari H, Rieu JP, Sano M. 4D traction force microscopy reveals asymmetric cortical forces in migrating Dictyostelium cells. Phys Rev Lett 2010;105(24):248103.

[194] Fidler IJ. The relationship of embolic homogeneity, number, size and viability to the incidence of experimental metastasis. Eur J Cancer 1973;9(3):223–7. Available from: http://www.ncbi.nlm.nih.gov/pubmed/4787857.

# Index

Note: Page numbers followed by *f* indicate figures, and *t* indicate tables.

## A

Accommodation
  defined, 355–356
  forces of, 357–358
Accommodative apparatus, 355–356, 356*f*, 358–359, 358*f*
Achilles tendon, 132
Achilles tendon stem cells (ATSCs), 135, 136*f*
Acid-sensing ion channels (ASICs), 391–392
Actin cytoskeleton, 111–112
Actin stress fibers (ASFs), 60–61
Adamlysins (ADAMs), 238
ADAMTS13, 291–293, 292*f*
Adenosine triphosphate (ATP), 173
Adhesion junctions, 179
Adipocyte volume fraction (AVF), 183
Afferent fibers, 380
AFM. *See* Atomic force microscopy (AFM)
Age-associated dermal microenvironment (AADM), 434
Aging
  of articular cartilage, 105
  tendons, 141–145, 142–144*f*
  vascular mechanics, 242–243
Aging-induced tendinopathy, 144–145
Ahesions, focal, 169
Anabolic responses, 133–135
Aneurysm, mechanosensory cilia, 312, 313*f*
AngII-induced hypertension, 241
Animal loading models
  in ovo immobilization, 13
  mammalian models, 13–14
  tibial four-point bend model, 12, 12*f*
  ulnar compression model, 12–13, 12*f*
  zebrafish models, 14
Annulus of Zinn, 364
Antigen-presenting cells (APCs), 230
Aortic valve (AV), 250–251
Aortic valve interstitial cells (AVICs), 253, 259, 259*f*, 267*f*
Appendicular limb development, 82–85
Aqueous humor, 359–362, 360*f*
Arterial mechanics, vascular
  ex vivo
    assumptions, 222–223
    experimental approaches, 223
    mechanotargets, 226
    stress/strain determination, 223–225, 225*f*
  in vivo, 221–222
Arterial stiffness, 221
Articular cartilage
  aging of, 105
  chondrocyte mechanoreceptors
    integrins, 110–111
    ion channels, 109
    primary cilia, 110
  chondrocyte mechanotransduction
    cytoskeleton, 111–112
    miRNAs, 114–115
    mitogen-activated protein kinases, 113
    Wingless-type family, 113–114
  collagen fibrils, 100
  extracellular matrix, 100–104, 102*f*
  mechanical loading, 105–106
  micromechanical environment, 106–107
  structural overview, 100–106, 101*f*
  zonal characteristics, 103–104, 104*f*
Astrocytes, 363
Atherosclerosis, mechanosensory cilia, 312–313
Atomic force microscopy (AFM), 4, 255, 256*f*, 287–288, 288*f*, 342, 466, 466*f*, 468
Attractive forces, 351–352
AVICs. *See* Aortic valve interstitial cells (AVICs)
Axoneme, 306, 307*f*

## B

Beam theory, 8
β-catenin, 113–114, 179
Biaxial tensile testing, 260
Biglycan, 132
Bimodular model, 264
Biomechanical changes, osteocytes, 180–185, 184*f*
Biomembrane force probe (BFP), 287–288, 288*f*
Bioreactors, 8–11, 9*f*
Birefringence, 352–354
BK-like channels, 393
Blebbistatin, 64–65
B lymphocytes, 230
Bone
  biomechanics, 158–159

495

Bone (Continued)
  cytoskeleton, 172
  fluid flow, 159–161
  growth
    endochondral ossification, 165–166
    intramembranous ossification, 166
  hierarchical structure, 159, 160f
  mechanical environment, 176–178
  mechanobiological therapy, 187
  mechanosensation, 169–173
  mechanosensors
    cytoskeleton, 172
    gap junctions, 172–173
    G-protein-coupled membrane receptors, 173
    hemichannels, 172–173
    integrins, 169–170
    ion channels, 171
    primary cilia, 170–171
  mechanotransduction
    calcium signaling, 175
    G-protein-related signaling, 175–176
    mitogen-activated protein kinases, 175
    nitric oxide, 176
    Wnt signaling, 175
  mesenchymal stem cells, 161
  metastatic bone disease, 185–187
  modeling, 166
  osteoarthritis, 185
  osteoblasts, 161–163
  osteoclasts, 165
  osteocytes, 163–165, 164f
  osteoporosis
    biomechanical changes, 180–185, 184f
    mechanical environment, 180–185
    mechanosensors, 178–179
    mechanotransduction, 179–180
  osteoprogenitors, 161
  osteosarcoma, 185–187
  porosity, 159–161
  remodeling, 166–167, 167f
Bone metastases, 186
Bone modeling, 166
Bone morphogenic proteins (BMPs), 147
Bone regeneration, 187–188
Bone remodeling, 166–167, 167f
Borges' lines, 418f, 433–435
Boundary conditions, 57–58
Buckling forces, 336

# C

$Ca^{2+}$ channels, 109
*Caenorhabditis elegans*, 314
Calcific aortic valve disease (CAVD), 251–252
Calcium signaling, 175, 308–309
Canaliculi, 163, 164f
Cartilage oligomeric matrix protein (COMP), 103
Cas genes, 25–26
Cataract, 359
Cathepsins, 238
Cauchy biaxial stress, 224
Cell division, 78
Cell fate, tracking, 85–86
Cell growth, 78–80, 79f
Cell hypertrophy, 78–80
Cell mechanics assays
  atomic force microscopy and cytoindentation, 466f, 468
  intracellular particle tracking, 466f, 467
  magnetic rotational cytometry, 466f, 468–469
  microconstrictions and microchannels, 466f, 467
  micropipette aspiration, 466–467, 466f
  optical (laser) tweezers, 466f, 468
Cell migration, 71
  collective cell, 459–460, 459f
  gap closure exclusion zone assay, 469–470, 469f
  gap closure scratching/crushing assays, 469, 469f
  microfluidic assays, 469f, 470
  single cell, 458–459, 459f
  transwell/modified Boyden chamber assay, 469f, 470
Cell shape index (CSI), 60–61
Cellular interactions
  micropillar arrays, 8
  traction force microscopy, 7
Cerebrospinal fluid (CSF), 362
Chemotaxis, 460
*Chlamydomonas*, 314
Chloride ($Cl^-$), 81–82
Chondrocyte
  mechanoreceptors
    integrins, 110–111
    ion channels, 109
    primary cilia, 110
  mechanosensitivity, 107
  mechanotransduction
    cytoskeleton, 111–112
    miRNAs, 114–115
    mitogen-activated protein kinases, 113
    Wingless-type family, 113–114
Chondroitin sulfate chains, 105
Chondron, 100–103, 106–107
Cilia, 81
Cilia bending, 308–309
Ciliary necklace, 306, 307f

Ciliopathy, 308
Cilioplasm, 306, 307f
Circulating tumor cells (CTCs), 450
Clustered regularly interspaced short palindromic repeats (CRISPR), 25–26
Cohesotaxis, 461
Collagenases, 237
Collagen fiber architecture (CFA), 271–272, 271f
Collagen fibrils, 100, 103, 351–353, 352f
Complementary DNA (cDNA), 18–20
Compression bioreactors, 9f, 10
Compressive forces, 10, 12–13, 392–393
Computational fluid dynamics (CFD), 27–28
Computational modeling
    computational fluid dynamics, 27–28
    finite element analysis, 28–29
    multiscale and multiphysics modeling, 29–31, 30f
Confocal laser scanning microscopy, 17
Contractile cellular forces, 6
Convergent extension, 78–80
Cornea, 350–355, 351f
Corneal stress, 353, 354t
Corneal transparency, 351–352
Cortical bone, 158–159
Cortical elasticity, 357
Cramer/Helmholtz theory, 355–356
Cutaneous wound healing
    in aging, 433–434
    diabetic ulcers, 429–433
Cyclic pressurization, 113
Cyclooxygenase-2 (Cox-2) expression, 169
Cystic fibrosis transmembrane conductance regulator (CFTR), 135
Cytochalasin D, 111–112
Cytoskeletal elements
    actin, 336–338
    axons, 337–338
    buckling forces, 336
    depolymerization, 336
    G-actin, 336
    hippocampus, 337–338
    intracellular forces, 336
    microtubules, 338–339
    polymerization, 336
    presynaptic vesicle clustering, 337–338
    stalling force, 336
    synapses, 337–338
    tensegrity, 335–336
    viscoelasticity of dendritic spines, 338
Cytoskeletal forces, 61
Cytoskeleton
    actin filament network, 452
    bone, 172
    chondrocyte mechanotransduction, 111–112
    force application, 453–455
    intermediate filaments, 453
    mechanical structure of, 452–453
    motility, 453–455
    in nuclear fluctuations, 66–68
    in nuclear rotation, 63–65
    in nuclear translation, 60–61
    structural localization of, 455
    tumor microenvironment role, 456–457
Cytotoxic T cells, 235

# D

Deformation, 61, 69, 71, 251
Dendritic cells (DCs), 235–236
Dendritic spines, viscoelasticity of, 338
Diabetic ulcers
    abnormal architecture of, 430–431
    complicating wound healing, 429
    mechanotransduction factors, 431–432, 432f
    pathological pathways, 433
    proper clinical management, 430–431
Differentiation, 77
Digital image correlation (DIC), 31–32, 32f
Dogiel type I, 392
Dogiel type II, 392
Donnan osmotic pressure, 106
Dopamine receptors, 310–311, 311t
Dopamine receptor type 5 (DR5), 310
Dorsal root ganglia (DRG), 380
Dupuytren's disease, 427–428
Durotaxis, 461
Dynamic force spectroscopy (DFS), 288–289
Dynein, 61, 63–64
Dysfunctional mechanobiology
    biochemical and pharmacological inhibitors, 436–438, 437f
    mechanomodulatory devices, 435–436
    preoperative surgical techniques, 435

# E

ECM. *See* Extracellular matrix (ECM)
Elastic fibers, 218, 222–223
Elastomeric polymer technology, 435–436
Electrodeformation, 330–333
Electrowetting, 332
Endochondral ossification, 165–166
Endocrine cells, gastrointestinal (GI) tract, 398–399, 399f
Endotenon, 128
Endothelial cells (ECs), 217, 227
Endothelin-1, 217–218, 252

## Index

Enteric nervous system (ENS), 392
Enterochromaffin cells, 381, 398, 399f
Enzyme-linked immunosorbent assay
    (ELISA), 24–25
Epiboly, 78–80
Epigastric pain syndrome, 385
Epitenon, 128
Epithelial airways, 81–82
Epithelial cells, 355–356
Epithelium, 351
Estrogen, 178–180
Extracellular matrix (ECM), 100–104, 102f,
    163–165, 238–239, 250–251
Extracellular-regulated kinase (ERK1/2),
    113
Extra ocular system
    muscle functions, 364–365, 366f
    nystagmus, 366–367
Extrinsic innervation
    anatomy, 380–381
    ion channels, 389–392
    spinal afferents, 386–389
    vagal afferents, 381–386, 381f, 383–384f
Eye
    anterior segment
        accommodation forces, 357–358, 358f
        accommodative apparatus, 358–359
        aqueous humor, 359–362, 360f
        capsule, 355–357, 356f
        cornea, 350–355, 351f
        glaucoma, 360–361
        lens, 355–357, 356f
        tear film, 350–351
        zonules, 355–357, 356f
    axial section, 361f
    extra ocular system
        muscle functions, 364–365, 366f
        nystagmus, 366–367
    function, 349–350
    human, 350f
    intraocular pressure, 350
    posterior segment, 362–364
    retina, 349–350
Eye movements, 364
    muscle functions, 364–365

## F

Fate mapping techniques, 85–86
Fibroblast growth factor 2 (FGF-2), 107
Fibroblasts, mechanosensing in, 229–230
Fibromodulin, 132
Fibronectin, 100–103

Fibrotic skin disorders
    Dupuytren's disease, 427–428
    hypertrophic scars, 424–425
    keloids, 425–426
    prevalence of, 428
    scleroderma, 426–427
Finite element (FE) analysis, 28–29
5-hydroxytryptamine (5-HT), 381, 398–399
Flexcell system, 133
Flexoelectricity, 330–333
Flexural deformation tests, 260
Flow perfusion bioreactors, 8–10
Fluid pressure, 106
Fluid-structure interaction (FSI) model, 30, 30f
Fluorescence recovery after photobleaching
    (FRAP), 69–70
Fluorescent proteins (FPS)
    confocal and two-photon microscopy, 16–17
    fluorescent recovery after photobleaching, 16
    fluorescent resonance energy transfer, 16
    live cell imaging, 15
    as markers in mechanobiology, 15
Fluorophore, 16
Focal adhesions, 169
Focused ion beam scanning electron microscope
    (FIB-SEM), 279
Force application techniques
    atomic force microscopy, 4
    micropipette aspiration, 4–6
    optical tweezers, 4, 5f
Formin, 66
Förster resonance energy transfer (FRET), 341–342
Four-step model
    mechanopresentation, 296, 297f
    mechanoreception, 296, 297f
    mechanotransduction, 297, 297f
    mechanotransmission, 297, 297f
Free energy, 334
FrzB expression, 113–114
FSI. See Fluid-structure interaction (FSI)
    model
Functional dyspepsia, 385–386

## G

G-actin, 336
Gap junctions, bone, 172–173
Gastrointestinal (GI) tract
    endocrine cells, 398–399, 399f
    extrinsic innervation
        anatomy, 380–381
        ion channels, 389–392
        spinal afferents, 386–389

vagal afferents, 381–386, 381f, 383–384f
interstitial cells of Cajal
  motility disorders, 396
  structure and organisation, 394, 395t
intrinsic innervation
  anatomy, 392
  intrinsic neurones, 392–394
  mechanosensation, 378–380, 379f
  smooth muscle cells, 397–398
Gelatinases, 237
Glaucoma, 360–361
Global gene expression profiles, 433–434
Glycocalyx, 176–178
Glycoprotein Ibα (GPIbα), 285–287
  force-induced domain unfolding, 293
  mechanotransduction, 295–296
Glycosaminoglycans (GAGs), 103–104
G-protein-coupled receptors (GPCRs), 16, 173, 419
Groove of Ranvier, 83

# H

Haptotaxis, 460
Heart valve interstitial cells (VICs)
  finite element models, 273–277
  in situ tissue level evaluation
    aortic valve leaflet, 261–264, 262f, 271f
    bidirectional valve leaflet bending response, 260–261
    downscale model, 265–268
    3D hydrogel, 265–268
    VIC-ECM coupling, 260–261
  isolated cell studies
    mechanical models, 256–259, 257f
    microindentation, 255–256
    micropipette aspiration, 253–255, 254f
  uniaxial planar stretch bioreactors
    clinical relevance, 273, 274f
    collagen fiber architecture, 271–272, 271f
    design, 269–271
    mitral valve interstitial cells activation, 272, 272f
Heat shock protein 47 (HSP47), 255
Helmholtz equation, 331
Hematopoietic stem cell (HSC), 179
Hemichannels, bone, 172–173
Heparin affin regulatory protein (HARP), 138–140
Hering's law of equal innervations, 364
Hertz model, 4
High-fat diet (HFD)-induced obesity, 384–385
Hip joint, 84–85, 85f
Hydrostatic pressure, 331
Hydrostatic pressure bioreactors, 9f, 10

Hypertension, 240–242
Hypertrophic scars, 424–425

# I

ICCs. See Interstitial cells of Cajal (ICCs)
Image analysis
  digital image correlation, 31–32, 32f
  particle image velocimetry, 32–33, 33f
Immunohistochemistry (IHC), 23–24
Inertial forces, 363
Inflammatory bowel disease (IBD), 387–388
Inflammatory response, vascular
  alternatively activated macrophages, 233
  classically activated macrophages, 231–233
  dendritic cells, 235–236
  monocytes/macrophages, 231–233
  T cells, 234–235
Inflation tests, 353
In ovo/ex ovo manipulation, 26
In ovo immobilization, 13
In situ hybridization, 19f, 22–23, 22f
Integrins, 110–111
  bone, 169–170
  chondrocyte mechanoreceptors, 110–111
  vascular smooth muscle, 228–229
Intensive treadmill running (ITR), 138, 139f
Interstitial cells of Cajal (ICCs)
  intramuscular, 394–396
  motility disorders, 396
  myenteric plexus, 394
  structure and organisation, 394, 395t
Intracellular calcium, 175
Intracellular forces, 336, 342–343
Intracranial pressure (ICP), 362
Intraganglionic laminar endings (IGLEs), 381–382
Intraganglionic varicose endings (IGVEs), 386
Intramembrane cavitation, 340–341
Intramembranous ossification, 166
Intramural cells, 227
Intramuscular arrays (IMAs), 382
Intraocular pressure (IOP), 350, 353–354, 359–361, 361f
Intrinsic innervation
  anatomy, 392
  intrinsic neurones, 392–394
Intrinsic neurones, 392–394
Intrinsic primary afferent neurons (IPANs), 393, 398
Ion channels, 109
  acid-sensing ion channels, 391–392
  bone, 171
  chondrocyte mechanoreceptors, 109

Ion channels *(Continued)*
    extrinsic afferents, 389–392
    in neuronal function, 333–335
    transient receptor potential channels, 389–391
IOP. *See* Intraocular pressure (IOP)
Irritable bowel syndrome (IBS), 385–386, 388–389

## J

Jun N-terminal kinase (JNK), 113

## K

Keloids, 425–426
keratoconus (KC), 354–355
Kinesin-1, 61, 63
Kraissl's lines, 418$f$, 433–434

## L

Lacunar-canalicular porosity, 159–161
Lamina, 361–362
Lamin-deficient cells, 61
Langer's lines, 417–418, 418$f$, 435
Lens, 355–357, 356$f$
Leucine-rich repeat domain (LRRD), 285–287, 293–295
Leukocytes, 217
Limb patterning, 83
Lipid membranes, 329
Live cell imaging, 15
Loading-induced tendinopathy, 140–141
Low-magnitude high-frequency (LMHF) signals, 187
LRRD. *See* Leucine-rich repeat domain (LRRD)
Lung, morphogenesis, 81–82

## M

Macro-micro finite element methodology, 265, 265$f$
Magnetic resonance elastography (MRE), 340–341
Magnetic tweezer (MT), 287–288, 288$f$
Mammalian models, 13–14
Matrilin (MATNs), 106–107
Matrix metalloproteinases (MMPs), 172, 219–220, 236–237
Mean squared displacement, nuclear translation, 59–60, 59$f$
Mechanical behavior, of developing tissues, 86–88, 87$t$
Mechanical buckling, 82
Mechanical forces, 15, 305–306, 306$t$, 327–328, 342–343
Mechanical interfacing, with nervous system, 339–342

Mechanical loading
    of articular cartilage, 105–106
    on tendon, 137
Mechanical stimulation, 2–3
Mechanical stimuli, 28–29, 392–393
Mechanobiological responses
    in vitro, 133–137, 134$f$, 136$f$
    in vivo, 137–140, 137$f$, 139$f$
Mechanogrowth factor (MGF), 137–138
Mechanosensing
    in endothelial cells, 227–230
    in fibroblasts, 229–230
    gastrointestinal tract, 378–380, 379$f$
    in vascular smooth muscle cells, 227
Mechanosensitive domain (MSD) unfolding, 293–296
Mechanosensitive proteins, 308
Mechanosensors, 167–169
    in metastatic bone disease, 186
    osteoporosis, 178–179
Mechanosensory cilia
    in renal system
        kidney functions, 315
        polycystic kidney disease, 314–315
    in vascular system
        aneurysm, 312, 313$f$
        atherosclerosis, 312–313
Mechanosignaling, in TSC differentiation, 145–147
Mechanotargets, 226
Mechanotransduction, 145, 167–169
    bone
        calcium signaling, 175
        G-protein-related signaling, 175–176
        mitogen-activated protein kinases, 175
        nitric oxide, 176
        Wnt signaling, 175
    diseases, 309$t$
    gastrointestinal tract, 378–380, 379$f$
    glycoprotein Ibα, 295–296
    osteoporosis, 179–180
Memory T cells, 235
Merkel cells, 398
Mesenchymal stem cells (MSCs), 130, 161
    endochondral ossification, 165–166
    intramembranous ossification, 166
Mesenteric afferents, 387
Messenger RNA (mRNA) expression analysis
    in situ hybridization, 22–23, 22$f$
    microarray analysis, 18–20, 19$f$
    quantitative real time PCR, 21
    transcriptome sequencing, 19$f$, 20–21
Metalloelastases, 237

Metastatic cancer
  cell-microenvironment mechanical interactions
    adhesive forces, 477–479
    cell adhesion, 477–479
    mechanical invasiveness, 479–480
  cell migration, assays
    cell lattice models, 471
    cell motility, 470
    collective cell, 459–460, 459$f$
    continuous functions, 472
    experimental methods, 469–470
    motility model, 472
    single cell, 458–459, 459$f$
    subcellular attributes, 471
  cell strength evaluation
    micropillars, 473$f$, 474
    psuedo 3D mechanical invasiveness, 473$f$, 475
    3D traction microscopy, 473$f$, 475
    2D traction force microscopy, 473–474, 473$f$
  cytomechanobiology methods
    cell mechanics assays, 466–469, 466$f$
  directional motility
    chemical/biochemical cues, 460
    environmental topology, 461
    intercellular mechanics, 461
    substrate mechanics/mechanotaxis, 461
  gel-indentation, 480, 481$f$
  mechanical invasiveness, 480, 481$f$
  metastasis steps
    blood stream, 464
    extravasation, 464–465
    intravasation phase, 463–464
    invasive capabilities, 461–463
    mechanical properties of, 462$f$
  tumor staging, 481
Microarray analysis, 18–20, 19$f$
Microglia, 363
Microindentation, 255–256
Micropillar arrays, 8
Micropipette aspiration (MA), 4–6
Microtubule-associated proteins (MAPs), 338–339
Microtubule organizing center (MTOC), 63, 65
Microtubules, 338–339
miRNAs, 114–115
Mitogen-activated protein kinases (MAPKs), 113, 175
Mitral valve (MV), 250–251
Mitral valve anterior leaflets (MVALs), 269–271, 270$f$, 273–275, 276$f$
Mitral valve interstitial cells (MVICs), 253, 272–275, 275$f$

MLO-Y4 osteocytes, 179–180, 181$f$
MMPs. See Matrix metalloproteinases (MMPs)
Moderate mechanical stretching, 133–135
  Achilles tendon stem cells, 135, 136$f$
  of aging TSCs, 142
  patellar tendon stem cells, 135, 136$f$
Moderate treadmill running (MTR), 137–138, 137$f$
  aging tendons, 142–143, 144$f$
Moens–Korteweg equation, 221
Mohawk (Mkx), 135, 138, 146
Monocytes, 231
Morphogenesis, 77–78
  appendicular limb development, 82–85
  of human embryo, 78, 78$f$
  lung, 81–82
  neural tube, 80–81
Motility disorders, interstitial cells of Cajal and, 396
Mucosal afferents, 381, 381$f$, 386
Müller cells, 363
Multiscale modeling techniques, 29–31
Muscle loading, 83–85
Muscle paralysis, 84–85, 85$f$
Muscular afferents, 386
MVALs. See Mitral valve anterior leaflets (MVALs)
MVICs. See Mitral valve interstitial cells (MVICs)
Myenteric plexus (MP), 394
Myosin, 66
Myxomatous MV degeneration, 251–252

# N

Nanoindentation, 86–88
NanoNewton force measurement device (nNFMD), 86–88
NAR. See Nuclear aspect ratio (NAR)
Natural killer T (NKT) cells, 235
N-cadherin, 179
Nesprin3, 64
Neural tube, 80–81
Neuromechanobiology
  cytoskeletal elements, 335–339
  electrodeformation, 330–333
  flexoelectricity, 330–333
  ion channels, 333–335
  mechanical interfacing, 339–342
  viscoelastic plasma membranes, 328–330
Neuroscience, 327–328
NIH3T3 fibroblasts, 66, 68–69
Nitric oxide (NO), 176, 217–218
Nocodazole, 63–64

Noncoding RNAs (ncRNAs), 115
Normalized orientation index (NOI), 271, 271f
Nuclear aspect ratio (NAR), 252, 279–280
Nuclear displacement, 59–60
Nuclear dynamics, 55–56
Nuclear fluctuations
  cell geometric regulation, 66
  computational methods, 66
  cytoskeleton role, 66–68
  image processing, 66, 67f
Nuclear lamina, 68–69
Nuclear positioning, 61
Nuclear rotation
  cell geometric regulation, 63
  in cellular functions, 65
  computational methods, 62–63
  cytoskeleton role, 63–65
  image processing, 62–63, 62f
Nuclear translation
  computational methods, 58–60
  crossover timescale, 60
  cytoskeleton role, 60–61
  geometric control, 60
  image processing, 58–60
  mean squared displacement, 59–60, 59f
  translational motion, 58, 59f
Nucleus
  active force, 57
  actomyosin, 56
  cytoskeletal forces, 56–57
  cytoskeleton, 56
  eukaryotic cell, 56
  microrheology, 57
  passive force, 57
Nystagmus, 366–367

## O

Ocular response analyzer (ORA), 353
Optical tweezer (OT), 4, 5f, 287–288, 288f
Oscillatory fluid flow (OFF), 179–180, 182f
Osteoarthritis (OA), 185
Osteoblasts, 161–163
Osteocalcin (OCN), 173
Osteoclastic lineage, 161, 162f
Osteoclasts, 165
Osteocytes, 163–165, 164f
Osteogenic lineage, 161, 162f
Osteopontin (OPN), 169–170, 173
Osteoporosis
  biomechanical changes, 180–185, 184f
  mechanical environment, 180–185

mechanosensors, 178–179
mechanotransduction, 179–180
Osteoprogenitors, 161
Osteosarcoma, 185–187
OT. *See* Optical tweezer (OT)

## P

Papilledema, 363
Parathyroid hormone (PTH), 171
Particle image velocimetry (PIV), 32–33, 33f
Patellar tendon stem cells (PTSCs), 135, 136f
Pathogenassociated molecular patterns (PAMPs), 230
Pattern recognition receptors (PRRs), 230
P-cadherin, 179
PCM. *See* Pericellular matrix (PCM)
Pelvic afferents, 387
Pericellular matrix (PCM), 100–103, 106–107, 163–165, 172
Perichondrium, 83
Perifibrillar adapter proteins, 100–103
Peterson's elastic modulus, 221–222
Phosphoinositide 3-kinase (PI3K) signaling cascade, 147
Phospholipid bilayers, 329
Physiological loading, on tendon, 137–138
Piezo2 channel, 398, 399f
Piezoelectric nanoribbons, 340
Piola–Kirchhoff stress, 224
Planar biaxial tests, 223
Planar membrane tension, 260
Plasmacytoid dendritic cells, 236
Platelet and endothelial cell adhesion molecule (PECAM-1), 217–218
Platelets, 285–287
  four-step model, 296–298, 297f
  glycoprotein Ibα
    force-induced domain unfolding, 293
    mechanotransduction, 295–296
  mechanosensitive domain unfolding, 295–296
  ultrasensitive force techniques, 287–289, 288f
  unbinding kinetics, 293–295
  unfolding mechanics, 293–295
  von Willebrand factor
    force-induced activation, 289–290
    force-induced cleavage, 291–293, 292f
    GPIBα binding kinetics, 290–291
Plithotaxis, 461
Poisson's ratio, 329
Polycystic kidney disease, 314–315
Polycystins, 171, 314

Polydimethylsiloxane (PDMS), 133
Poly(ethylene glycol) (PEG) hydrogels, 268
Polymer device, 435–436
Porosity, bone, 159–161, 160f
Postaglandin E2 (PGE$_2$), 146–147, 169, 172, 179–180
Postprandial distress syndrome, 385
Presbyopia, 358–359
Primary cilia, 110, 145–146, 305
   axoneme, 306, 307f
   basal body, 306, 307f
   bone, 170–171
   chondrocyte mechanoreceptors, 110
   ciliary membrane, 306, 307f
   ciliary necklace, 306, 307f
   ciliary tip, 306, 307f
   ciliopathy, 308
   cilioplasm, 306, 307f
   ciliotherapy in vascular hypertension, 310–311
   length regulation, 309–315
   as mechanical sensors, 306–308, 307f
   mechanobiology, 308–309
   mechanosensory cilia
      in renal system, 314–315
      in vascular system, 311–313
Protease-activated receptor-2 (PAR$_2$), 391
Proteases, 238
Protein level
   enzyme-linked immunosorbent assay, 24–25
   immunohistochemistry, 23–24
   Western blotting, 24
Protein Op18, 61
Proteoglycan, 106
Pseudoelastic response, 222–223
Pulmonary valve (PV), 250
Pulmonary valve interstitial cells (PVICs), 253, 259, 259f
Pulse wave velocity (PWV), 221
Pursuit movement, 364
PVICs. *See* Pulmonary valve interstitial cells (PVICs)

## Q
Quantitative real time PCR (qRT-PCR), 21
Quiescent osteoblasts, 163. *See also* Osteoblasts

## R
Reactive oxygen species (ROS), 236
Repulsive forces, 351–352
Retina, 349–350, 362–364
Ribonucleic acid interference (RNAi), 69–70
RNA-induced silencing complex (RISC), 114

## S
Saccades, 364
Saint Venant–Kirchhoff (SVK) material model, 273
Schlemm's canal, 359–361
Scleral buckling, 364
Scleraxis promoter (ScxGFP), 138
Scleroderma, 426–427
Sclerostin, 187
Senescence, 141–142, 358–359
Senescence-associated β-galactosidase (SA β-gal), 141
Senescence-associated secretory phenotypes (SASPs), 141
Serosal afferents, 386
Shape fluctuations, 66
Shear force, 392–393
Shear stress, 289–290, 306, 310, 314, 397
Sherrington's law of reciprocal innervations, 364–365
Single cell analysis
   atomic force microscopy, 4
   micropipette aspiration, 4–6
   optical tweezers, 4, 5f
Single-cell RNA-sequencing technology (scRNA-seq), 20–21
Skin mechanobiology
   Borges' lines, 418f, 433–435
   diseases and wound healing, 422
   homeostasis, 417–418
   Kraissl's lines, 418f, 433–434
   Langer's lines, 417–418, 418f, 435
   mechanical homeostasis, 435–436, 437f
   mechanical stress, 421–422
   mechanoresponsive cells, 418–419
   mediate mechanotransduction in, 419–420
   molecular pathways, 419–420
   stretch-activated ion channels, 419
Small-angle light scattering (SALS), 271
Smooth muscle actin (SMA), 255
Smooth muscle cells, 397–398
Sonocrystals, 275, 277f
Spinal afferents
   inflammatory bowel disease, 387–388
   intraganglionic varicose endings, 386
   irritable bowel syndrome, 388–389
   mesenteric afferents, 387
   mucosal afferents, 386
   muscular afferents, 386
   serosal afferents, 386
   signalling, 387–389

Splanchnic afferents, 388
Spongiosa, 250–251
Squint, 366
Stalling force, 336
STB-140 system (STREX), 133
Stem cells, 129–132
Stiffness
    arterial, 221
    lens, 357, 359
    valve interstitial cells, 255–256
Strabismus, 366–367
Stress/strain determination, 223–225, 225f
Stretch-dependent $K^+$ (SDK), 397–398
Stretching apparatus, 133, 134f
Stretching, tendon stem/progenitor cells, 133
Strip extension tests, 353
Subchondral bone remodeling, 185
Sulci, 80–81
Survivin, 308
Systemic blood pressure, 240

## T

T cells, 234–235, 241
Tear film, 350–351
Tendons, 128
    Achilles, 129
    aging, 141–145, 142–144f
    matrix, 128–129
    patellar, 129
    structure, 128f
    tenocytes, 128–129
Tendon stem/progenitor cells (TSCs)
    in aging-induced tendinopathy development, 144–145
    aging tendons, 141–145, 142–144f
    in culture, 130–132, 131f
    differentiation, mechanosignaling in, 145–147
    discovery, 129–130
    influencing factors, 132–133
    in loading-induced tendinopathy development, 140–141
    mechanobiological responses
        in vitro, 133–137, 134f, 136f
        in vivo, 137–140, 137f, 139f
Tenocytes, 128–129
    in culture, 130–132, 131f
    vs. tendon stem/progenitor cells, 130–132
Tenomodulin, 133–135
Tensegrity, 172, 335–336
Tensile forces, 10
Tensile stress, 392–393

Tension sensitive afferents
    intraganglionic laminar endings, 382
    intramuscular arrays, 382
Tensotaxis, 461
Territorial matrix (TM), 100–103
Tetrodotoxin, 396
T helper cells, 234–235
3D hydrogel, 268–269, 270f
Thrombospondin type 1 motif (TSP), 291
Thrombotic thrombocytopenic purpura (TTP), 291
Thymocytes, 234
Tibial four-point bend model, 12, 12f
Tissue inhibitors of metalloproteinases (TIMPs), 238
T lymphocytes, 230
Tonic fluid pressure, 81–82
Trabecular bone, 158–159
Trabecular meshwork (TM), 359–361
Trabecular porosity, 159–161
Traction force microscopy (TFM), 7
Traction forces, 7
Transcriptome sequencing, 19f, 20–21
Transforming growth factor-β (TGF-β), 179–180
Transient receptor potential (TRP) channel, 171, 335, 389–391
Transient receptor potential vanilloid 1 (TRPV1), 384–385
Transient receptor potential vanilloid 4 (TRPV4), 109
Transvalvular pressures (TVPs), 250, 252–253
Treadmill running, of rodents, 137, 137f
Triaxial tests, 223
Tricuspid valve (TV), 250
Tricuspid valve interstitial cells (TVICs), 253
Trinitrobenzenesulfonic acid (TNBS), 394
TRPV4-mediated $Ca^{2+}$ signaling, 109
TSCs. See Tendon stem/progenitor cells (TSCs)
Two-photon excitation microscopy, 17
Two-pore-domain $K^+$ ($K_{2P}$) channel, 335
Type I collagen, 128–129

## U

Ulnar compression model, 12–13, 12f
Ultrasensitive force techniques, 287–289, 288f
Ultrasound, 340–341
Uniaxial planar stretch bioreactors
    clinical relevance, 273, 274f
    collagen fiber architecture, 271–272, 271f
    design, 269–271
    mitral valve anterior leaflet, 269–271, 270f
    mitral valve interstitial cells activation, 272, 272f

## V

Vagal afferents
  appetite regulation, 384–385
  esophageal, 383, 386
  functional dyspepsia, 385–386
  mechanosensory properties, 381–386
  mucosal afferents, 381, 381$f$
  obesity, 384–385
  tension sensitive afferents, 382, 383$f$
Valve endothelial cells (VECs), 252
Valve interstitial cells (VICs), 250–252
Vascular
  adventitia, 219–220
  arterial mechanics
    ex vivo, 222–226
    in vivo, 221–222
  arteries, 216–217
  function, 216–220
  inflammatory response
    alternatively activated macrophages, 233
    classically activated macrophages, 231–233
    dendritic cells, 235–236
    monocytes/macrophages, 231–233
    T cells, 234–235
  intima, 217–218
  matrix G&R in cardiovascular disease, 238–243, 239$f$
  matrix metalloproteinases, 236–237
  mechanosensing
    in endothelial cells, 227–230
    in vascular smooth muscle cells, 227
  media, 218–219
  proteases, 235–236
  structure, 216–220
  tissue inhibitors of metalloproteinases, 238
Vascular porosity, 159–161
Vascular smooth muscle cells, 228–230
Vergence movement, 364
Vimentin, 64, 112
Viscoelasticity, of dendritic spines, 338
Viscoelastic plasma membranes, 328–330
von Willebrand factor (VWF)
  force-induced activation, 289–290
  force-induced cleavage, 291–293, 292$f$

## W

Western blotting, 24
Whole-body vibration, 187
Wide-field imaging, 339
Wingless-type family (Wnt) signaling, 113–114, 147, 175
Wound repair, 422, 423$f$

## Y

Young–Laplace equation, 332
Young–Lippmann equation, 332–333
Young's modulus, 4–6

## Z

Zebrafish models, 14
Zebra fish, paralysis of, 14
Zonules, 355–357, 356$f$